HANDBOOK
FOR
ELECTRICITY METERING

NINTH EDITION

EDISON ELECTRIC INSTITUTE

701 Pennsylvania Avenue, N.W.
Washington, D.C. 20004-2696

First, Second, and Third editions entitled
Electrical Meterman's Handbook
copyright 1912, 1915, 1917
by the National Electric Light Association
Fourth edition entitled
Handbook for Electrical Metermen
copyright 1924
by the National Electric Light Association
Fifth, Sixth, and Seventh editions entitled
Electrical Metermen's Handbook
copyright 1940, 1950, and 1965
by the
Edison Electric Institute

First edition, 1912 .... 5,000 copies
Second edition, 1915 .... 2,500 copies
Third edition, 1917 .... 5,000 copies
Fourth edition, 1923 .... 21,300 copies
Fifth edition, 1940 .... 15,000 copies
Sixth edition, 1950 .... 25,000 copies
Seventh edition, 1965 .... 20,000 copies
Eighth edition, 1981 .... 17,000 copies
Ninth edition, 1992 .... 10,000 copies

EEI Publication No. 06-92-01

LIBRARY OF CONGRESS CATALOGING-IN-PUBLICATION DATA

Handbook for Electricity Metering.—Ninth Edition
p. cm.
Includes index.
ISBN 0-931032-30-X : $75.00
1. Electric meters. I. Edison Electric Institute
TK301.H428 1992
621.37'45—dc20 91-46517
CIP

PREFACE

TO THE NINTH EDITION

The first edition of the *Electrical Meterman's Handbook*, now the *Handbook for Electricity Metering*, was first published in 1912. Seven revisions have since been published; the eighth edition appeared in 1981. As in the previous editions, the emphasis has been on fulfilling the needs of the meter tester and installer.

In the ninth edition, the chapters on mathematics and electrical theory continue to serve as refresher or reference. The chapter on basic electronics has expanded material on digital logic circuits. Where applicable, tables have been changed to reflect the latest national standards. The appearance of the entire text and many illustrations are improved in an effort to make the presentation easier to understand.

To provide room for more information on solid-state electronics, former Chapters 10 and 12 are combined into Chapter 10 and former Chapter 19 is eliminated. The companion publication, the Appendices, is greatly expanded and provides specific product information that complements the theory presented in the *Handbook*.

To make the *Handbook* convenient either as a reference or textbook, a great deal of duplication has been permitted. Thus, although phasor diagrams are treated in Chapter 3, "Mathematics for Metering," similar phasor diagrams are given and explained in succeeding chapters where the subject matter requires their use. Formulas have also been repeated.

In some cases, information concerning devices that were recently phased out of current production and are slowly disappearing has been dropped, with a reference to an earlier edition.

For advanced metering personnel, such subjects as compensating metering and automatic meter reading have been expanded with text intended to meet the specific problems of design and maintenance of these complex metering installations. It is hoped that the treatment of both elementary and advanced material will be useful.

In the preparation of this *Handbook*, the Task Force wishes to make grateful acknowledgment for all help received. Above all, credit must be given to the editors and committees responsible for previous editions of the *Handbook*. Although the ninth edition has been rewritten and rearranged, the eighth edition provided most of the material that made this rewriting possible. Changes have been dictated by advances in technology rather than by any belief that the material presented could be improved.

The contribution by the manufacturers has been far greater than that of providing material for the chapters concerned with their products. In other chapters of the *Handbook* that are more general, manufacturers have freely provided both illustrations and text. Although borrowed text is used without quotation marks, major participation is acknowledged by showing manufacturers as contributors.

It is hoped that future editions will be prepared as new developments make them necessary. If any users of this handbook have any suggestions which they believe would make future editions more useful, such suggestions, comments, or criticisms would be welcomed. They should be sent to the Meter and Service Committee, Edison Electric Institute, 701 Pennsylvania Avenue, N.W., Washington, D.C. 20004-2696.

STAFF

William Mambert ....Project Director
Richard D. Alexander ....Managing Editor
Greg C. Tice ....Graphics and Text Production
William Bickel ....Art Director
David E. Soffrin ....Director, Engineering & Standards
Sondra L. Rubin ....Coordinator, Engineering Programs

EEI METER AND SERVICE COMMITTEE

Al Wilkinson, *EEI Chairman* ....Central Vermont Public Service

Clifford Hubbard, *AEIC Chairman* ....Pacific Gas & Electric

James Mining, *EEI Vice Chairman* ....TU Electric

Arlin A. Rummel, *AEIC Vice Chairman* ....Duquesne Lighting

Donald W. Eggleston, *Past EEI Chairman* ....Duke Power

James P. Suttle, *Past EEI Chairman* ....Ohio Edison

James R. Christensen, *Meter & Service Subcommitte* ....Nevada Power

William A. England, *Research & Development Subcommittee* ....Carolina Power & Light

Leon Harris, *Codes and Standards Subcommittee* ....Rochester Gas & Electric Company

Mario Natividad, *Management Subcommittee* ....Southern California Edison

EDITORIAL COMMITTEE

Kenneth E. Hauzer, *Chairman* ....Centerior Energy Corporation
Robert D. Adams ....Indianapolis Power & Light
Gordon Belcher ....Northeast Utilities
Joseph D. Blackmer ....Consumers Power
Richard C. Guenther ....Union Electric Company
Shannon L. Killinger ....Virginia Power
Gerald Martin ....South Carolina Electric & Gas
Glen Powers ....Public Service Electric & Gas
Arlin A. Rummel ....Duquesne Light
David Scott ....Northeast Utilities
Clark Smith ....Pacificorp
Rod K. Tipton ....Public Service Indiana Energy, Inc.
Tim C. Vahlstrom ....Pacific Gas & Electric

CONTRIBUTORS

Harley Gilleland, Paul Johnston,
Mark L. Munday, Kate A. Shiring .... ABB Power T&D Company

Bill Smith .... Anchor Electric Company

Del Weers .... Associated Engineering

Bill Doherty .... Astra Corporation

Michael D. Patena, Gary W. Proch .... Centerior Energy

John Williams .... Ekstrom Industries

Kathy Bergstrom, Dale M. Bickford,
Joe Maida, Les J. Rosenau .... General Electric

Harley L. Friend .... Jemtec Electronics

John Argy .... Landis & Gyr Metering Inc.

Richard Schwartz .... Pennsylvania Power & Light

Lee West .... Process Systems, Inc.

John Burney, Jr., Stephanie Jones,
George H. Stephens, Ray Stevens .... Schlumberger Industries

M.V. Mathis .... Spinlab Utility Instrument

John Carr .... Utility Test Equipment

W. Earl Garthright, Paul J. Ireland,
Daniel C. Pillar .... Virginia Power

CONTENTS

CHAPTER 1
INTRODUCTION TO THE METER DEPARTMENT

CHAPTER 2
COMMON TERMS USED IN METERING

CHAPTER 3
MATHEMATICS FOR METERING

CHAPTER 4
ELECTRICAL CIRCUITS

CHAPTER 5
SOLID-STATE ELECTRONICS

CHAPTER 6
INSTRUMENTS

CHAPTER 7
THE WATTHOUR METER

CHAPTER 8
DEMAND METERS

CHAPTER 9
KILOVAR AND KILOVOLT-AMPERE METERING

CHAPTER 10
SPECIAL METERING

CHAPTER 11
INSTRUMENT TRANSFORMERS

CHAPTER 12
METER WIRING DIAGRAMS

CHAPTER 13

THE CUSTOMERS' PREMISES — SERVICES AND INSTALLATION

CHAPTER 14

WATTHOUR METER TESTING AND MAINTENANCE

CHAPTER 15
DEMAND METER TESTING AND MAINTENANCE

CHAPTER 16
THE METER LABORATORY

CHAPTER 17
METER READING

LIST OF TABLES

CHAPTER 3
MATHEMATICS FOR METERING

CHAPTER 4
ELECTRICAL CIRCUITS

CHAPTER 9
KILOVAR AND KILOVOLT-AMPERE METERING

CHAPTER 11
INSTRUMENT TRANSFORMERS

CHAPTER 1

INTRODUCTION TO THE METER DEPARTMENT

THE ELECTRIC UTILITY AND THE COMMUNITY

THE ELECTRIC COMPANY and the community which it serves are permanently interdependent. An electric company, by the nature of its business, cannot pick up its generating plant, transmission, and distribution system and move to some other community. It is firmly rooted where it is located. Its progress depends to a large extent upon the progress of the area it serves; also, it depends greatly upon the respect and active support of its customers. It makes good sense for the electric company to work cordially and cooperatively with its customers toward the improvement of economic and civic conditions. Because of this, the meter reader or meter tester calling at a customer's house is in a different position than, for example, a person making a package delivery.

What the electric company sells has become essential to the point that loss of electric power causes more than inconvenience; it can mean real hardship, even tragedy. In addition, large quantities of electricity cannot be produced and stored and so must be immediately available in sufficient quantities upon demand. What this means is that we sell not only the commodity of electric energy but a very valuable service as well.

The service performed by the electric company and its employees should be so well done that every member of the company and the community can be proud of it.

THE DUTIES OF THE METER DEPARTMENT

The primary function of the meter department is to maintain revenue metering installations at the high level of accuracy specified by company policy and regulatory requirements. This involves the installation, testing, and maintenance of meters, instrument transformers and other associated equipment, and verification of the complete installation to assure accurate metering.

Secondary functions which vary with an individual company may include a broad range of responsibilities: appliance repair, connection of services, testing of rubber protective equipment, operation of standardizing laboratories, meter reading, chart and magnetic tape changes, acceptance testing of material and equipment, instrument calibration and repair, investigation of customer complaints, installation and maintenance of load survey and load management equipment, relay testing, and high-voltage testing. Although possibly quite removed from metering, these and many similar functions may become the responsibility of the meter department predominantly for two reasons: first, the direct association of the work with metering, as in the case of meter reading, and, second, the characteristic ability of meter personnel to translate their knowledge and techniques to other fields requiring detailed electrical knowledge and specialized skills, as in the case of operation of standardizing laboratories, instrument repair, or load surveys.

The electric meter, since it serves as the basis for customer billing, must be installed, maintained, tested, and calibrated to assure accuracy of registration. To accomplish this, the accuracy of all test equipment must be traceable through suitable intermediate standards to the basic and legal standards of electrical measurement maintained by the National Institute of Standards and Technology. Quality of workmanship must be consistently maintained at a level which will achieve this desired accuracy. Poor workmanship can have a serious effect on both the customer and the company. Standards, procedures, and instructions are essential to insure uniformity of operations, to prevent errors, and for overall economy.

CUSTOMER CONTACTS

Because of the electric company's place in the community, and because members of the meter department so frequently meet customers, it is important that all meter personnel exemplify those qualities of integrity and courtesy which generate confidence in the company. Day-to-day contacts with customers provide these employees with exceptional opportunities to serve as good-will ambassadors and may earn public appreciation, for the services they and their company perform. To achieve this appreciation, employees must demonstrate a sincere desire to be helpful, as well as a high ethical standard in the performance of their work.

In many companies the increase in outdoor meters, as well as certain other factors, has resulted in a decrease in the meetings between customers and company employees. Thus the friendly association with people which meter personnel often enjoy presents opportunities for building good will that should not be neglected.

First impressions are often lasting impressions. It is desirable that the meter personnel look their best so that a pleasant picture of the company they represent will be left in the customer's mind. Neatness and cleanliness are of utmost importance. The little things which customers notice may have considerable influence on the company's reputation.

Visits to a customer's premises for meter reading and testing or for other reasons afford opportunities for personnel to demonstrate the company's interest in the customer's welfare. Courteous consideration of every request will create satisfaction and appreciation of the efforts made by the company to render good service. However, customers should be referred to the local office for answers to all questions on meter accuracy, billing, and any matter about which the meter employee is not certain of the answer. Promises requiring action beyond the employee's instructions should be avoided. In practically all cases, assurance that any request will be conveyed to the proper party will satisfy the customer.

Upon entering a customer's premises, meter personnel should make their presence and business known and should cheerfully present identification card, badge, or other credentials when requested. All work done on customers' premises should be planned carefully and carried out promptly. While on customers' premises, conversations between company personnel should be about the work at hand and should not be argumentative.

If utility personnel notice any unusual conditions on the customer's premises or in the immediate vicinity which might affect the company's system or the customer's service, they should report them promptly to their immediate supervisors.

Telephone conversations with customers, like visits, can go far toward expressing the company's interest in the customer, if they are conducted intelligently and sympathetically. Sometimes considerable patience may be required, but even then, as at all times, a courteous tone of voice will prove most helpful.

OPPORTUNITIES FOR METER PERSONNEL

Metering of electric energy is a dynamic enterprise and offers a challenge to utility personnel who are interested. More than a century ago, Ayrton, Siemens, Edison and others started the art of electric energy measurement. Soon thereafter, Shallenberger, Thomson, Gutmann, and Duncan conceived their first AC watthour meters, which have evolved into the fine products of the four major United States meter manufacturers. Few production line products can compare with the watthour meter in accuracy, long life, economy, and design. As the use of electricity grows, the need to conserve natural resources fosters efficient use of energy sources. These conditions will provide a continuing need for new and improved metering devices based on new technologies and methods. The meter personnel of tomorrow will have opportunities to help devise, apply, and evaluate these new aids and establish new heights of performance.

KNOWLEDGE REQUIRED IN METERING

The theory of metering is technical. To understand their jobs, meter personnel must have a working knowledge of instruments and meters, elementary electricity, elementary mathematics, and certain practical aspects of customers' services. A good understanding of electronics is rapidly becoming a requirement for work on electronic metering equipment, such as time-of-day meters, magnetic tape recorders, and electronic meters. A knowledge of the following subjects is essential:

- The use of fractions and decimals necessary to calculate meter constants and register ratios.
- A knowledge of AC and DC circuits with particular reference to Ohm's Law and Kirchhoff's Law.
- A basic understanding of inductance, capacitance, power factor, and vector analysis.

- A basic understanding of electronic components and circuits.
- A general knowledge of the current-carrying capacity of wire, the relationship between electricity and heat, and the causes and effects of voltage drops.
- The principles of indicating instruments.
- The principles of watthour meters, the effects of the various adjustments, and a good understanding of how to test and calibrate watthour meters.
- Single and polyphase circuits and how to meter them.
- Blondel's Theorem.
- Principles of power, current, and voltage transformers and how to interconnect them.
- The reasons for, and correct methods of, grounding.
- The reasons for fuses or circuit breakers.

Various books on metering which can be studied to attain technical knowledge are generally made available within the company. There are also many instructional pamphlets issued by the manufacturers which are excellent and have clear introductions to their subjects.

Besides the technical subjects mentioned before, effective meter personnel must be familiar with company policies, and the methods and practices of their department. They should attain such additional knowledge of electrical engineering, self improvement, and the utility business in general as opportunities provide. Above all, they must be willing to study and to learn.

METER SECURITY

As the cost of electricity rises to become a significant portion of the cost of living, the temptation to violate the security of metering equipment for the purpose of energy theft becomes irresistible for some customers. In addition, the possibility of an organized effort to tamper with metering equipment increases with the increased cost of energy. Therefore, the meter employee must be aware of the various techniques of energy theft and be constantly on the lookout for such violations. Since meter security systems vary throughout the industry, it becomes necessary for meter employees to completely familiarize themselves with their company's devices and procedures for sealing meters and associated devices, and to keep a constant vigil against their violation. Incidents of tampering should be reported immediately in accordance with company instructions, taking care to preserve all evidence and to submit complete, well written, and brief reports.

It is imperative to bear in mind that circumstantial evidence of tampering should not be interpreted as guilt until all evidence has been examined by those designated to do so. Therefore, courtesy towards all customers, even in strained circumstances, will speak well for you, your department, and your company.

Meter security begins with the seal that secures the glass cover to the base of the meter. This seal is applied without a tool and offers no interference when installing the meter. After the meter is installed, a seal must be applied to secure the meter mounting device whether it is the ring-type or ringless. Ring-type sockets are secured by sealing the ring that holds the meter in place. Various seals are available for this purpose ranging from the lead and wire seal and the plastic and wire padlock seals to other specially designed sealing devices, all of which must be destroyed in order to remove the ring. Ringless sockets are secured by installing the socket cover after the meter is in place, then using any of the various sealing devices described above to seal the cover hasp.

The demand reset mechanism is another area which needs to be secured with a seal to prevent tampering. It should be sealed each time the demand is reset. If a different color seal is used each reading cycle, there is proof that the demand was reset at the end of the last cycle.

To be sure your company's sealing program maintains its integrity, seals should be treated as security items. Only authorized personnel should have access to seals, and they should not be left where unauthorized people would come in contact with them.

The most important part of the sealing procedure is the follow-up. Every time the meter is read, the seal should be inspected, not just visually but physically. This seal should be tugged on and visually inspected to make sure that it has not been violated and is the proper seal for that meter. Evidence of tampering, should be reported immediately. Steps may then be taken to securely lock the meter with various types of hardware to assure that the system is not tampered with again. Even if the service is locked, a seal should still be applied to seal off the key-way of the lock to avoid removal without detection.

There are a wide variety of seals available for all of these applications. Some require tools for installation, some do not. Some are all metal, some all plastic and some a combination of both. Whatever seal is used, however, it should offer the following benefits:

- Be unique to your company and readily identifiable.
- Be impossible to remove without leaving visible signs of tampering.
- Be numbered so that particular seals can be identified with the location.

SAFETY

Safety is a full-time business and requires the hard work and full cooperation of every meter employee. Safety suggestions are not to be considered as arbitrary procedures, but measures which, if followed, will enable personnel to work without injury to themselves or others and without damage to property.

Simply issuing safety suggestions, rules, or regulations does not guarantee safe work practices or produce good safety records. Meter employees must learn the safety rules of their company, apply them daily, and become safety-minded.

The meter personnel owe it to themselves, their families, and their company to do each step of every job the safe way. Careful planning of every job is essential. Nothing should be taken for granted. The meter employee must develop the "feel" for safety. Constant awareness of safety, coupled with training, experience, and knowledge of what to do and how to do it, will prevent many accidents.

Every meter employee's attention is directed to the following general suggestions, which are almost without exception incorporated in company safety rules:

- Horseplay and practical jokes are dangerous. Work safely, consider each act, and do nothing to cause an accident.
- Knowledge of safe practices and methods, first aid, and resuscitation is a must for meter personnel.
- Report unsafe conditions or defective equipment to immediate supervisor.
- Have injuries treated immediately.
- Report all accidents as prescribed by company safety regulations.

- Exercise general care and orderliness in performance of work.
- The right way is the safe way.
- Study the job! Plan ahead! Prevent accidents!
- Select the right tools for the job.
- Keep tools in good working order.
- Use tools properly.
- Exercise good housekeeping at all times.
- Handle material with care. Lift and carry properly.
- Respect low voltage. It can be fatal.
- Never substitute assumptions for facts.
- The importance of working safely cannot be over-emphasized. Safety pays dividends in happiness to meter personnel and their families.
- **Remember, there is no job so important that it cannot be done in a safe manner.**

CHAPTER 2

COMMON TERMS USED IN METERING

THE FOLLOWING PARAGRAPHS are to be considered as practical common terms. In order to keep the explanations as clear and simple as possible, occasional departures from exact definitions have been permitted. The explanations given are intended to be useful for meter personnel rather than for scientists. For additional definitions see the current version of ANSI C12.1 *Code for Electricity Metering*— Definitions Section, and ANSI/IEEE 100-1988 *Standard Dictionary of Electrical and Electronics Terms.*

Accuracy—The extent to which a given measurement agrees with the defined value.

Ammeter—An instrument to measure current flow, usually indicating in amperes. Where indication is in milliamperes the instrument may be called a milliammeter.

Ampere—The practical unit of electric current. One ampere is the current caused to flow through a resistance of 1 ohm by 1 volt.

Ampere-Hour—The average quantity of electric current flowing in a circuit for one hour.

Ampere-Turn—A unit of magnetomotive force equal to that produced by one ampere flowing in a single turn of wire.

Annunciator—A label that is displayed to identify a particular quantity being displayed.

Automatic Meter Reading (AMR)—The reading of meters from a location remote from where the meter is installed. Telephone, radio, and electric power lines are used to communicate meter readings to remote locations.

Autotransformer—A transformer in which a part of the winding is common to both the input and output circuits. Thus, there is no electrical insulation between input and output as in the usual transformer. Because of this interconnection, care must be exercised in using autotransformers.

Balanced Load—The term balanced load is used to indicate equal currents in all phases and relatively equal voltages between phases and between each phase and neutral (if one exists), with approximately equal watts in each phase of the load.

Base Load—The normal minimum load of a utility system; the load which is carried 24 hours a day. Plants supplying this load, and operating day and night are spoken of as "base-load plants."

Basic Lightning Impulse Insulation Level (BIL)—A specific insulation level expressed in kilovolts of the crest value of a standard lightning impulse (1.2×50 microsecond wave).

Blondel's Theorem—In a system of N conductors, N-1 meter elements, properly connected, will measure the power or energy taken. The connection must be such that all voltage coils have a common tie to the conductor in which there is no current coil.

Bottom-Connected Meter—A meter having a bottom connection terminal assembly. Also referred to as an A-base electricity meter.

Bridge, Kelvin—An arrangement of six resistors, electromotive force and a galvanometer for measuring low values of resistance. In this bridge a large current is passed through the unknown resistance and a known low resistance. The galvanometer compares the voltage drops across these two resistors in a high-resistance double ratio circuit made up of the other four resistors. Hence, the bridge is often called a "double bridge."

Bridge, Wheatstone—An arrangement of four resistances, one of which may be unknown and one generally adjustable, to which is applied an electromotive force. A galvanometer is used for continually comparing the voltage drops, thereby indicating the resistance values.

BTU (British Thermal Unit)—A unit of heat. One kilowatthour is equivalent to 3,413 British Thermal Units.

Burden—The load, usually expressed in volt-amperes at a specified power factor, placed on instrument transformer secondaries by the associated meter coils, leads, and other connected devices.

Calibration—Comparison of the indication of the instrument under test, or registration of meter under test, with an appropriate standard.

Capacitance—That property of an electric circuit which allows storage of energy and exists whenever two conductors are in close proximity but separated by an insulator or dielectric material. When direct voltage is impressed on the conductors, a current flows momentarily while energy is being stored in the dielectric material, but stops when electrical equilibrium is reached. With an alternating voltage between the conductors, the capacitive energy is transferred to and from the dielectric materials, resulting in an alternating current flow in the circuit.

Capacitive Reactance—Reactance due to capacitance. This is expressed in ohms. The capacitive reactance varies indirectly with frequency.

Central Station—Control equipment, typically a computer system, which can communicate with metering and load control devices. The equipment may also interpret and process data, accept input from other sources, and prepare reports.

Circuit, Three-Wire—A metallic circuit formed by three conductors insulated from each other. See Three-Wire System.

Circuit, Two-Wire—A metallic circuit formed by two adjacent conductors insulated from each other. When serving domestic loads one of these wires is usually grounded.

Circuit Breaker—A device, other than a fuse, designed to open a circuit when an overload or short circuit occurs. The circuit breaker may be reset after the conditions which caused the breaker to open have been corrected.

Circular Mil—The area of a circle whose diameter is one mil (1/1000 in). It is a unit of area equal to $\pi/4$ or 0.7854 square mil. The area of a circle in circular mils is, therefore, equal to the square of its diameter in mils.

Class Designation—The maximum of the watthour meter load range in amperes.

Clockwise Rotation—Motion in the same direction as that of the hands of a clock, front view.

Conductance—The ability of a substance or body to pass an electric current. Conductance is the reciprocal of resistance.

Conductor Losses—The watts consumed in the wires or conductors of an electric circuit. Such power only heats the wires, doing no useful work, so it is a loss. It may be calculated from I^2R where I is the conductor current and R is the circuit resistance.

Connected Load—The sum of the continuous ratings of the connected load-consuming apparatus.

Constant—A quantity used in an equation, the value of which remains the same regardless of the values of other quantities used in the equation.

Constant Kilowatthour, of a Meter (Register Constant, Dial Constant)—The multiplier applied to the register reading to obtain kilowatthours,

Constant, KYZ Output (K_e)—Pulse constant for the KYZ outputs of a solid-state meter, programmable in unit-hours per pulse.

Constant, Mass Memory (K_m)—The value, in unit quantities, of one increment (pulse period) of stored serial data. Example: K_m = 2.500 watthours/pulse.

Constant, Watthour—

(a) For an electromechanical meter (K_h): The number of watthours represented by one revolution of the disk. Determined by the design of the meter and not normally changed. Also called: Disk Constant.
(b) For a solid state meter (K_h or K_t): The number of watthours represented by one increment (pulse period) of serial data. Example: K_h or K_t = 1.8 watthours/pulse.

Core Losses—Core losses usually refer to a transformer and are the watts required in the excitation circuit to supply the heating in the core. Core heating is caused by magnetic hysteresis, a condition which occurs when iron is magnetized with alternating current, and by the eddy currents flowing in the iron. Core losses are often called iron losses.

Creep—For mechanical meters, a continuous motion of the rotor of a meter with normal operating voltage applied and the load terminals open-circuited. For electronic meters, a continuous accumulation of data in a consumption register when no power is being consumed.

Current Transformer—An instrument transformer designed for the measurement or control of current. Its primary winding, which may be a single turn or bus bar, is connected in series with the load. It is normally used to reduce primary current by a known ratio to within the range of a connected measuring device.

Current Transformer, Continuous Thermal Current Rating Factor—The factor by which the rated primary current is multiplied to obtain the maximum allowable primary current based on the maximum permissible temperature rise on a continuous basis.

Current Transformer Phase Angle—The angle between the current leaving the identified secondary terminal and the current entering the identified primary terminal. This angle is considered positive when the secondary current leads the primary current.

Cutout—A means of disconnecting an electric circuit. The cutout generally consists of a fuse block and latching device or switch.

Cycle—One complete set of positive and negative values of an alternating current or voltage. These values repeat themselves at regular intervals (See Hertz).

Damping of an Instrument—The term applied to its performance to denote the manner in which the pointer settles to its steady indication after a change in the value of the measured quantity. Two general classes of damped motion are distinguished as follows:

(a) **Under-Damped**—When a meter pointer oscillates about the final position before coming to rest.

(b) **Over-Damped**—When the pointer comes to rest without overshooting the rest position.

The point of change between under-damped and over-damped is called critical damping and is considered to be when the degree of pointer overshoot does not exceed an amount equal to one half the rated accuracy of the instrument.

Dead-Front—Equipment which under normal operating conditions has no live parts exposed is called dead-front.

Demand—The average value of power or related quantity over a specified interval of time. Demand is expressed in kilowatts, kilovolt-amperes, kilovars, or other suitable units. An interval may be 1, 5, 10, 15, 30, or 60 minutes.

Demand, Continuous Cumulative—The sum of the previous billing period maximum demands, and the present period maximum demand.

Demand, Cumulative—The sum of the previous billing period maximum demand readings. At time of billing period reset, the maximum demand for the most recent billing period is added to the previously accumulated total of all maximum demands.

Demand, Maximum—The highest demand measured over a selected period of time, such as one month. Also called peak demand.

Demand, Rolling Interval—A method of measuring power or other quantity by taking measurements within fixed intervals of the demand period. This method can be used to determine total demand, average demand, maximum demand, and average maximum demand during the full interval.

Demand, Sliding Window—See Demand, Rolling Interval

Demand, Threshold Alert—An output to indicate that a programmed value of demand has been exceeded.

Demand-Interval Deviation—The difference between the measured demand interval and the specified demand interval, expressed as a percentage of the specified demand interval.

Demand Constant (Pulse Receiver)—The value of the measured quantity for each received pulse, divided by the demand interval, expressed in kilowatts per pulse, kilovars per pulse, or other suitable units. The demand interval must be expressed in parts of an hour such as 1/4 for a 15 minute interval or 1/12 for a 5 minute interval.

Demand Delay—The programmable amount of time before demand calculations are restarted after a power outage. Also called Cold Load Pickup and Demand Forgiveness.

Demand Deviation—The difference between the indicated or recorded demand and the true demand, expressed as a percentage of the fullscale value of the demand meter or demand register.

Demand Factor—The ratio of the maximum demand to the connected load.

Demand Interval (Block-Interval Demand Meter)—The specified interval of time on which a demand measurement is based. Intervals such as 10, 15, or 60 minutes are commonly specified.

Demand Interval Synchronization—Physical linking of meters to synchronize the demand intervals of all meters. Also called Demand Timing Pulse.

Demand Meter—A metering device that indicates or records the demand, maximum demand, or both. Since demand involves both an electrical factor and a time factor, mechanisms responsive to each of these factors are required, as well as an indicating or recording mechanism. These mechanisms may be either separate from or structurally combined with one another.

Demand Meter, Indicating—A demand meter equipped with a readout that indicates demand, maximum demand, or both.

Demand Meter, Integrating (Block-Interval)—A meter that integrates power or a related quantity over a fixed time interval and indicates or records the average.

Demand Meter, Lagged—A meter that indicates demand by means of thermal or mechanical devices having an approximately exponential response.

Demand Meter, Time Characteristic (Lagged-Demand Meter)—The nominal time required for 90% of the final indication, with constant load suddenly applied. The time characteristic of lagged-demand meters describes the exponential response of the meter to the applied load. The response of the lagged-demand meter to the load is continuous and independent of selected discrete time intervals.

Demand Meter, Timing Deviation—The difference between the elapsed time indicated by the timing element and the true elapsed time, expressed as a percentage of the true elapsed time.

Demand Register—A mechanism, for use with an integrating electricity meter, that indicates maximum demand and also registers energy (or other integrated quantity).

Detent—A device installed in a meter to prevent reverse rotation.

Dial-Out Capability—Ability of a meter to initiate communications with a central station.

Disk Position Indicator, or "Caterpillar"—An indicator on the display of a solid state register that simulates rotation of a disk at a rate proportional to power.

Diversity—A result of variation in time of use of connected electrical equipment so that the total maximum demand is less than the sum of the maximum demands of the individual units.

Eddy Currents—Those currents resulting from voltages which are introduced in a conducting material by a variation of magnetic flux through the material.

Effective Resistance—Effective resistance is equal to watts divided by the square of the effective value of current.

Effective Value (Root-Mean-Square Value)—The effective value of a periodic quantity is the square root of the average of the squares of the instantaneous value of the quantity taken throughout one period. This value is also called the root-mean-square value and is the value normally reported by AC instruments.

Electrical Degree—The 360th part of one complete alternating current cycle.

Electricity Meter—A device that measures and records the summation of an electrical quantity over a period of time.

Electromagnet—A magnet in which the magnetic field is produced by an electric current. A common form of electromagnet is a coil of wire wound on a laminated iron core, such as the voltage coil of a watthour meter stator.

Electromotive Force (EMF)—The force which tends to produce an electric current in a circuit. The common unit of electromotive force is the volt.

Element—A combination of a voltage-sensing unit and a current-sensing unit, which provides an output proportional to the quantities measured.

Embedded Coil—A coil in close proximity to, and nested within, a current circuit loop of a meter, to measure the strength of a magnetic field and develop a voltage proportional to the flow of current.

Embedded System—A microcomputer system including microprocessor, memory, power supply, and supporting input and output devices, usually designed for a dedicated application.

Energy—The integral of active power with respect to time.

Farad—The practical unit of capacitance. The common unit of capacitance is the microfarad.

Field, Magnetic—A region of magnetic influence surrounding a magnet or a conductor carrying electric current.

Field, Stray—Usually a disturbing magnetic field produced by sources external or foreign to any given apparatus.

Firmware—Computer programs used by embedded systems and typically stored in read-only memories. See Memory.

Galvanometer—An instrument for indicating a small electric current.

Gear Ratio—The number of revolutions of the rotating element of a meter for one revolution of the first dial pointer.

Ground—A conducting connection, whether intentional or accidental, between an electric circuit or equipment and earth.

Ground Return Circuit—A current in which the earth is utilized to complete the circuit.

Grounding Conductor—A conductor used to connect any equipment device, or wiring system, with a grounding electrode or electrodes.

Grounding Electrode—A conductor embedded in the earth, used for maintaining ground potential on conductors connected to it and for dissipating into the earth current conducted to it.

Henry—The practical unit of inductance. The millihenry is commonly encountered.

Hertz (Cycles per Second)—The practical unit of frequency of an alternating current or voltage. It is the number of cycles (sets of positive and negative values) occurring in one second.

Horsepower—A commercial unit of power equal to the average rate of doing work when 33,000 pounds are raised one foot in one minute. One horsepower is approximately equal to 746 watts.

Hot-Wire Instrument—An electrothermic instrument which depends for its operation on the expansion by heat of a wire carrying the current which produces the heat.

Hysteresis Loss—The energy lost in a magnetic core due to the variation of magnetic flux within the core.

Impedance—The total opposing effect to the flow of current in an alternating current circuit. It may be determined in ohms from the effective value of the total circuit voltage divided by the effective value of total circuit current. Impedance may consist of resistance or resistance and reactance.

Induced Current—A current flow resulting from an electromotive force induced in a conductor by changing the number of lines of magnetic force linking the conductor.

Inductance—That property of an electric circuit which opposes any change of current through the circuit. In a direct current circuit, where current does not change, there is no inductive effect except at the instant of turn-on and turn-off. However, in alternating current circuits the current is constantly changing, so the inductive effect is appreciable. Changing current produces changing flux which, in turn, produces induced voltage. The induced voltage opposes the change in applied voltage, hence the opposition to the change in current. Since the current changes more rapidly with increasing frequency, the inductive effect also increases with frequency.

Inductance, Mutual—If the current change causes induced voltage and an opposing effect in a second conductor, we have mutual inductance.

Inductance, Self—If the preceding effect occurs in the same conductor as that carrying the current, we have self-inductance. The self-inductance of a straight conductor at power frequency is almost negligible because the changing flux will not induce any appreciable voltage, but self-inductance increases rapidly if the conductor is in the form of a coil and more so if the coil is wound on iron.

Inductive—Having inductance, e.g., inductive circuit, inductive load. Circuits containing iron or steel that is magnetized by the passage of current are highly inductive.

Inductive Reactance—Reactance due to inductance. This is expressed in ohms. The inductive reactance varies directly with the frequency.

Instrument Transformer—A transformer that reproduces in its secondary circuit, in a definite and known proportion, the voltage or current of its primary circuit, with the phase relation substantially preserved.

Instrument Transformer, Accuracy Class—The limits of transformer correction factor, in terms of percent error, that have been established to cover specific performance ranges for line power factor conditions between 1.0 and 0.6 lag.

Instrument Transformer, Accuracy Rating for Metering—The accuracy class, together with the standard burden for which the accuracy class applies.

Instrument Transformer, Burden—The impedance of the circuit connected to the secondary winding. For voltage transformers it is convenient to express the burden in terms of the equivalent volt-amperes and power factor at its specified voltage and frequency.

Instrument Transformer, Correction Factor—The factor by which the reading of a wattmeter or the registration of a watthour meter must be multiplied to correct for the effects of the error in ratio and the phase angle of the instrument transformer. This factor is the product of the ratio and phase-angle correction factors for the existing conditions of operation.

Instrument Transformer, Marked Ratio—The ratio of the rated primary value to the rated secondary value as stated on the name-plate.

Instrument Transformer, Phase Angle—The angle between the current or voltage leaving the identified secondary terminal and the current or voltage entering the identified primary terminal. This angle is considered positive when the secondary circuit or voltage leads the primary current or voltage.

Instrument Transformer, Phase Angle Correction Factor—The factor by which the reading of a wattmeter or the registration of a watthour meter, operated from the secondary of a current transformer, or a voltage transformer, or both, must be multiplied to correct for the effect of phase displacement of secondary current, or voltage, or both, with respect to primary values. This factor equals the ratio of true power factor to apparent power factor and is a function of both the phase angle of the instrument transformer and the power factor of the primary circuit being measured.

Instrument Transformer, Ratio Correction Factor—The factor by which the marked ratio of a current transformer or a voltage transformer must be multiplied to obtain the true ratio. This factor is expressed as the ratio of true ratio to marked ratio. If both the current transformer and the voltage transformer are used in conjunction with a wattmeter or watthour meter, the resultant ratio correction factor is the product of the individual ratio correction factors.

Instrument Transformer, True Ratio—The ratio of the magnitude of the primary quantity (voltage or current) to the magnitude of the corresponding secondary quantity.

Joule's Law—The rate at which heat is produced in an electric circuit of constant resistance is proportional to the square of the current.

Kilo—A prefix meaning one thousand of a specified unit (kilovolt, kilowatt). 1000 watts = 1 kilowatt.

KVA—The common abbreviation for kilovolt-ampere (equal to 1000 voltamperes).

KYZ Output—A three-wire pulse output from a metering device to drive external control or recording equipment. Each pulse or transition represents a predetermined increment of power or other quantity.

Lagging Current—An alternating current which, in each half-cycle, reaches its maximum value a fraction of a cycle later than the maximum value of the voltage which produces it.

Laminated Core—An iron core composed of sheets stacked in planes parallel to its magnetic flux paths in order to minimize eddy currents.

Leading Current—An alternating current which, in each half-cycle, reaches its maximum value a fraction of a cycle sooner than the maximum value of the voltage which produces it.

Lenz's Law—The induced voltage and resultant current flow in a conductor as a result of its motion in a magnetic field is in such a direction as to exert a mechanical force opposing the motion.

Load, Artificial—See Phantom Load.

Load, System—The load of an electric system is the demand in kilowatts.

Load Compensation—That portion of the design of a watthour meter which provides good performance and accuracy over a wide range of loads. In modern, self-contained meters, this load range extends from load currents under 10% of the rated meter test amperes to 667% of the test amperes for class 200 meters.

Load Control—Procedure for turning off portions of customers' loads based on predetermined time schedules, system demand thresholds, or other circumstances.

Load Factor—The ratio of average load over a designated time period to the maximum demand occurring in that period.

Loading Transformer—A transformer of low secondary voltage, usually provided with means for obtaining various definite values of current, whereby the current circuit of the device under test and of the test standard can be energized.

Loss Compensation—A means for correcting the reading of a meter when the metering point and point of service are physically separated resulting in measureable losses including I^2R losses in conductors and transformers, and iron-core losses. These losses may be added to, or subtracted from, the meter registration.

Magnetomotive Force—The force which produces magnetic flux. The magnetomotive force resulting from a current is directly proportional to the current.

Mega—A prefix meaning one million of a specified unit (megawatt, megohm). 1 megohm = 1,000,000 ohms.

Memory—Electronic devices which store computer instructions and data.

(a) Volatile memories can be written to, and read from, repeatedly. Random access memories (RAMs) require uninterrupted power to retain their contents.

(b) Read Only Memories (ROMs) are programmed once, may be read (only) repeatedly, and do not require constant power to retain their contents. ROMs are typically used to store firmware in dedicated systems.

Meter, Excess—A meter that records, either exclusively or separately, that portion of the energy consumption taken at a demand in excess of a predetermined demand.

Meter Sequence—Refers to the order in which meter, service switch, and fuses are connected. Meter-switch-fuse is a common modern sequence. The switch-fuse-meter sequence is also used.

Micro—A prefix meaning one millionth part of a specified unit (microfarad, microhm). 1 microhm = 0.000001 ohm.

Mil—A unit of length equal to one thousandth of an inch.

Milli—A prefix meaning one thousandth part of a specified unit (milliampere, millihenry, millivolt). 1 millivolt = 0.001 volt.

Multifunction Meter—A meter that displays more than one electricity-related quantity. Typically an electronic meter.

N.E.C. (National Electrical Code)—A regulation covering the electric wiring systems on the customer's premises particularly in regard to safety. The code represents the consensus of expert opinion as to the practical method and materials of installation to provide for the safety of person and property in the use of electrical equipment.

Ohm—The practical unit of electrical resistance. It is the resistance which allows one ampere to flow when the impressed electromotive force is one volt.

Ohm's Law—Ohm's Law states that the current which flows in an electrical circuit is directly proportional to the electromotive force impressed on the circuit and inversely proportional to the resistance in a direct current circuit or the impedance in an alternating current circuit.

Optical Port—A communications interface on metering products which allows the transfer of information, while providing electrical isolation and metering security. The communications medium is typically infrared light transmitted and received through the meter cover.

Optical Probe—An interface device which mates with the optical port of the meter, to read data or to program the meter.

Peak Load—The maximum demand on an electric system during any particular period. Units may be kilowatts or megawatts.

Percent Error—The percent error of a meter is the difference between its percent registration and one hundred percent.

Percent Registration—Percent registration of a meter is the ratio of the actual registration of the meter to the true value of the quantity measured in a given time, expressed as a percentage. Percent registration is also sometimes referred to as the accuracy of the meter.

Phantom Load—A device which supplies the various load currents for meter testing, used in portable form for field testing. The power source is usually the service voltage which is transformed to a low value. The load currents are obtained by suitable resistors switched in series with the isolated low voltage secondary and output terminals. The same principle is used in most meter test boards.

Phase Angle—The phase angle or phase difference between a sinusoidal voltage and a sinusoidal current is defined as the number of electrical degrees between the beginning of the cycle of voltage and the beginning of the cycle of current.

Phase Sequence—The order in which the instantaneous values of the voltages or currents of a polyphase system reach their maximum positive values.

Phase Shifter—A device for creating a phase difference between alternating currents and voltages or between voltages.

Phasor (Vector)—A quantity which has magnitude, direction, and time relationship. Phasors are used to represent sinusoidal voltages and currents by plotting on rectangular coordinates. The time relationship is present in that if the phasors were allowed to rotate about the origin, and a plot made of ordinates against rotation time, the instantaneous sinusoidal wave form would be represented by the phasor.

Phasor Diagram—A phasor diagram contains two or more phasors drawn to scale showing the relative magnitude and phase, or time, relationships among the various voltages and currents.

Photoelectric Tester (or Counter)—This device is used in the shop testing of meters to compare the revolutions of a watthour meter standard with a meter under test. The device receives pulses from a photoelectric pickup which is actuated by the anti-creep holes in the meter disk or the black spots on the disk. These pulses are used to control the standard meter revolutions on an accuracy indicator by means of various relay and electronic circuits.

Power, Active—The time average of the instantaneous power over one period of the wave. For sinusoidal quantities in a two-wire circuit, it is the product of the voltage, the current, and the cosine of the phase angle between them. For nonsinusoidal quantities, it is the sum of all the harmonic components, each determined as above. In a polyphase circuit it is the sum of the active powers of the individual phases.

Power, Apparent—The product of the root-mean-square current and root-mean-square voltage for any waveform. For sinusoidal quantities, apparent power is the square root of the sum of the squares of the active and reactive powers.

Power, Reactive—For sinusoidal quantities in a two-wire circuit, reactive power is the product of the voltage, the current, and the sine of the phase angle between them with the current taken as reference. With nonsinusoidal quantities, it is the sum of all the harmonic components, each determined as above. In a polyphase circuit, it is the sum of the reactive powers of the individual phases.

Power Factor—The ratio of the active power to the apparent power.

Power Line Carrier—A type of communication where signals are sent through existing electrical power lines.

Primary-Secondary—In distribution and meter work, primary and secondary are relative terms. The primary circuit usually operates at the higher voltage. For example, a distribution transformer may be rated at 14,400 volts to 2,400 volts, in which case the 14,400-volt winding is the primary and the 2,400-volt winding is the secondary. Another transformer may be rated 2,400 volts to 240 volts, in which case the 2,400-volt winding is the primary. Thus, in one case the 2,400-volt rating is secondary while in the latter case it is a primary value.

Pulse—An electrical signal which departs from an initial level for a limited duration of time and returns to the original level. Example: a sudden change in voltage or current produced by the opening or closing of a contact.

Pulse Device (for Electricity Metering)—The functional unit for initiating, transmitting, retransmitting, or receiving electric pulses, representing finite quantities, such as energy, normally transmitted from some form of electricity meter to a receiver unit.

Pulse Initiator—Any device, mechanical or electrical, used with a meter to initiate pulses, the number of which are proportional to the quantity being measured. It may include an external amplifier or auxiliary relay or both.

Q-Hour Meter—An electricity meter that measures the quantity obtained by lagging the applied voltage to a watthour meter by 60 degrees, or for electronic meters by delaying the digitized voltage samples by a time equivalent to 60 electrical degrees.

Reactance—The measure of opposition to current flow in an electric circuit caused by the circuit properties of the inductance and capacitance. Reactance is normally expressed in ohms.

Reactiformer—A phase-shifting auto transformer used to shift the voltages of a watthour meter 90 degrees when reactive volt-ampere measurement is wanted.

Reactive Volt-Amperes—The out-of-phase component of the total volt-amperes in a circuit which includes inductive or capacitive reactance. In an AC circuit, reactive volt-amperes are the product of the total volt-amperes and the sine of the angle between the current and voltage. The unit of reactive volt-amperes is the var.

Reactor—A device used for introducing reactance into a circuit for purposes such as motor starting, paralleling transformers, and controlling currents.

Rectifier—A device which permits current to flow in one direction only, thus converting alternating current into unidirectional current.

Register Constant—The number by which the register reading is multiplied to obtain kilowatt-hours. The register constant on a particular meter is directly proportional to the register ratio, so any change in ratio will change the register constant.

Register Ratio—The number of revolutions of the gear meshing with the worm or pinion on the rotating element for one revolution of the first dial pointer.

Register Freeze—The function of a meter or register to make a copy of its data, and perhaps reset its demand, at a pre-programmed time after a certain event (such as demand reset) or upon receipt of an external signal. Also called: Self-Read, Auto-Read, and Data Copy.

Registration—The registration of the meter is equal to the product of the register reading and the register constant. The registration during a given period of time is equal to the product of the register constant and the difference between the register readings at the beginning and the end of the period.

Resistance—The opposition offered by a substance or body to the passage through it of an electric current. Resistance is the reciprocal of conductance.

Rheostat—An adjustable resistor so constructed that its resistance may be changed without opening the circuit in which it is connected.

SE Cable—Service entrance cable usually consists of two conductors, with conventional insulation, laid parallel and overlaid with a heavy insulating wrap about which is wrapped a stranded, bare neutral conductor. The final covering is a flame retarding and waterproof braid. ASE cable is a variant of the SE cable in which a flat steel strip is inserted between the neutral conductor and the outside braid.

Service—The conductors and equipment for delivering electric energy from a street distribution system to, and including, the service equipment of the premises served.

Service Conductors—The conductors which extend from a street distribution system, transformers on private property, or a private generating plant outside the building served, to the point of connection with the service equipment.

Service Drop—That portion of the overhead service conductors between the last pole or other aerial support and the first point of attachment to the building or structure.

Service Entrance Conductors—For an overhead service, that portion of the service conductors which connect the service drop to the service equipment. The service entrance conductors for an underground service are that portion of the service conductors between the terminal box located on either the inside or outside building wall, or the point of entrance in the building if no terminal box is installed, and the service equipment.

Service Equipment—The necessary equipment, usually consisting of one or more circuit breakers or switches and fuses, and their accessories, intended to constitute the main control and means of disconnecting the load from the supply source.

Shaft Reduction (Spindle Reduction, First Reduction)—The gear reduction between the shaft, or spindle, of the rotating element and the first gear of the register.

Shop, Meter—A place where meters are inspected, repaired, tested, and adjusted.

Short Circuit—A fault in an electric circuit, instrument, or utilization equipment such that the current follows a low resistance by-pass instead of its intended course.

Socket or **Trough**—The mounting device consisting of jaws, connectors, and enclosure for socket-type meters. A mounting device may be either a single socket or a trough. The socket may have a cast or drawn enclosure, the trough an assembled enclosure which may be extensible to accommodate more than one mounting unit.

Standard, Basic Reference—Those standards with which the value of electrical units are maintained in the laboratory, and which serve as the starting point of the chain of sequential measurements carried out in the laboratory.

Stator—The unit which provides the driving torque in a watthour meter. It contains a voltage coil, one or more current coils, and the necessary steel to provide the required magnetic paths. Other names used for stator are element or driving element.

Sub-Metering—The metering of individual loads within a building for billing or load control purposes. For billing applications, usually the building is metered by a master meter and the property owner desires to meter and charge individual tenants for their portion of the electricity consumed.

Switchboard-Mount Meter—A meter mounted in a drawn-out case where the meter may be removed as a functional module, with provisions to properly shunt current paths before meter disconnection, leaving behind the outer case to which service connections are permanently made.

Synchronism—It expresses the phase relationship between two or more periodic quantities of the same period when the phase difference between them is zero. A generator must be in synchronism with the system before it is connected to the system.

Temperature Compensation—In reference to a watthour meter, refers to the factors included in the design and construction of the meter which make it perform with good accuracy over a wide range of temperatures. In modern meters this range may extend from -20°F to +140°F.

Testing, Statistical Sample—A testing method which conforms to accepted principles of statistical sampling based on either the variables or attributes method. The following expressions are associated with statistical sample testing:

(a) **Method of Attributes**—A statistical sample testing method in which only the percentage of meters tested found outside certain accuracy limits is used for determining the quality or accuracy of the entire group of meters.

(b) **Method of Variables**—A statistical sample testing method in which the accuracy of each meter tested is used in the total results for determining the quality or accuracy of the entire group of meters.

(c) **Bar X**—A mathematical term used to indicate the average accuracy of a group of meters tested.

(d) **Sigma**—A mathematical term used to indicate the dispersion of the test results about the average accuracy (Bar X) of a group of meters tested.

Thermocouple—A pair of dissimilar conductors so joined that two junctions are formed. An electromotive force is developed by the thermoelectric effect when the two junctions are at different temperatures.

Thermoelectric Effect (Seebeck Effect)—One in which an electromotive force results from a difference of temperature between two junctions of dissimilar metals in the same circuit.

Thermoelectric Laws—(1) The thermoelectromotive force is, for the same pair of metals, proportional through a considerable range of temperature to the excess of temperature of the junction over the rest of the circuit. (2) The total thermoelectromotive force in a circuit is the algebraic sum of all the separate thermoelectromotive forces at the various junctions.

Three-Wire System (Direct Current, Single-Phase, or Network Alternating Current)—A system of electric supply comprising three conductors, one of which (known as the neutral wire) is generally grounded and has the same approximate voltage between it and either of the other two wires (referred to as the outer or "hot" conductors). Part of the load may be connected directly between the outer conductors; the remainder being divided as evenly as possible into two parts, each of which is connected between the neutral and one outer conductor.

Time-of-Use Metering—A metering method which records demand during selected periods of time so consumption during different time periods can be billed at different rates.

Time Division Multiplication—An electronic measuring technique which produces an output signal proportional to two inputs, for example voltage and current. The width or duration of the output signal is proportional to one of the input quantities; the height is proportional to the other. The area of the signal is then proportional to the product of the two inputs, for example voltage and current.

Torque of an Instrument—The turning moment produced by the electric quantity to be measured acting through the mechanism.

Transducer—A device to receive energy from one system and supply energy, of either the same or of a different kind, to another system, in such a manner that the desired characteristics of the energy input appear at the output.

Transformer—An electric device without moving parts which transfers energy from one circuit to one or more other circuits by means of electromagnetic fields. The name implies, unless otherwise described, that there is complete electrical isolation among all windings of a transformer, as contrasted to an auto-transformer.

Var—The term commonly used for volt-amperes reactive.

Varhour Meter—An electricity meter that measures and registers the integral, with respect to time, of the reactive power of the circuit in which it is connected. The unit in which this integral is measured is usually the kilovarhour.

Volt—The practical unit of electromotive force, or potential difference. One volt will cause one ampere to flow when impressed across a one ohm resistor.

Volt-Ampere—Volt-amperes are the product of volts and the total current which flows because of the voltage. In DC circuits, and AC circuits with unity power factor, the volt-amperes and the watts are equal. In AC circuits at other than unity power factor, the volt-amperes equal the square root of watts squared plus reactive volt-amperes squared.

Voltage Transformer—An instrument transformer intended for measurement or control purposes which is designed to have its primary winding connected in parallel with a circuit, the voltage of which is to be measured or controlled.

Watt—The practical unit of active power which is defined as the rate at which energy is delivered to a circuit. It is the power expended when a current of one ampere flows through a resistance of one ohm.

Watthour—The practical unit of electric energy which is expended in one hour when the average power during the hour is one watt.

Watthour Meter—An electricity meter that measures and registers the integral, with respect to time, of the active power of the circuit in which it is connected. This power integral is the energy delivered to the circuit during the interval over which the integration extends, and the unit in which it is measured is usually the kilowatthour.

Watthour Meter Portable Standard (Rotating Standard)—A special watthour meter used as the reference for tests of other meters. The standard has multiple current and voltage coils or electronic equivalents, so a single unit may be used in the field or in the shop for tests of any normally rated meter. The portable standard watthour meter is designed and constructed to provide better accuracy and stability than would normally be required in customer meters.

CHAPTER 3

MATHEMATICS FOR METERING
(A Brief Review)

BASIC LAWS OF EQUATIONS

AN EQUATION is a statement of equality in mathematical form. In the study of electricity one of the most familiar equations is the expression of Ohm's law for DC circuits:

$$I = \frac{E}{R}$$

Where: I = current in amperes
E = potential in volts
R = resistance in ohms

This law, in equation form, states that the amperes are equal to the volts divided by the ohms.

To understand equations and to make them useful, certain rules must be remembered. One important rule states that when the identical operation is performed on both sides of the equal sign, the equation remains true.

If both sides of the equation are multiplied by the same quantity, or divided by the same quantity, the equation is still true, as shown in the following examples.

With the equation $I = \frac{E}{R}$, multiply both sides by R to give

$$I \times R = \frac{E}{R} \times R$$

which simplifies to $IR = E$.

With the equation $IR = E$, if the value of R is wanted, divide both sides by I to give

$$\frac{IR}{I} = \frac{E}{I}$$

which simplifies to

$$R = \frac{E}{I}$$

The same quantity may be added to, or subtracted from, both sides of an equation without violating the state of equality.

For example, in a parallel circuit the total current is equal to the sum of the currents in the branches. With three resistors connected in parallel, the equation for the total circuit current may be expressed as follows:

$$I_{Total} = I_1 + I_2 + I_3$$

To determine the value of I_2, subtract I_1 and I_3 from both sides of the equation:

$$I_{Total} - I_1 - I_3 = I_1 + I_2 + I_3 - I_1 - I_3$$

which simplifies to

$$I_{Total} - I_1 - I_3 = I_2$$

The preceding example illustrates another general rule which states that any complete term may be shifted from one side of an equation to the other by changing its sign. This must be a complete term or the equation is no longer true. In the example, the $+I_1$ and $+I_3$ terms were shifted from the right to left side where they became minus.

To summarize:

If $x = 2y$, and C is any constant except zero, then the following equations are also true:

$$x + C = 2y + C$$
$$x - C = 2y - C$$
$$Cx = 2Cy$$
$$\frac{x}{C} = \frac{2y}{C}$$

Parentheses

In an expression such as $IR_1 + IR_2 + IR_3$, the subscripts 1, 2, and 3 after the symbol R indicate that R does not necessarily have the same value in each term. Since the symbol I does not have subscripts, it does have the same value in each term. Such a series can be written $I(R_1 + R_2 + R_3)$ which means that the total quantity inside the parentheses is to be multiplied by I.

The equation $E = IR_1 + IR_2 + IR_3$ states that the voltage across three resistors in series is equal to the sum of the products of the current times each of the resistances. If E, R_1, R_2 *and* R_3 are all known and the value of I is wanted, the equation may be rearranged as follows:

$$E = IR_1 + IR_2 + IR_3$$

$$E = I(R_1 + R_2 + R_3)$$

Dividing both sides by $(R_1 + R_2 + R_3)$ gives

$$\frac{E}{(R_1 + R_2 + R_3)} = I$$

When removing parentheses, pay attention to the laws of signs which are summarized as follows:

$$a \times b = ab$$
$$a \times (-b) = -ab$$
$$-a \times b = -ab$$
$$-a \times (-b) = ab$$

No sign before a term implies a + sign.

A minus sign before a parenthetical expression is equivalent to multiplication by −1 and means all signs within the parentheses must be changed when the parentheses are removed.

THE GRAPH

A graph is a pictorial representation of the relationship between the magnitudes of two quantities. A graph may represent a mathematical equation, or the relationship between the quantities may be such that it cannot be expressed by any simple equation.

An example of a graph which may be used in metering is the calibration curve for an indicating instrument. Figure 3-1 shows a typical graph of voltmeter corrections. When the voltmeter reads 120 volts, reference to the correction curve shows that at this point, marked X in Fig. 3-1, the correction to be applied is + 1 volt and the true voltage is 121 volts. With a scale reading of 100 volts, the correction is +0.5 volts, shown by ○ and the true voltage is 100.5. With a scale reading of 70 volts, the correction is −0.5 volts, shown by □, to give 69.5 true volts.

The sine wave has important applications in alternating-current circuit theory. The equation of a sine wave, $y = \sin x$, is shown graphically in Fig. 3-2. The x quantity is commonly expressed in angular degrees and the y values which are plotted on the graph are the sine values of the corresponding angles. Thus, for any particular angular value, it is possible to use the graph to determine its sine. For example, the sine of 30 degrees is equal to 0.5, as shown by the dotted lines on Fig. 3-2.

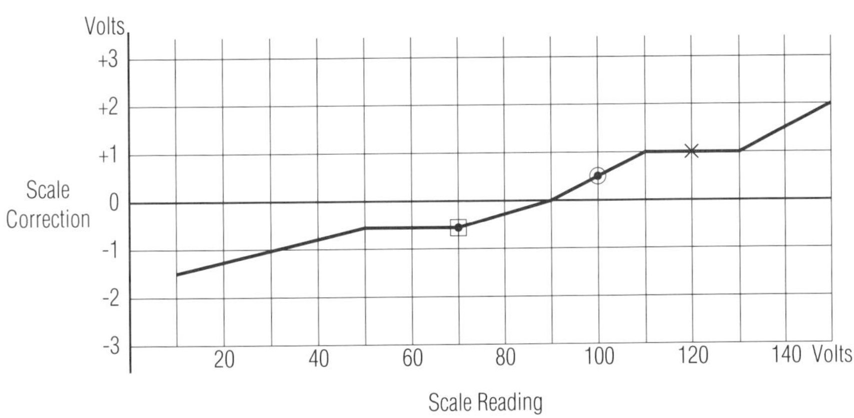

Figure 3-1. Graph of Voltmeter Corrections.

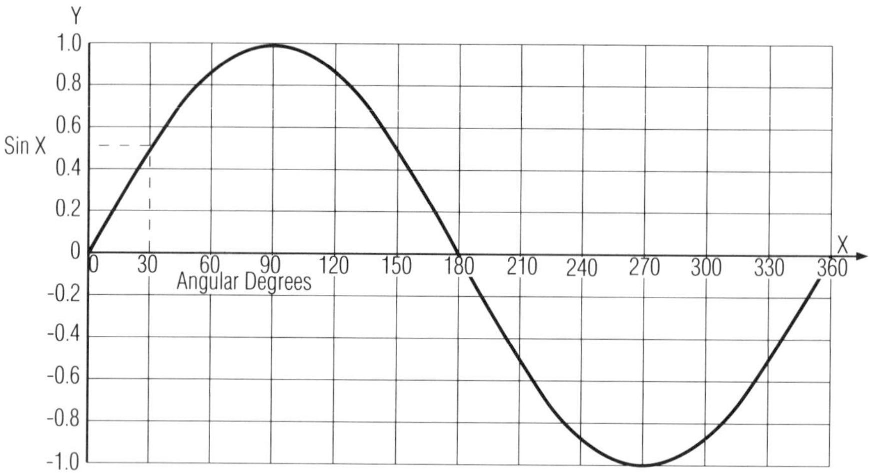

Figure 3-2. Graph of Sine Wave.

THE RIGHT TRIANGLE

A right triangle is a triangle having one right (90°) angle. In every right triangle a definite relation exists between the sides of the triangle, so that when the lengths of two of the sides are known, the length of the third can be calculated using the right triangle formula. The side opposite the right angle is termed the hypotenuse of the triangle and the two sides forming the right angle are known as the legs of the triangle. Mathematically this relationship is stated in the formula:

$$c^2 = a^2 + b^2$$

The relationship between the sides of the right triangle in Fig. 3-3 are:

$$a^2 = c^2 - b^2 \text{ or } a = \sqrt{c^2 - b^2}$$

$$b^2 = c^2 - a^2 \text{ or } b = \sqrt{c^2 - a^2}$$

$$c^2 = a^2 + b^2 \text{ or } c = \sqrt{a^2 + b^2}$$

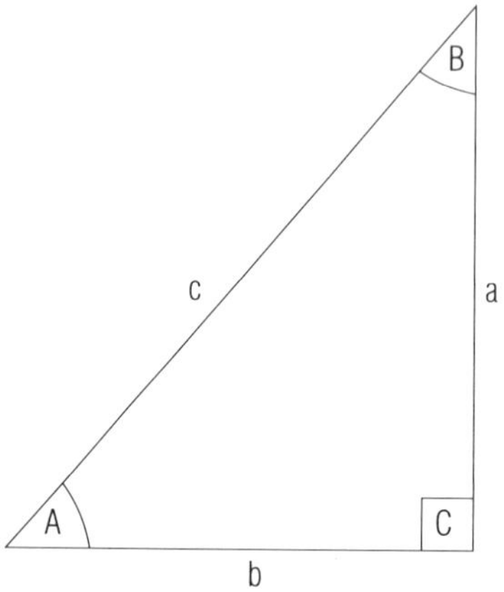

Figure 3-3. The Right Triangle.

TRIGONOMETRIC FUNCTIONS

Sine, cosine, and tangent are three of the relationships existing between the sides of a right triangle. The six possible ratios between the sides of a right triangle are called trigonometric functions. In the solution of alternating-current problems requiring the use of trigonometric functions, the following tabulation of their definitions and relationships will be useful.

In the right triangle of Fig. 3-3, side *a* is opposite angle *A*; side *b* is adjacent to angle *A*; side *a* is adjacent to angle *B*; and side *b* is opposite angle *B*. The ratios between the length of sides of the triangle determine the trigonometric functions of the angle *A* follows:

By definition, the ratios are named:

$$\frac{a}{c} = \frac{\text{opposite}}{\text{hypotenuse}} = \text{sine } A \textit{ or } \sin A$$

$$\frac{b}{c} = \frac{\text{adjacent}}{\text{hypotenuse}} = \text{cosine } A \textit{ or } \cos A$$

$$\frac{a}{b} = \frac{\text{opposite}}{\text{adjacent}} = \text{tangent } A \textit{ or } \tan A$$

The reciprocals of these functions are:

$$\frac{b}{a} = \frac{\text{adjacent}}{\text{opposite}} = \text{cotangent } A \textit{ or } \cot A$$

$$\frac{c}{b} = \frac{\text{hypotenuse}}{\text{adjacent}} = \text{secant } A \textit{ or } \sec A$$

$$\frac{c}{a} = \frac{\text{hypotenuse}}{\text{opposite}} = \text{cosecant } A \textit{ or } \csc A$$

From before:

$$\sin A = \frac{a}{c} = \frac{1}{\frac{c}{a}} = \frac{1}{\csc A} = \frac{\frac{b}{c}}{\frac{b}{a}} = \frac{\cos A}{\cot A}$$

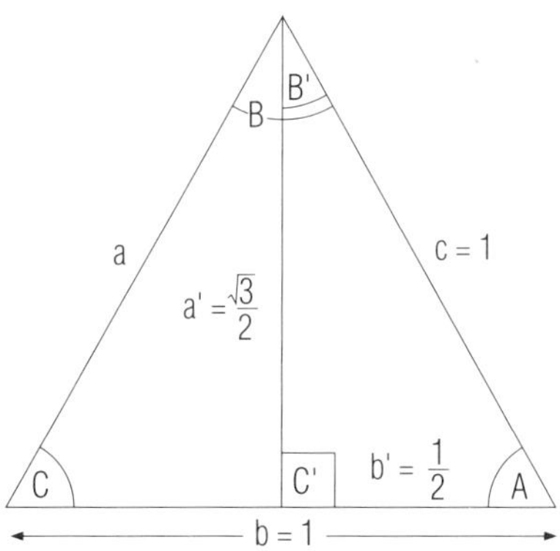

Figure 3-4. Functions of 30° and 60°.

Similarly:

$$\cos A = \frac{1}{\sec A} = \frac{\sin A}{\tan A}$$

$$\tan A = \frac{1}{\cot A} = \frac{\sin A}{\cos A}$$

$$\cot A = \frac{1}{\tan A} = \frac{\cos A}{\sin A}$$

$$\sec A = \frac{1}{\cos A} = \frac{\tan A}{\sin A}$$

$$\csc A = \frac{1}{\sin A} = \frac{\cot A}{\cos A}$$

While $\frac{a}{c}$ is the sine of *A*, it is also the cosine of *B*, since *a* is adjacent to angle *B*. Therefore, it will be seen that:

$$\sin A = \cos B \qquad \cot A = \tan B$$
$$\cos A = \sin B \qquad \sec A = \csc B$$
$$\tan A = \cot B \qquad \csc A = \sec B$$

Numerical values for the functions of every angle are computed from the ratios of the sides of a right triangle containing that angle. In a right triangle, which is a triangle containing one right (90-degree) angle, the sum of the other two angles must equal 90 degrees, since the sum of the three angles of any triangle must be 180 degrees. Also, the sum of the squares of the two shorter sides must equal the square of the longer side or hypotenuse (the side opposite the 90-degree angle). *Regardless of the values which may be assigned to the sides of a right triangle, the ratio of any two sides for any given angle is always the same.*

The functions of 30 degrees and 60 degrees may be derived from Fig. 3-4. The triangle ABC of Fig. 3-4 is equilateral (having equal sides). Therefore, it is also equiangular (having equal angles), so that each angle equals 60 degrees.

If a line is drawn from the midpoint of the base to the vertex as shown, then $b' = 1/2$ and angle $B' = 30$ degrees. In the triangle $AB'C'$:

$$\sin 30° = \cos 60° = \frac{b'}{1} = \frac{\frac{1}{2}}{1} = \frac{1}{2} = 0.500$$

$$\cos 30° = \sin 60° = \frac{a'}{1} = \frac{\frac{1}{2}\sqrt{3}}{1} = \frac{1}{2}\sqrt{3} = 0.866$$

$$\tan 30° = \cot 60° = \frac{b'}{a'} = \frac{\frac{1}{2}}{\frac{1}{2}\sqrt{3}} = \frac{1}{\sqrt{3}} = 0.577$$

The functions of 45 degrees may be derived from Fig. 3-5. In this triangle, $a = b = 1$, and $c = \sqrt{a^2 + b^2} = \sqrt{2}$. From geometry, angle A must equal angle B, or one-half of 90 degrees, or 45 degrees.

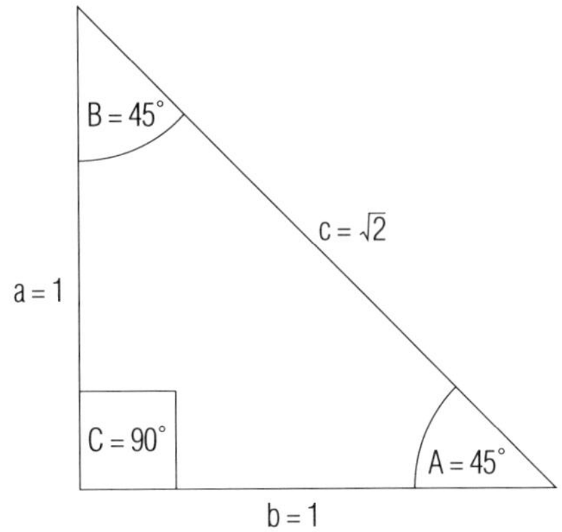

Figure 3-5. Functions of 45°.

$$\sin A = \sin B = \sin 45° = \frac{1}{\sqrt{2}} = \frac{\sqrt{2}}{2} = 0.707$$

$$\cos A = \cos B = \cos 45° = \frac{1}{\sqrt{2}} = \frac{\sqrt{2}}{2} = 0.707$$

$$\tan A = \tan B = \tan 45° = \frac{1}{1} = 1 = 1.000$$

To determine the functions of an angle greater than 90 degrees and less than 180 degrees, subtract the angle from 180 degrees and refer to a table of functions. For an angle greater than 180 degrees and less than 270 degrees, subtract 180 degrees from the angle and refer to the table. For an angle greater than 270 degrees and less than 360 degrees, subtract the angle from 360 degrees and refer to the table.

The algebraic signs of the functions of all angles between 0 degrees and 90 degrees are +; beyond 90 degrees the signs can be determined from Table 3-1.

Table 3-1. Signs of the Functions of Angles.

	sin cos	tan cot	sec csc
1st Quadrant (0° to 90°)	+ +	+ +	+ +
2nd Quadrant (90° to 180°)	+ –	– –	– +
3rd Quadrant (180° to 270°)	– –	+ +	– –
4th Quadrant (270° to 360°)	– +	– –	+ –

From the preceding formulas, it is evident that if two sides of a right triangle are known, the third side and the angles can be calculated. Also, if one side and either angle *A* or *B* are known, the other sides and angle can be calculated.

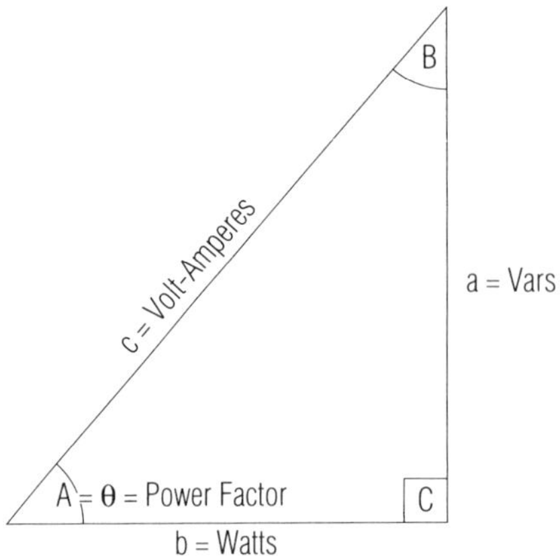

Figure 3-6. Relationship of Right Triangle to Volt-Amperes, Watts, and Vars.

Example:

To find a, given c and b: $a = \sqrt{c^2 - b^2}$

To find a, given c and A: $a = c \times \sin A$

To find b, given a and A: $b = \dfrac{a}{\tan A}$

To find B, given a and c: B = angle whose cosine is $\dfrac{a}{c}$

Since the volt-amperes, watts, and vars in a circuit are in proportion to the sides of a right triangle they may be represented as shown in Fig. 3-6. Trigonometry may also be used in calculating these quantities.

$$\text{Power factor} = \text{cosine of phase angle } \theta = \frac{\text{watts}}{\text{volt-amperes}}$$

Example: Voltmeter reads 120, ammeter 5, wattmeter 300.

$$\text{pf} = \frac{300}{120 \times 5} = 0.5$$

Power factor = 50%.

From Fig. 3-6:

$$\text{Tangent of phase angle: } \tan\theta = \frac{\text{vars}}{\text{watts}} = \frac{\text{varhours}}{\text{watthours}}$$

Example: Varhour meter reads 3733, watthour meter 9395, what is the power factor?

$$\frac{3733}{9395} = 0.3973 = \text{tangent of phase angle } \theta = \text{tangent of } 21°40'.$$

Power factor = cosine of phase angle $\theta = \cos 21°40' = 0.93$ or 93%.

To find watts, given volt-amperes and vars:

$$\text{watts} = \sqrt{(\text{volt-amp})^2 - (\text{vars})^2}$$

To find power factor, given vars and volt-amperes:

$$\text{Power factor} = \text{cosine of angle whose sine} = \frac{\text{vars}}{\text{volt-amp}}$$

To find vars, given power factor and watts:

$$\theta = \text{angle whose cosine equals the power factor}$$

$$\text{vars} = \text{watts} \times \tan\,\theta$$

To find volt-amperes, given power factor and vars:

$$\theta = \text{angle whose cosine equals the power factor}$$

$$\text{Volt-amperes} = \frac{\text{vars}}{\sin\,\theta}$$

SCIENTIFIC NOTATION

Scientific notation is a form of mathematical shorthand. It is a method of indicating a number having a large number of zeros before or after the decimal point and it is based on the theory of exponents. Some powers of ten are shown in Table 3-2.

Any number may be expressed as a power of ten by applying the following rules:

1. To express a decimal fraction as a whole number times a power of ten, move the decimal point to the right and count the number of places back to the original position of the decimal point. The number of places moved is the correct negative power of ten.

Examples:

$$0.00756 = 7.56 \times 10^{-3}$$
$$0.000095 = 9.5 \times 10^{-5}$$
$$0.866 = 86.6 \times 10^{-2}$$
$$0.0866 = 86.6 \times 10^{-3}$$

2. To express a large number as a smaller number times a power of ten, move the decimal point to the left and count the number of places back to the original position of the decimal point. The number of places moved is the correct positive power of ten.

Table 3-2. Powers of Ten.

Number	Power of Ten	Expressed in English
0.000001	10^{-6}	ten to the negative sixth power
0.00001	10^{-5}	ten to the negative fifth power
0.0001	10^{-4}	ten to the negative fourth power
0.001	10^{-3}	ten to the negative third power
0.01	10^{-2}	ten to the negative second power
0.1	10^{-1}	ten to the negative first power
1.0	10^{-0}	ten to the zero power
10.0	10^{1}	ten to the first power
100.0	10^{2}	ten to the second power
1000.0	10^{3}	ten to the third power
10000.0	10^{4}	ten to the fourth power
100000.0	10^{5}	ten to the fifth power
1000000.0	10^{6}	ten to the sixth power

Examples:

746.	=	7.46	×	10^2
95.	=	9.5	×	10^1
866.	=	86.6	×	10^1
8,660.	=	8.66	×	10^3

PHASORS

Sine Wave Representation

"Phasor" is the name adopted by the electrical industry for what was formerly called a vector. Alternating current electrical quantities can be represented by either sine waves or by phasor diagrams. A sine wave represents the cycle of an electrical quantity from zero to maximum positive, through zero to maximum negative, and back to zero in repeated cycles having equal intervals of time.

The electrical value may be represented by a vertical line. This line shows magnitude only.

If while tracing such a line, starting at and returning to the center point, you pull the paper smoothly to the left, you will draw an approximate sine wave as shown in Fig. 3-8.

The movement of the paper introduces the factor of time. This identical principle is used in the light-beam oscillograph. The light beam reacting to an alternating voltage oscillates back and forth in a rectangular aperture. When viewed with the eye, only a straight line is visible. To obtain a record, a film is drawn past the aperture at constant speed by a motor. This introduces the time factor so that when the film is developed, the trace on the film is in the same form as the original alternating voltage—a sine wave.

The sine wave in electrical work is associated with the rotation of a generator and hence can be measured in degrees, like a circle. A complete cycle of the voltage waveform is 360 degrees.

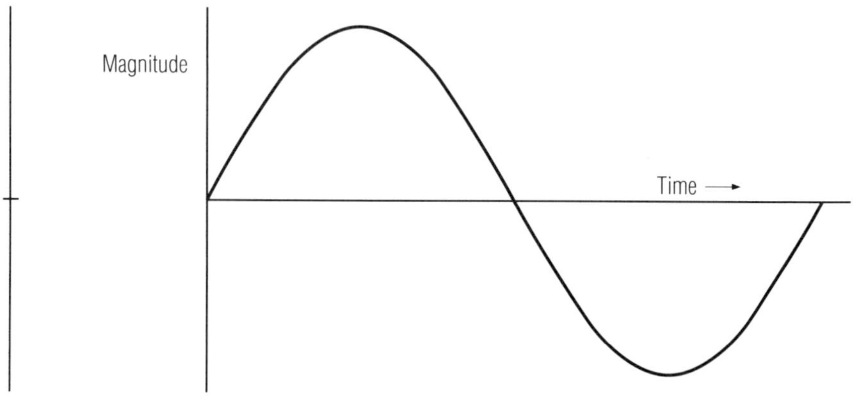

Figures 3-7. (left) and 3-8. Variation in Magnitude of Electrical Quantity without (left) and with Reference to Time.

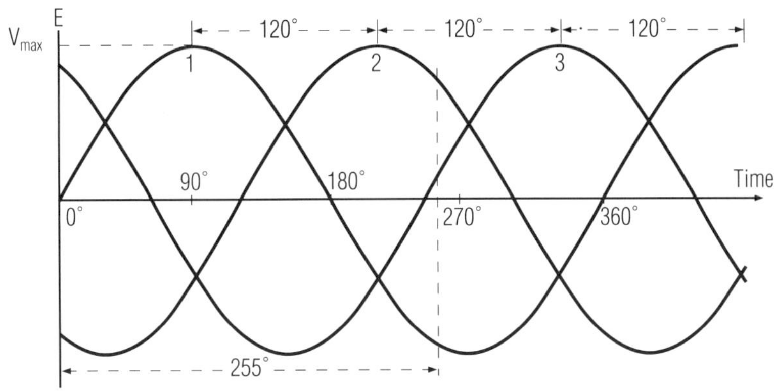

Figure 3-9. Three Equal Sine-Wave Voltages Symmetrically Displaced in Time.

Electrical calculations can be performed by the use of sine waves, but this method is unwieldy. For example, to determine the sum of instantaneous voltages in a three-phase circuit requires three sine waves placed 120 degrees apart, such as in Fig. 3-9.

It can be seen by inspection of Fig. 3-9 that at any instant the three voltages add up to zero. At the time indicated by the dotted line at 255 degrees there are the following sine values:

Curve 1:	sin 255°	=	−sin 75° =			−0.97
Curve 2:	sin (255° − 120°)	=	sin 135° =	sin 45°	=	0.71
Curve 3:	sin (255° − 240°)	=	sin 15° =			0.26
						0.00

In a phasor diagram, arrows are substituted for sine waves, such as in Fig. 3-10. These arrows are considered to rotate one complete revolution, or 360 degrees, while the sine wave passes through one cycle, or 360 degrees. The length

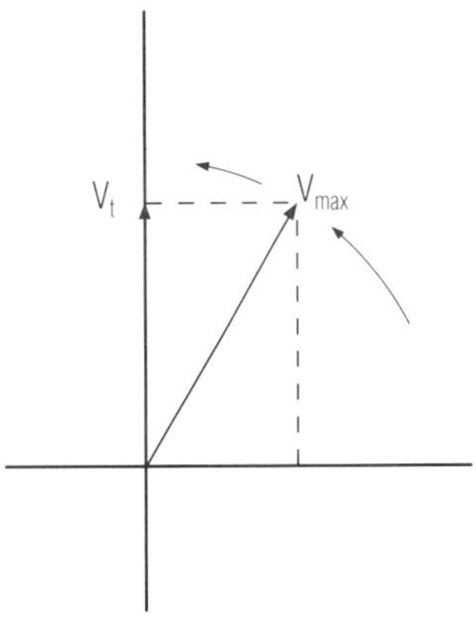

Figure 3-10. Phasor Representation.

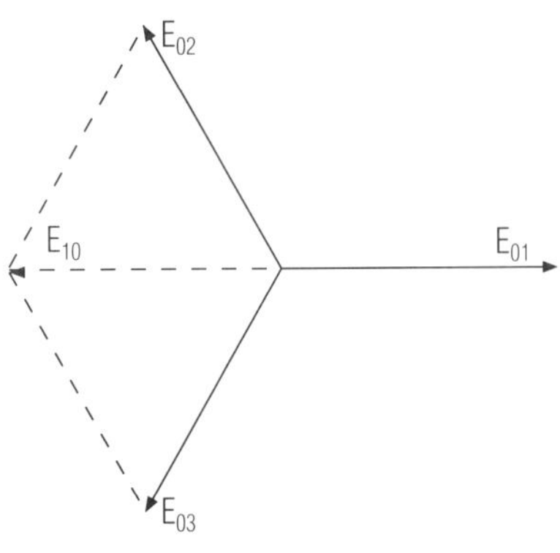

Figure 3-11. Voltage Phasors for Three-Phase Circuits.

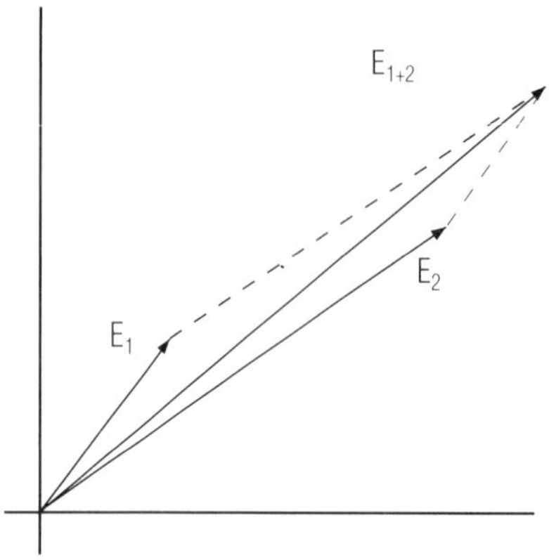

Figure 3-12a. Adding Phasors by Completing the Parallelogram.

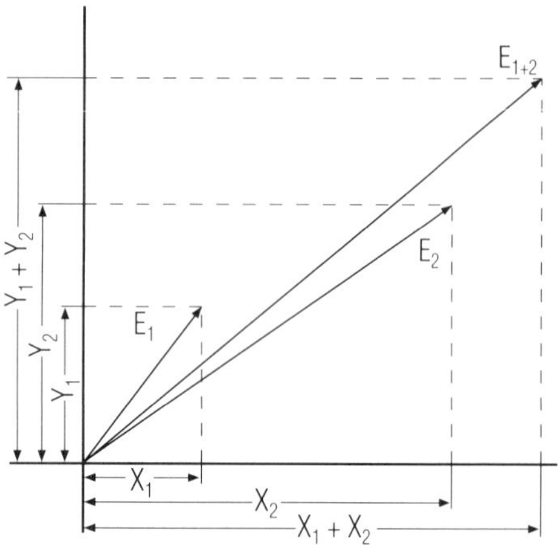

Figure 3-12b. Adding Phasors by Adding Vertical and Horizontal Projections.

of the arrow represents the distance V_{max} in the sine-wave diagram or the peak value of the wave. The position of the arrow represents the time element in the sine wave. Since the length of the arrow does not change, the instantaneous values of the electrical quantity represented by the phasor are obtained by the projection or shadow of the arrow on a vertical line drawn through the center of rotation. Thus in Fig. 3-10, V_{max} represents the peak value of the sine wave, and V_1 represents the instantaneous value of the voltage at this point in the rotation. Note that this projection, or shadow, increases through the same values that the sine wave passes through.

If we wish to determine the sum of the voltages in a three-phase circuit, as was done before by the sine wave method, by using phasors, the three phasors representing the three sine waves, 1, 2, and 3, are drawn as in Fig. 3-11, 120 degrees apart. Any two of these phasors may be added. It is then immediately apparent that the sum of these two phasors is equal to, but in the opposite direction from, the third phasor. The sum of the three must then be equal to zero.

Plotting the vertical projections of phasors against time will produce sine waves *with the correct magnitudes* only if the phasor lengths are equal to the peak values of the waves. The common practice is to use the effective values of voltages and currents for phasor lengths rather than peak values, since the relative phasor relationships are not changed. This procedure greatly simplifies working with phasor diagrams because the deflection of most alternating-current ammeters and voltmeters is proportional to the effective value.

Phasors can be added to, or subtracted from, other phasors. This can be done either by completing the parallelogram and drawing the diagonal, as in Fig. 3-12a, or by adding both vertical and horizontal projections, as in Fig. 3-12b.

In all discussions about alternating-current power, the concept of power factor must be included. Power factor results from the phase difference between the voltage and current in the circuit considered. This phase difference can be shown by two sine waves *E* and *I*, as in Fig. 3-13a. It is possible to compute average power from such sine-wave representation, but it is difficult. The maximum

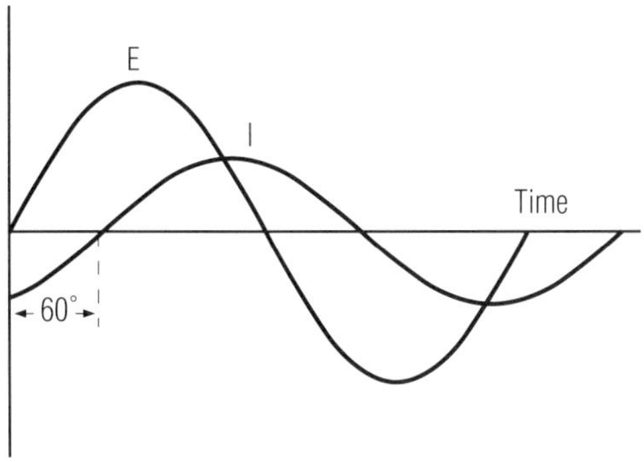

Figure 3-13a. Sine-Wave Representation of 50 Percent Power Factor Lagging Current.

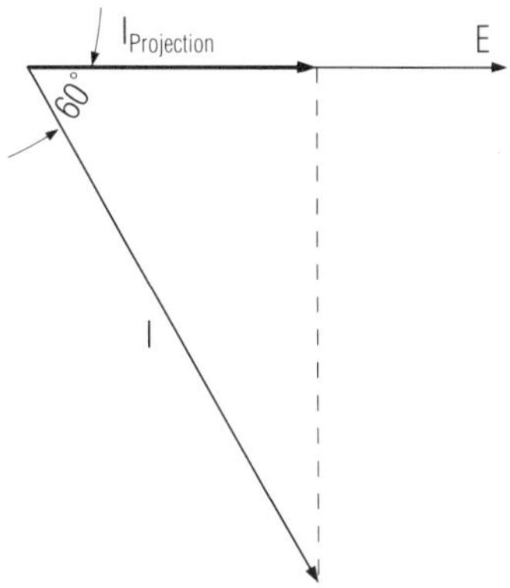

Figure 3-13b. Phasor Representation of 50 Percent Power Factor Lagging Current

voltage acts not with the maximum current but with the instantaneous current value occurring at the time that the maximum voltage occurs. A phasor diagram simplifies the calculation. Figure 3-13b is the phasor diagram for the sine-wave diagram in Fig. 3-13a, the phasors *E* and *I* representing the effective values of the sine waves. The arrow shows the direction of rotation of the phasors. It is seen that the *I* phasor lags the *E* phasor by 60 degrees. The active *I* component is the projection of the *I* phasor on the *E* phasor. This is equal to *I* times the cosine of the angle between the two phasors, in this case 60 degrees. Thus, the power is equal to *EI* cos 60 degrees.

In the use of phasor diagrams, several rules or conventions must be observed. These are:

1. A phasor can be moved to any part of the diagram if its magnitude and direction are maintained. As an example, the phasor diagram of a voltage transformer shows the induced voltage *ES*, the voltage losses *IX* and *IR*, and the terminal voltage *VS*. All the phasors can have a common origin, as in Fig. 3-14a, or be strung in a chain, as in Fig. 3-14b. Both phasor diagrams are identical in meaning and are correct. However, the concept of phasors rotating to generate sine waves would apply only in Fig. 3-14a, which has a common origin.
2. Double-subscript notation (for example V_{12}) is used to designate between which two points and the direction, or instantaneous polarity, of the voltage. However, a phasor diagram is not a circuit diagram and no attempt should be made to draw phasor diagrams in the same schematic configurations as their corresponding circuits.
3. Reversing the order of the double subscripts on the phasor reverses the direction of the phasor. Thus, in Fig. 3-15, E_{01} and E_{10} represent one circuit voltage. Only the instantaneous polarity, or circuit direction, is different.
4. The direction of rotation of all phasors in a single diagram must be the same.
5. Only phasors having the same frequency can be shown in one diagram.
6. The direction of rotation of a phasor should be indicated by a separate arrow near the diagram or, if no arrow is shown, it will be assumed the rotation is counterclockwise.

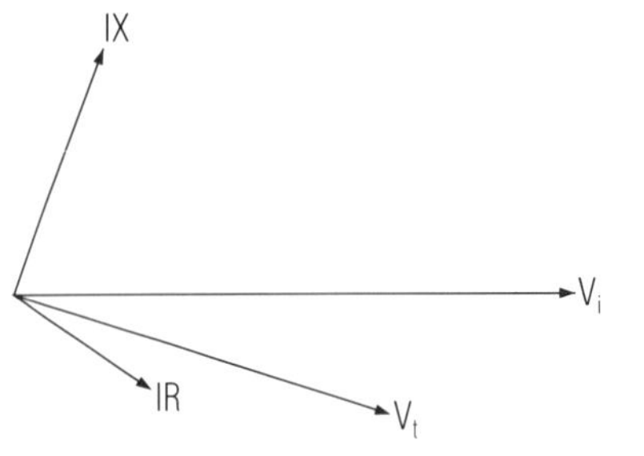

Figure 3-14a. Phasors Having a Common Origin.

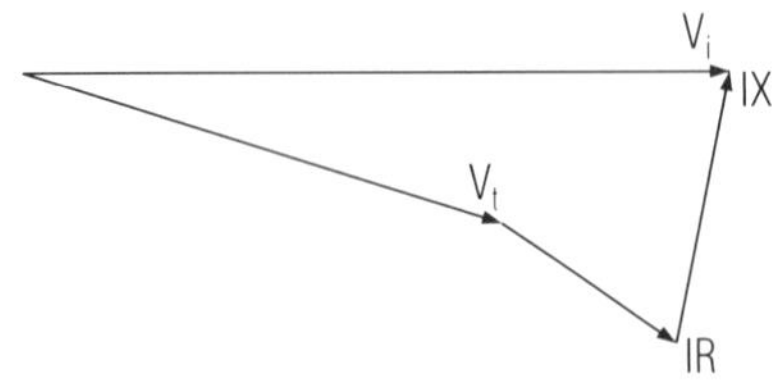

Figure 3-14b. Phasors Drawn in Chain.

7. The addition or subtraction of two phasors is accomplished by adding or subtracting the projections of these phasors that are in the same direction. Usually the horizontal and vertical projections are used.
8. Addition and subtraction of phasors can be accomplished on a single phasor diagram.
9. Multiplication and division of two phasors has the effect of changing frequency and cannot be represented on one phasor diagram. Thus volts and amperes can be shown on one diagram but power ($EI \cos \theta$), vars ($EI \sin \theta$), and volt-amperes (EI) cannot be presented on the same diagram.
10. Phasors represent sine waves only. *Non-sinusoidal waveforms cannot be represented by a phasor.* A non-sinusoidal waveform cannot be generated by the rotation at a constant rate of a phasor of constant length which is a basic principle in phasor representation.
11. Only phasors representing like quantities can be added or subtracted. One can add or subtract voltage phasors or current phasors, but one cannot add a voltage phasor to a current phasor.

To illustrate again the advantages of a phasor diagram over a sine-wave diagram, refer to Figs. 3-16a, 3-16b, and 3-16c. Here a series circuit is shown which contains a generator with its associated voltage rise (E_{12}) supplying a

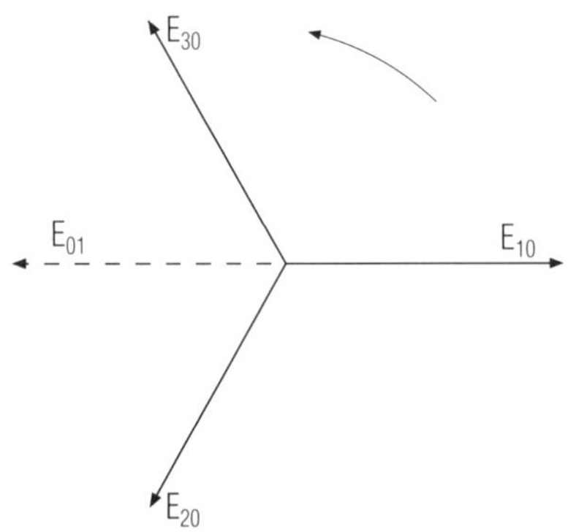

Figure 3-15. Relation of Phasor Subscript Order and Phasor Direction.

resistance and inductive load between points 3 and 5. The phasor diagram representing all voltages and currents is shown in Fig. 3-16b. The sine-wave diagram for the same circuit is shown in Fig. 3-16c. This sine-wave diagram also shows the power absorbed by the resistor R as a product of V_{34} and I_{23}, the power wave being drawn one-half scale. Note that the power wave has two cycles that are displaced from the voltage and current waves. This double-frequency power wave cannot be represented in the phasor diagram of Fig. 3-16b.

The phasor diagram is basically a picture of voltages and currents at one instant, while the sine-wave diagram shows the relations for a period of time and thus looks more complicated. The phasor diagram is extremely useful for metering application. It lends itself readily to illustrating the relations between currents and voltages in a single or multiphase meter installation showing which current

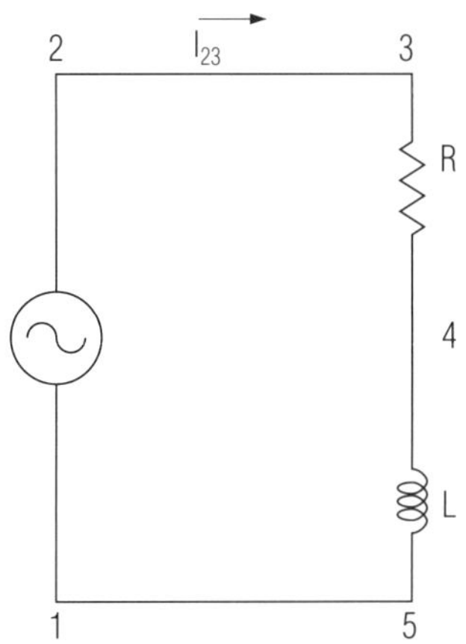

Figure 3-16a. Series Circuit.

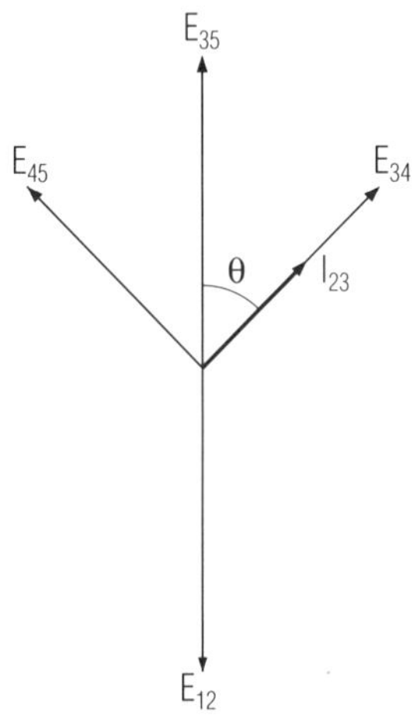

Figure 3-16b. Phasor Diagram of Series Circuit of Fig. 3-16a.

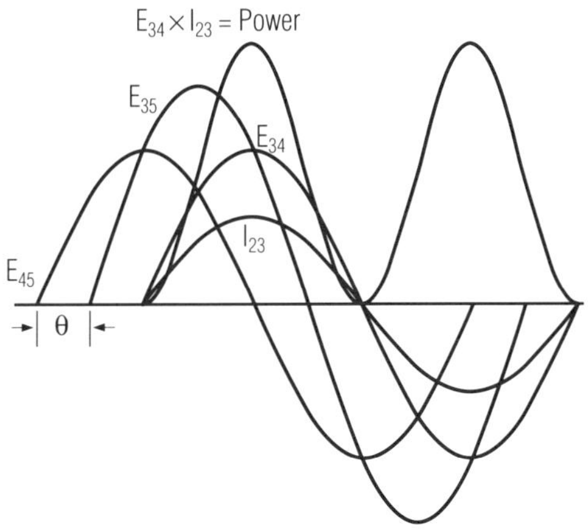

Figure 3-16c. Sine-Wave Diagram for Series Circuit of Fig. 3-16a.

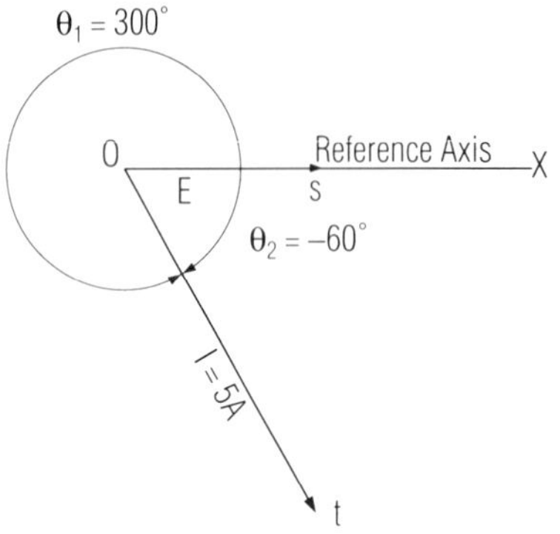

Figure 3-17a. Phasors in Polar Coordinate System.

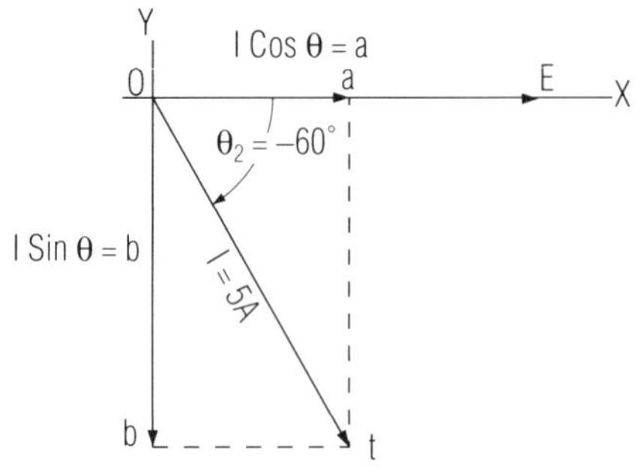

Figure 3-17b. Phasors in Rectangular Coordinate System.

coils act with which voltage coil. The phase angle between a current and voltage is easily determined from a phasor diagram where the units of measure already are in angular degrees or radians.

A sine-wave diagram, though awkward for metering problems, contains much basic information about alternating-current theory that should not be overlooked. It is useful for showing the difference between absolute value and rate of change. For example, when a magnetizing current waveform crosses the zero line, its absolute value is zero, but its rate of change is at a maximum. Induced voltage, which is proportional to rate of change of magnetizing current is, therefore, at maximum at the time magnetizing current is zero, thereby establishing a 90-degree phase difference between these two waveforms. Both sine-wave diagrams and phasor diagrams should be firmly remembered.

Phasor Calculations

This section is not intended for the beginner. However, advanced metering personnel who may have occasion to perform calculations with phasor quantities will find it helpful.

Phasor magnitude is represented by a straight line of a length proportional to the magnitude of the quantity represented. Its direction can be graphically represented by its position in either a polar or rectangular coordinate system. Its position can be described mathematically by either polar or rectangular number notation. Phasor direction is graphically represented by an arrowhead on the line, and it is common practice to use open arrowheads on voltage phasors and closed arrowheads on current phasors.

In Figs. 3-17a and 3-17b, the reference axis is placed at a position commonly labeled as 0°. Measurement of angles from the reference axis in the counterclockwise direction generates a positive (+) angle. Measurement in the clockwise direction will generate a negative (–) angle, as shown by θ_2 in Fig. 3-17a and θ in Fig. 3-17b. Phasor rotation is conventionally taken as counterclockwise unless otherwise indicated by an arrow.

In the polar coordinate system of Fig. 3-17a, the length of line $\overline{Ot}$ represents magnitude of the current I, and the closed arrowhead the direction. When $\overline{Ot}$ is used to express a phasor length it is written with a bar above it, as $\overline{Ot}$. If the voltage E in the diagrams was used in a phasor equation, such as $P = \dot{E}\dot{I}$, it would be written with a dot above it as shown to signify that it was a phasor quantity and not an algebraic expression. The phasor $\overline{Os}$ indicates that the voltage E is taken as the reference by placing it on the OX reference. The position of the phasor I is expressed mathematically either by $I\underline{/-60°}$, which is read "current I at an angle of minus 60 degrees," or by $I\underline{/300°}$, which is read "current I at an angle of 300 degrees." If in the diagram $I = 5$ amperes, the phasor I represents a current of 5 amperes at an angle of –60 degrees, or lagging voltage E by 60 degrees.

Figure 3-17b shows the same phasors on a rectangular coordinate system. Here the phasor is described by its projections on the X and Y axes, both of which may be either positive or negative depending upon the direction from the center (origin) O. X is positive to the right and Y is positive above the X axis. The letter j is used in electrical problems to designate the projection on the Y axis. The mathematical description of a phasor on a rectangular coordinate system has the general form of $a + jb$, where a is the projection on the X axis and b is the projection on the Y axis. Either a or b may be positive or negative depending upon the sign of the X or Y axis on which the projection falls. Expressions in the form of $a \pm jb$ are called complex numbers.

The numerical value of the X axis projection a is the scalar value of the phasor times the cosine of the angle of rotation θ of the phasor from the reference axis X. The numerical value of the Y axis projection b is the product of the scalar value of the phasor times the sine of the angle of rotation θ of the phasor from the reference axis.

The phasor in Fig.3-17b is therefore written:

$$\begin{gathered} I\cos\theta + jI\sin\theta \text{ or} \\ 5\cos(-60°) + j5\sin(-60°) \text{ or} \\ 5 \times 0.5 + j5 \times (-0.866) \text{ or} \\ 2.5 - j4.33 \end{gathered}$$

In Fig. 3-17b line $\overline{Ot}$ is the hypotenuse of the triangle having legs $\overline{Oa}$ and $\overline{at}$. From trigonometry, $\overline{Ot}^2 = \overline{Oa}^2 + \overline{at}^2$. From this relationship there can be developed a means of changing from a rectangular coordinate system to a polar system or the reverse. In Fig. 3-17b, $\overline{Oa}$ is the projection a of I on the X axis. Also, $\overline{at}$ is the projection b of I on the Y axis.

From the preceding:

$$\begin{aligned} \overline{Ot} &= I \text{ and } \overline{Ot}^2 = I^2 \\ \overline{Oa} &= a \text{ and } \overline{Oa}^2 = a^2 \\ \overline{at} &= b \text{ and } \overline{at}^2 = b^2 \\ \overline{Ot} &= \overline{Oa}^2 + \overline{at}^2 \text{ and } I^2 = a^2 + b^2 \\ \text{or } I &= \sqrt{a^2 + b^2} \end{aligned}$$

$$\text{The tangent of } \theta = \frac{\overline{at}}{\overline{Oa}} = \frac{b}{a} \text{ or } \theta = \underline{/\tan^{-1}\frac{b}{a}}$$

$$\left(\text{read "}\theta \text{ equals the angle whose tangent is } \frac{b}{a}\text{"}\right)$$

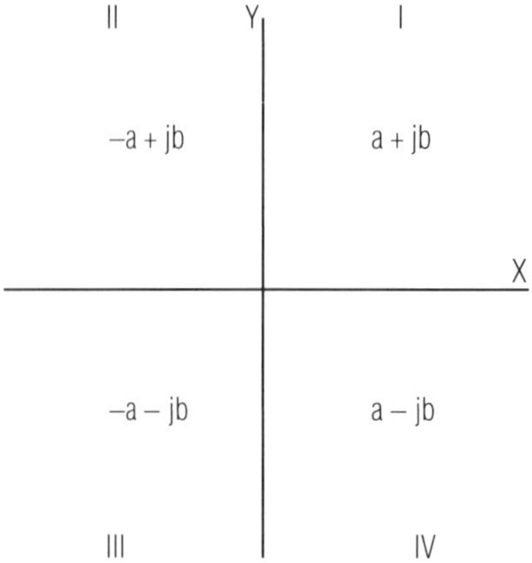

Figure 3-18. Algebraic Signs in Rectangular Coordinate System.

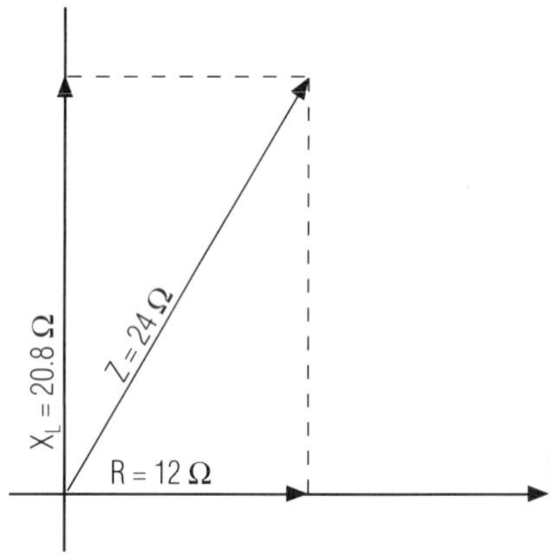

Figure 3-19. Impedance Diagram.

The phasor in Fig. 3-17a was written $I\underline{/\theta}$. Since I can be expressed as $\sqrt{a^2 + b^2}$ and θ as $\underline{/\tan^{-1} \frac{b}{a}}$, the equation $I\underline{/\theta} = \sqrt{a^2 + b^2} \; \underline{/\tan^{-1} \frac{b}{a}}$ may be used to convert from one system to the other.

Figure 3-18 shows the relationship between the signs of a and jb to the position of the phasor.

From Fig. 3-18 it is seen that the relative signs of a and jb provide the information needed to locate the phasor. That is:

$a + jb$ must be in first quadrant
$-a + jb$ must be in second quadrant
$-a - jb$ must be in third quadrant
$a - jb$ must be in fourth quadrant

Complex numbers may be used to describe impedance in a series circuit. It must be kept in mind that the signs of a and jb are determined by the quadrant in which the impressed voltage phasor falls, assuming that the current in a series circuit is taken as the reference in the phasor diagram. As an example:

Assuming E in Fig. 3-17b is 120 volts, the impedance is

$$Z = \frac{E}{I} = \frac{120}{5} = 24 \text{ ohms}$$

The impedance diagram would be as in Fig. 3-19.

In an impedance diagram for a series circuit, the resistance R is placed on the X axis as the reference, since the voltage drop across a resistor is in phase with the current through it. Since the current through an inductance lags the voltage across it by 90 degrees, the inductor voltage-drop phasor leads the current phasor by 90 degrees so that the inductive reactance leads the resistance by 90 degrees. Similarly, the capacitive reactance lags the resistance by 90 degrees since the current in a capacitance leads the voltage drop across it.

In Fig. 3-17b, the current phasor lagged the voltage phasor by 60 degrees. In Fig. 3-19 the impedance leads the resistance by 60 degrees, so that

$$\begin{aligned} Z\angle\theta &= 24\angle 60^\circ \\ &= 24\cos 60^\circ + j24\sin 60^\circ \\ &= 24\times 0.5 + j24\times 0.866 \\ &= 12 + j20.8 \text{ ohms} \end{aligned}$$

Phasors expressed in rectangular form are easily added or subtracted. Multiplication or division can be done with the rectangular form, but is more difficult than when expressed in the polar form. Phasors expressed in the polar form cannot be added or subtracted.

Phasor multiplication may be performed in the easiest manner by the following procedure:

1. Express phasors in polar form.
2. Multiply their magnitudes.
3. Add their angles.
4. The result is the polar form of the product.

Example:

Multiply $24\angle 60^\circ$ ohms by $5\angle -60^\circ$ amperes

$$24 \times 5\angle 60^\circ - 60^\circ = 120\angle 0^\circ \text{ volts}$$

Division of phasors is easily accomplished as follows:

1. Express phasors in polar form.
2. Divide their magnitudes.
3. Subtract the angle of the divisor from the angle of the dividend.
4. The result is the polar form of the quotient.

Example:

Divide $120\angle 0^\circ$ by $24\angle 60^\circ$ ohms

$$\frac{120\angle 0^\circ}{24\angle 60^\circ} = \frac{120}{24}\angle 0^\circ - 60^\circ = 5\angle -60^\circ \text{ amperes}$$

Examples of series and parallel circuit calculations using complex numbers are given below:

Series circuits:

1. Express circuit impedance components in rectangular number form.
2. Add or subtract the *a* and *jb* parts of the expressions.
3. The resulting expression is a complex number representing the net impedance.

Example:

Add $5\angle 60^\circ$ ohms to $5\angle -30^\circ$ ohms

$$5\angle 60^\circ = 5\cos 60^\circ + j5\sin 60^\circ = 2.5 + j4.33$$
$$5\angle -30^\circ = 5\cos -30^\circ + j5\sin -30^\circ = 4.33 - j2.50$$
$$6.83 + j1.83$$

$$Z = \sqrt{(6.83)^2 + (1.83)^2} = 7.07 \angle \tan^{-1}\frac{1.83}{6.83} \text{ ohms}$$

$$\text{where } \tan\theta = \frac{1.83}{6.83} = 0.2678$$
$$\theta = 15^\circ$$
$$Z = 7.07\angle 15^\circ \text{ ohms.}$$

Parallel circuits:

1. Express currents in rectangular form.
2. Add or subtract the a and jb parts of the expressions.
3. The resulting expression is a complex number representing the net current.

Example:

Subtract $10\angle -30^\circ$ amperes from $19.8\angle -37^\circ 30'$ amperes

$$19.8\angle -37^\circ 30' = 15.73 - j12.07 \text{ amp}$$
$$10\angle -30^\circ = 8.66 - j\,5.00 \text{ amp}$$
$$19.8\angle -37^\circ 30' - 10\angle -30^\circ = 7.07 - j\,7.07 \text{ amp} = 10\angle -45^\circ \text{ amp}$$

Table 3-3 shows the polar and rectangular forms of notation for impedance in series circuits and Table 3-4 shows complex notation for currents in parallel circuits.

Phasor analysis is a mathematical tool that can be used to obtain precise solutions to electrical circuit problems. Phasors are also used to represent visually the voltages and currents in a circuit or meter as an aid in understanding their phase and quantitative relationship.

COMPUTATIONS USED IN METERING

Before presenting typical metering computations, the following definitions should be reviewed.

The percentage registration of a meter is the ratio, expressed as a percentage, of the registration in a given time to the true kilowatt-hours.

The percentage error of a meter is the difference between its percentage registration and one hundred percent.

The correction factor is the number by which the registered kilowatt hours must be multiplied to obtain the true kilowatt-hours.

Table 3-5 illustrates the numerical relationship of these quantities.

Calculating Percentage Registration Using Rotating Standard

When no correction is to be applied to the rotating standard readings, the percentage registration of the watthour meter under test is calculated as follows:

$$\text{Percentage registration} = \frac{k_h \times r \times 100}{K_h \times R}$$

Table 3-3. Polar and Rectangular Representation of Impedance in Series Circuits.

Circuit	Phasor	Rectangular Form	Polar Form
$R = 5$  R	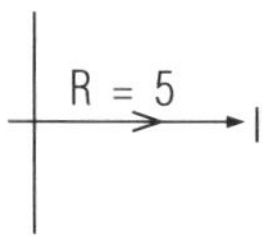 $R = 5$, I	$Z = 5 + j0$ $\tan\theta = \frac{0}{5} = 0$ $\theta = 0^\circ$	$Z = R\angle 0^\circ$ $Z = 5\angle 0^\circ$
$X_L = 5$  L	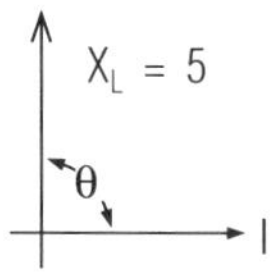 $X_L = 5$, θ, I	$Z = 0 + j5$ $\tan\theta = \frac{5}{0} = \infty$ $\theta = 90^\circ$	$Z = X_L\angle 90^\circ$ $Z = 5\angle 90^\circ$
$X_C = 5$  C	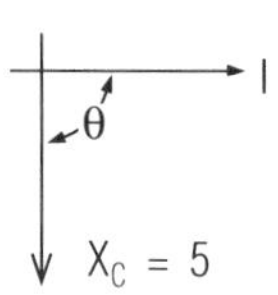 I, θ, $X_C = 5$	$Z = 0 - j5$ $\tan\theta = \frac{-5}{0} = -\infty$ $\theta = -90^\circ$	$Z = X_C\angle -90^\circ$ $Z = 5\angle -90^\circ$
$R = 5$ $X_L = 5$  RL	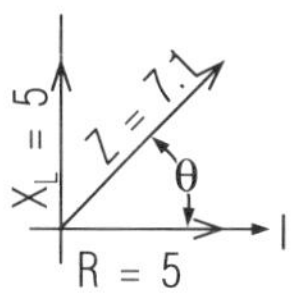 $X_L = 5$, $Z = 7.1$, θ, $R = 5$, I	$R = 5 + j0$ $X_L = 0 + j5$ $Z = 5 + j5$ $\tan\theta = \frac{5}{5} = 1$ $\theta = 45^\circ$	$Z = \sqrt{R^2 + X_L^2}$ $\tan\theta = \frac{X_L}{R}$ $Z = \sqrt{5^2 + 5^2} = 7.1$ $\tan\theta = \frac{5}{5} = 1;\ \theta = 45^\circ$ $Z = 7.1\angle 45^\circ$
$R = 5$ $X_C = 5$  RC	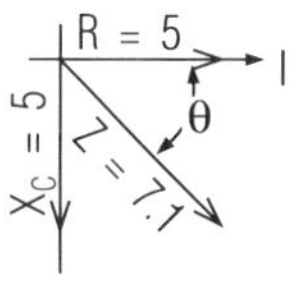 $R = 5$, I, θ, $X_C = 5$, $Z = 7.1$	$R = 5 + j0$ $X_C = 0 - j5$ $Z = 5 - j5$ $\tan\theta = \frac{-5}{5} = -1$ $\theta = -45^\circ$	$Z = \sqrt{R^2 + X_C^2}$ $\tan\theta = \frac{-X_C}{R}$ $Z = \sqrt{5^2 + 5^2} = 7.1$ $\tan\theta = \frac{-5}{5} = 1;\ \theta = -45^\circ$ $Z = 7.1\angle -45^\circ$
$R = 5$ $X_C = 5$ $X_L = 5$  RLC	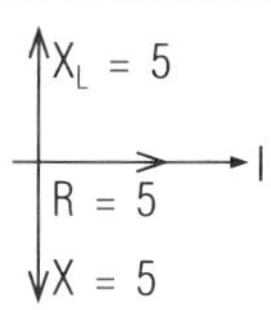 $X_L = 5$, I, $R = 5$, $X = 5$	$R = 5 + j0$ $X_C = 0 - j5$ $X_L = 0 + j5$ $Z = 5 + j0$ $\tan\theta = \frac{0}{5} = 0$ $\theta = 0^\circ$	$Z = \sqrt{R^2 + (X_L - X_C)^2}$ $\tan\theta = \frac{X_L - X_C}{R}$ $Z = \sqrt{5^2 + (5-5)^2} = 5$ $\tan\theta = \frac{5-5}{5} = 0;\ \theta = 0^\circ$ $Z = 5\angle 0^\circ$

Table 3-3 shows the polar and rectangular forms of notation for impedance in series circuits.

---

where: r = revolutions of meter under test
R = revolutions of standard
k_h = watthour constant of meter under test
K_h = watthour constant of standard

Table 3-4. Polar and Rectangular Representation of Currents in Parallel Circuits.

Circuit	Phasor	Rectangular Form	Polar Form
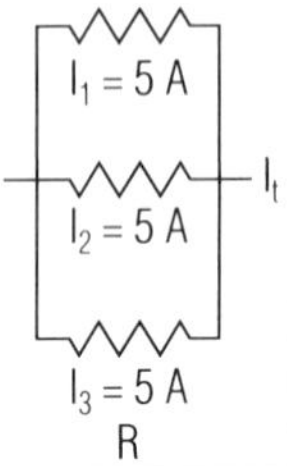	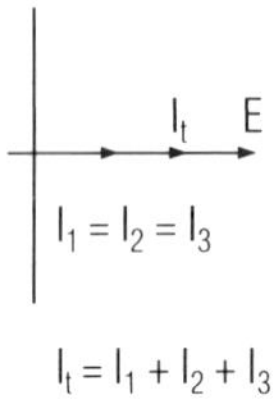	$I_1 = 5 + j0$ A $I_2 = 5 + j0$ A $I_3 = 5 + j0$ A $I_t = 15 + j0$ A $\tan\theta = \frac{0}{15} = 0$ $\theta = 0°$	$I_t = 15\angle 0°$ A
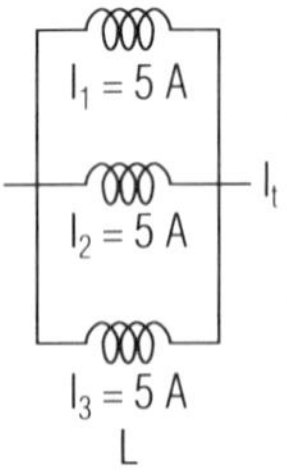	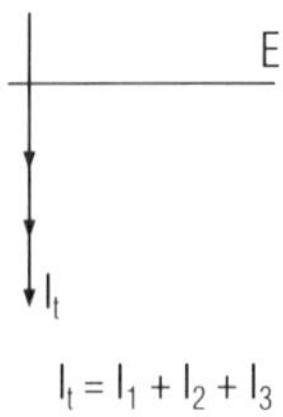	$I_1 = 0 - j5$ A $I_2 = 0 - j5$ A $I_3 = 0 - j5$ A $I_t = 0 - j15$ A $\tan\theta = \frac{-15}{0} = -\infty$ $\theta = -90°$	$I_t = 15\angle -90°$ A
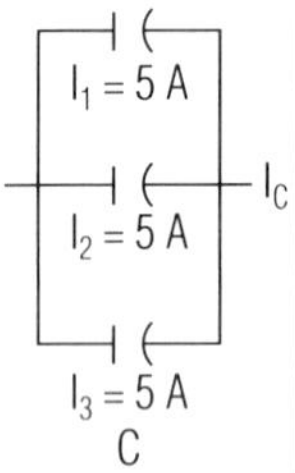	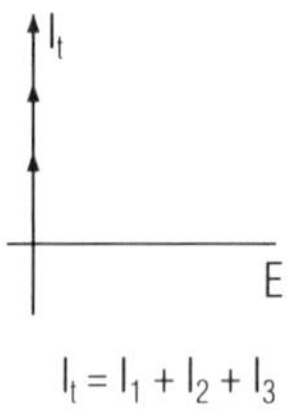	$I_1 = 0 + j5$ A $I_2 = 0 + j5$ A $I_3 = 0 + j5$ A $I_t = 0 + j15$ A $\tan\theta = \frac{15}{0} = \infty$ $\theta = 90°$	$I_t = 15\angle 90°$ A
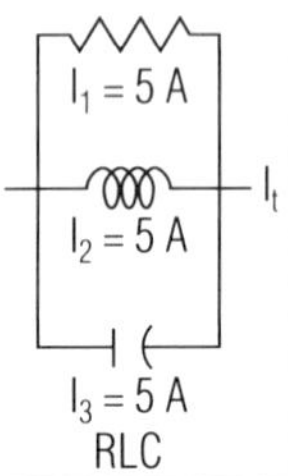	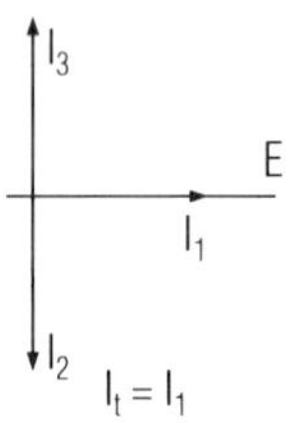	$I_1 = 5 + j0$ A $I_2 = 0 - j5$ A $I_3 = 0 + j5$ A $I_t = 5 + j0$ A $\tan\theta = \frac{0}{5} = 0$ $\theta = 0°$	$I_t = 5\angle 0°$ A

Table 3-4 shows complex notation for currents in parallel circuits.

Example: $r = 10$
$R = 10.87$
$k_h = 7.2$
$K_h = 6^2/_3$

$$\text{Percentage registration} = \frac{7.2 \times 10}{6^2/_3 \times 10.87} \times 100 = 99.4\%$$

Table 3-5. Relationship of Percent Registration, Percent Error, and Correction Factor.

% Registration	% Error	Correction Factor
100.4	+0.4	0.996
100.2	+0.2	0.998
100.0	0.0	1.000
99.8	−0.2	1.002
99.6	−0.4	1.004

The procedure may be simplified by introducing an additional symbol "R_o" where R_o is the number of revolutions the standard should make when the meter under test is correct. The values of R_o may be given to metering personnel in tabular form.

The number of revolutions of two watthour meters on a given load vary inversely as their disk constants.

$$\frac{R_o}{r} = \frac{k_h}{K_h} \qquad R_o = \frac{k_h \times r}{K_h}$$

Substituting R_o in the equation for percentage registration:

$$\text{Percentage registration} = \frac{R_o}{R} \times 100$$

Example:

The watthour meter under test and the standard have the following constants:

Meter $k_h = 7.2$
Standard $K_h = 6\,^2/_3$

The number of revolutions of the meter under test, r, equals 10.

$$\text{Then } R_o = \frac{7.2 \times 10}{6^2/_3} = 10.80 \text{ revolutions}$$

That is, for ten revolutions of the meter under test, the standard should make 10.80 revolutions. Assume the standard actually registered 10.87 revolutions, then:

$$\text{Percentage registration} = \frac{10.80}{10.87} \times 100 = 99.4\%$$

It is frequently easier to calculate mentally the percentage error of the meter, then algebraically add it to 100 percent to determine the percentage registration of the meter.

$$\text{Percentage error} = \frac{R_o - R}{R} \times 100$$

Using the same values as in the preceding example, then,

$$\text{Percentage error} = \frac{10.80 - 10.87}{10.87} \times 100 = -0.6\% \text{ and}$$

Percentage registration = 100.0 −0.6 = 99.4%.

When a correction is to be applied to the readings of the standard, the percentage registration is calculated as follows:

If A = percentage registration of the standard, then

$$\text{Percentage meter registration} = \frac{k_h \times r \times A}{K_h \times R}.$$

Using the same values as in the preceding examples, and assuming the percentage registration of the standard is 99.5 percent, then

$$\text{Percentage meter registration} = \frac{7.2 \times 10 \times 99.5}{6^2/_3 \times 10.87} = 98.9\%$$

To save time, the percentage error of the rotating standard is calculated and applied to the indicated percentage of meter registration to determine the true percentage meter registration.

Percentage meter registration = indicated percentage meter registration + percentage standard error. That is, the percentage standard error is added to the apparent percentage registration of the meter under test if the percentage standard error is positive, and subtracted if negative. Then, for this example,

$$\text{Percentage standard error} = 99.5\% - 100\% = -0.5\%.$$

Referring to computations made earlier, if a rotating standard with a –0.5 percentage error is used, then percentage registration of the meter is 99.4% – 0.5% = 98.9%. This method is usable where the percentage error does not exceed 3%.

Calculating Percentage Registration Using Indicating Instruments

Where P = true watts (corrected readings of instruments)
k_h = watthour constant of self-contained watthour meter
r = number of revolutions of meter disk
s = time in seconds for r revolutions

$$\text{Percentage registration} = \frac{k_h \times r \times 3{,}600 \times 100}{P \times s}$$

Example:

Where P = 7,200
k_h = 7.2
r = 10
s = 36.23

$$\text{Percentage registration} = \frac{7.2 \times 10 \times 3{,}600 \times 100}{7{,}200 \times 36.23} = 99.4\%$$

The seconds for 100 percent accuracy (S_s) may be determined from:

$$S_s = \frac{3{,}600 \times r \times k_h}{P}$$

For 100 percent accuracy

$$S_s = \frac{3{,}600 \times 10 \times 7.2}{7{,}200} = 36.00 \text{ and}$$

$$\text{Since percentage registration} = \frac{S_s \times 100}{s}$$

$$\text{Percentage registration} = \frac{36.00 \times 100}{36.23} = 99.4\%$$

The correction for instrument error may be applied similarly to the correction for rotating standards, where *P* equals the observed reading of the wattmeter.

Assume the observed reading of a wattmeter is 7,200 watts and true watts are 7,236. Then the wattmeter indicates:

$$\frac{7200}{7236} = 99.5\% \text{ of true watts.}$$

This percentage indication may be used for A in the formula:

$$\text{Percentage registration} = \frac{k_h \times r \times 3{,}600 \times A}{P \times S}$$

$$\text{Percentage registration} = \frac{7.2 \times 10 \times 3{,}600 \times 99.5}{7{,}200 \times 36.23} = 98.9\%$$

If the preceding meter had been a direct-current meter, the test could have been made with a voltmeter and ammeter.

Assume an observed ammeter reading of 30 amperes and true current is 30.2 amperes. Then:

$$\text{Percentage indication} = \frac{30}{30.2} \times 100 = 99.3\%$$

Assume an observed voltmeter reading of a 240 volts and true voltage is 239.5 volts. Then:

$$\text{Percentage indication} = \frac{240}{239.5} \times 100 = 100.2\%$$

$$\text{Indicated watts} = 240 \times 30 = 7{,}200 \text{ watts}$$

$$\text{Actual watts} = 239.5 \times 30.2 = 7{,}233 \text{ watts}$$

$$\text{Percentage indication} = \frac{7{,}200}{7{,}233} \times 100 = 99.5\%$$

Again, the 99.5 percent calculated could be used for A in the formula the same as the 99.5 percent obtained as the percentage indication of the wattmeter.

In either case, the percent registration is the sum of the apparent percentage indication and the percentage error. The percentage registration = 99.4% + (–0.5%) = 98.9%.

Register Formulas and Their Application

R_r = register ratio

K_h = watthour constant

R_s = gear reduction between worm or spur gear on disk shaft and meshing gear wheel of register

K_r = register constant

R_g = gear ratio = $R_r \times R_s$

CTR = current transformer ratio

VTR = voltage transformer ratio

TR = Transformer ratio ($CTR \times VTR$)

PK_h = primary watthour constant = $K_h \times TR$

Example 1:

Self-contained meter, K_h = 7.2, 100 teeth on first wheel or register, 1 pitch worm on shaft, register constant 10.

To find the Register Ratio:

$$R_r = \frac{10{,}000 \times K_r}{K_h \times R_s \times TR} = \frac{10{,}000 \times 10}{7.2 \times 100 \times 1} = 138\,{}^{8}/{}_{9}$$

To check the Register Constant:

$$K_r = \frac{K_h \times R_r \times R_s \times TR}{10{,}000} = \frac{7.2 \times 138\,{}^{8}/{}_{9} \times 100 \times 1}{10{,}000} = 10$$

To determine the Gear Reduction:

$$R_g = R_r \times R_s$$

$$R_g = 138\,{}^{8}/{}_{9} \times \frac{100}{1} = \frac{1{,}250}{9} \times 100 = 13{,}888\,{}^{8}/{}_{9}$$

Example 2:

Transformer-rated meter installed with 400/5 (80/1) CTR, register constant (K_r) 100, K_h = 1.8, 100 teeth on first wheel, 2 pitch worm on shaft.

$$R_r = \frac{10{,}000 \times K_r}{PK_h \times R_s} = \frac{10{,}000 \times K_r}{(K_h \times TR) \times R_s} = \frac{10{,}000 \times 100}{(1.8 \times 80) \times \frac{100}{2}} = 138\,{}^{8}/{}_{9}$$

$$K_r = \frac{PK_h \times R_r \times R_s}{10{,}000} = \frac{(K_h \times TR) \times R_r \times R_s}{10{,}000}$$

$$= \frac{(1.8 \times 80) \times 138\,{}^{8}/{}_{9} \times \frac{100}{2}}{10{,}000} = 100$$

$$R_g = 138\,{}^{8}/{}_{9} \times \frac{100}{2} = \frac{1{,}250}{9} \times 50 = 6{,}944\,{}^{4}/{}_{9}$$

Example 3:

Transformer rated meter installed with 50/5 (10/1) CTR, 14,400/120 (120/1) VTR, register constant (K_r) 1,000, K_h= 0.6, and 100 teeth on first wheel, with 1 pitch worm on shaft.

$$R_r = \frac{10{,}000 \times 1{,}000}{0.6 \times 10 \times 120 \times 100} = 138\,{}^{8}/{}_{9}$$

$$K_r = \frac{0.6 \times 10 \times 120 \times 138\,{}^{8}/{}_{9} \times 100}{10{,}000} = 1{,}000$$

$$R_g = 138\,{}^{8}/{}_{9} \times 100 = \frac{1{,}250}{9} \times 100 = 13{,}888\,{}^{8}/{}_{9}$$

CHAPTER

4

ELECTRICAL CIRCUITS

PART 1 — DIRECT CURRENT

INTRODUCTION

DIRECT CURRENT DISTRIBUTION serves a very minor role in electric distribution in this country. However, the fundamentals of direct current circuits are included in this text because the laws of direct current are basic to all electric circuits. These laws, if understood for direct current, can be easily modified for a better understanding of alternating-current circuits.

A direct current is defined as an electric current flowing in one direction only. The values of rate of current flow (amperes), the pressure (volts), and the resistance (ohms) of a DC circuit can be calculated by the use of formulas based on Ohm's law or Kirchhoff's laws for direct-current circuits.

OHM'S LAW

Ohm's law states that the current flowing in a direct-current circuit is directly proportional to the total voltage applied to the circuit and inversely proportional to the total circuit resistance.

$$\text{(Amperes)}\quad I = \frac{E}{R} = \frac{\text{volts}}{\text{ohms}}$$

$$\text{(Volts)}\quad E = IR = \text{amperes} \times \text{ohms}$$

$$\text{(Ohms)}\quad R = \frac{E}{I} = \frac{\text{volts}}{\text{amperes}}$$

KIRCHHOFF'S LAWS

1. Kirchhoff's voltage law states that the sum of the voltage drops around a circuit is equal to the supply voltage or voltages.
2. Kirchhoff's current law states that the sum of the currents flowing into a junction point is equal to the sum of the currents flowing away from the junction point.

GENERAL CIRCUIT CHARACTERISTICS

Resistances in a direct-current circuit are found connected in one of four ways: series, parallel, series-parallel, or a network of series and parallel circuits. Equivalent resistance is the total effect of all the resistances in a circuit opposing the flow of current in the circuit.

Resistances Connected in Series

1. The current in a series circuit is the same in all parts of the circuit.
2. The input voltage to a series circuit is the sum of the voltage drops across all resistances in the circuit (Kirchhoff's voltage law).
3. In a series circuit, the equivalent resistance is the total resistance and is equal to the sum of the individual resistances.

Resistances Connected in Parallel

1. The total current in a parallel circuit is the sum of the currents through the branches (Kirchhoff's current law).
2. The voltage in a parallel circuit is the same across each parallel branch.
3. Equivalent resistance of a parallel circuit is always less than that of the smallest resistance branch.
4. Equivalent resistance of resistances connected in parallel equals one divided by the sum of the conductances connected in parallel. Conductance is the reciprocal of resistance, or one divided by the resistance. As an example: if the resistance is 10 ohms, the conductance is 1/10 mho (ohm spelled backward).
5. Another method of computing the equivalent resistance of a circuit composed of many parallel branches is to calculate the current through each branch, add the currents, and determine the equivalent resistance by Ohm's law.

Resistances Connected Series-Parallel

The equivalent resistance of a series-parallel circuit equals the equivalent resistances of each group or branch of parallel resistances, added to the resistances connected in series. Computation to determine the equivalent resistance of a series-parallel circuit can be simplified if, by inspection, it can be determined which resistances are connected in parallel and which resistances are connected in series.

USEFUL DIRECT CURRENT FORMULAS

The following formulas are useful in DC circuit calculations:

$$\text{Current: } I = \frac{E}{R} \text{ or } \frac{W}{E} \text{ or } \sqrt{\frac{W}{R}}$$

$$\text{Resistance: } R = \frac{E}{I} \text{ or } \frac{E^2}{W} \text{ or } \frac{W}{I^2}$$

$$\text{Voltage: } E = IR \text{ or } \frac{W}{I} \text{ or } \sqrt{WR}$$

$$\text{Watts (power): } W = EI \text{ or } I^2R \text{ or } \frac{E^2}{R}$$

$$\text{Energy} = Wt = EIt \text{ or } I^2Rt \text{ or } \frac{E^2t}{R}, \text{ where } t = \text{time.}$$

APPLICATION OF OHM'S LAW

As previously expressed, Ohm's law is as follows:

$I = \frac{E}{R}$; $E = IR$; $R = \frac{E}{I}$; where I is current, E is voltage, and R is resistance.

Example 1:

With a voltage of 112 V and a resistance of 8 ohms, what current would flow?

$$I = \frac{E}{R} \text{ or } \frac{112}{8} = 14\text{ A}$$

Example 2:

What resistance is necessary to obtain a current of 14 amperes at 112 volts?

$$R = \frac{E}{I} \text{ or } \frac{112}{14} = 8 \text{ ohms}$$

Example 3:

What voltage is required to produce a flow of 14 amperes through a resistance of 8 ohms?

$$E = IR = 14 \times 8 = 112\text{ V}$$

Example 4:

Ohm's law may be used to calculate the effect of conductor resistance on the operation of a customer's appliance.

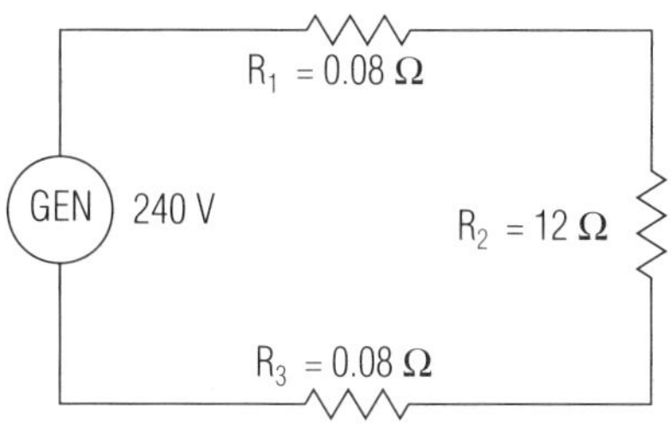

Figure 4-1. Solving for Conductor Losses Using Ohm's Law.

$$
\begin{aligned}
R_1 + R_3 &= 0.16 \text{ ohm or approximate resistance of 100 feet of No. 12 wire} \\
R_2 &= 12 \text{ ohms, the resistance of a 4800 watt, 240 volt heating element} \\
R_{tot} &= R_1 + R_2 + R_3 = 0.08 + 12 + 0.08 = 12.16 \text{ ohms} \\
I &= \frac{240}{12.16} = 19.74 \text{ amperes} \\
W_1 &= E \times I = 240 \times 19.74 = 4{,}737.6 \text{ watts (input power)} \\
W_{Heater} &= I^2 R_{Heater} = (19.74)^2 \times 12 = 4{,}676.0 \text{ watts}
\end{aligned}
$$

In the preceding example the heating element will not develop the heat for which it was designed, due to the voltage drop in the wire.

Conductor Voltage Drop = $E_{cond} = I(R_1 + R_3) = 19.74(0.08 + 0.08) = 3.16\text{ V}$

The heat loss in the No. 12 wire which is paid for by the customer, but serves no useful purpose, is:

$$W_2 = E \times I = I(R_1 + R_3) \times I = I^2(R_1 + R_3) = 19.74^2 \times 0.16 = 62.35 \text{ watts heat loss}$$

$$\text{Efficiency (percent)} = 100 \times \frac{\text{useful output watts}}{\text{input watts}}$$

$$\text{Efficiency} = 100 \times \frac{4{,}737.6 - 62.35}{4{,}737.6} = 98.7\%$$

To reduce line losses and improve efficiency, the unit should be wired with a larger size wire.

The resistance of metals and most alloys used in electrical conductors has a positive temperature coefficient, which means the resistance of the conductor increases with a rise in temperature.

Tables are available which give the resistance in ohms per thousand feet of conductor and the current-carrying capacity of different size conductors for a specific temperature. Using these tables and Ohm's law, it is possible to calculate the losses for a conductor at various current loadings. By applying Ohm's law and data from these resistance tables, distribution circuits can be designed for minimum losses for a specified circuit loading. In electric heating, the resistance of high-resistance conductors can be calculated so that the watts dissipated will be such as to generate a predetermined heat, expressed in BTUs.

A fuse link is a conductor designed with sufficient resistance to melt at a given current rating.

Example 5:

The resistance of a copper wire 1 foot long and 1 circular mil in cross section is 10.371 ohms at 20° C (National Bureau of Standards). 10.4 is used for practical calculations.

In Ohm's law $I = \frac{E}{R}$, where R, the resistance, is equal to the length of the conductor in feet times 10.4, and divided by the sectional area in circular mils of the conductor or:

$$R = \frac{2 \times \text{ft (length of circuit)} \times 10.4}{\text{cir mil}}$$

Using Ohm's law, $E = IR$,

$$E = \frac{I \times 2 \times \text{ft} \times 10.4}{\text{cir mil}}$$

where here the term "ft" indicates the length of the circuit. Note that the number of feet of wire in the circuit is double the length of the circuit.

What would be the volts lost in a circuit of No. 12 wire carrying 20 amperes a distance of 50 feet? (The area of #12 copper wire is 6530 circular mils).

$$E = \frac{20 \times 2 \times 50 \times 10.4}{\text{cir mil}}$$

or approximately 3% on a 110 volt circuit.

Example 6:

What size conductor would be necessary to give a 3 percent drop on a 110 V circuit carrying 20 amperes a distance of 50 ft?

$$\text{cir mil} = \frac{I \times 2 \times \text{ft} \times 10.4}{E}$$

$$\text{cir mil} = \frac{20 \times 2 \times 50 \times 10.4}{3.3} = 6{,}303 \text{ cir mil,}$$

or approximately a No. 12 wire.

Example 7:

What current can a No. 12 wire carry on a 50-ft circuit with a voltage drop of 3.3 V?

$$I = \frac{\text{cir mil} \times E}{2 \times \text{ft} \times 10.4}$$

$$I = \frac{6{,}530 \times 3.3}{50 \times 2 \times 10.4} = 20.7 \text{ A}$$

APPLICATION OF KIRCHHOFF'S LAWS

In solving for the various unknown values of current, voltage, and resistance of complex direct-current circuits, Kirchhoff's laws of voltage and current are often used in conjunction with Ohm's law. Complex circuits are defined as circuits consisting of parallel and series branches having one or more sources of voltage supply. The voltage law is adapted for use with series circuits and the current law with parallel circuits.

As previously expressed, Kirchhoff's voltage law states that the sum of the voltage drops around a circuit is equal to the supply voltage or voltages. Kirchhoff's voltage law also states that the algebraic sum of the voltages around a circuit is equal to zero, that is:

$$E_s = E_1 + E_2 + E_3$$
$$E_s - E_1 - E_2 - E_3 = 0$$

This law can be illustrated by referring to Fig. 4-2 and using Ohm's law. E_s = source voltage, with E_1, E_2, and E_3 the voltage drops across R_1, R_2, and R_3, respectively.

$$\text{Current in circuit} = I = \frac{120}{10 + 15 + 15} = \frac{120}{40} = 3 \text{ A}$$

Voltage drop E_1 across $R_1 = IR_1 = 3 \times 10 = 30$ V
Voltage drop E_2 across $R_2 = IR_2 = 3 \times 15 = 45$ V
Voltage drop E_3 across $R_3 = IR_3 = 3 \times 15 = 45$ V

$$E_s = E_1 + E_2 + E_3 = 30 + 45 + 45 = 120 \text{ V}$$
$$E_s - E_1 - E_2 - E_3 = 120 - 30 - 45 - 45 = 0 \text{ V}$$

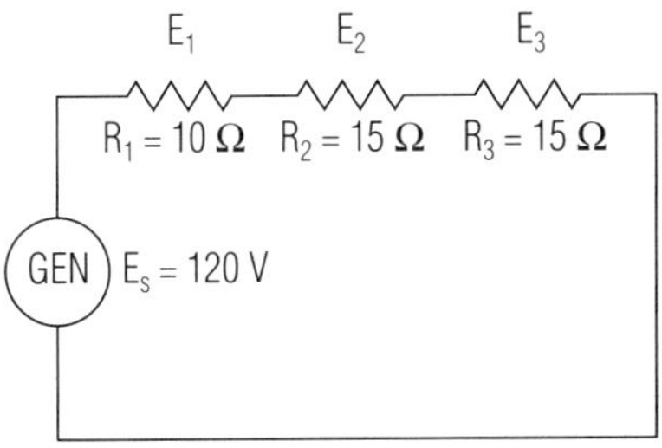

Figure 4-2. Proving Kirchhoff's Voltage Law Using Ohm's Law.

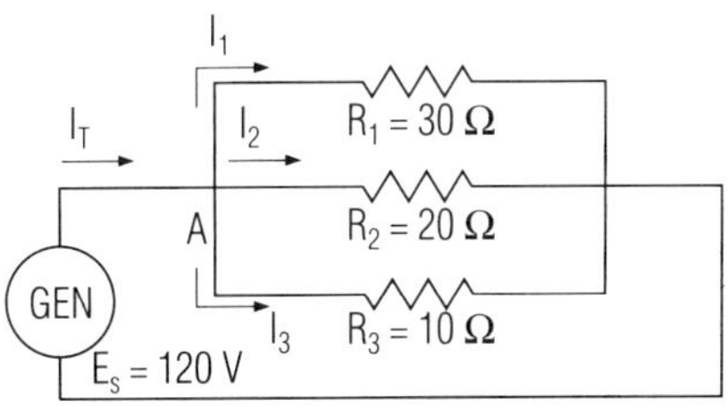

Figure 4-3. Proving Kirchhoff's Current Law Using Ohm's Law.

Kirchhoff's current law states that the sum of the currents flowing into a junction point is equal to the sum of the currents flowing away from the junction point. Kirchhoff's current law also states that at any junction of conductors, the algebraic sum of the currents is zero. That is:

$$I_{tot} = I_1 + I_2 + I_3$$
$$I_{tot} - I_1 - I_2 - I_3 = 0$$

Referring to junction A in Fig. 4-3, the law states that as much current flows away from the junction of the conductors in the branch circuits as flows into the junction. This law can also be illustrated by referring to Fig. 4-3 and using Ohm's law.

$$R_{eq} = \frac{1}{\frac{1}{R_1} + \frac{1}{R_2} + \frac{1}{R_3}} = \frac{1}{\frac{1}{30} + \frac{1}{20} + \frac{1}{10}}$$

$$R_{eq} = \frac{1}{\frac{2}{60} + \frac{3}{60} + \frac{6}{60}} = \frac{1}{\frac{11}{60}} = \frac{60}{11} = 5.45 \text{ ohms}$$

$$I_{tot} = \frac{E}{R} = \frac{120}{\frac{60}{11}} = \frac{120 \times 11}{60} = 22 \text{ A}$$

$$I_1 = \frac{E}{R_1} = \frac{120}{30} = 4 \text{ A}$$

$$I_2 = \frac{E}{R_2} = \frac{120}{20} = 6 \text{ A}$$

$$I_3 = \frac{E}{R_3} = \frac{120}{10} = 12 \text{ A}$$

$$I_{tot} = I_1 + I_2 + I_3 = 4 + 6 + 12 = 22 \text{ A}$$

Aids in understanding Kirchhoff's laws:

1. The sum of the currents flowing into a junction must equal the sum of the currents flowing away. If this were not so, current would collect at the junction. Since we know from experiment that current cannot be continuously stored at, or removed from, a junction, the law is true.
2. The sum of the voltages around a closed circuit must equal zero. If this were not so, a single point on a circuit could be at two different voltages at the same time relative to the same fixed reference point. Since experiment shows that each point can have only one voltage at any instant relative to a fixed reference point, the law is true.

CIRCUIT CALCULATIONS

Series Circuit

Example 8:

A series circuit consisting of three resistors as shown in Fig. 4-4.

Formula:

$$R_{eq} = R_{tot} = R_1 + R_2 + R_3$$
$$R_{eq} = 10 + 20 + 30 = 60 \text{ ohms}$$

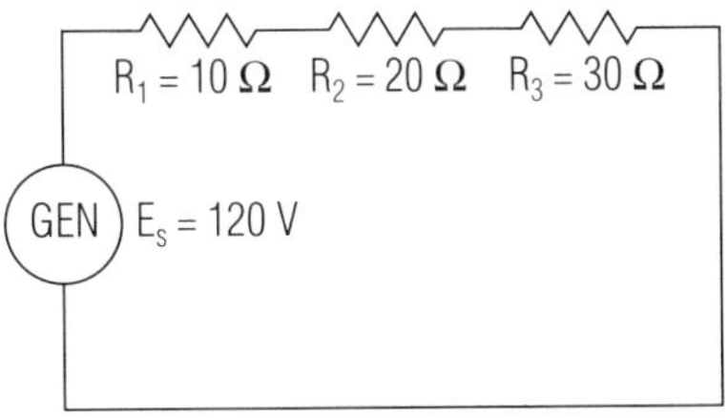

Figure 4-4. Resistances Connected in Series.

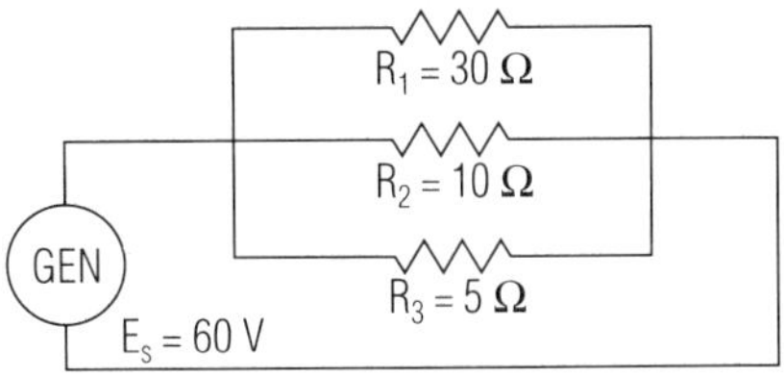

Figure 4-5. Resistances Connected in Parallel.

Formula:

$$I = \frac{E}{R}$$

$$I = \frac{120}{60} = 2\text{ A}$$

Formula:

$$E_S = IR_1 + IR_2 + IR_3$$
$$E_S = (2 \times 10) + (2 \times 20) + (2 \times 30) = 20 + 40 + 60 = 120\text{ V}$$

Parallel Circuit

Example 9:

The lights of a house circuit are connected in parallel. The current at the fuse panel is the sum of the currents through the lamps. Each lamp on this circuit has the same voltage impressed on it, nominally 120 volts.

Example 10:

A parallel circuit consisting of three resistors connected in parallel as shown in Fig. 4-5.

Formula:

$$R_{eq} = \frac{1}{\frac{1}{R_1} + \frac{1}{R_2} + \frac{1}{R_3}} = \frac{1}{\frac{1}{30} + \frac{1}{10} + \frac{1}{5}}$$

$$= \frac{1}{\frac{1}{30} + \frac{3}{30} + \frac{6}{30}} = \frac{1}{\frac{10}{30}} = \frac{30}{10} = 3\text{ ohms}$$

Also:

$$\text{30-ohm branch: } I = \frac{60\text{ V}}{30\text{ ohms}} = 2\text{ A}$$

$$\text{10-ohm branch: } I = \frac{60\text{ V}}{10\text{ ohms}} = 6\text{ A}$$

$$\text{5-ohm branch: } I = \frac{60\text{ V}}{5\text{ ohms}} = 12\text{ A}$$

$$\text{Total current} = 20\text{ A}$$

$$\text{Equivalent resistance} = \frac{60\text{ V}}{20\text{ A}} = 3\text{ ohms}$$

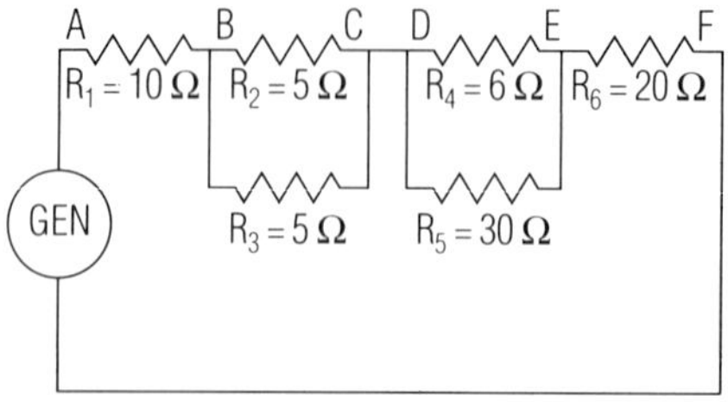

Figure 4-6. Circuit Connected in Series-Parallel.

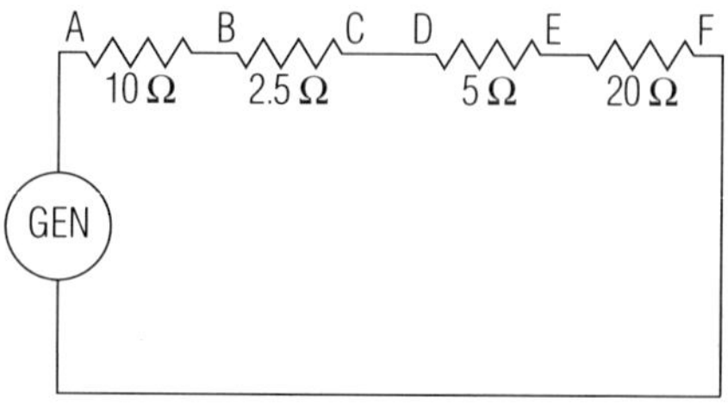

Figure 4-6a. Circuit Redrawn as Series Circuit.

Series-Parallel Circuit

Example 11:

A series-parallel circuit consisting of two parallel branches and two series resistors is shown in Fig. 4-6. Fig. 4-6 shows the actual circuit and Fig. 4-6a shows the circuit redrawn as a series circuit with the equivalent resistances between each point calculated.

Formula:

$$R_{eq} = R_1 + \frac{1}{\frac{1}{R_2} + \frac{1}{R_3}} + \frac{1}{\frac{1}{R_4} + \frac{1}{R_5}} + R_6$$

Resistance from A to B = R_1 = 10 ohms

$$\text{Resistance from B to C} = \frac{1}{\frac{1}{R_2} + \frac{1}{R_3}} + \frac{1}{\frac{1}{5} + \frac{1}{5}} = \frac{1}{\frac{2}{5}} = \frac{5}{2} = 2.5 \text{ ohms}$$

$$\text{Resistance from D to E} = \frac{1}{\frac{1}{R_4} + \frac{1}{R_5}} + \frac{1}{\frac{1}{6} + \frac{1}{30}} = \frac{1}{\frac{2}{10}} = \frac{10}{2} = 5 \text{ ohms}$$

Resistance from E to F = R_6 = 20 ohms
Resistance from A to F = R_{eq} = 10 + 2.5 + 5 + 20 = 37.5 ohms

Resistance Network

Example 12:

Ohm's law and Kirchhoff's laws are used to determine the value of the current I_2 in a circuit having two sources of voltage supply, as shown in Fig. 4-7.

(a) $I_1R_1 + I_2R_2 = 20$ or $4I_1 + 6I_2 = 20$

(b) $I_2R_2 + I_3R_3 = 14$ or $6I_2 + 8I_3 = 14$

(c) $I_1 + I_3 = I_2$ or $I_3 = I_2 - I_1$

Combining (b) and (c):

(d) $6I_2 + 8(I_2 - I_1) = 14$

$6I_2 + 8I_2 - 8I_1 = 14$ or $14I_2 - 8I_1 = 14$

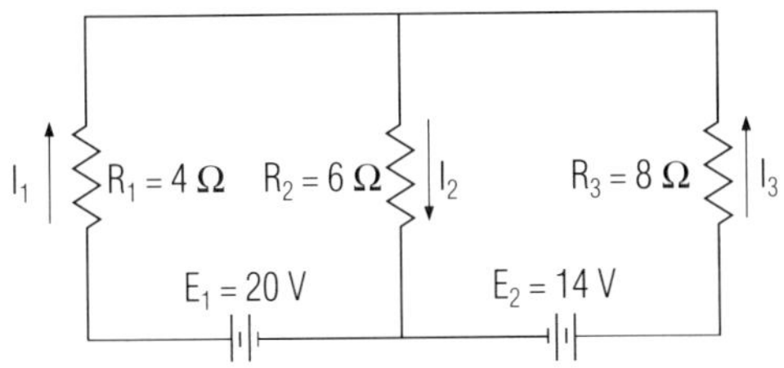

Figure 4-7. Solving for Currents in Resistances Connected in Network with Two Voltage Supplies.

(e) Multiply (a) by 2:
$12I_2 + 8I_1 = 40$

(f) Add (d) and (e):
$26I_2 = 54$
$I_2 = {}^{54}/_{26} = {}^{27}/_{13}$ A

(g) Substitute (f) in (a):
$4I_1 + 6 \times {}^{27}/_{13} = 20$
$I_1 = {}^{49}/_{26}$ A

(h) Substitute (f) in (b):
$6 \times {}^{27}/_{13} + 8I_3 = 14$
$I_3 = {}^{5}/_{26}$ A

(i) Check: Substitute (f), (g), and (h) in (c):
${}^{5}/_{26} = {}^{27}/_{13} - {}^{49}/_{26} = {}^{54}/_{26} - {}^{49}/_{26} = {}^{5}/_{26}$

THREE-WIRE EDISON DISTRIBUTION SYSTEM

Edison discovered that if the positive wire of one generator and the negative wire of another generator having an equal output voltage were combined, one wire of the four from the two generators could be eliminated between the station and the customer. This system resulted in a saving of copper and a reduction of distribution losses.

On balanced load between the outside wires and the common wire, no current flows in the common wire or neutral, as shown in Fig. 4-8. If the load is unbalanced, the neutral will carry the amount of unbalanced current to or away from the generator, depending on which side of the system is more heavily loaded. This is illustrated in Figs. 4-8a and 4-8b.

When the neutral is carrying current due to an unbalanced load condition, the opening of the neutral wire results in higher voltage at the load from one outside wire to the neutral and lower voltage from the other outside wire to the neutral, the higher voltage being on the lightly loaded side. This condition is explained by the use of Ohm's law in Fig. 4-8c.

Figure 4-8c is the same as Fig. 4-8b except that the neutral is open and the 5-ohm and l0-ohm resistances are in series with 240 volts across the line wires.

The current through the total 15 ohms is:

$$I = \frac{E}{R} \text{ or } \frac{240}{15} = 16 \text{ A}$$

The voltage from A to B = $E_1 = IR_{AE} = 16 \times 5 = 80$ V

The voltage from B to C = $E_2 = IR_{EC} = 16 \times 10 = 160$ V

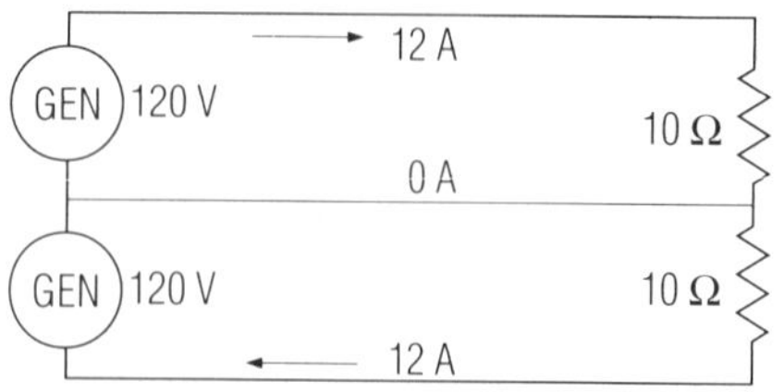

Figure 4-8. Edison System - Currents and Voltages Balanced.

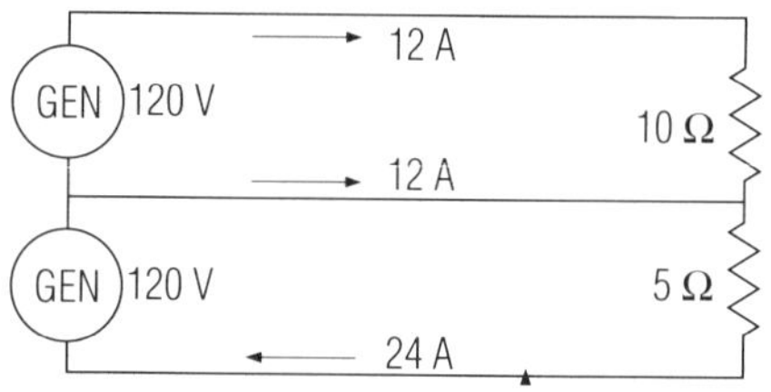

Figure 4-8a. Edison System - Current in Neutral Flowing from Generator with Unbalanced Load.

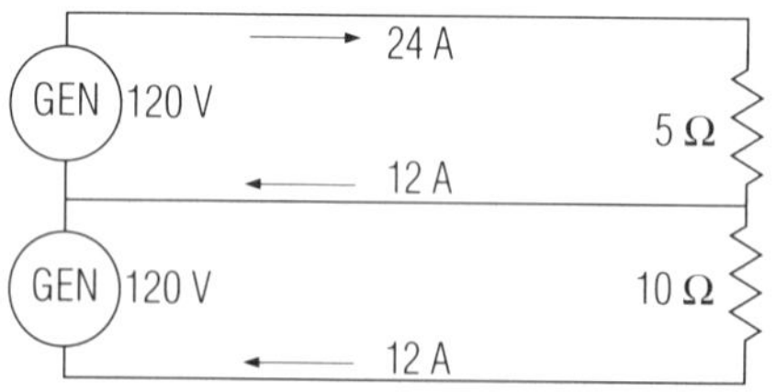

Figure 4-8b. Edison System - Current in Neutral Flowing Toward Generator with Unbalanced Load.

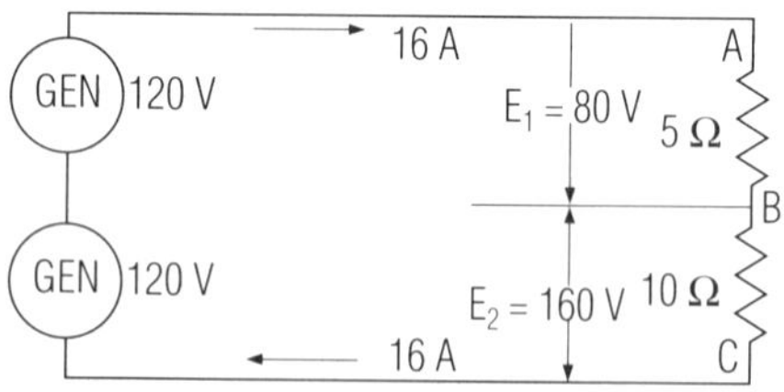

Figure 4-8c. Edison System - Current and Voltages, Unbalanced System with Open Neutral.

PART 2—ALTERNATING CURRENT

INTRODUCTION

Although direct current is necessary for some industrial purposes such as electrolytic processes and for all digital logic circuits, practically all electric energy today is generated and transmitted as alternating current. Alternating current permits the use of static transformers by which voltages can readily be raised or lowered, thus allowing the transmission of energy at high voltages and the usage of energy at low voltages. Transformers operate on the principle of induction. They transfer energy using both magnetic circuits and electric circuits.

It is the influence of the magnetic circuit on the electric circuit with which it is associated that causes the major differences between the AC circuit and the DC circuit.

In an alternating-current circuit, voltage and current vary from instant to instant and power cannot generally be determined by the product of voltages and amperes alone. To understand the phenomena occurring in an AC circuit frequency, time relations and wave shapes must be studied.

ALTERNATING CURRENT

An alternating current is an electric current which periodically passes through a regular succession of changing values, positive and negative, with the total positive and negative values of current equal.

FREQUENCY

A cycle consists of one complete pattern of change of the AC wave; that is, the period from any point on an AC wave to the next point of the same magnitude and direction at which the wave pattern begins to repeat itself (Fig. 4-9).

The number of cycles completed per second is the frequency of the wave expressed in hertz (Hz). Higher frequencies may be expressed in kilohertz (kHz) or megahertz (MHz).

Typical Electrical Frequencies

Application	Frequency
Direct current	0
Standard AC power (USA)	60 Hz
Audio sound	16 to 16,000 Hz
Radio (AM broadcast band)	535 to 1,605 kHz
Radio (FM broadcast band)	88 to 105 MHz
Television (channels 2-13)	55 to 216 MHz

The time it takes to complete one cycle is called the period, and the symbol is T. A 60 Hz wave has a period of 1/60 second.

$$T = \frac{1}{f} \quad \text{or} \quad f = \frac{1}{T}$$

where T is in seconds and f is the frequency in hertz.

SINE WAVES

If the usual alternating voltage is plotted against time, it looks like the curve in Fig. 4-9.

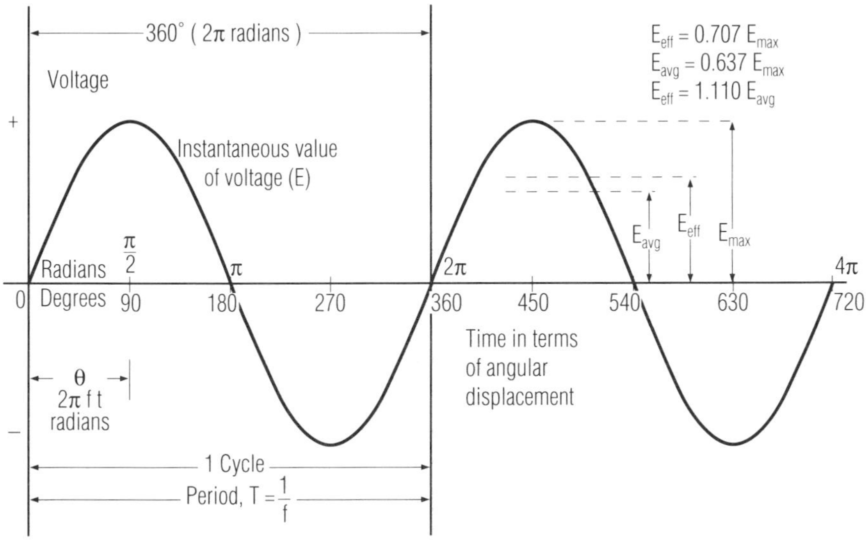

Figure 4-9. Sine-Wave Relationships.

This curve is called a sine wave because it has exactly the same shape as the trigonometric curve described by the equation:

$$e = E_{max} \sin \theta$$

where e is the instantaneous voltage and θ is the angle corresponding to the instantaneous voltage.

Figure 4-9 shows that a cycle covers a definite period of time and is completed in 360 degrees. Therefore it follows that time may be expressed in terms of an angle.

$$\theta = 2 \pi f t \text{ radians}$$

Note that for t = T, one period or 360°, $\theta = 2 \pi$ radians.

So the equation of the sine wave in Fig. 4-9 may be expressed as:

$$e = E_{max} \sin 2 \pi f t$$

$$\text{or } e = E_{max} \sin \omega t \qquad \text{where } \omega = 2 \pi f$$

$$\text{or } e = E_{max} \sin \theta$$

A sine-wave voltage can be assigned a value in three ways:

1. By the maximum or peak value (E_{max}). This value is used in insulation stress calculations.
2. By the average value (E_{avg}) which is equal to the average value of e for the positive half (or negative half) of the cycle. This value is often used in rectification problems.
3. By the root-mean-square (rms) or effective value (E_{eff}). This is the value most generally used. In electricity, the effective value of an alternating current is that value of current which gives the same heating effect in a given resistor as the same value of direct current. Unless some other description is specified, when alternating currents or voltages are mentioned, it is always the "effective" value that is meant.

HARMONICS

In the ideal case, alternating voltages and currents follow a true sine wave, and most notations and formulas are based on a true sine wave.

However, in actual practice, there are effects that may cause the sine wave to be distorted to some extent.

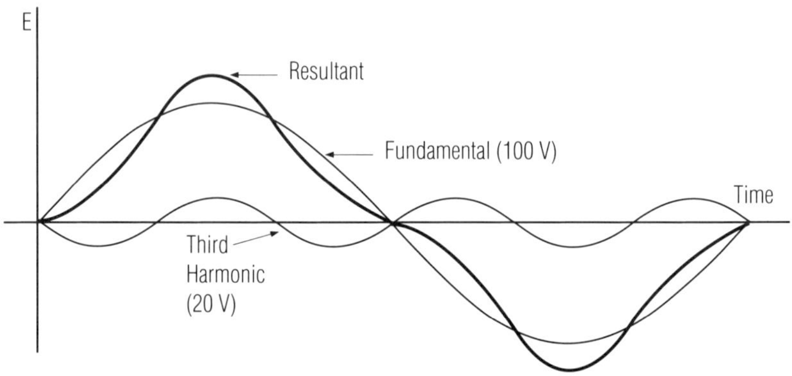

Figure 4-10. Sine Wave with 20 Percent Third Harmonic.

Any AC wave, no matter how distorted, may be represented by a combination of a wave of the fundamental frequency plus one or more harmonics or multiples of the original frequency. The relative magnitude of the fundamental wave and the number, magnitude, and phase displacement of the harmonics determine the resultant wave form.

For instance, in Fig. 4-10, a voltage wave is shown having a 20 percent third harmonic. This harmonic is shown such that it crosses the zero line at the same instant as the fundamental wave. Harmonics, however, can be displaced in time from the fundamental, depending on circuit characteristics.

In this wave there is present a 100-volt "fundamental" 60 hertz wave and a 20 volt third harmonic, or 180 hertz wave.

Under certain circumstances, harmonics can be quite important and also quite troublesome. Harmonics may be caused by various circuit loads such as fluorescent lights and by certain three-phase transformer bank connections. They cause heating in wiring and will not be detected by most digital test meters unless they are true RMS measuring devices.

PHASE RELATIONS – POWER FACTOR

If two sine waves of the same frequency do not coincide with respect to time, they are said to be out of phase with each other. In Fig. 4-11, the current wave *I*, is θ degrees out of phase with the voltage wave. As shown, it is behind or lagging the voltage wave by the angle θ. It reaches its peak value θ degrees of angular displacement after the voltage wave reaches its peak. The trigonometric cosine of this angle of lag (or lead) is the power factor of the circuit.

TRIGONOMETRY FOR ALTERNATING CURRENT

A basic knowledge of right triangles is necessary for an understanding of alternating current, particularly those relationships shown in Table 4-1.

PHASORS

The representation of electrical quantities by sine waves is unwieldy and time consuming. A much preferred method is to represent currents and voltages by phasor diagrams in which arrows are substituted for sine waves. These arrows or phasors are considered to rotate one complete revolution, or 360 degrees, while the sine wave passes through one cycle, or 360 degrees.

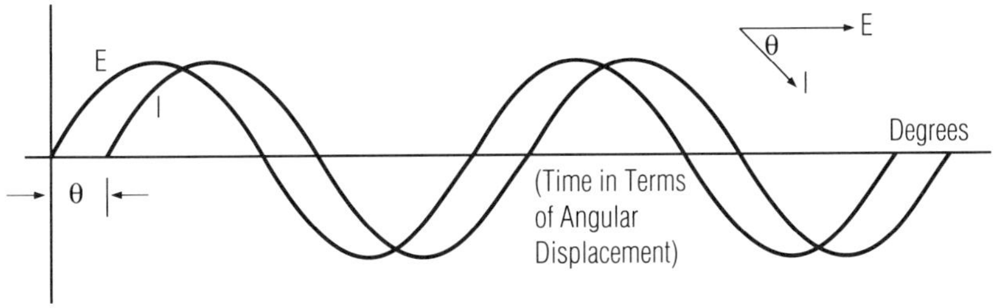

Figure 4-11. Current Wave Lagging Voltage Wave.

Table 4-1. Right Triangle Relationships.

$\sin\theta =$	$\frac{a}{c}$	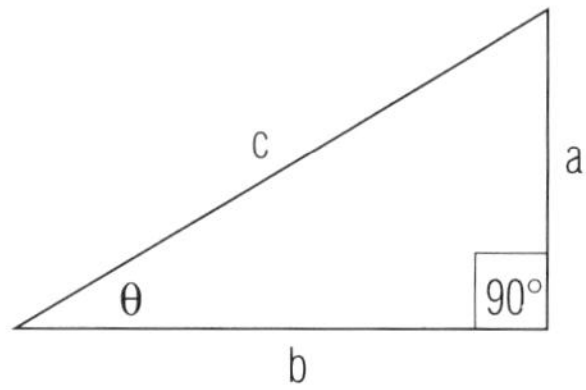	
$\cos\theta =$	$\frac{b}{c}$		
$\tan\theta =$	$\frac{a}{b}$		
$a =$	$c \sin\theta$	$b \tan\theta$	$\sqrt{c^2 - b^2}$
$b =$	$c \cos\theta$	$\frac{a}{\tan\theta}$	$\sqrt{c^2 - a^2}$
$c =$	$\frac{b}{\cos\theta}$	$\frac{a}{\sin\theta}$	$\sqrt{a^2 + b^2}$

The length of the phasor is proportional to the effective value of the current or voltage, while the angle indicates the position of the phasor relative to another phasor, the zero referenceline usually horizontal and to the right of the origin (Fig. 4-12).

One method of phasor notation is $I\underline{/\theta}$, meaning the phasor I is at an angle of θ degrees counterclockwise from the horizontal reference (Fig. 4-12).

Phasors representing currents or voltages can be resolved into vertical and horizontal components using right-triangle relationships (Fig. 4-13).

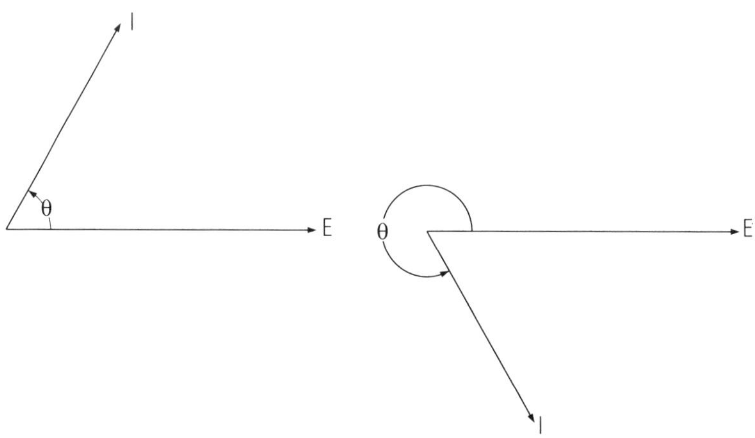

Figure 4-12. Current and Voltage Phasors.

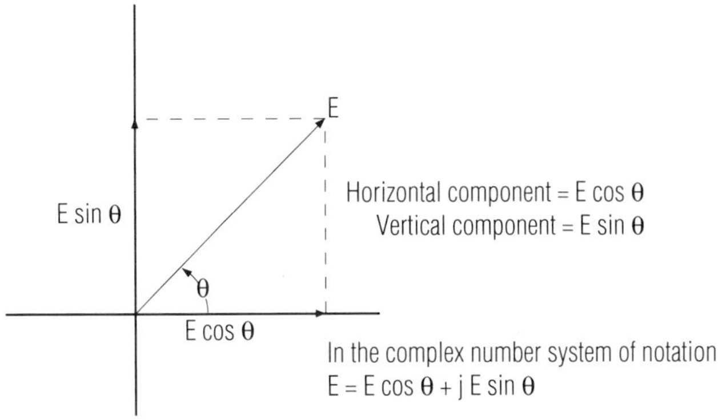

Figure 4-13. Phasor Voltage Resolved into Components.

ADDITION AND SUBTRACTION OF ALTERNATING VOLTAGES AND CURRENTS

Addition and subtraction of alternating voltages or currents cannot be done algebraically except in the case where the voltages or currents are in phase with each other. At all other times they must be added or subtracted phasorially. This can be done on a phasor diagram by resolving each phasor into its horizontal and vertical components and then adding or subtracting the components which are in the same direction. The resulting summations or subtractions are then recombined into the new phasor. This is shown in Fig. 4-14.

A second method of adding two phasors is a graphic method, "completing the parallelogram," as shown in Fig. 4-15. This construction gives both the value and the angle of the new phasor. The value only of the summation can be obtained from the law of cosines:

$$I_s = \sqrt{(I_A)^2 + (I_B)^2 + 2I_AI_B \cos\theta}$$

and the angle can be calculated by:

$$\theta = \cos^{-1}\left(\frac{I_B^2 + I_S^2 - I_A^2}{2I_BI_S}\right)$$

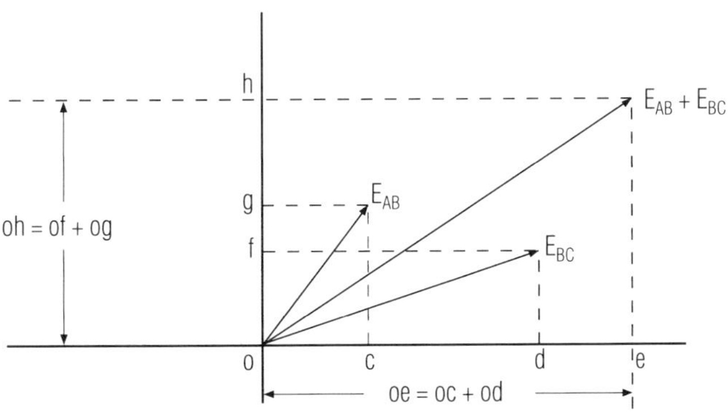

Figure 4-14. Addition of Phasors by Components.

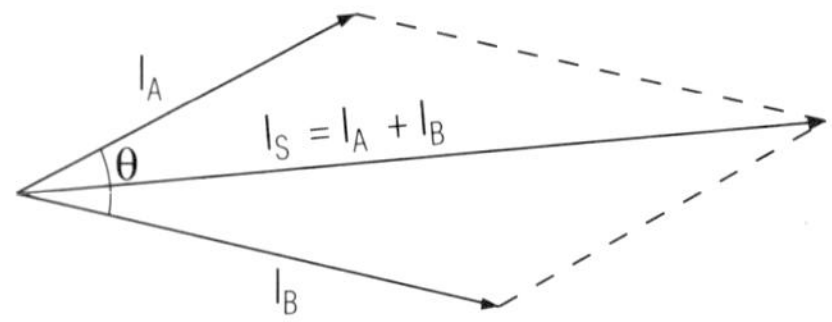

Figure 4-15. Graphical Method of Adding Phasors.

RESISTANCE

Resistance in an AC circuit has the same effect as it has in a DC circuit. An alternating current flowing through a resistance results in a power loss in the resistor. This loss is expressed the same as in direct current and is equal to I^2R.

With alternating current, a given resistor or coil may have a higher equivalent AC resistance than its DC resistance. This is especially true in coils with iron cores. Here, there is not only a power loss in the winding itself, but there is also a heat loss in the iron core. The total loss is represented by I^2R, where R is now the equivalent AC resistance of the coil.

INDUCTANCE

Any conductor which is carrying current is cut by the flux of its own field when the current changes in value. A voltage is thereby induced in the conductor which, by Lenz's law, opposes the *change* in current in the conductor. If the current is decreasing, the induced voltage is of such polarity as to try to maintain the current; if the current is increasing, the induced voltage tends to keep the current down. The amount of induced voltage depends upon the number of flux lines which cut the conductor which, in turn, depends upon the rate of change of current in the conductor. The proportionality factor between the induced voltage and the rate of change of current is the inductance, L, of the circuit.

The inductance of a circuit is generally independent of the current magnitude in the circuit. It is dependent mainly on the physical characteristics of the circuit. A conductor in the form of a coil exhibits greater inductance than a straight conductor, and iron in the flux path further increases inductance.

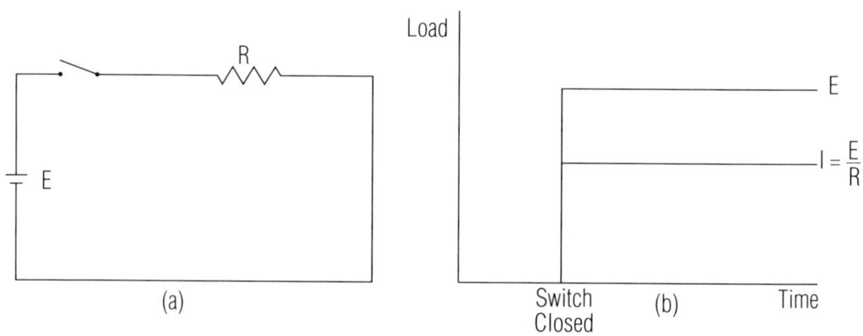

Figure 4-16. Direct Current in a Resistance Circuit.

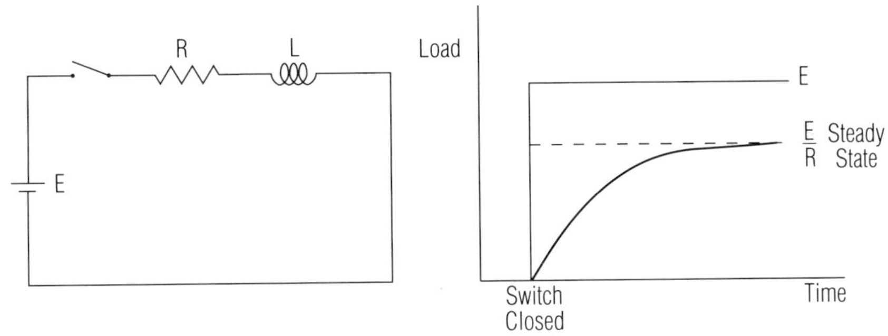

Figure 4-17. Direct Current in an Inductive Circuit.

Inductance is expressed in henrys. A more common, smaller unit is the millihenry which is one-thousandth of a henry.

In direct-current circuits inductance has no effect except when current is changing. Consider a pure resistance DC circuit (Fig. 4-16a). When a voltage is impressed, the current instantly assumes its steady-state value determined by $\frac{E}{R}$, as shown in Fig. 4-16b.

If an inductance is inserted in series with the same resistor, the current does not increase instantly to its steady-state value when the switch is closed. Instead, there is a time delay before the current reaches the same steady-state value as before, as shown in Fig. 4-17. This effect is caused by the induced voltage in the inductor opposing the rising current.

The larger the value of the circuit inductance, the longer the time required for the current to reach its steady-state value. However, once this value has been reached, the inductance has no further effect and only the resistance limits the magnitude of circuit current.

With alternating current the instantaneous current is always changing, so in an inductive circuit the inductive effect is always present.

INDUCTIVE REACTANCE

Inductance has a very definite current-limiting effect on alternating current as contrasted with steady-state direct current. This effect is directly proportional to the magnitude of the inductance L. It is also proportional to the rate of change

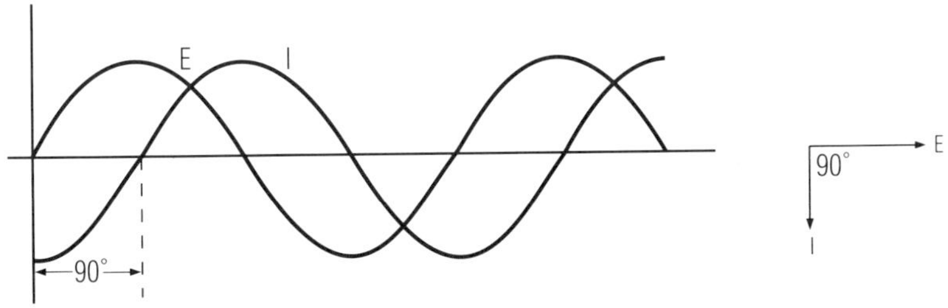

Figure 4-18. Phase Relationships in a Circuit of Pure Inductance.

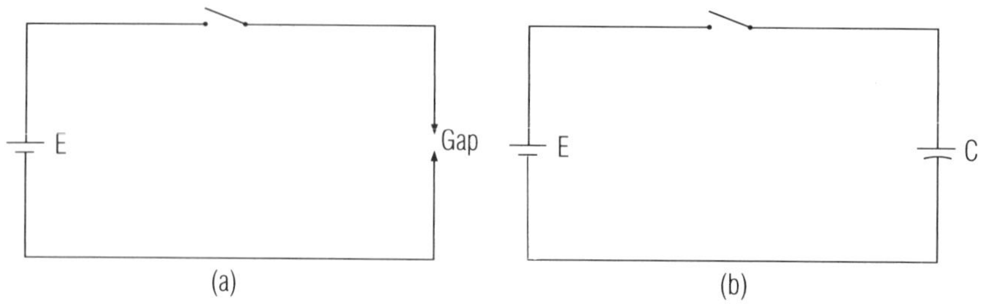

Figure 4-19. Capacitive Circuits.

of current which is a function of the frequency of the supply voltage. The total opposing, or limiting, effect of inductance on current may be calculated according to the following equation, and is called the inductive reactance.

$$X_L = 2\pi f L$$

where X_L = inductive reactance in ohms
f = frequency in hertz
L = inductance in henrys

In an AC circuit with pure inductance and a sine-wave current, the maximum rate of change of current occurs when the current passes through the zero axis. It is at this instant of zero-magnitude current, but maximum change, that there is maximum induced voltage and the voltage wave is at its peak value. When the sine-wave current reaches its peak value, the rate of change of current is zero and the induced voltage is zero. As shown in Fig. 4-18, the current wave *lags* the voltage wave by 90 degrees.

CAPACITANCE

Electric current flow is generally considered to be a movement of negative charges, or electrons, in a conductor. In conducting materials, some of the electrons are loosely attached to the atoms so that when a voltage is applied to a closed circuit, these electrons are separated from the atoms and their movement constitutes a current flow.

In an insulator the electrons are much more firmly bonded to the atoms than in a conductor. When a voltage is applied to an insulator, the electrons tend to leave the atoms, but cannot do so. However, the electrons are displaced an amount dependent upon the force applied, the voltage difference. When voltage changes, the displacement also changes. When this electron motion takes place, a displacement current flows through the dielectric and there is a charging-current flow throughout the entire circuit.

Consider the circuit of Fig. 4-19a, which has a small insulating gap between the ends of the wires. When the switch is closed there is no continuous current flow in the circuit. However, a very small current may be measured with an extremely sensitive instrument for a short time. What occurs is that electrons move through the circuit to build up an electrical charge across the gap which is equal to the impressed voltage *E*. Once the charge has been established there is no further electron movement.

If, instead of a small gap, the area is enlarged by connecting plates to each of the conductors, as in Fig. 4-19b, the current required to raise the charge to a given level is increased because a greater movement of electrons is required. Such devices consisting of large conducting areas separated by thin insulating materials (such as air, mica, glass, etc.) are called capacitors. Any two conductors separated by insulation constitute a capacitor, but normally the capacitance effect is negligible unless the components and their arrangement have been specifically designed to provide capacitance.

The capacitance, C, then, is a function of the physical characteristics of the capacitor, such as the plate area, the distance, and the type of insulation between the plates. Capacitance is expressed in farads. A more common, smaller unit is the microfarad, which is one-millionth of a farad.

Capacitors may be connected in parallel or in series. The total capacitance of capacitors connected in parallel is simply the sum of the individual capacitances.

$$\text{Parallel: } C = C_1 + C_2 + C_3$$

For a series connection, the net capacitance is found by a formula similar to that for parallel resistances.

$$\text{Series: } \frac{1}{C} = \frac{1}{C_1} + \frac{1}{C_2} + \frac{1}{C_3}$$

In a DC circuit, current flows through a capacitor only when the voltage across it changes. In an AC circuit, the voltage is continually changing and current flows in a capacitor as long as the voltage is applied. The current magnitude is proportional to the rate of change of voltage. With a sine-wave voltage, the maximum rate of change occurs when the voltage wave crosses the zero axis, and peak value of current occurs at this instant. When the voltage wave is at its peak, its rate of change is zero and current magnitude is zero. Therefore, there is a 90-degree phase displacement between current and voltage in a capacitor. When rate of change of voltage is positive, the current must be in the positive direction to supply the increasing positive charge. Therefore, the current *leads* the voltage in a capacitor. These relationships are shown in Fig. 4-20.

CAPACITIVE REACTANCE

The current-limiting effect of a capacitor, its reactance, is dependent on two quantities: capacitance and frequency. Charging current increases with increasing capacitance, so with a given voltage the reactance must be inversely propor-

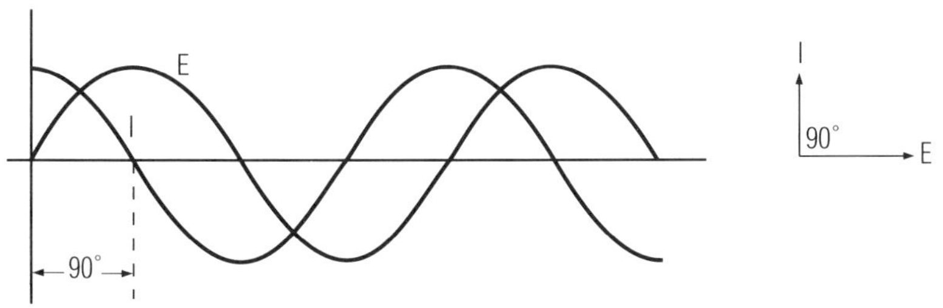

Figure 4-20. Phase Relationship in a Circuit of Pure Capacitance.

tional to capacitance. Rate of change of voltage is proportional to frequency, hence charging current is also proportional to frequency and reactance is inversely proportional. Capacitive reactance may be calculated from the following equation:

$$X_C = \frac{1}{2\pi f C}$$

where X_C = capacitive reactance in ohms
f = frequency in hertz
C = capacitance in farads

As brought out previously, capacitive reactance causes leading current and leading power factor while inductive reactance causes lagging current and lagging power factor. Capacitors are often used to balance out some of the inductive reactance of a circuit and therefore to increase the circuit power factor. They are also used to balance out some of the inductive voltage drop in a circuit and therefore increase the available voltage.

IMPEDANCE

In alternating currents, then, there are three quantities that limit the flow of current: resistance (R), inductive reactance (X_L) and capacitive reactance (X_C). Combined they become the impedance (Z) of the circuit.

In a series circuit the relationship of these quantities is shown by the formula:

$$Z = \sqrt{R^2 + (X_L - X_C)^2}$$

or by the complex number, phasor notation

$$Z = R + j(X_L - X_C)$$

Impedance may also be represented by the impedance triangles (Fig. 4-21).

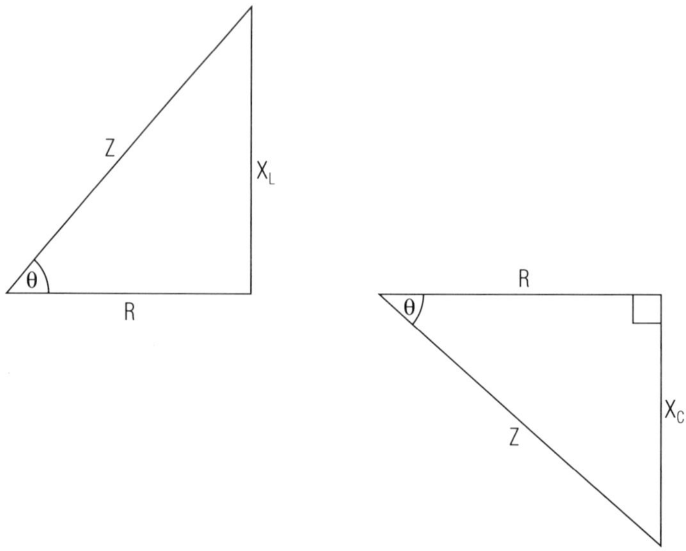

Figure 4-21. Impedance Triangles for Series Circuits.

Table 4-2. Formulas for Single-Phase AC Series Circuits.

E Volts		$\frac{W}{I \text{ Cos } \theta}$	IZ	$\frac{\sqrt{WR}}{\text{Cos } \theta}$	$\sqrt{\frac{WZ}{\text{Cos } \theta}}$	
I Amperes	$\frac{W}{E \text{ Cos } \theta}$		$\frac{E}{Z}$	$\sqrt{\frac{W}{R}}$	$\sqrt{\frac{W}{Z \text{ Cos } \theta}}$	
Z Ohms	$\frac{E}{I}$	$\frac{W}{I^2 \text{ Cos } \theta}$		$\frac{R}{\text{Cos } \theta}$	$\frac{E^2 \text{ Cos } \theta}{W}$	$\sqrt{R^2 + X^2}$
R Ohms	$\frac{E^2 \text{ Cos}^2 \theta}{W}$	$\frac{E}{I} \text{ Cos } \theta$	$Z \text{ Cos } \theta$		$\frac{W}{I^2}$	$\sqrt{Z^2 - X^2}$
W Watts	$\frac{E^2 \text{ Cos } \theta}{Z}$	$EI \text{ Cos } \theta$	$I^2Z \text{ Cos } \theta$	$I^2 R$		
Cos θ Power Factor	$\frac{IR}{E}$	$\frac{W}{I^2 Z}$	$\frac{WZ}{E^2}$	$\frac{R}{Z}$	$\frac{W}{EI}$	$\frac{R}{\sqrt{R^2 + X^2}}$
X Ohms	$(X_L - X_C)$		$\left(2\pi fL - \frac{1}{2\pi fC}\right)$			$\sqrt{Z^2 - R^2}$

From these triangles, various other trigonometric relationships between *Z*, *R*, and *X* can be obtained. See Table 4-2.

The various components of the impedance *Z* determine not only the amount of current flowing in a circuit, but also the phase relationship between the voltage and the current.

If the circuit has only resistance, the current is "in phase" with the voltage and the circuit is said to have unity power factor.

If the inductive reactance exceeds the capacitive reactance in a series circuit, the current lags the voltage and the circuit has lagging power factor.

If the capacitive reactance exceeds the inductive reactance in a series circuit, the current leads the voltage and the circuit has leading power factor.If the inductive reactance and the capacitive reactance are equal to each other in a series circuit, the circuit is said to be resonant; the current flow is limited only by the resistance and the circuit power factor is again unity.

OHM'S LAW — AC CIRCUITS

Ohm's law as applied to AC circuits takes the form:

$$E = IZ \quad \text{or} \quad I = \frac{E}{Z}$$

In this form, only the magnitude of the voltage or current is obtained, but the same formula can be applied phasorially to obtain both magnitude and direction.

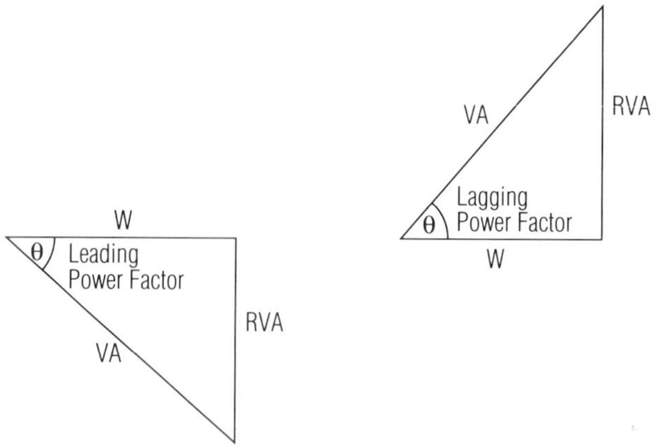

Figure 4-22. Power Triangles (Single-Phase or Three-Phase).

THE POWER EQUATIONS

The commonly used power equations for single-phase AC circuits are:

$$W = I^2R$$
$$W = EI\cos\theta$$

where $\cos\theta$ is equal to the power factor of the circuit.

Ohm's law and the power equations are combined to give the various formulas for single-phase AC series circuits shown in Table 4-2.

THE POWER TRIANGLE

In alternating current, two other terms are important: volt-amperes, and reactive-volt-amperes or vars. In formula form these are:

$$\text{Volt-amperes (VA)} = E \times I$$
$$\text{Reactive-volt-amperes (RVA)} = \text{Vars} = E \times I\sin\theta$$

The relationship between these two terms and the power is best shown by the power triangle, Fig. 4-22.

From these triangles, the following relationships are obtained:

$$VA = \sqrt{W^2 + (RVA)^2}$$
$$W = (VA)\ \cos\theta$$
$$RVA = (VA)\ \sin\theta$$

$$\text{Power factor} = \cos\theta = \frac{W}{VA} \quad \text{(percent power factor = power factor} \times 100)$$

$$\text{Reactive factor} = \sin\theta = \frac{RVA}{VA}$$

In complex number notation:

$$VA = W + jRVA \text{ for lagging power factor}$$
$$VA = W - jRVA \text{ for leading power factor}$$

Watts is the "in phase" component, while reactive-volt-amperes, or vars, is the "out-of-phase" or quadrature component.

MAGNETISM

1. *Permanent Magnets* — Certain elements and alloys have the property that, once magnetized, they will retain a large part of that magnetism. "Permanent magnets" are the result, and they maintain a continuous magnetic field or flux which comes out of their north magnetic pole and goes into their south magnetic pole. Iron, nickel, and cobalt are the elements that can be magnetized. Alloys containing one or more of these magnetic elements, plus certain non-magnetic elements, can be made into strong permanent magnets. Aluminum-nickel-cobalt alloys have very strong permanent magnet properties.
2. *An electromagnet* is made by simply winding a coil around an iron core. When a current flows in the winding, the core is magnetized, a magnetic flux is set up, and north and south poles are established. However, when the current is removed, the core loses most of its magnetism. The strength of an electromagnet depends on the current and the number of turns in the coil — the ampere-turns. It also depends on the type iron in the core.
3. *The magnetic field* equation is similar to Ohm's law for the electric circuit.

For the electric circuit: $I = \dfrac{E}{R}$

For the magnetic circuit: $\Phi = \dfrac{\mathfrak{F}}{\mathfrak{R}}$

where Φ = flux in core

$\mathfrak{F}$ = magnetomotive force (ampere-turns)

$\mathfrak{R}$ = reluctance of flux path

Whenever a current flows, a flux or magnetic field is set up.

4. *Induced Voltage* — Whenever a flux about a conductor changes, an induced voltage is set up in that conductor in such a direction as to oppose the change in flux.

In alternating current, where the current is changing constantly, the flux or field is constantly changing and voltage is constantly being induced in the original conductor (self-induction) and in any surrounding conductors. Also, a voltage is induced in any conductor moving through a magnetic field (generator action). The direction of induced voltage and resulting current will be such that the field set up by the current will oppose the original field. Similar generator action results if the conductor is stationary but the magnetic field is moving.

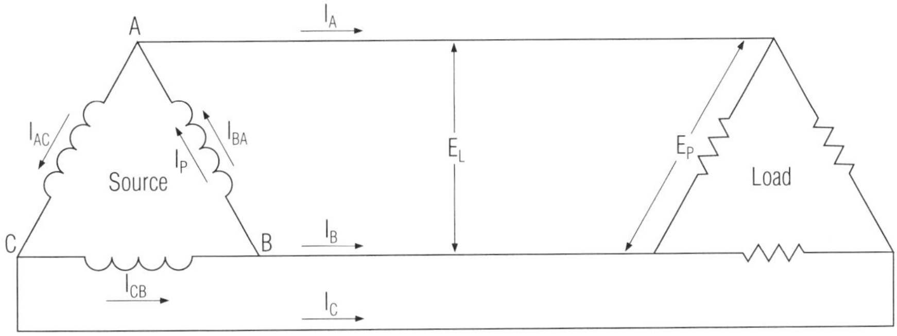

Figure 4-23. Three-Phase Delta Connection.

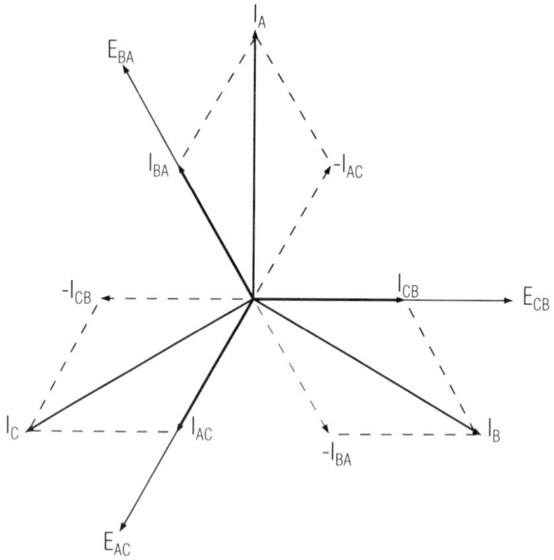

Figure 4-24. Phasor Diagram for Three-Phase Delta Connection for a Balanced 100 Percent Power Factor Load.

POLYPHASE DELTA CONNECTION

In electric generators, the magnetic field is set up by current flowing in conductors or coils called the "field winding." The field winding may be either the stationary member (stator) or the rotating member (rotor) of the generator. Voltage is induced in the "armature winding" which is then the opposite member.

In modern large generators, because of the heavy currents involved, the armature is the stationary winding or stator. The magnetic field is set up by the field current flowing in the coils of the rotating winding or rotor.

The armature coils may be wound so that only one voltage is obtained, as in a single-phase generator. But the armature may also be wound and taps brought out so that several voltages displaced in time from each other may be obtained.

If three voltages are brought out, displaced from each other by 120 degrees, a three-phase system is obtained. By proper connection, this may be either a delta or wye system. In a delta system the connections are made as shown in Fig. 4-23.

In a balanced delta system the three phase currents are equal in magnitude and are spaced 120 degrees apart. The same relationships exist among the three line currents.

Therefore (in magnitude):

$$I_{BA} = I_{AC} = I_{CB} = I_{Phase} = I_P$$

$$I_A = I_B = I_C = I_{Line} = I_L$$

From the phasor diagram in Fig. 4-24 it can be proven that:
Line current = $\sqrt{3}$ phase current or $I_L = \sqrt{3}\ I_P$
Also, by inspection, line-to-line voltage = phase voltage, or $E_L = E_P$

POLYPHASE WYE CONNECTION

In a wye system the connections are made as shown in Fig. 4-25.

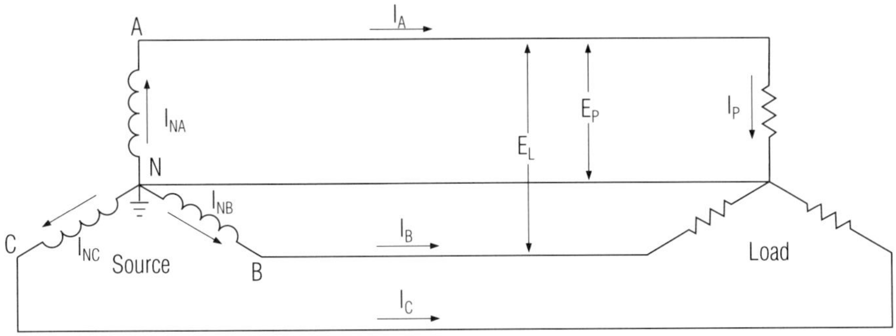

Figure 4-25. Three-Phase Wye Connection.

In a balanced wye system the three phase currents are equal in magnitude and are spaced 120 degrees apart. The same relationship holds for the three phase voltages and the three line voltages. Therefore (in magnitude):

$$I_{NA} = I_{NB} = I_{NC} = I_{Phase} = I_P$$
$$E_{NA} = E_{NB} = E_{NC} = E_{Phase} = E_P$$
$$E_{BA} = E_{AC} = E_{CB} = E_{Line} = E_L$$

Also, by inspection, line current = phase current, or $I_L = I_P$

From the phasor diagram in Fig. 4-26, it can be proven that:

Line-to-line voltage = $\sqrt{3}$ phase voltage, or $E_L = \sqrt{3}\ E_P$

POWER EQUATION FOR POLYPHASE SYSTEMS

In a balanced polyphase system, the three line currents are equal in magnitude to each other and are 120 degrees apart. In a balanced wye system, there is no current in the neutral wire. Likewise, the voltages are equal to each other and 120 degrees apart. For the balanced polyphase system, either delta or wye, the power equation is as shown in Table 4-3.

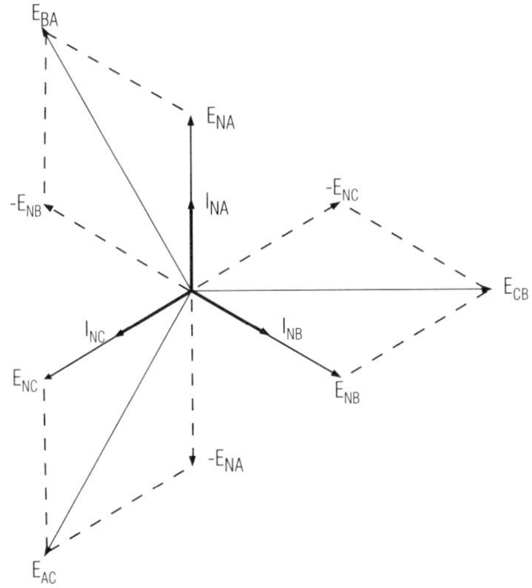

Figure 4-26. Phasor Diagram for Three-Phase Wye System for Balanced 100 Percent Power Factor Load.

Table 4-3. Common Formulas for Three-Phase AC Circuits

	For Balanced Systems Only	For Both Balanced and Unbalanced Systems	
W Watts	$\sqrt{3}\ E_L\ I_L \cos\theta$	$\sqrt{(VA)^2 - (RVA)^2}$	$(VA) \times (PF)$
RVA Reactive Voltamperes	$\sqrt{3}\ E_L\ I_L \sin\theta$	$\sqrt{(VA)^2 - (W)^2}$	$(VA) \times (RF)$
VA Voltamperes	$\sqrt{3}\ E_L\ I_L$	$\sqrt{(W)^2 + (RVA)^2}$	$\frac{W}{PF}$
PF Power Factor	$\cos\theta$		$\frac{W}{VA}$
RF Reactive Factor	$\sin\theta$		$\frac{RVA}{VA}$
E_L Line to Line Voltage	$\sqrt{3}\ E_P$ (Wye) E_P (Delta)	$\sqrt{3}\ E_P$ (Wye) (If voltages are balanced)	E_P (Delta)
I_L Line Current	I_P (Wye) $\sqrt{3}\ I_P$ (Delta)		

$$\text{Total watts} = W = \sqrt{3}\ E_L I_L \cos\theta$$

$$\text{Total reactive-volt-amperes (vars)} = RVA = \sqrt{3}\ E_L I_L \sin\theta$$

$$\text{Total volt-amperes } (VA) = \sqrt{3}\ E_L I_L$$

In these equations E_L and I_L are the line voltage and current, θ is the angle between the phase current and the phase voltage, and $\cos\theta$ is the power factor of the load. For a *balanced wye system only,* the following equations are correct and are often used:

$$W = 3E_P I_L \cos\theta$$

$$RVA = 3E_P I_L \sin\theta$$

$$VA = 3E_P I_L$$

For an unbalanced polyphase system the power must be calculated for each phase and these added to obtain total power.

$$W = W_A + W_B + W_C$$

$$RVA = (RVA)_A + (RVA)_B + (RVA)_C$$

$$VA = \sqrt{W^2 + (RVA)^2}$$

TRANSFORMERS

One of the greatest advantages of alternating current is that the voltage level can be stepped up or down readily through the use of transformers. In simplest form, a transformer consists of two windings wound on a common laminated-iron core. See Fig. 4-27.

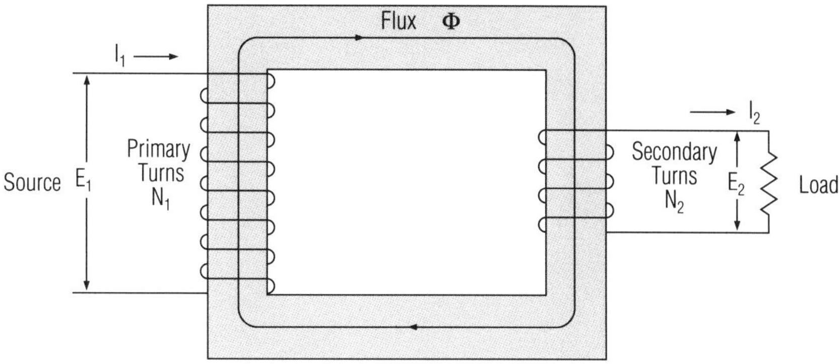

Figure 4-27. Simple Form of Transformer.

The alternating current flowing in the primary winding sets up an alternating flux in the common iron core. A voltage is induced in the secondary winding which in turn causes an alternating current to flow in the secondary circuit.

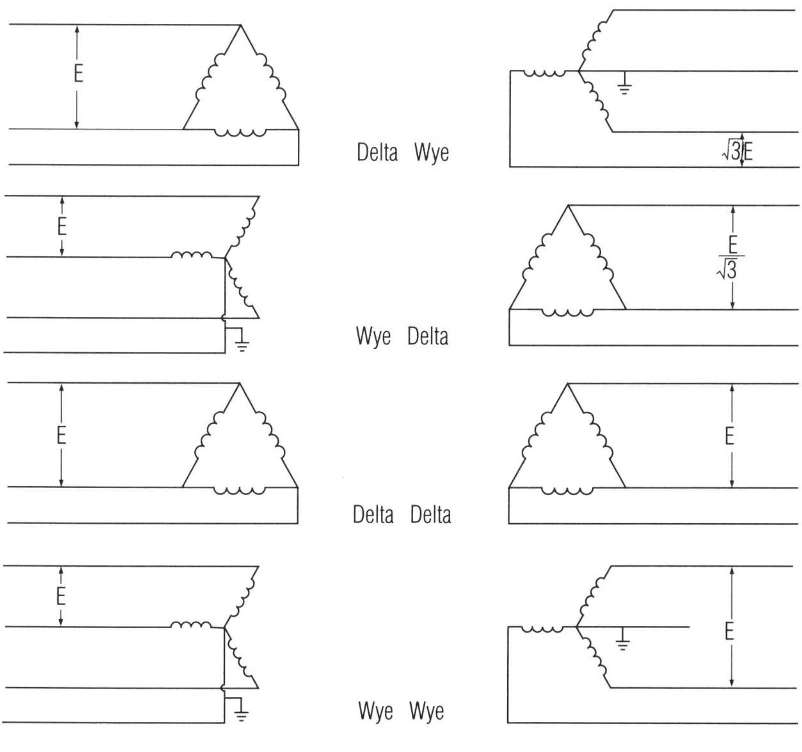

Figure 4-28. Various Transformer Bank Connections with Line Voltages Shown for a 1:1 Ratio of Transformation for the Individual Transformers.

Primary and secondary relationships in an ideal transformer are:

$$\frac{E_2}{E_1} = \frac{N_2}{N_1}$$

$$\frac{I_2}{I_1} = \frac{N_1}{N_2}$$

$$W_1 = W_2$$

TRANSFORMER CONNECTIONS

Banks of transformers may be connected in such a way as to set up various types of distribution systems. Some of these are shown in the diagrams in Fig. 4-28.

DISTRIBUTION CIRCUITS

Common distribution circuits used in the electrical industry are shown in Fig. 4-29.

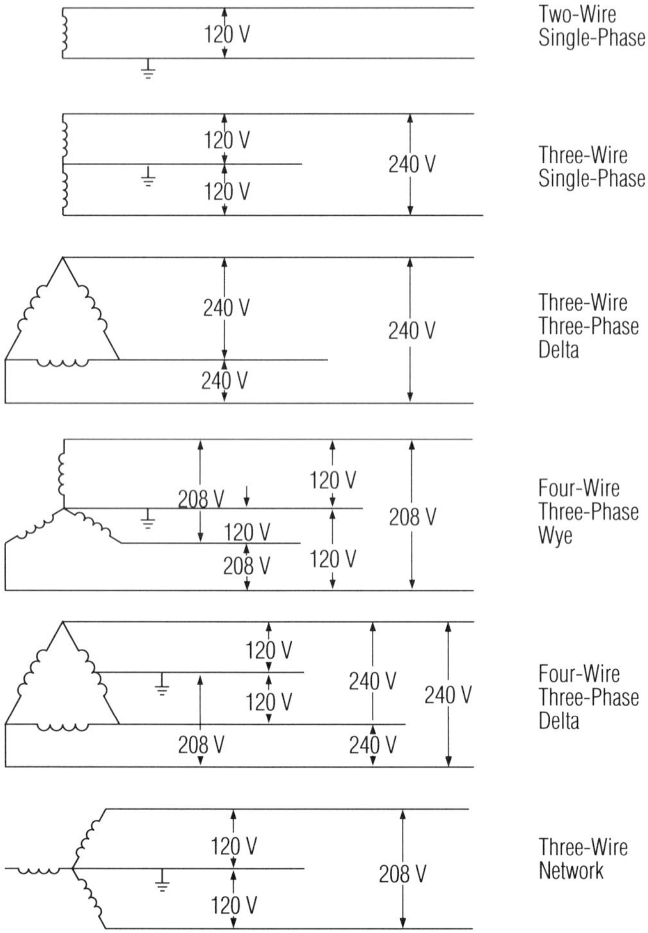

Figure 4-29. Common Distribution Circuits.

CHAPTER 5
SOLID STATE ELECTRONICS

THIS CHAPTER DEALS WITH basic solid state electronics as applied to modern metering devices. The information contained here is intended as a review for metering personnel with a background in electronics and as an introduction for those yet unfamiliar with the subject, with the intention of stimulating further study.

In the study of solid state electronics it is necessary to understand the effects of combining semiconductors of differing atomic structures. For this reason the chapter begins with a discussion of the atom in order to introduce the concept of current flow across the semiconductor junction. Finally, the chapter introduces digital electronics including the microprocessor.

PART 1—THE ATOM

Atomic structure is best demonstrated by the hydrogen atom, which is composed of a nucleus (center core) containing one proton and a single orbiting electron. As the electron revolves around the nucleus it is held in this orbit by two counteracting forces. One of these forces is the centrifugal force, which tends to cause the electron to fly outward as it travels around its circular orbit. The second force is centripetal force, which tends to pull the electron in towards the nucleus and is provided by the mutual attraction between the positive nucleus and negative electron. At some given radius the two forces will exactly balance each other providing a stable path for the electron. By virtue of its motion, the electron in the hydrogen atom contains kinetic energy. Due to its position it also contains potential energy.

The total energy contained by the electron (kinetic plus potential) is the factor which determines the radius of the electron orbit around the nucleus. The orbit shown in Fig. 5-1 is the smallest possible orbit the hydrogen electron can have. For the electron to remain in this orbit is must neither gain nor lose energy.

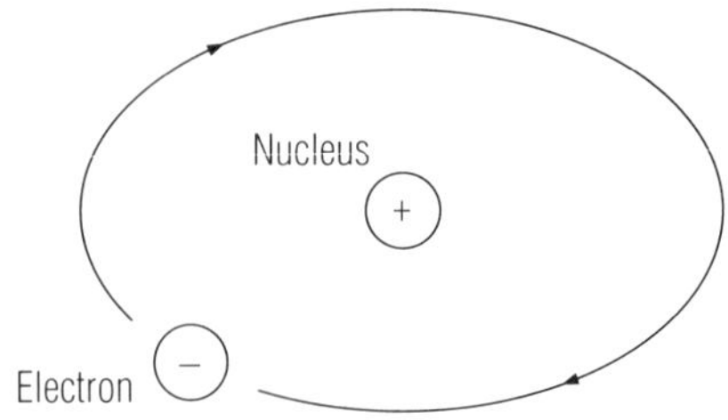

Figure 5-1. Hydrogen Atom.

The electron will remain in its lowest orbit until a sufficient amount of energy is available, at which time the electron will accept the energy and jump to one of a series of permissible orbits. An electron cannot exist in the space between permissible orbits or energy levels. This indicates that the electron will not accept energy unless it is great enough to elevate the electron to one of the allowed energy levels. Light and heat energy as well as collisions with other particles can cause the electron to jump orbit.

Once the electron has been elevated to an energy level higher than the lowest possible energy level, the atom is said to be in an excited state. The electron will not remain in this excited condition for more than a fraction of a second before it will radiate the excess energy and return to a lower energy orbit.

A second alternative would be for the electron to return to the lower level in two jumps; from the third to the second, and then from the second to the first. In this case the electron would emit energy twice, once from each jump. Each emission would have less energy than the original amount which originally excited the electron.

Although hydrogen has the simplest of all atoms, the principles just developed apply equally well to the atoms of more complex elements. The manner in which the orbits are established in an atom containing more than one electron is somewhat complicated and is part of a science known as Quantum Mechanics. In an atom containing two or more electrons, the electrons interact with each other and the exact path of any one electron is very difficult to predict. However, each electron will lie in a specific energy band and the above-mentioned orbits will be considered as an average of the electron positions. Also, the various electron orbits found in large atoms are grouped into shells which correspond to fixed energy levels.

The number of electrons in the outermost orbit group or shell determines the valence of the atom and, therefore, is called the valence shell. The valence of an atom determines its ability to gain or lose an electron which, in turn, determines the chemical and electrical properties of the atom. An atom that is lacking only one or two electrons from its outer shell will easily gain electrons to complete its shell, but a large amount of energy is required to free any other electrons. An atom having a relatively small number of electrons in its outer shell in comparison to the number of electrons required to fill the shell will easily lose its valence electrons. Gaining or losing electrons in valence shells is called ionization. Atoms gaining electrons are negative ions and atoms losing electrons are positive ions.

PART 2—SEMICONDUCTOR ELECTRONICS

Any element can be categorized as either a conductor, semiconductor, or insulator. Conductors are elements, such as copper or silver, which will conduct electricity readily. Insulators (non-conductors) do not conduct electricity to any great degree and are therefore used to prevent a flow of electricity. Rubber and glass are good insulators. Material such as germanium and silicon are not good conductors, but cannot be used as insulators either, since their electrical characteristics fall between those of conductors and insulators. These are called semiconductors.

The electrical conductivity of matter is ultimately dependent upon the energy levels of the atoms of which the material is constructed. In any solid material such as copper, the atoms which make up the molecular structure are bound together in a crystal lattice which is a rigid structure of copper atoms. Since the atoms of copper are firmly fixed in position within the lattice structure, they are not free to migrate through the material and therefore cannot carry the electricity through the conductor without application of some external force. However, by ionization, electrons could be removed from the influence of the parent atom and made to move through the copper lattice under the influence of external forces. It is by virtue of the movement of these free electrons that electrical energy is transported within the copper material. Since copper is a good conductor, it must contain vast numbers of free electrons.

HOLE CURRENT AND ELECTRON CURRENT

The degree of difficulty in freeing valence electrons from the nucleus of an atom determines whether the element is a conductor, semiconductor, or an insulator. When an electron is freed in a block of pure semiconductor material, it creates a hole which acts as a positively charged current carrier. Thus, an electron liberation creates two currents which are known as electron current and hole current.

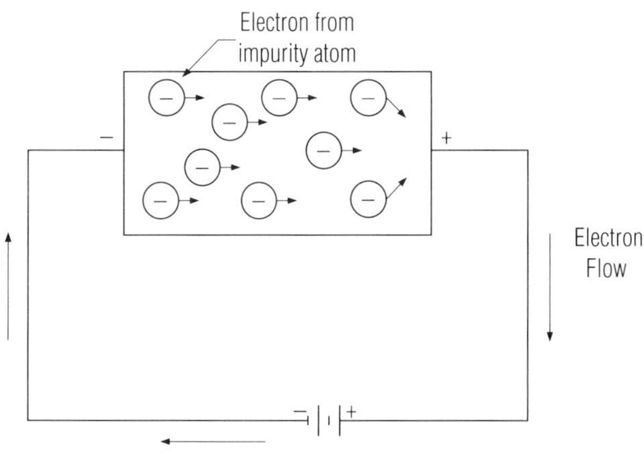

Figure 5-2. Electron Flow in N-Type Material.

When an electric field is applied, holes and electrons are accelerated in opposite directions. The life spans (time until recombination) of the hole and the free electron in a given semiconductor sample are not necessarily the same. Hole conduction may be thought of as the unfilled tracks of a moving electron. Because the hole is a region of net positive charge, the apparent motion is like the flow of particles having a positive charge.

If suitable impurity is added to the semiconductor, the resulting mixture can be made to have either an excess of electrons, causing more electron current, or an excess of holes, causing more hole current.

Depending upon the kind of impurity added to a semiconductor, it will have more (or less) free electrons than holes. Both electron current and hole current will be present, but a majority carrier will dominate. The holes are called positive carriers and the electrons negative carriers. The one present in the greatest quantity is called the majority carrier; the other is called the minority carrier. The quality and quantity of the impurity are carefully controlled by the doping process.

N AND P TYPE MATERIALS

When an impurity like arsenic is added to germanium it will change the germanium crystal lattice in such a way as to leave one electron relatively free in the crystal structure. Because this type of material conducts by electron movement, it is called a negative-carrier or N-type semiconductor. Pure germanium may be converted into an N-type semiconductor by doping it with a donor impurity consisting of any element containing five electrons in its outer shell. The amount of the impurity added is very small.

An impurity element can also be added to pure germanium to dope the material so as to leave one electron lacking in the crystal lattice, thereby creating a hole in the lattice. Because this semiconductor material conducts by the movement of holes which are positive charges, it is called a positive carrier or P-type semiconductor. When an electron fills a hole, the hole appears to move to the spot previously occupied by the electron.

As stated previously, both holes and electrons are involved in conduction. In N-type material the electrons are the majority carriers and holes are the minority carriers. In P-type material the holes are the majority carriers and the electrons are the minority carriers.

Current flow through an N-type material is illustrated in Fig. 5-2. Conduction in this type of semiconductor is similar to conduction in a copper conductor. That is, an application of voltage across the material will cause the loosely bound electron to be released from the impurity atom and move towards the positive potential point.

Current flow through a P-type material is illustrated in Fig. 5-3. Conduction in this material is by positive carrier (holes) from the positive to the negative terminal. Electrons from the negative terminal cancel holes in the vicinity of the terminal, while, at the positive terminal, electrons are being removed from the crystal lattice, thus creating new holes. The new holes then move toward the negative terminal (the electrons shifting to the positive terminal) and are canceled by more electrons emitted into the material from the negative terminal. This process continues as a steady stream of holes (hole current) move toward the negative terminal.

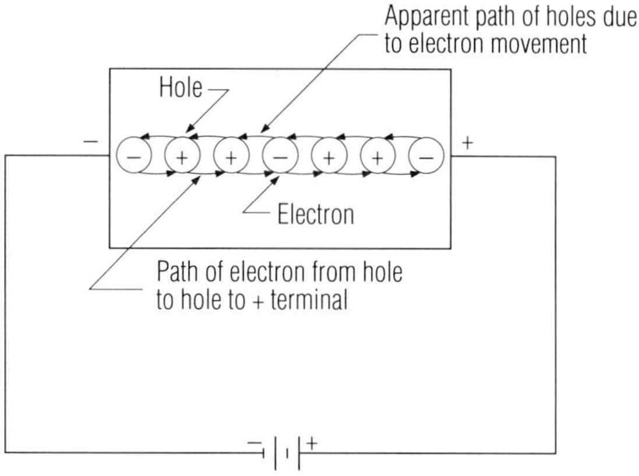

Figure 5-3. Electron Flow in P-Type Material.

P-N JUNCTION

Both N-type and P-type semiconductor material are electrically neutral. However, a block of semiconductor material may be doped with impurities so as to make half the crystals N-type material and the other half P-type material. A force will then exist across the thin junction of the N-type and P-type material. The force is an electro-chemical attraction by the P-type material for electrons in the N-type material. Due to this force, electrons will be caused to leave the N-type material and enter the P-type material. This will make the N-type material near to the junction positive with respect to the remainder of the N-type material. Also the P-type material near to the junction will be negative with respect to the remaining P-type material.

After the initial movement of charges, further migration of electrons ceases, due to the equalization of electron concentration in the immediate vicinity of the junction. The charged areas on either side of the junction constitute a potential barrier, or junction barrier, which prevents further current flow. This region is also called a depletion region. The device thus formed is called a semiconductor diode.

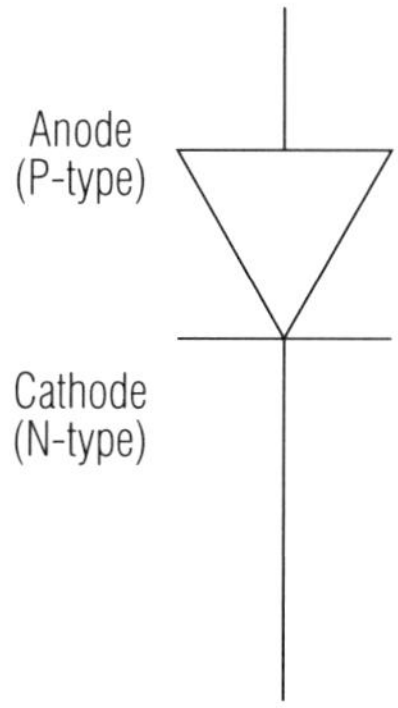

Figure 5-4. Semiconductor Diode Symbol.

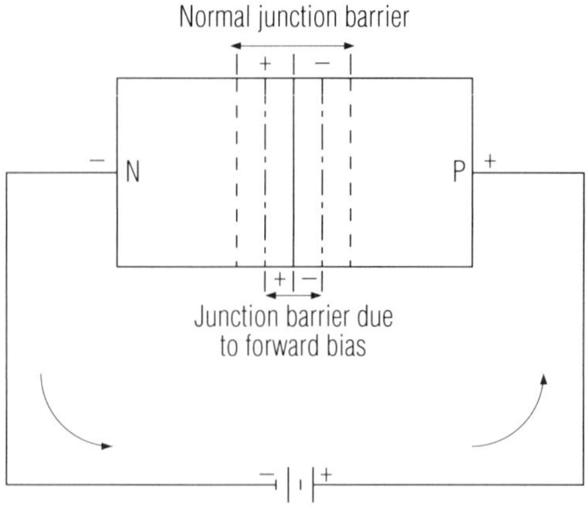

Figure 5-5. Semiconductor Diode with Forward Bias.

SEMICONDUCTOR DIODE

The schematic symbol for semiconductor diode is illustrated in Fig. 5-4. The N-type material section of the device is called the cathode and the P-type material section the anode. The device permits electron current flow from cathode to anode, and restricts electron current flow from anode to cathode.

Consider the case where a potential is placed externally across the diode, positive on the anode with respect to the cathode. This is depicted in Fig. 5-5. This polarity of voltage (anode positive with respect to the cathode) is called forward bias since it decreases the junction barrier and causes the device to conduct appreciable current. Next, consider the case where the anode is made negative with respect to the cathode. Figure 5-6 illustrates this reverse bias condition.

Theoretically, no current flow should be possible with reverse bias applied across the junction, due to the increase in the junction barrier.

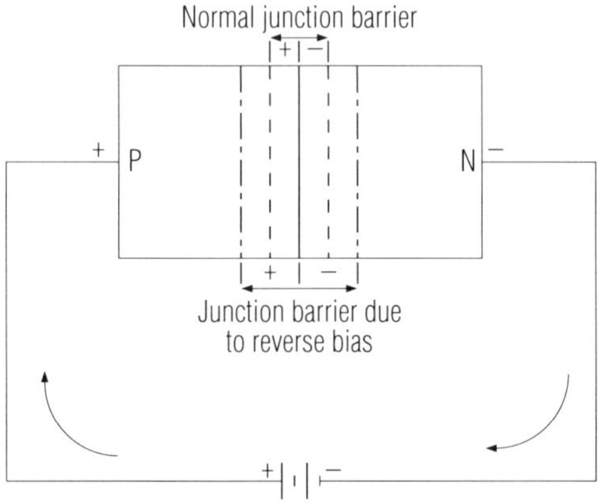

Figure 5-6. Semiconductor Diode with Reverse Bias.

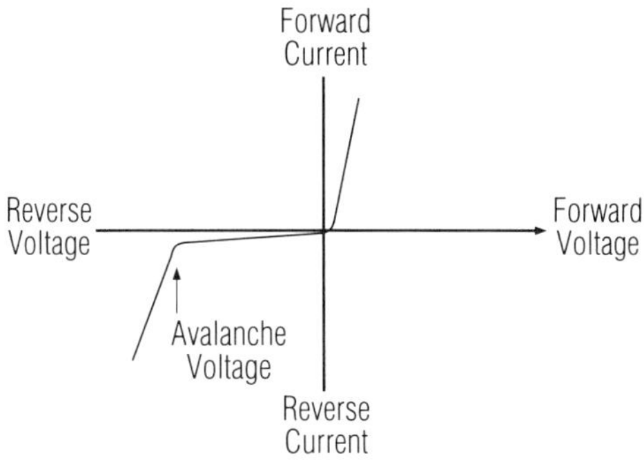

Figure 5-7. Semiconductor Diode Characteristic Curve.

However, since the block of semiconductor material is not a perfect insulator, a very small reverse or leakage current will flow. At normal operating temperatures this current may be neglected. It is noteworthy, however, that leakage current increases with an increase in temperature. The characteristic curve of the typical diode is shown in Fig. 5-7. Note that excessive forward bias results in a rapid increase of forward current and could destroy the diode. By the same token, excess reverse bias could cause a breakdown in the junction due to the stress of the electric field. The reverse bias point at which breakdown occurs is called the breakdown or avalanche voltage.

Some semiconductor diodes are made to operate in the breakdown or avalanche region, the most common being the zener diode which is discussed later.

TRANSISTORS

By connecting together two P-N junctions, either at their N sides or their P sides, and appropriately applying forward bias to one junction while reverse biasing the other junction, an interesting phenomenon occurs. If we call the thin

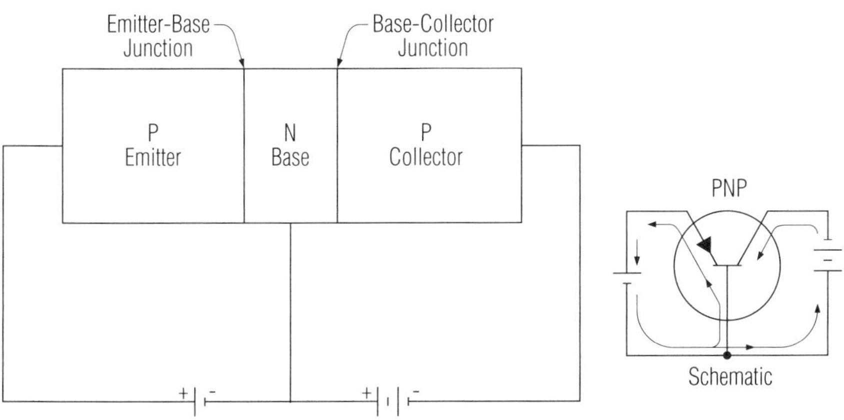

Figure 5-8. Basic PNP Transistor Circuit.

connecting section of material the base, and the sections on either end of the junction, the emitter and collector respectively, we have the device shown in Fig. 5-8. Reverse bias applied to the base-collector junction causes a small reverse current as shown in Fig. 5-7 for a typical P-N junction. By forward biasing the emitter-base junction, the base-collector junction is driven further into the breakdown, or avalanche region, resulting in a much larger collector current. What, then, is the difference between the simple junction diode and the transistor? Well, if a small, varying signal is applied between the emitter and base, the bias across the base-emitter junction can be used to control the large current flow in the collector circuit, and if the bias is reversed, current flow ceases. What we have here, then, is the means for controlling a large current by varying a smaller one, which is the basis for amplification.

PART 3—DIGITAL ELECTRONICS

While analog circuits operate on a continuous range of signals, digital electronic circuits have only two states: on and off. Digital circuits use electronic components which can be controlled to operate in either of two electrical conditions for example, conducting and non-conducting and these two conditions represent the on and off states.

NUMBER SYSTEMS

Decimal Numbers

The decimal number system that we all use is comprised of ten digits, zero through nine. It is based on units of ten, and is called a base-ten system. For example, 4,732 is immediately recognized as four thousand, seven hundred, thirty-two. What the digits 4732 actually mean is the sum of four thousands, seven hundreds, three tens, and two ones. In the decimal number system, the right-most digit is multiplied by 10^0, the next digit by 10^1, the next digit by 10^2, and so on across, for as many digits as there are in the number.

$$
\begin{array}{rcrcrcrcr}
4 & \times & 1{,}000 & = & 4 & \times & 10^3 & = & 4{,}000 \\
7 & \times & 100 & = & 7 & \times & 10^2 & = & 700 \\
3 & \times & 10 & = & 3 & \times & 10^1 & = & 30 \\
2 & \times & 1 & = & 2 & \times & 10^0 & = & 2 \\
\hline
 & & & & & & & & 4{,}732
\end{array}
$$

Binary Numbers

The binary number system uses two digits, one and zero. It is called a base-two system. The binary system fits well with the way digital circuits operate. For example, a transistor can be controlled either to conduct or not conduct current, and can quickly switch from one state to the other. Whether the transistor is conducting or not controls whether its output is high or low, signifying a one or a zero.

The two-state digit is called a bit, a contraction of the words "binary digit." An example of a binary number is 100101 (read as one-zero-zero-one-zero-one). This number is actually the sum of six binary numbers, and totals 37 in the decimal number system. In the binary number system, the right-most digit is multiplied by 2^0, the next digit by 2^1, the next digit by 2^2, and so on across, for as many digits as there are in the number.

$$
\begin{array}{ccccccccc}
1 & \times & 32 & = & 1 & \times & 2^5 & = & 32 \\
0 & \times & 16 & = & 0 & \times & 2^4 & = & 0 \\
0 & \times & 8 & = & 0 & \times & 2^3 & = & 0 \\
1 & \times & 4 & = & 1 & \times & 2^2 & = & 4 \\
0 & \times & 2 & = & 0 & \times & 2^1 & = & 0 \\
1 & \times & 1 & = & 1 & \times & 2^0 & = & \underline{1} \\
 & & & & & & & & 37
\end{array}
$$

In computers, a combination of several bits is used to represent standard characters. Most computer products handle information in bytes, where a byte is a combination of eight bits. One byte can represent a number, a letter, or a special character.

Hexadecimal Numbers

Hexadecimal numbers are sometimes easier to manage when working with large binary numbers. Since it takes a long string of 1's and 0's to represent a number of any size, it might be convenient to group together binary numbers to form another value with the same meaning. Since computers operate mostly with 4-, 8-, and 16-bit numbers, a grouping of four bits is useful. The hexadecimal system, a base-16 system, is a useful choice because 4 bits can be arranged 16 different ways, and 4 is a common factor of 4, 8, and 16. When working with bytes, two hexadecimal digits define one byte rather than the eight 1's and 0's which would be necessary in the binary number system.

The relationship between the decimal, binary, and hexadecimal number systems, counting up through the first 16 numbers, is as follows:

Decimal	Binary	Hexadecimal
0	0	0
1	1	1
2	10	2
3	11	3
4	100	4
5	101	5
6	110	6
7	111	7
8	1000	8
9	1001	9
10	1010	A
11	1011	B
12	1100	C
13	1101	D
14	1110	E
15	1111	F

Binary Digital Logic Circuits

Figure 5-9 shows a schematic diagram of a circuit with a battery, a light bulb, and a switch. When the switch is closed, the light bulb is on. When the switch is open, the light bulb is off. In the truth table, the binary numbers 1 and 0 are used to represent on and off, or that the switch is open and closed.

The NOT function is illustrated in Fig. 5-10. The light is on when the switch is not closed, and off when the switch is not open. The NOT function is an inverting function, and the circuit element is called an inverter. If the input is A, the output is $\overline{A}$, which is read: "A bar", or "Not A."

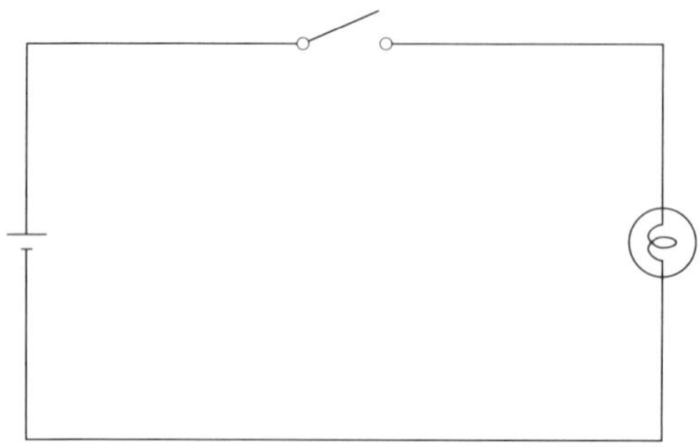

Switch	Light
0	0
1	1

Figure 5-9. Logic Circuit and Truth Table.

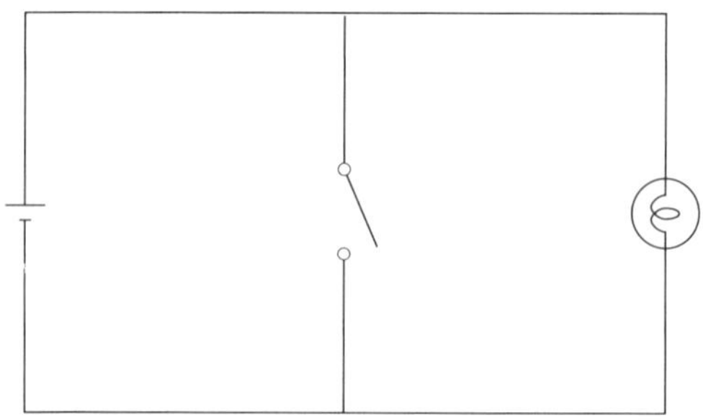

Switch	Light
0	1
1	0

Figure 5-10. NOT Function.

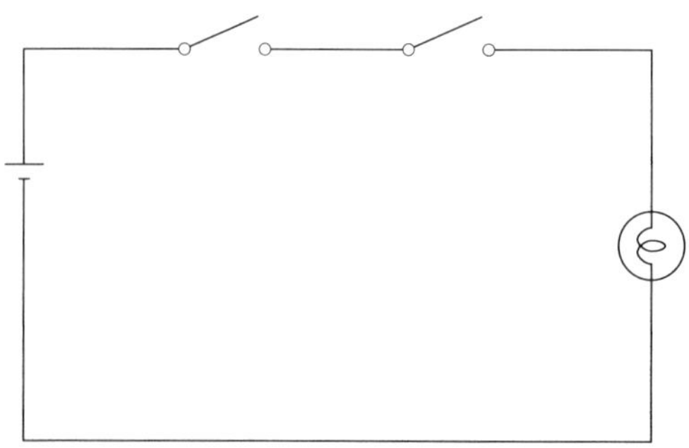

S_1	S_2	Light
0	0	0
0	1	0
1	0	0
1	1	1

Figure 5-11. AND Function.

Certain digital circuits are called logic circuits because they perform like the logic functions AND, OR, NAND, and NOR. Figure 5-11 shows the AND function: the light is on only when both switches are closed. Figure 5-12 shows the OR function: the light is on when one switch or the other is closed. Figure 5-13 shows the NAND function: the light is not on only when both switches are closed. Figure 5-14 shows the NOR function: the light is not on when either switch is closed.

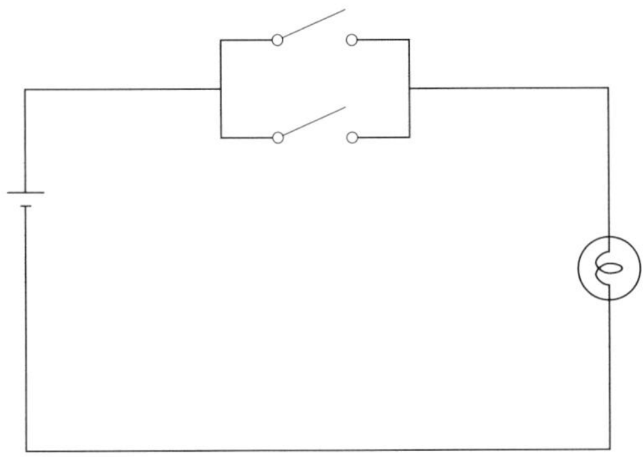

S_1	S_2	Light
0	0	0
0	1	1
1	0	1
1	1	1

Figure 5-12. OR Function.

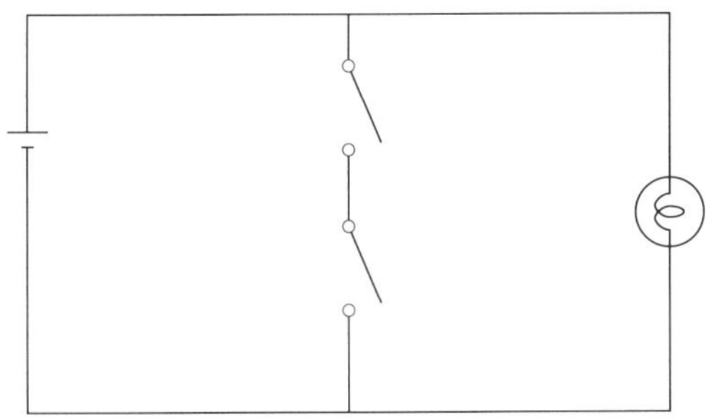

S_1	S_2	Light
0	0	1
0	1	1
1	0	1
1	1	0

Figure 5-13. NAND Function.

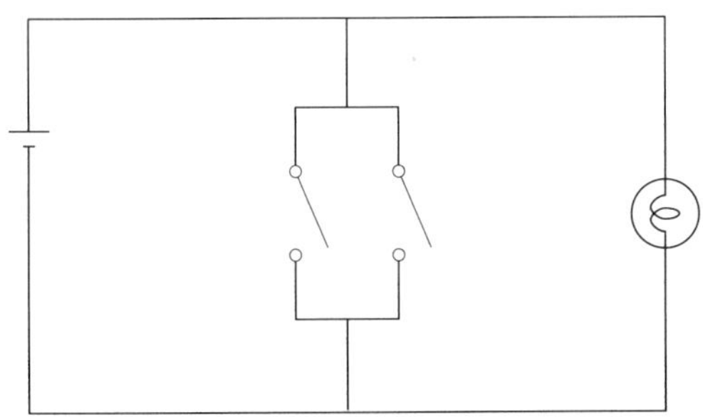

S_1	S_2	Light
0	0	1
0	1	0
1	0	0
1	1	0

Figure 5-14. NOR Function.

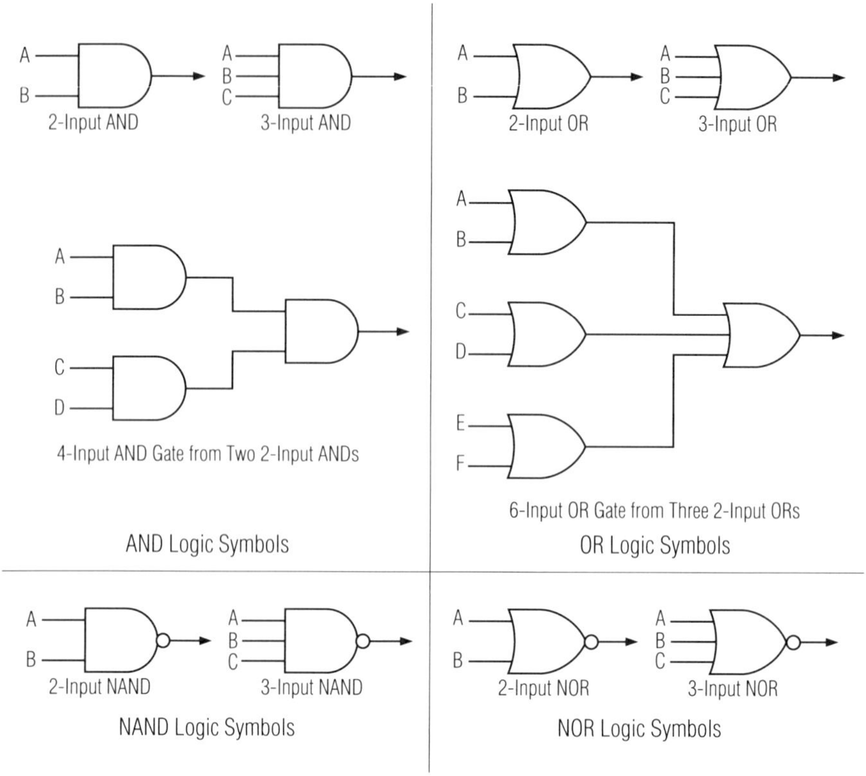

Figure 5-15. Logic Symbols.

Memory Circuits

In digital electronics, logic elements are used to make decisions. The decisions are then stored in memory elements, whose basic building block is the flip-flop. A flip-flop is a one-bit circuit which remembers 1's and 0's.

The logic diagram for a J-K flip-flop is shown in Fig. 5-16. The truth table lists possible values for the inputs J and K, and the output Q which will appear after the next clock pulse. The value of Q can change only each time a clock pulse appears. The new value of Q depends on the inputs J and K. If J and K are both 0, the output Q keeps the same value (0 or 1) it had before the new clock pulse. If J and K are both 1's, the output Q changes to the opposite of the value it had before the clock pulse. If J and K are 0 and 1, the output Q is set to is 0; if J and K are 1 and 0, the output Q is set to 1. The other output $\overline{Q}$ is always the inverse of Q.

While with AND, OR, NAND, and NOR circuits the outputs change immediately when inputs change, flip-flop outputs change only when a clock pulse arrives. So a flip-flop is a memory circuit which remembers the input status from the last clock pulse.

Most digital systems and all computers need to remember thousands of bits of information. Large memories are made by integrating thousands of flip-flops

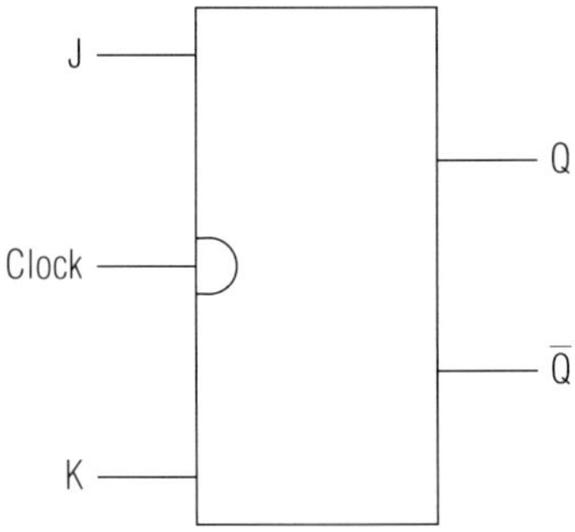

J	K	$Q_{(n+1)}$
0	0	1
0	1	0
1	0	0
1	1	0

Figure 5-16. J-K Flip-Flop.

onto a single piece of silicon, forming a single integrated circuit. Since most computers operate with 8-bit bytes, integrated circuits are designed to store bytes, by the thousands.

Several construction techniques exist for memory circuits, to satisfy different needs. The most popular electronic memory types are Random Access Memory (RAM), Read-Only Memory (ROM), and Programmable Read-Only Memory (PROM). All forms of memory can be read over and over again without changing the contents.

Random Access Memory (RAM) is used for the temporary storage of data. It is used for applications where information is stored and retrieved quickly and frequently. A disadvantage of RAM memory is that all data is lost when power is removed from the circuit. RAM memory is also called volatile memory. RAM can be supported with batteries (called battery back-up) to power the RAM in the event the main power supply fails.

Read-Only-Memory (ROM) is made at an integrated circuit factory to a set of specifications called a mask. ROMs can be programmed to perform like many individual gates, or to store data which can be accessed as needed. When making a masked ROM, permanent changes are made to the silicon inside the integrated circuit (IC) package. Once a ROM is programmed at the factory, the contents of its memory can never be re-written. ROM memory is permanent and remains intact even when power is removed. It is called non-volatile memory.

Programmable Read-Only-Memory (PROM) stores information which can be re-written, but only under special conditions. PROMs therefore can be programmed by the user at his laboratory or manufacturing facility. And once programmed, PROMs can be erased and re-programmed. One form of PROM is erased and programmed by exposing the silicon, through a glass window in the package, to intense ultra-violet light while entering new data. PROMs are also called EPROMs (Erasable Programmable Read-Only-Memories), or EEPROMS (Electrically Erasable Programmable Read-Only-Memories). PROMs are useful for prototyping, for low volume requirements, and for applications where the data stored might be changed periodically. The software programs stored in PROMs are sometimes called firmware.

MICROPROCESSORS

Microprocessors get instructions and data from memory, perform arithmetic and logical functions on the data, and store the results back in memory. The instructions are in a specific sequence, written uniquely for each application by a programmer. Microprocessors typically operate on data organized in groups of 8 or 16 bits. Usually the data path used by the microprocessor defines the data path of the memory chips in the circuit. For example, 8-bit microprocessors use memory circuits with 8-bit data paths. Microprocessors are packaged as integrated circuits.

CHAPTER 6
INSTRUMENTS

ELECTRICAL MEASURING INSTRUMENTS are necessary because the very nature of most electrical phenomena is beyond the reach of man's physical senses. Only by measurement of the invisible electrical quantities has it become possible to design, manufacture, and maintain the innumerable electrical devices now in use.

The main purpose of any electrical instrument is to measure and indicate the value of an electrical quantity. The measurement may be indicated by a numeric value on a digital readout, by an appropriate position of a pointer with respect to a scale, or by recording the value of the electrical quantity on a moving chart. The devices commonly used for such measurements are voltmeters, ammeters, and wattmeters.

Since the field of instrumentation is extensive and includes many classifications of instruments according to portability, type of indication or record, accuracy, design features, etc., we shall discuss briefly only those instruments commonly used in meter departments. These instruments are typically displaying, indicating, electronic digital, or recording measuring devices.

PERMANENT-MAGNET, MOVING-COIL INSTRUMENT

Fig. 6-1 represents the mechanism of a permanent-magnet, moving-coil instrument. Here, the field produced by the direct current in the moving coil reacts with the field of the permanent magnet to produce torque.

Essentially, the permanent-magnet, moving-coil instrument, often called a d'Arsonval instrument, consists of a very light-weight, rigid coil of fine wire suspended in the field of a permanent magnet. The moving coil in most instruments consists of a very light-weight frame of aluminum, flanged for strength and to retain the windings, which, in turn, consist generally of several layers of fine enameled wire. To the ends of the coil frame are cemented pivot bases carrying the hardened steel pivots on which the coil turns as well as the inner ends of the control and current-carrying springs. In addition, the upper pivot base mounts

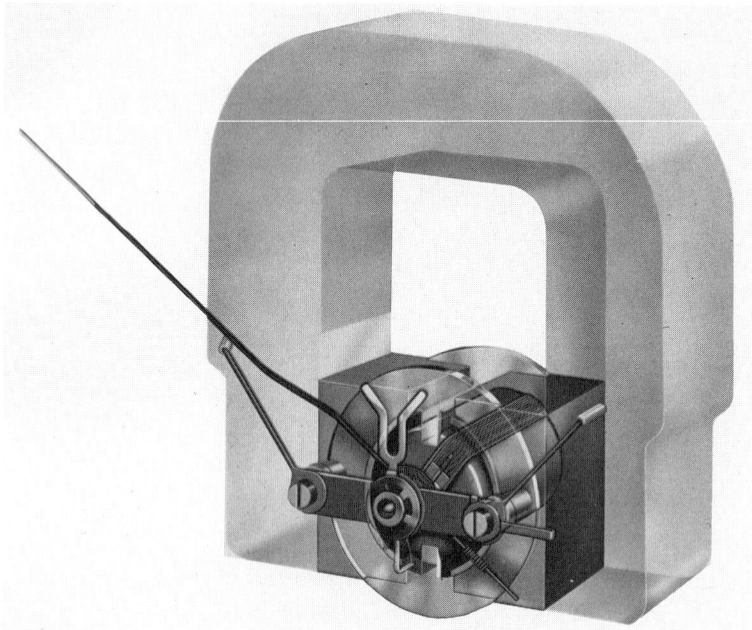

Figure 6-1. Mechanism of Permanent-Magnet, Moving-Coil Instrument.

the pointer and the balance cross. Threaded balance weights, or their equivalent, are adjusted on the balance cross to balance the moving element in its bearing system. The pivots ride in jewel bearings to keep friction at a minimum. A taut band suspension may be used in place of the pivot and jewel-bearing system. Here the moving coil is supported by two metal ribbons under tension sustained by springs. Either bearing system allows a properly balanced instrument to be used in any position with little error.

Current is carried to the coil by two springs. These control and current-carrying springs oppose the torque of the moving coil and are the calibrating means of the instrument. These springs are generally made of carefully selected phosphor bronze or beryllium copper specially manufactured to provide stability so that the instrument accuracy will not be changed by time and use.

The torque developed by current flowing through the moving coil is a function of the field strength of the permanent magnet and of the current in the moving coil, as well as of dimensional factors of both magnet and coil. The torque T in dyne-centimeters is given by this equation:

$$T = \frac{B \times A \times I \times N}{10}$$

where:

B = flux density in lines per square centimeter in the air gap
A = coil area in square centimeters
I = moving-coil current in amperes
N = turns of wire in moving coil

The characteristics of this type instrument are very desirable. It has high accuracy, high sensitivity, low cost, and uniform scale. It can measure extremely small currents, because of the fine wire in the moving coil. The instrument is unique in the variety of accessories that can be used in conjunction with it. The four most commonly used are the series resistor, the shunt, the thermocouple, and the rectifier.

Rectifier-Type Instruments

Rectifier-type instruments may be used to measure AC milliamperes or AC volts. Since the DC mechanism is available by disconnecting the rectifier, this type is widely used in compact test sets where, by suitable switching, both alternating and direct current can be indicated on the same instrument. When used on alternating current, the rectifier-type instrument is suitable for audio-frequency current measurement. It is subject to errors due to waveform distortion if used on waveforms differing substantially from that with which the instrument was calibrated. See Fig. 6-2.

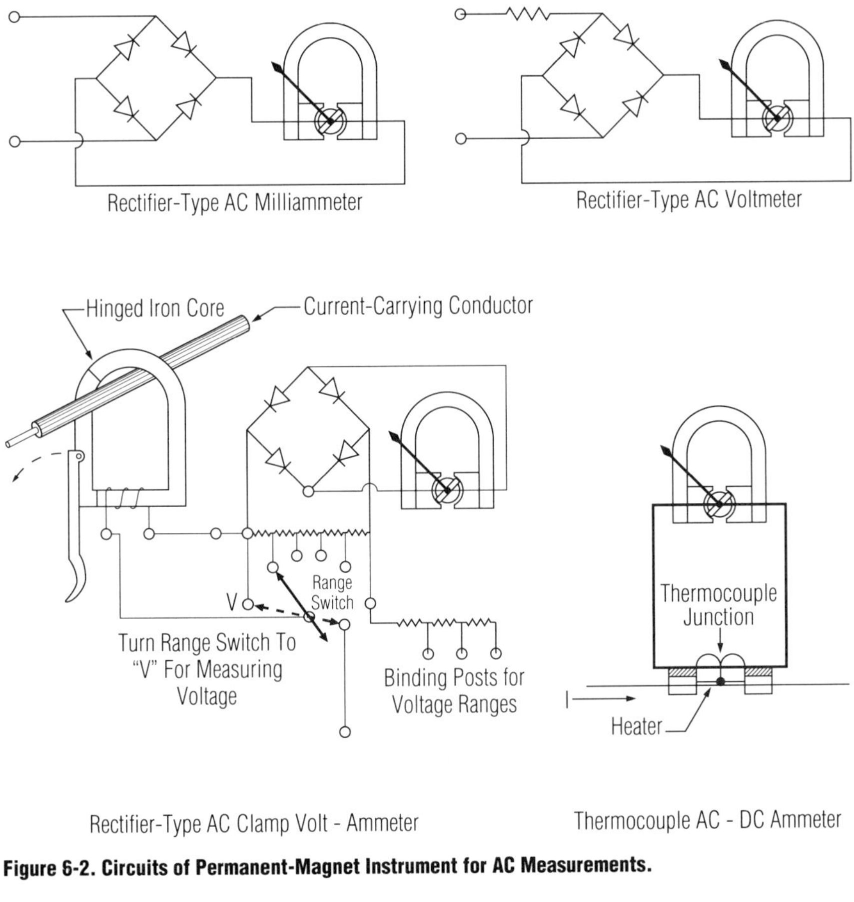

Figure 6-2. Circuits of Permanent-Magnet Instrument for AC Measurements.

Clamp Volt-ammeter

A commonly used development of the rectifier-type instrument is the clamp volt-ammeter. The principal use of this instrument is the measurement of AC current without interrupting the circuit. Provision is also made for AC voltage measurements. The circuit arrangement of the instrument is shown in Fig. 6-2.

The line current is measured through the use of a hinged-core current transformer, the secondary current of which is suitably divided by a multiple-range series shunt for several ranges of line current. The secondary current is then rectified and applied to the permanent-magnet instrument mechanism. Voltage is measured by short-circuiting the transformer secondary and making direct connection to the rectifier with sufficient resistance added in series to produce the desired range. This instrument is subject to the waveform and frequency errors which are characteristic of rectifier-type instruments.

Thermocouple Instrument

This instrument is a combination of a permanent-magnet, moving-coil mechanism and a thermocouple or thermal converter. The latter consists of a heater, which is a short wire or tube of platinum alloy, to the center of which is welded the junction of a thermocouple of constantan and platinum or other non-corroding alloys. The cold ends of the thermocouple are soldered to copper strips thermally in contact with, but insulated from, the heavy end terminals. This construction is necessary to reduce temperature errors. The copper strips in turn are connected to a sensitive moving-coil instrument.

The current to be measured passes through the heater causing a temperature rise of the thermocouple junction over the cold ends, and the resultant voltage is proportional to the temperature differential. Since the temperature rise of the hot junction is proportional to the square of the heater current, the instrument reading is also proportional to the square of the heater current.

With suitable conversion and circuit components, the thermocouple instrument may be used as a millivoltmeter, ammeter, milliammeter, or voltmeter. It indicates true root-mean-square (RMS) values on all waveforms and shows little error over a frequency range from direct-current to 20 kHz or more. Its disadvantages are its low overload capacity, its scale distribution, and its relatively slow response.

In Fig. 6-2 are shown some representative circuits of the permanent-magnet, moving-coil mechanism for AC measurements.

There are two inherent shortcomings of the permanent-magnet, moving-coil mechanism: except in a specially scaled instrument with rectifiers or thermocouples, it cannot measure AC quantities, and without auxiliary shunts or multipliers it can measure only small electrical quantities.

MULTIFUNCTIONAL DIGITAL INSTRUMENTS

A multifunctional digital instrument is an electronic device that measures voltage, current, and resistance by converting the measured analog input signal into a digital representation which is then displayed as a digital readout by the instrument.

Advances in technology have lead to digital instruments which are capable of high degrees of accuracy in the measurement of voltages, currents, and resistances over a wide range of values. Before the development of digital instruments, analog instruments indicated measured quantities by the deflection of a pointer on a scale that required the user to "eye up" the reading. Today, modern digital multimeters display the measured quantities as discrete numbers rather than as a pointer deflection on a scale as in analog meters.

The central component of a digital instrument is the digital DC voltmeter that uses electronic circuits to sense, process, and display a measured quantity. Input quantities other than DC voltages are converted to DC through the use of transducers. Examples of transducers include internal shunts used to measure current and AC-to-DC converters to measure AC quantities. The transduced analog quantity (the DC voltage) is then converted to an equivalent digital signal through the use of active electronic components such as transistors and diodes, or complete analog-to-digital modules which contain these components. These active elements are never found in analog instruments.

Several different analog-to-digital (A/D) conversion processes are in use in multifunctional instruments. These include dual-slope integration, ramp-and-counter, successive approximation, and voltage-to-frequency conversion. Each of these techniques produces a digital output which is equivalent to the measured analog input, and can be shown by a digital display. The display may be a liquid crystal display (LCD), light emitting diode (LED), or a gas discharge type display.

The digital output can also be transmitted to other devices such as printers, and received by computers through an interface such as a RS-232-C serial communications port, or the IEEE-488 bus standard which is used for controlling processes.

One of the simplest versions of the A/D converter is shown in Figure 6-3. In this example, a binary counter increments one count with each Clock pulse, until V_{out} equals V_{in}. This type A/D converter is called a "digital ramp and counter" because

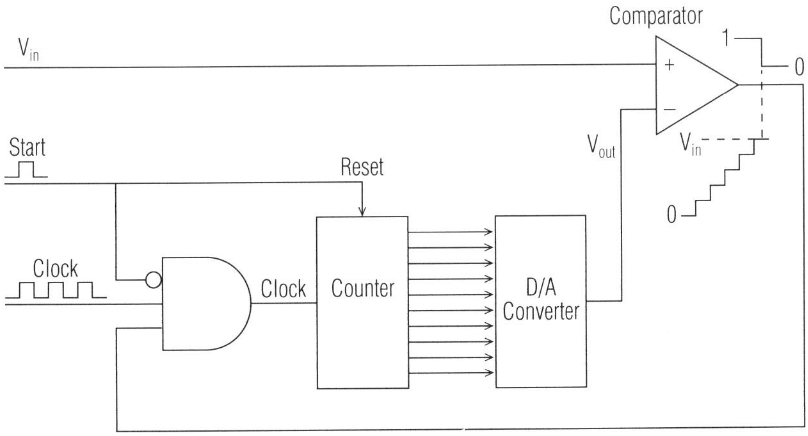

Figure 6-3. Analog-to-Digital Converter.

the waveform at V_{out} ramps up step-by-step, like a staircase. It operates as follows:

1. A positive Start pulse is applied which resets the counter to zero. It also inhibits the AND gate so no Clock pulses get through to the counter while the Start pulse is High.
2. With the counter at zero, $V_{out} = 0$, so the comparator output is High.
3. When the Start pulse goes Low, the AND gate is enabled, allowing pulses to enter the counter.
4. As the counter advances, the digital-to-analog (D/A) output (V_{out}) increases one step at a time, with the size of each step equal to its resolution.
5. This continues until V_{out} reaches a step that exceeds V_{in}. At this point, the comparator output goes Low, stopping the pulses to the counter, stopping the counter at the digital equivalent of the analog input of V_{in}. The conversion process (analog-to-digital) is now complete.

The following example illustrates the resolution and accuracy of the digital-ramp A-to-D converter.

Assume the following values for the A-to-D converter of Fig. 6-3: D-to-A converter has a 10-bit input and a full scale analog output of 10.23 volts; the comparator can detect a voltage difference of 1 millivolt or greater; V_{in} is 3.728 volts.

Since the digital-to-analog converter has a 10-bit input, the maximum number of steps possible is $(2^{10} - 1) = 1023$. With a full-scale output of 10.23 volts reached in 1023 steps, the step size is 10 millivolts. This means V_{out} increases in steps of 10 mV as the counter counts up from zero. Since $V_{in} = 3.728$ volts and the comparator threshold is 1 mV, then V_{out} has to reach 3.729 volts or greater before the comparator switches Low. At 10 mV per step, this requires 373 steps.

At the end of the conversion, the counter holds the binary equivalent of 373, which is 0101110101. This is the digital equivalent of the analog input of $V_{in} = 3.728$ volts. The resolution of this A-to-D converter is equal to the step size of the D-to-A converter which is 10 mV, or approximately 0.1 percent (.010/10.23 x 100 = 0.1%).

The resolution of an A-to-D converter is equal to the resolution of the D-to-A converter which it contains. The D-to-A output voltage V_{out} is a staircase waveform (digital ramp) that goes up in discrete steps until it exceeds V_{in}. Thus, V_{out} is an approximation of the value of V_{in} and when the resolution (step size) is 10 mV, the best we can expect is that V_{out} is within 10 mV of V_{in}. The resolution of the D-to-A converter is an inherent error, often referred to as a quantizing error. This quantizing error can be reduced by increasing the number of bits in the counter and in the D-to-A converter, and is specified as an error ±1 least significant bit (LSB), indicating that the result can be off by that much due to the step size.

From another point of view, the input voltage V_{in} can take on an infinite number of values, from 0 to full scale. However, the output voltage V_{out} can take on only a finite number of discrete values. This means that similar values of V_{in} within a small range could have the same digital representation. For example, if the counter goes through 1,000 steps from zero to full scale, any value of V_{in} from 3.720 to 3.729 will require 373 steps, thus resulting in the same digital representation. In other words, V_{in} must change by 10 millivolts (the resolution) to produce a change in the digital output.

The D/A converter accuracy is not related to the resolution. It is related to the accuracy of the components in its circuit such as the resistors in the D/A network, comparator, level amplifiers, and the reference power supply. If a D/A has an accuracy of 0.01 percent full scale, the A/D converter may be off by 0.01 percent full scale owing to non-perfect components. This error is in addition to the quantizing error due to resolution. These two sources of error are usually specified separately, and for a given A/D converter are usually of the same order of magnitude.

In addition to the inherent errors noted above, the accuracy of a multifunction instrument depends on proper selection of the meter range. Normally, the uncertainty of measurements is expressed as a percentage of the reading, plus the number of counts of the least significant digit (LSD) displayed for that range. For example, for a three-and-a-half digit digital multimeter with a nameplate accuracy of ± 0.5% of input voltage ± 1 LSD, if the 1,000 volt DC range is selected to measure a 2-volt signal, this setup would result in a meter accuracy of ± 50.5%, as shown below.

Given: Meter Range Accuracy (MRA) is ± 0.5% of input voltage + 1 LSD
Meter range set to 1,000 volts DC
Input voltage is 2 volts DC

Then: Meter Accuracy $= \pm[(\text{MRA} \times \text{input V} + \text{LSD})/\text{Input Voltage}] \times 100$
$= \pm[(0.5\% \times 2 + 1)/2] \times 100$
$= \pm 50.5\%$

However, selecting a meter range of 2 volts DC on the same digital multimeter would result in an accuracy of 0.60%, nearly 100 times better, as shown below.

Given: Meter Range Accuracy (MRA) is ± 0.5% of input ± 1 LSD
Meter range set to 2 volts DC
Input voltage is 2 volts DC

Then: Meter Accuracy $= \pm[(\text{MRA} \times \text{input V} + \text{LSD})/\text{Input Voltage}] \times 100$
$= \pm[(0.5\% \times 2 + 0.002)/2] \times 100$
$= \pm 0.60\%$

The application and the decision of when to use digital instruments versus analog instruments will depend on several considerations including accuracy, resolution, speed of measurement and reading, size and portability, environmental considerations, and cost.

THE MOVING-IRON INSTRUMENT

The measurement of alternating current (or voltage) is the measurement of a quantity which is continuously reversing direction. The permanent-magnet, moving-coil instrument movement cannot be used since the alternating-current field of the moving coil reacting with the unidirectional permanent-magnet field will produce a torque reversing in direction at line frequency. Because of its inertia the moving element will be unable to respond to this rapidly reversing torque and the pointer will only vibrate at zero. A different type of meter movement is therefore required.

The moving-iron instrument is specifically designed to operate on alternating-current circuits. This instrument is called the moving-iron type because its moving member is a piece of soft iron in which magnetism induced from a field

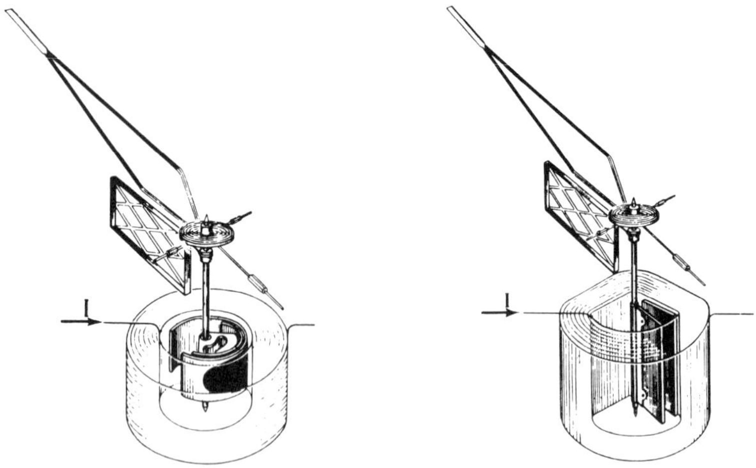

Concentric Vane Mechanism   Radial Vane Mechanism

Figure 6-4. Mechanism of Moving-Iron Instrument.

coil interacts with the magnetic field of a fixed piece of soft iron to produce torque.

The mechanism of this instrument, shown in Fig. 6-4, consists essentially of a stationary field coil, within the magnetic field of which are two soft iron pieces. One is fixed while the other, commonly called the moving vane, is attached to a pivoted shaft provided with a pointer and is free to rotate. When current flows through the field coil the two pieces are magnetized with the same polarity, since they are both under the influence of the same field and, hence, repel each other, causing the pivoted member to rotate. The angular deflection of the moving unit stops at the point of equilibrium between the actuating torque and the counter torque of the spiral control spring.

It is evident from the illustration that the operating current flows through a stationary winding. Depending upon the use for which it is designed, the coil may be wound with fine or heavy wire, giving this type of instrument a wide range of capacities. The instrument will tolerate overloads with less damage to springs and pointer than will most other types of instruments, since, with excess current, the iron vanes tend to become saturated and limit the torque. Damping is provided by either a light aluminum vane fixed to the shaft and moving in a closed air chamber, or by a segment of an aluminum disc moving between poles of small permanent magnets.

The bearing system may consist of a pivoted shaft turning in jeweled bearings or may be of the taut-band suspension type where the moving element is supported by two metal ribbons under tension sustained by springs.

APPLICATION OF MOVING-IRON INSTRUMENT

Measurement of Current

Since the actuating coil may be wound with a choice of many wire sizes, the instrument may be constructed to measure current from a few milliamperes up to 100 or 200 amperes in self-contained ratings. For measurement of currents beyond this range, a 5-ampere instrument may be used with a current transformer.

Current Transformer Field Test Set

A special application of the moving-iron ammeter is the current transformer field test set.

The circuit of this instrument is shown in Fig. 6-5. It is used to check current transformer installations in service, on the secondary side, for possible defects such as short-circuited primary or secondary turns, high-resistance connections in the secondary circuit, or inadvertent grounds, any of which conditions could cause incorrect metering.

It is essentially a multi-range, moving-iron-type ammeter with a built-in burden which is normally shunted out, but which can be put in series with the meter by the push button.

In the typical instrument illustrated here, ammeter current ranges of 1.25, 2.5, 5, and 10 amperes are obtained from the tapped primary winding of a small internal current transformer, the secondary winding of which is connected to the ammeter which has corresponding multiple scales. It is thus possible to obtain a reading well up-scale on the ammeter for most load conditions under which the current transformer being checked is operating. The rotary burden switch permits the addition of 0.25, 0.5, 1, 2, or 4 ohms to the secondary circuit as desired.

The imposition of an additional secondary burden on a current transformer having the defects previously mentioned will result in an abnormal decrease in the secondary current. The extent of this decrease and the ohms burden required to effect it depend on the characteristics of the transformer under test. The check on the current transformer consists of inserting the field test set in series with the current transformer secondary circuit and comparing the ammeter readings under normal operating conditions with the readings after the additional field test set burden is added.

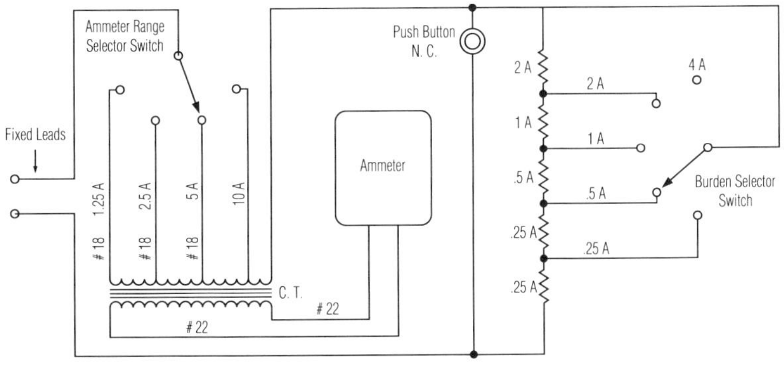

Figure 6-5. Circuit of Current Transformer Field Test Set.

The use of this device under field conditions is discussed in Chapter 10, "Instrument Transformers."

Measurement of Voltage

By the use of an actuating coil of many turns of fine wire in series with a resistor, the moving-iron instrument may be used to measure voltage. Such a voltmeter may have an operating current of around 15 milliamperes with a range up to 750 volts. External multipliers may be used to extend this range. These voltmeters are used in applications where sensitivities lower than those of the rectifier d'Arsonval instrument are satisfactory. The moving-iron voltmeter may be used on direct current with some loss in accuracy. The best accuracy is obtained by using the average of readings taken before and after reversal of the leads to the instrument terminals. This instrument will not indicate the polarity of direct current.

ELECTRODYNAMOMETER INSTRUMENTS

The electrodynamometer-type mechanism, shown in Fig. 6-6, is adaptable to a greater variety of measurements than any of the instruments previously described and is especially useful in AC measurement and as a DC-to-AC transfer instrument.

In this instrument, both stator and rotor are coils. Current flowing through the stationary or field coil winding produces a field in proportion to the current. As current is applied to the moving coil, the coil moves by reason of the reaction on a current-carrying conductor in a magnetic field. The torque actuating the moving element is a function of the product of the two magnetic fields and their angular displacement.

This instrument can be used for measurement of volts, amperes, watts, either alternating or direct current, as well as power factor and frequency. Figure 6-7 shows the coil arrangements for various applications.

APPLICATION OF ELECTRO-DYNAMOMETER INSTRUMENT

Measurement of Power

The most important use of the electrodynamometer mechanism is as a wattmeter. In this construction the moving coil is in series with a resistance and is connected across the circuit as a voltmeter, while the other, the field coil, is connected in series with the load as an ammeter coil. The torque between these coils is proportional on direct current to the product of volts and amperes, or watts. On alternating-current, the instrument recognizes the phase difference between volts and amperes, which is the power factor. Its readings then are proportional to the product of volts, amperes, and power factor.

When used as a wattmeter, the moving coil is wound with fine wire, while the field coil may be wound with large-size wire, the nominal rating of the latter being usually 5 amperes.

By superimposing two complete wattmeter elements with the two moving coils on the same shaft, power in a three-wire circuit may be measured by one instrument. The torques developed by the two elements add algebraically to give an indication of total power. By using phasing transformers to shift the voltages to the moving coils 90 degrees, a three-phase varmeter is obtained.

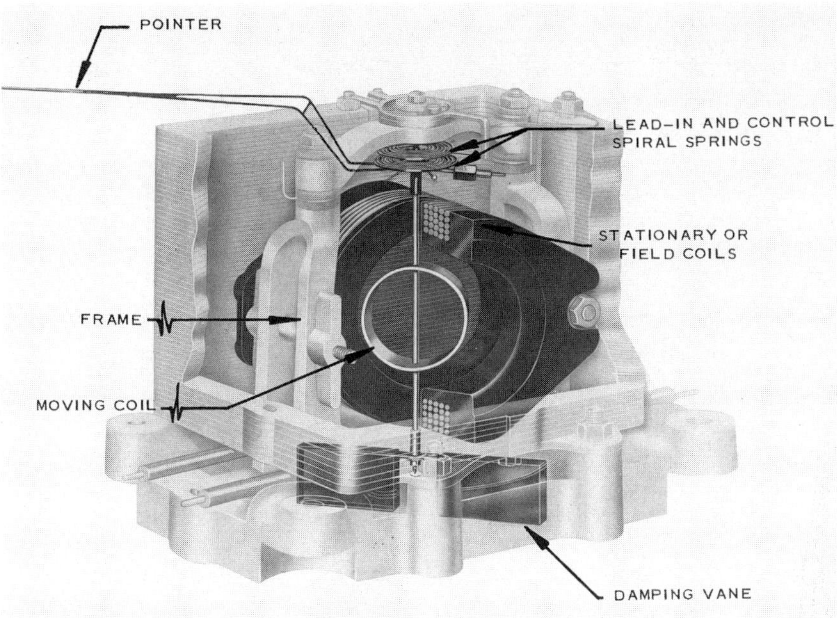

Figure 6-6. Mechanism of Electrodynamometer Instrument.

Measurement of Current

Electrodynamometer ammeters have the field and moving coils connected in series. Since the moving coil is connected to the circuit by the rather fragile lead-in spirals, it is evident that the current-carrying capacity of that part of the instrument is limited. For this reason the moving coil is shunted in instruments above 100 milliamperes capacity. The full line current passes through the field coil and the shunt.

Measurement of Voltage

In the electrodynamometer voltmeter, the field coil is connected in series with the moving coil and a resistance across the line. The sensitivity of this instrument is less than that of a DC voltmeter because of the greater current required by the dynamometer mechanism. It is, however, more accurate than the moving-iron voltmeter and is better adapted to precise voltage measurements.

Measurement of Power Factor and Phase Angle

A variation of the fundamental electrodynamometer instrument is used to measure power factor or phase angle, and is called the crossed-coil type. See Fig. 6-7. In this design the moving element consists of two separate coils, instead of one, which are mounted on the same shaft and set at an angle with each other. The lead-in springs or spirals to the crossed coils are made as light or weak as possible so as to exert practically no torque. In the single-phase instrument, one of the crossed moving coils is connected in series with a resistor across the line while the other is

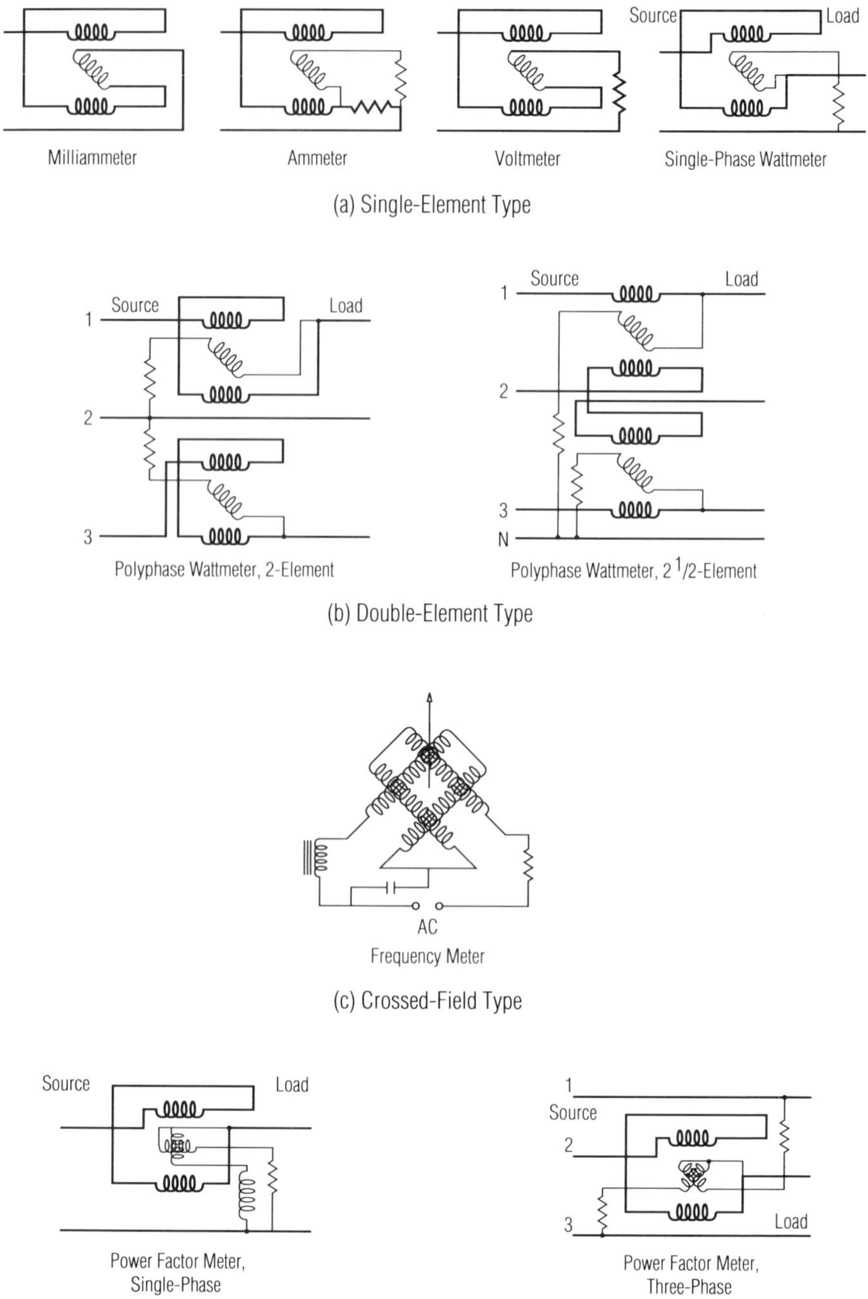

Figure 6-7. Circuits of Electrodynamometer Instruments.

connected in series with a reactor across the line. The current flowing through the reactor-connected coil is approximately 90 degrees out of phase with the line voltage. The field coil is connected in series with the line as an ammeter coil.

In operation, the moving system assumes a position depending upon the phase relationship between the line current and the line voltage. If the line current is in phase with the line voltage, the reactor-connected moving coil will exert no torque and the resistor-connected coil will align its polarities with those of the fixed-coil field. If the line current is out of phase with the line voltage, the reactor-connected moving coil will exert a restraining or counter torque and the moving element will assume a position in the field of the fixed coil where the two torques are in balance.

This instrument may be calibrated to indicate either power factor or the phase angle between the line voltage and current.

In the three-phase power factor instrument, the crossed moving coils are connected to opposite legs of a three-phase system. The fixed coils are connected in series with the line used as a common for the moving-coil connection. This instrument will give correct indication on balanced load only.

When these instruments are not energized, the pointer has no definite zero or rest position, as do instruments whose restraining torque is a spring. They are therefore known as free-balance instruments.

Power factor meters may also be of the induction type. In one such type, for single-phase use, the fixed element consists of three stationary coils and the moving element comprises an indicator shaft bearing an iron armature. As in the electrodynamometer type, the operation is based on the interaction of a rotating and an alternating magnetic field.

Measurement of Frequency

Another variation of the electrodynamometer instrument, called the crossed-field type, is used to measure frequency. Crossed field, or stationary, coils are connected to the line through inductive and capacitive circuit elements so that the relative strengths of the fields become a function of the frequency. See Fig. 6-7. An iron vane attached to a freely rotatable pointer shaft will align itself with the direction of the resultant field and the instrument will thus indicate the frequency. This frequency meter is also a free-balance instrument.

THERMAL AMPERE DEMAND METERS

Thermal ampere demand meters differ from instruments previously discussed in that the moving-element deflection does not result from the electromagnetic interaction between fixed and moving instrument components, but is due entirely to the mechanical torque exerted on a shaft by the distortion by heat of a bimetallic strip or coil. These instruments are generally not designed for precise measurements. On the other hand they are simple, inexpensive, and rugged measuring units which are easily adapted to a wide variety of applications.

Lincoln Ampere Demand Meter

The moving element of a Lincoln ampere demand meter consists of a horizontal shaft mounted in bearings to which are attached an indicating pointer and two bimetallic coils which are temperature sensitive. The inner ends of both bimetallic coils are attached to the shaft while the outer ends are attached to the instrument frame. These two bimetallic coils are carefully matched and are wound in opposite directions.

One bimetallic coil is placed in an enclosure of heat-insulating material with, and adjacent to, a noninductive nichrome or manganin resistor used as a heater. See Fig. 6-8. When a current I passes through the heater R, the heater and its enclosure are heated at a rate proportional to I^2R, thus producing a temperature difference between the bimetallic coils which results in motion of the shaft and pointer. This motion is proportional to the current squared. This heater current may be either line current or a smaller current from the secondary of an internal current transformer.

When both coils are at the same temperature, the instrument pointer reads zero. Changes in external temperature affect both bimetallic coils equally and therefore cause no change in pointer position. Since this arrangement is equivalent to a differential thermometer, the shaft motion is proportional to the temperature difference between the two coils.

The restraining torque is largely supplied by the bimetallic coil which is not heated. Additional restraining torque is supplied by a coiled spring and a third weak spiral spring. The former aids in improving scale distribution while the latter provides for a slight zero adjustment.

Thermal capacity of the system is large and so the pointer responds very slowly, reaching 90 percent of the current value in approximately 15 minutes under steady current, 99 percent in 30 minutes, and 99.9 percent in 45 minutes.

To obtain maximum indication over a long period of time, the regular pointer pushes a second pointer which is not attached to the shaft and which has sufficient friction so that it will stay at the maximum point to which it is pushed. Provision is made for setting the free pointer to the pusher pointer when desired.

Typical movements of this type require 3 to 6 amperes through the heater resistor for full-scale deflection.

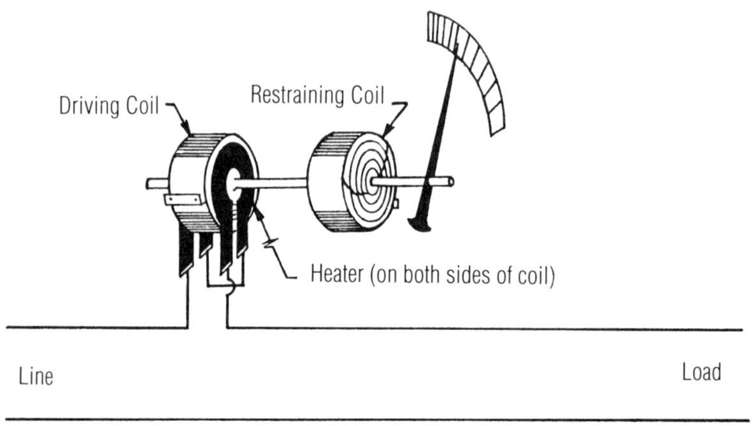

Figure 6-8. Mechanism of Thermal Demand Ammeter.

INSTRUMENT SCALES

In Fig. 6-9a is shown a sample instrument scale of the kind most easy to read. This type of scale is characteristic of DC, permanent-magnet, moving-coil instruments in that the divisions are of equal size from zero to full scale. Since each major division is equal to 0.1 and each minor division to 0.01 of full scale, reading errors are minimized and rather precise readings may be made.

In Fig. 6-9b is shown the more complicated scale of the three-range voltmeter. Note that the divisions are not uniform and are smaller or cramped near the zero end of the scale. This scale is characteristic of moving-iron instruments. Since the values of each major and minor division are different for each of the

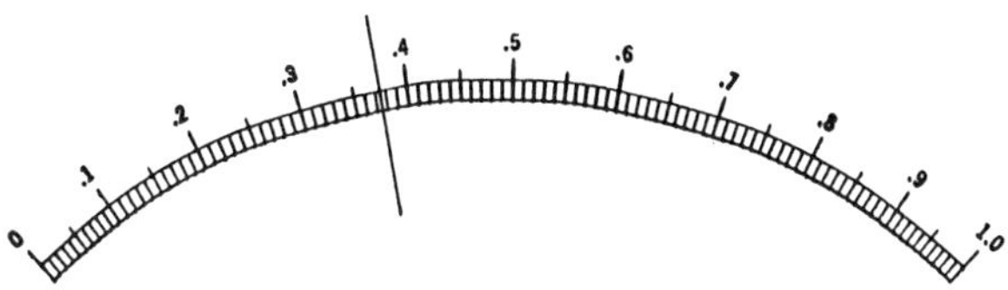

(a) Linear Instrument Scale

Instrument Reads 0.374

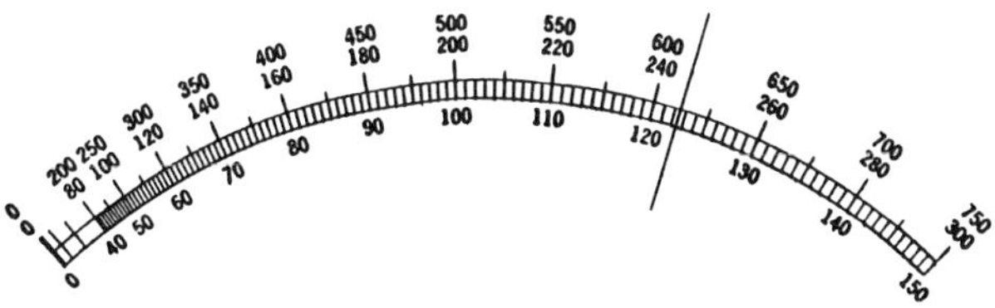

(b) Scale of Triple Range Voltmeter

On 300 Volt Scale, Instrument Reads 245

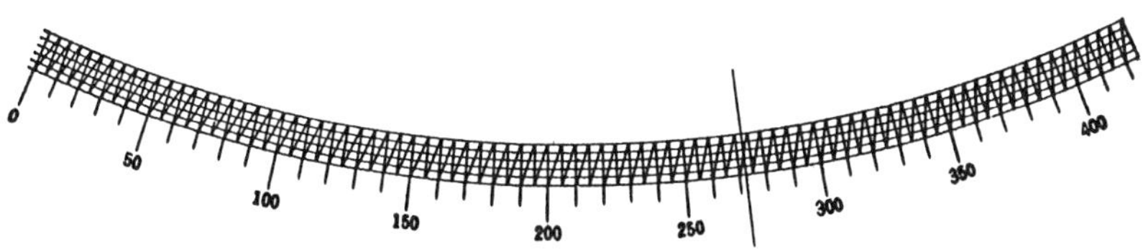

(c) Scale of Precision Wattmeter

Instrument Reads 272.5

Figure 6-9. Typical Instrument Scales.

three ranges, it is important to know on what range the meter is connected in order to determine the correct reading.

In Fig. 6-9c is shown the type of scale often used in high-accuracy instruments, such as secondary standards in the meter laboratory. The diagonals connecting each minor division plus the additional lines parallel to the arc of the scale permit close readings to fractions of a division. Where four intermediate arcs are drawn, each intersection of a diagonal with an arc line is 0.2 of the marked division.

When reading instrument indications, one must be careful to have one's eye directly above the knife edge of the pointer to avoid errors due to parallax. In many instruments and especially those of high accuracy, the scale is equipped with a mirror so that the pointer may be lined up with its reflection, thus assuring that the eye is in the proper position for accurately reading the scale.

MEASUREMENT OF RESISTANCE

Voltmeter-Ammeter Method

The simplest approach to resistance measurement is simultaneous measurement of both the current through a circuit or component and the voltage across it. This method is shown in Fig. 6-10. From the current and voltage readings we can calculate the resistance by the application of Ohm's law:

$$R = \frac{E}{I}$$

For accurate results in measuring resistance by this method, it is necessary to keep in mind the relative resistances of the voltmeter and ammeter themselves. For low values of the resistance R, the resistance of the ammeter must be deducted from the value of the circuit resistance determined by the preceding equation. For relatively high values of R (compared to the resistance of the ammeter), the value of the ammeter resistance may be neglected.

Ohmmeters

These are self-contained instruments with a source of low DC voltage that measure within reasonable accuracy limits the resistances of a circuit or component and indicate the value of resistance on a meter scale calibrated in ohms.

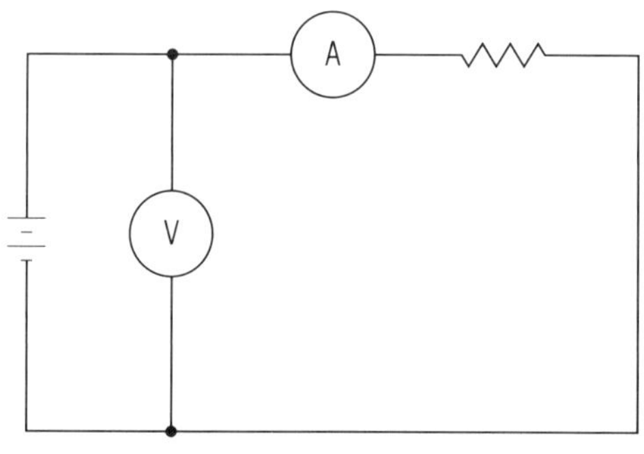

Figure 6-10. Circuit for Measurement of Resistance by Voltmeter-Ammeter Method.

In application, the ohmmeter is one of the most useful test instruments, being used also to locate open circuits and to check circuit continuity. This instrument should not be used on energized circuits. Figure 6-11 shows the circuit arrangement of a typical ohmmeter.

The megger is a special variety of ohmmeter. It is a portable test set for measuring the extremely high values of resistance, up to 100,000 megohms or more, which are encountered during testing of the insulation resistance of cables; the resistance between conductors in multiple cables; between windings; or from windings to ground in transformers, motors, or other forms of electrical equipment. These instruments have built-in, hand-driven or electronic DC generators to supply 500 volts or more so that measurable currents can be produced through the high resistances encountered. Fig. 6-12 shows a typical megger for measuring insulation resistance.

Ground-resistance ohmmeters are used for measuring resistance to earth of ground connections, such as substation or transmission tower and lightning arrester grounds. These instruments have built-in AC generators instead of DC units and also differ from the type used for insulation resistance measurements by their range and the fact that the resistance of the leads is electrically removed from the indicated reading by the nature of the test connections. Whereas the former has a full-scale range of 100,000 megohms or more, the ground-resistance

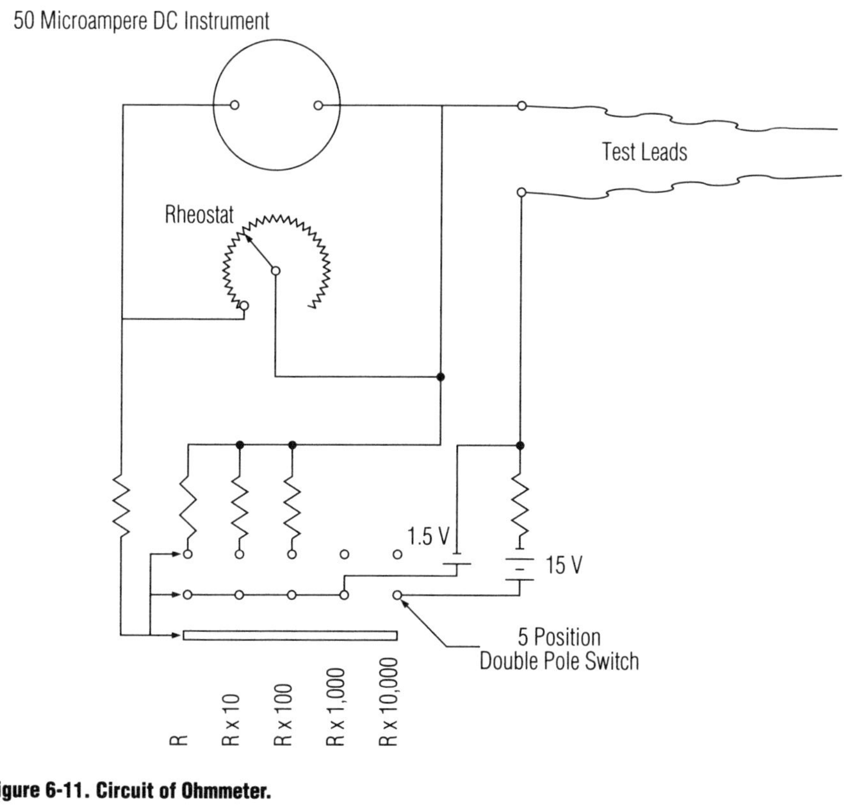

Figure 6-11. Circuit of Ohmmeter.

Figure 6-12. Megger Insulation Tester.

megger shown in Fig. 6-13 may have several ranges in one instrument and has the capability of measuring ground-resistance values from fractions of an ohm to several thousand ohms.

SELECTION OF INSTRUMENTS

The selection of instruments and related equipment best adapted to meet the requirements of a particular use is very important and should receive careful consideration. The most suitable types and ranges must be determined by the nature of the work for which they are to be used and the degree of accuracy required.

Consideration must be given to the choice between ruggedness and accuracy in an instrument. This choice is usually determined by service and economic requirements. The majority of all field test work is done with instruments of the general-purpose type. These are rugged, moderately priced instruments with accuracies in the range of 0.5-1.0%. On field tests or investigations where high accuracy is required, portable standard instruments of the 0.25% accuracy class may be used. These instruments are considerably more expensive, but are less rugged and consequently must be handled with greater care.

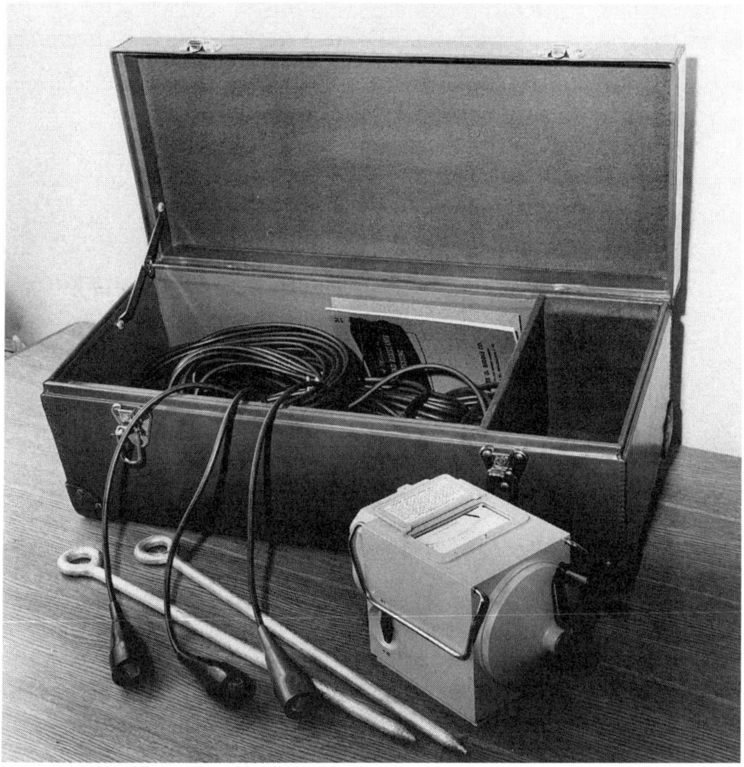

Figure 6-13. Megger Ground Tester.

Technically, high accuracy is desirable but may not be justified by economic consideration.

All instruments should be as free as possible from the effects of temperature changes, stray fields, frequency variation, waveform, spring set, pivot roll, and vibration. Indicating instruments should be equipped with accurate, legible scales having plainly marked divisions from which intermediate values can readily be determined.

CARE OF INSTRUMENTS

Instruments should be handled with great care and should receive no shock or blow from contact with table, bench, or other instruments in order to avoid damaging the fine points of the pivots of the moving elements or the polished surfaces of the jewel bearings. Instruments being transported should be carried in a vertical position so that the shafts of the moving elements are in a horizontal position. Chances of damage to the bearing systems will then be lessened, since shocks are taken on the sides of the pivots and jewels. Instruments should be carried in well padded, shock-mounted cases to further protect them. In placing an instrument in its case, the case should be laid flat and the instrument slowly slid into it.

Precaution should be taken to insure that the current or voltage to be measured is within the range of the instrument about to be used. If the range of quantities to be measured is not already known, a high-range instrument should be first connected to get an approximate indication of the value, after which an instrument with the proper range may be used.

Although fuses can be used in current-measuring instruments to prevent instrument windings from being burned out, they will not prevent the mechanical shock to the moving element when subject to sudden overloads. Fuses should not be used in instruments that might be connected in the secondary circuit of current transformers.

Precision instruments should be properly leveled before use. Zero should be adjusted before the instrument is energized. After the pointer has been deflected for a long time, a small zero shift may be noted. This shift is seldom permanent and should not be corrected until the instrument has been de-energized for some time. Zero shift caused by a bent pointer should be corrected by straightening the pointer rather than by the zero adjustment. In this connection, it should be emphasized that the case or cover should not be removed from the instrument for any reason except in the laboratory.

INFLUENCE OF TEMPERATURE

Temperature effects are inherent in almost all electrical instruments, although, through care in design and the use of special materials, the effect of temperature in most modern instruments has been greatly minimized. Temperature errors may be caused by exposure of the instrument to an environment of temperature extremes or by self-heating during use. If the instrument has been exposed to unusual heat or cold for some time, it should be exposed to room temperature until the temperature of the instrument is approximately normal. If it is necessary to use the instrument under extreme conditions of temperature, the test should be made as quickly as possible while the temperature of the instrument is near normal. If this is not practical, temperature-correction factors may be applied.

Temperature errors due to self-heating may be minimized by leaving the instrument in the circuit for as short a time as possible. Most portable voltage-measuring instruments are designed for intermittent use only and should not be left in the circuit indefinitely.

Temperature errors may also result from exposing the instrument to localized hot spots or temperature inequalities such as may be caused by a lamp close to the instrument scale. Such heating may affect the instrument but not the temperature compensation circuit, and thereby cause errors. A cold light should be used.

INFLUENCE OF STRAY FIELDS

Stray fields may cause appreciable errors in instruments. Such stray fields may be produced by other instruments, conductors carrying heavy currents, generators or motors, and even non-magnetized masses of iron. Since it is often not known if strong stray fields are present in the test area, the instrument should be read, then rotated 180 degrees, read again, and an average taken of the test readings. If an instrument is to be permanently used on magnetic panels (iron or steel), the instrument should be calibrated in location or with an equivalent panel.

MECHANICAL EFFECTS

Mechanical faults causing errors may include pivot friction, defective springs, unbalance of the moving coil, and, in some lower-grade instruments, incorrectly marked scales. Correction of the errors should be made only in the meter laboratory.

Another effect which may lead to serious errors in instrument readings is that due to electrostatic action. Rubbing the instrument scale window in order to clean it will often, in a cold dry atmosphere, cause the pointer to move from its zero position due to the action of an electric charge produced on the glass. Generally the static charge can be dissipated by breathing on the glass, the moisture in the breath causing the charge to leak away. In the laboratory, when calibrating wattmeters with separate sources of current and voltage, there may be an electrostatic force exerted between the fixed coil and the moving coil sufficient to introduce errors into the readings. The remedy is to arrange a common connection between the current and voltage circuits at some point.

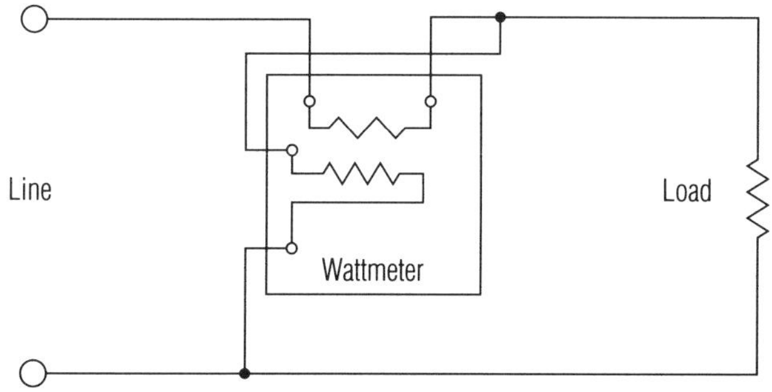

a. Wattmeter reading includes voltage circuit loss.

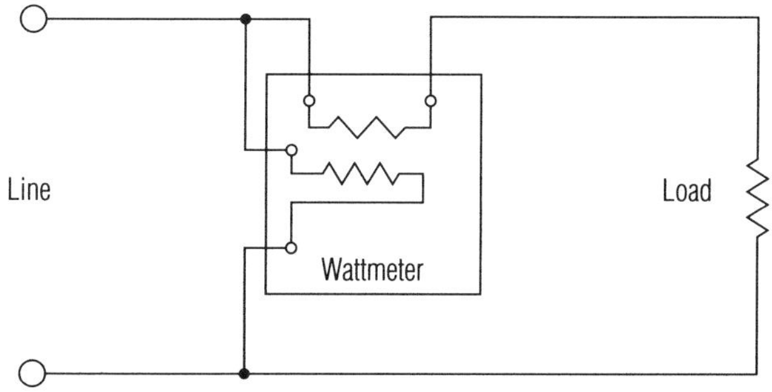

b. Wattmeter reading includes current coil loss.

Figure 6-14. Wattmeter Connections.

The matter of good electrical contact is important in connection with the use of electric instruments. Contact surfaces where connections are made must be clean and binding posts must be tightened. This is particularly important when using a millivoltmeter with shunt leads, since small voltage drops due to poor contact can have significant effects on accuracy.

INFLUENCE OF INSTRUMENTS ON CIRCUITS

All instruments of the types under discussion require energy to actuate them. This energy must come from the circuit under test, and, to some degree, this energy requirement will affect the circuit. In general, the more sensitive the instrument, the less it will modify circuit conditions. However, where measurements of a high degree of accuracy are desired, the effect of the energy requirements by the instrument must be considered. A voltmeter, for instance, connected to measure the voltage drop across a resistor, provides a parallel path for the current so that the total resistance of the circuit and, hence, the total current is not the same with the voltmeter connected as it was before such connection. In order to keep such circuit changes to a minimum, it is necessary in certain applications to use a high-resistance instrument.

An ammeter also has some resistance and will, when inserted in series with a load, change the total resistance of the circuit. Low-range ammeters of 1 ampere capacity or less may have resistances exceeding 1 ohm.

In the wattmeter there is an error (unless compensated for) due to the method of connecting the voltage element. See Fig. 6-14.

In the connection shown in Fig. 6-14a, the current coil of the instrument measures not only the load current but also that taken by the voltage coil, since the latter is connected on the load side of the current coil. In the connection shown in Fig. 6-14b, the voltage coil measures the drop across the load plus the drop across the current coil of this instrument. Where precise measurements are to be made, it is necessary to apply corrections. In such cases, the connections shown in Fig. 6-14a should be used, since it is easy to calculate the comparatively constant loss in the voltage circuit and to subtract this value from the wattmeter reading. Where the error due to connections is not considered to be significant, the connections shown in Fig. 6-14b may be used.

ACCURACY RATING OF INSTRUMENTS

The accuracy rating of an indicating instrument is a figure expressing the maximum deviation from true value of measured quantity, and is expressed as a percentage of the full-scale rating of the instrument. Thus, an instrument of the 0.5% accuracy class, when new, can be depended upon to indicate the true value at any point of the scale with an error not more than 0.5% of the full-scale reading on the instrument. An AC voltmeter of 0.5% accuracy class and having a full-scale range of 150 V should, when new, be within 0.75 volts (0.5% of 150 volts) at any point on its scale. In general, the more accurate an instrument, the more fragile its construction and the greater its cost, and the greater the care which must be exercised in its use. It is, therefore, logical that the instrument should be selected with an accuracy rating no greater than that needed by the practical requirements of the measurement.

MAINTENANCE OF INSTRUMENTS

In most public utilities a laboratory is maintained for calibration and minor repairs of instruments. It is good practice to return all instruments to the laboratory for recalibration on a routine scheduled basis. The length of time between laboratory inspections and calibrations will generally depend upon the accuracy class of the instrument and its use. Instruments should also be returned to the laboratory for recalibration if they have been subjected to any accidental electrical or mechanical shock, even though no apparent damage may have resulted. Instruments should also be returned to the laboratory for test whenever there may be any doubt concerning the accuracy of their indications.

CHAPTER 7

THE WATTHOUR METER

ENERGY IS NOT the same thing as power. Power is an instantaneous quantity, energy includes the time function or how long the power has been applied. Energy equals average power multiplied by time, so when we want to measure energy, it is necessary to have a meter that will measure the amount of power used over a period of time. The basic unit of measurement for electric energy is the watthour, and the instrument used to measure electric energy is called a watthour meter. The previous chapter explains that an indication of electric power may be obtained by use of a wattmeter. A wattmeter and a watthour meter have roughly the same relationship to each other as do the speedometer and the odometer on an automobile. A speedometer indicates miles per hour. An odometer shows the total number of miles traveled. A wattmeter indicates the instantaneous consumption of watts. A watthour meter measures the total watthours that have been used. Just as an odometer will indicate 60 miles, for example, after a car has traveled two hours at a speedometer indication of 30 miles per hour, a watthour meter will indicate 1000 watthours if connected for two hours in a circuit using 500 watts.

PART 1— SINGLE-STATOR METERS

Basically, the watthour meter consists of a motor whose torque is proportional to the power flowing through it, a magnetic brake to retard the speed of the motor in such a way that it is proportional to power (by making the braking effect proportional to the speed of the rotor), and a register to count the number of revolutions the motor makes. Figure 7-1 shows these parts. If the speed of the motor is proportional to the power, the number of revolutions will be proportional to the energy.

THE MOTOR IN A SINGLE-STATOR AC METER

The motor is made up of a stator with electrical connections, as shown in Fig. 7-2, and a rotor. The stator is an electromagnet energized by the line voltage and load current. That portion of the stator energized by the line voltage is known as

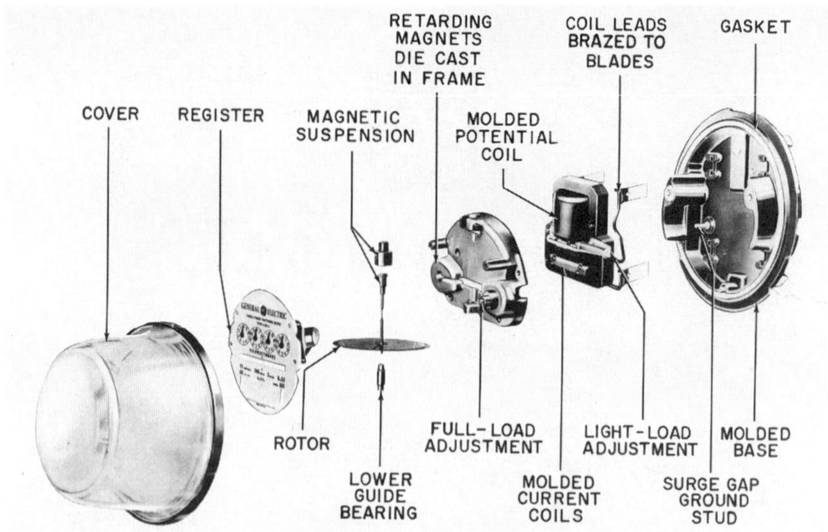

Figure 7-1. Basic Parts of a Watthour Meter.

the voltage coil. For meters built since 1960, the voltage coil consists of from about 2,400 turns of No. 29 AWG wire for a 120-volt coil to more than 9,600 turns of No. 35 AWG wire for a 480-volt coil. These coils are so compensated that the meter can be used within the range of 50 to 120 percent of nominal voltage, as explained later in this chapter. Because of the large number of turns, the voltage coil is highly reactive.

The portion of the stator energized by the load current is known as the current coil and usually consists of, for a Class 200 meter, two or four turns of wire equivalent to approximately 30,000 circular mils in size. The current coils are wound in reverse directions on the two current poles for correct meter operation.

Dr. Ferraris, in 1884, proved that torque could be produced electromagnetically by two alternating-current fluxes which have both a time displacement and a space displacement in the direction of proposed motion. The voltage coil is highly inductive, as mentioned before, so the current through the voltage coil (and hence the flux from it) lags almost 90 electrical degrees behind the line voltage. In modern meters, this angle is between 80 and 85 degrees. Although the current coil has very few turns, it is also wound on iron, so it is inductive. However, it is not as inductive as the voltage coil. The power factor of a modern meter current coil may be 0.5 to 0.7 or an angle of lag between 60 and 45 degrees. It is important to remember, however, that the meter current coils have negligible effect on the phase angle of the current flowing through them. This is true because the current coil impedance is extremely small in comparison with the load impedance which is connected in series. Hence, the load voltage and load impedance determine the phase position of the current through the meter. Thus, with a unity-power-factor load, the meter current will be in phase with the meter voltage. Since current through the voltage coil lags behind current through the current coil, flux from the voltage coil reaches the rotor after flux from the current coil, and a time displacement of fluxes exists. The stator is designed so that the current and voltage windings supply fluxes that are displaced in space. These two features combine to give us both the time and space displacement that Dr. Ferraris showed could be used to produce torque.

In order to understand why torque is produced, certain fundamental laws must be remembered. They are:

1. Around a current-carrying conductor there exists a magnetic field.
2. Like magnetic poles repel each other; unlike poles attract each other.
3. An electromotive force is induced in a conductor by electromagnetic action. This EMF is proportional to the rate at which the conductor cuts magnetic lines of force. The induced EMF lags 90 degrees behind the flux which produces it.
4. If a conducting material lies in an alternating-current magnetic field, the constantly changing or alternating magnetic lines of force induce EMFs in this material. Because of these EMFs, eddy currents circulate through the material and these eddy currents in turn produce magnetic fields of their own.
5. When a current is caused to flow through a conductor lying within a magnetic field, a mechanical force is set up which tends to move the current-carrying conductor out of the magnetic field.

The reason for this effect can be seen from Fig. 7-3. In Fig. 7-3(a), a conductor is indicated as carrying current from above into the plane of the paper, which establishes a magnetic field that is clockwise in direction. Figure 7-3(b) indicates an external magnetic field. When the current-carrying conductor is moved into the external field, as in Fig. 7-3(c), it reacts with the external field and causes a crowding of the flux lines on the left where the two fields are additive. On the right, where the fields are in opposition, the flux lines move apart. The flux lines may be considered as elastic bands acting on the conductor, causing a force which tends to move the conductor to the right.

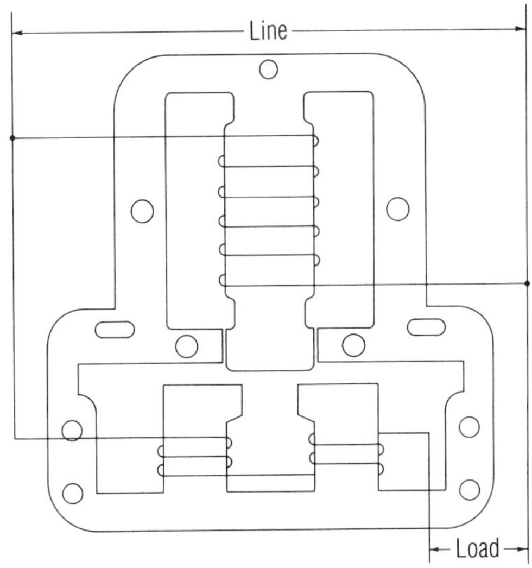

Figure 7-2. Basic Electromagnet (for Two-Wire Meter).

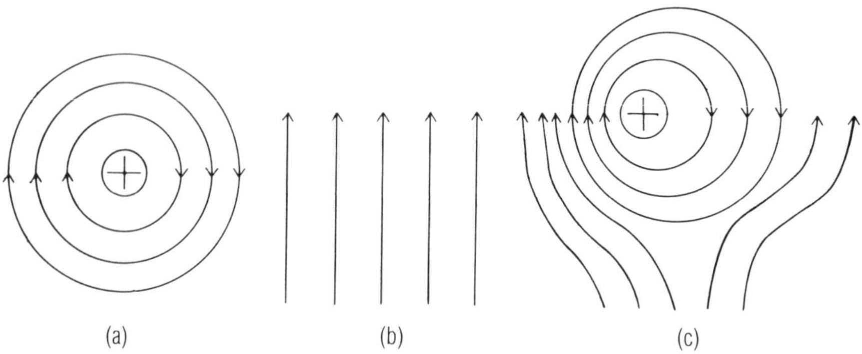

Figure 7-3. Effect of a Current-Carrying Conductor in an External Magnetic Field.

The rotor of the meter is an electrical conductor in the form of a disk which is placed between the pole faces of the stator as indicated in Fig. 7-4. The magnetic fluxes from the stator pass through a portion of the disk and, as the magnetic fields alternately build up and collapse, induced EMFs in the disk cause eddy currents which react with the alternating magnetic field, causing torque on the disk. The disk is free to turn, so it rotates.

Figure 7-5 shows the flux relationships and disk eddy currents in a meter at various instants of time during one cycle of supply voltage. It also indicates the space displacement which exists between the magnetic poles of the current coils and the voltage coil. The four conditions in Fig. 7-5 correspond to the similarly marked time points on the voltage and current flux waveforms of Fig. 7-6. This illustration also shows the time displacement existing between current and voltage coil fluxes. In relating the fluxes and eddy currents shown in Fig. 7-5 to the waveforms of Fig. 7-6, it must be remembered that when the flux waveforms

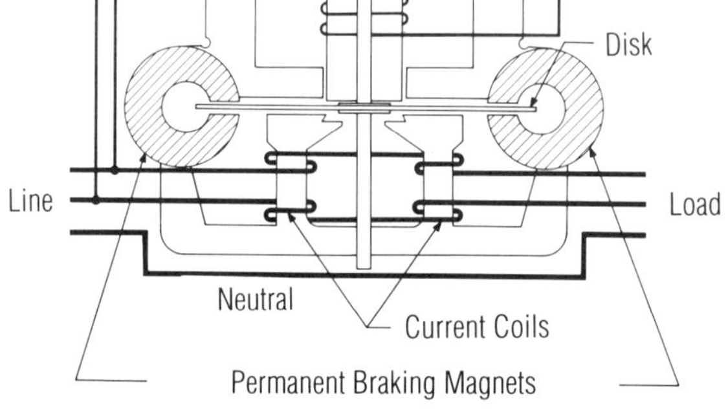

Figure 7-4. Schematic Diagram of a Three-Wire, Single-Phase Induction Watthour Meter.

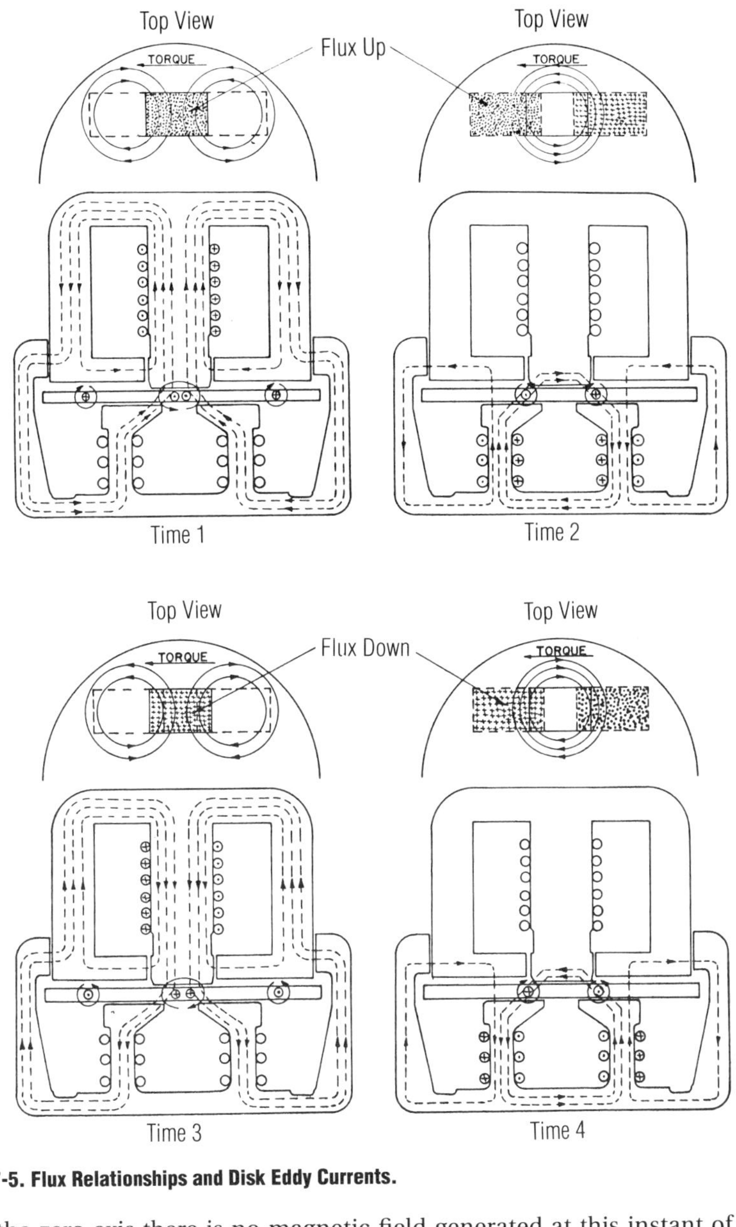

Figure 7-5. Flux Relationships and Disk Eddy Currents.

cross the zero axis there is no magnetic field generated at this instant of time. However, it is at this particular instant that the rate of change of flux is greatest, giving the maximum induced voltage in the disk and maximum resulting disk eddy currents. Thus, at Time 1, the voltage flux is at its maximum (negative) value as shown, but it causes no disk eddy currents because at this instant its rate of change is zero. At the same time, the current flux is zero, but its rate of change is a maximum, giving the greatest disk eddy currents. Consideration in Fig. 7-5 of the

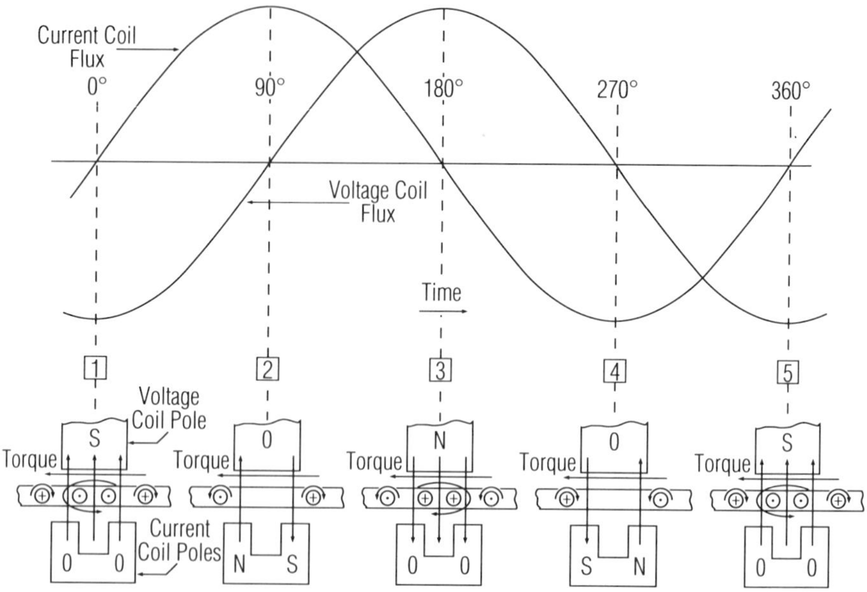

Figure 7-6. Voltage and Current Flux Waveforms.

directions of the fluxes created by the disk eddy currents and the air-gap fluxes from the voltage and current coils shows that, in accordance with Fig. 7-3, a force is developed with direction of the resultant torque as shown to the left.

In Fig. 7-6, below the flux waveforms, are enlarged views of the current and voltage poles and disk which may show more clearly the flux interactions which produce disk torque. It is assumed, of course, that an exact 90-degree phase relationship has been obtained between the two fluxes. Let us analyze each time condition separately:

1. The current coils are at the zero point of their flux curve; hence the rate of change of current flux is maximum, giving disk eddy currents as shown for the two current poles. The voltage-coil flux curve lags that of the current coil by 90 degrees. Since this curve is below the zero line, the voltage coil develops a south magnetic pole. Interaction of the disk eddy-current flux (in the central portion of the disk) and the voltage-coil flux develops a force to the left in the disk according to the principle shown in Fig. 7-3. The return paths of the disk eddy currents shown in the outer portions of the disk are too far removed from the voltage flux to have an appreciable effect on disk force.
2. At this time, 90 electrical degrees after Time 1, the voltage-coil flux has reached zero. Its rate of change is maximum, causing disk eddy current as shown. The current-coil flux has reached its maximum. North and south current poles are produced as indicated because the current coils are wound in reverse directions on the two poles. Again, the interaction of flux produced by the disk eddy currents with the current-coil flux creates disk force to the left.
3. This point is similar to that at Time 1 and occurs 180 electrical degrees later. Here, the voltage flux curve is above the zero line, producing a north pole.

The current-coil flux has again reached zero, but its rate of change is in the opposite direction so that the direction of disk eddy-current flow is reversed from that shown at 2. Since both the voltage flux and disk eddy-current flux are reversed in direction, the resultant disk force is still to the left.

4. This point, 270 electrical degrees after Time 1, is similar to that at Time 2. Here again the direction of both current-coil flux and disk eddy-current flux are reserved, giving resultant disk force to the left.
5. At this time the cycle of change is completed, producing the identical conditions of Time 1—360 electrical degrees later.

Summarizing the results, it is found that first a south pole, then a north pole, and then a south pole moves across the disk. At any position, the torque which causes the disk to turn is caused by interaction between flux from current in one coil and disk eddy currents caused by the changing flux from current through the other coil.

Because we want to measure watts, or active power, the force driving the disk must be proportional not only to the voltage and current, but also to the power factor of the load being metered. This means that, for a given voltage and current, the torque must be maximum when the load being metered is non-inductive and that it will be less as the power factor decreases.

When the successive values of the flux of a magnetic field follow a sine curve, the rate of flux change is greatest, as previously stated, at the instant of crossing the zero line and, thus, the induced electromotive force is greatest at this instant. Also, as previously mentioned, the magnetic field is greatest at the maximum point in the curve, but at this peak the rate of change is zero, so the induced EMF is zero. With two fields differing in phase relation, in order for one field to be at zero value while the other is at maximum, the phase difference must be 90 electrical degrees. The curves of Fig. 7-6 were shown this way. Since the torque on the disk depends on the interaction between the magnetic field and the disk eddy currents, the greatest torque occurs when the phase difference between the fields is 90 degrees. This is true at any instant throughout the cycle. When the two fields are in phase with each other, the disk eddy current produced by one field will be in a definite direction which will not change while the field changes from a maximum negative to a maximum positive value. The other field, in changing during the same period from a negative to a positive value and reacting with the disk eddy currents, tends to change the direction of rotation because direction of the torque changes. The change occurs every one-fourth cycle and the resultant average torque is zero. This is also true if the fields are 180 degrees apart.

It is apparent from previous discussion that if one field is proportional to the current in a power circuit and the other to the voltage across the circuit, the torque produced will be proportional to the product of these values. If an initial phase difference between the two fields is exactly 90 degrees when the line current and voltage are in phase, the torque produced on the rotating element when the current and voltage are not in phase will be proportional to the cosine of the angle of phase difference, which is the power factor. When the correct phase difference is obtained between the current and voltage flux (when the meter is properly "lagged"), the meter can be used to measure the active power in the circuit since the power is equal to the product of the voltage, current, and power factor.

THE PERMANENT MAGNET OR MAGNETIC BRAKE

As mentioned previously, another essential part of the meter is a magnetic brake. Torque on the disk caused by interaction of fluxes tends to cause constant acceleration. Without a brake, the speed of rotation would be limited by the supply frequency, by friction, and by certain counter torques at higher speeds which are discussed in those paragraphs concerning overload compensations, but the speed of rotation would be very high. Therefore, some method of making the speed proportional to power and also of reducing it to a usable value is needed. A permanent magnet performs these functions. As the disk moves through the field of the permanent magnet, eddy currents result in much the same manner as though the magnetic field were changing as described previously. These eddy currents remain fixed in space with respect to the magnet pole face as the rotor turns. Again, as in the case of eddy currents caused by fluxes from the voltage and current coils, the eddy currents are maximum when the rate of cutting flux lines is greatest. In this case the cutting of flux lines is caused by the motion of the disk, so the eddy currents are proportional to the speed of rotation of the disk. They react with the permanent-magnet flux, causing a retarding torque which is also proportional to the speed of the disk. This balances the driving torque from the stator so that the speed of the disk is proportional to the driving torque, which in turn is proportional to the power flowing through the meter. Thus, the number of revolutions made by the disk in any given time is proportional to the total energy flowing through the meter during that time interval. The strength of the permanent magnet is chosen so that the retarding torque will balance the driving torque at a certain speed. In this way the number of watthours represented by each revolution of the disk is established. This is known as the watthour constant (K_h) of the meter.

ADJUSTMENTS

On modern single-stator watthour meters there are three adjustments available to make the speed of the rotor agree with the watthour constant of the meter. They are the "Full-Load" adjustment, the "Light-Load" adjustment, and the "Power-Factor" adjustment.

Full-Load Adjustment

As noted previously, the eddy currents in the disk caused by the permanent magnets produce a retarding force on the disk. In order to adjust the rotor speed to the proper number of revolutions per minute at a given (or "rated") voltage and current at unity power factor, the full-load adjustment is provided.

Basically, there are two methods of making the full-load adjustment. One is to change the position of the permanent magnet. When the permanent magnet is moved, two effects result. As the magnet moves further away from the center of the disk, the "lever arm" becomes longer, which increases the retarding force. Also the rate at which the disk cuts the lines of flux from the permanent magnet increases and this too increases the retarding force.

The second method of making the full-load adjustment, by varying the amount of flux by means of a shunt, depends on the fact that flux tends to travel through the path of least reluctance. Reluctance in a magnetic circuit is resistance to magnetic lines of force, or flux. By changing the reluctance of the shunt, it is possible to vary the amount of flux that cuts the disk. One way of doing this is by means of a soft iron yoke used as a flux shunt, in which there is a movable iron screw. As the screw is moved into the yoke, the reluctance of this path decreases, more lines of flux from the permanent magnet flow through the yoke, and less through the disk, so the disk is subject to less retarding force and turns faster.

In either case, the retarding force is varied by the full-load adjustment and, by means of this adjustment, the rotor speed is varied until it is correct. Normally the full-load adjustment is made at unity power factor, at the voltage and test current (*TA*) shown on the nameplate of the watthour meter, but the effect of adjustment is the same, percentage-wise, at all loads within the class range of the meter.

Light-Load Adjustment

With no current in the current coil, any lack of symmetry in the voltage coil flux could produce a torque which might be either forward or reverse. Also, because electrical steels are not perfect conductors of magnetic flux, the flux produced by the current coils is not exactly proportional to the current, so that when a meter is carrying a small portion of its rated load it tends to run slow. In addition, a certain amount of friction is caused by the bearings and the register, which also tends to make the disk rotate at a slower speed than it should with small load currents. To compensate for these tendencies, a controlled driving torque which is dependent upon the voltage is added to the disk. This is done by means of a plate (or shading pole loop) mounted close to the voltage pole in the path of the voltage flux. As this plate is moved circumferentially with respect to the disk, the net driving torque is varied and the disk rotation speed changes accordingly. The plate is so designed that it can be adjusted to provide the necessary additional driving torque to make the disk revolve at the correct speed at 10 percent of the *TA* current marked on the nameplate of the meter. This torque is present under all conditions of loading. Since it is constant as long as applied voltage does not change, a change in the light-load adjustment at 10 percent of test amperes will also change full-load registration, but will change it only one-tenth as much as light-load registration is changed.

Inductive-Load or Power-Factor Adjustment

The theory behind the inductive load adjustment was presented by Shallenberger about 1890. The theory is that in order to have correct registration with varying load power factor, the voltage-coil flux must lag the current-coil flux exactly 90 degrees when the load on the meter is at unity power factor. This 90-degree relationship is essential to maintain a driving force on the disk proportional to the power at any load power-factor value. One way of doing this is to make the voltage-coil flux lag the current-coil flux by more than 90 degrees by means of a phasing band, or coil, around the core of the center leg of the voltage coil. It is then necessary to shift the current-coil flux toward the voltage-coil flux until the angle is exactly 90 degrees.

Figure 7-7 shows this phasorially. In Fig. 7-7, E is the voltage and ϕE is the flux caused by E. ϕE induces a voltage in the phasing band which causes a current to flow, creating the flux shown as ϕE_{PB}. This, added phasorially to ϕE, gives ϕE_T,

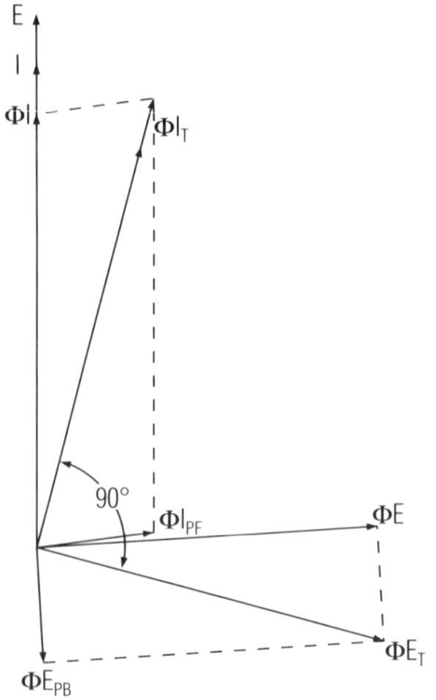

Figure 7-7. Phasor Diagram of Lag Adjustment.

which is the total resultant flux that acts on the disk and which lags E by more than 90 degrees. Since this analysis is for unity-power-factor load, the current I and its flux ϕI are in phase with E. But the flux ϕI must be shifted toward ϕE_T until the angle is exactly 90 degrees. A closed "figure-8" circuit loop is inserted on the current magnet. ϕI induces a voltage in this loop, which causes current to flow, creating the flux field ϕI_{PF}. This value can be changed by varying the resistance of the power-factor loop. Adding ϕI and ϕI_{PF} gives ϕI_T which is adjusted (by varying ϕI_{PF}) until it is exactly 90 degrees from ϕE_T. Also note that any reactance in the current coil which would cause ϕI to be slightly out of phase with I is compensated for at the same time.

As explained in the preceding discussion, the shift of resultant current-coil flux is done by means of a "figure-8" conducting loop on the current electromagnet. The coil usually consists of several turns of wire. The ends of this "lag coil" are twisted together and soldered at the point necessary to provide the 90-degree angle. A change in the length of the wire varies the resistance of the coil and the amount of current flowing, which results in a variation in the amount of compensating flux.

The means of adjusting the flux angle may be located on the voltage-coil pole instead of the current poles, in which case it would vary ϕE_{PB} instead of ϕI_{PF}. The adjustment may be in the form of a coil with soldered ends, so that loop resistance may be varied, or a lag plate. A lag plate would be movable under the voltage pole piece radially with respect to the disk. In this manner it would provide adjustable phase compensation with minimum effect on light-load characteristics.

Many modern meters use a fixed lag plate operating on voltage flux with the compensation permanently made by the manufacturer at the factory. Such plates may be located on the voltage coil pole or may form a single loop around both current poles.

For practical purposes, all modern meters leave the factories properly adjusted and, once made, this lag or power-factor adjustment seldom requires change, regardless of what type it is.

Once the proper phase relationship between the load-current flux and the voltage flux is attained, there will be no appreciable error at any power factor. If this adjustment is improperly made, an error will be present at all power factors other than unity and it will increase as the power factor decreases. This is calculated as follows:

$$\%\ \text{error} = 100\left(1 - \frac{\text{Meter watts}}{\text{True watts}}\right)$$

Using the information supplied and the method explained in Chapter 3, this can be developed into a formula which may be resolved into the following:

$$\%\ \text{error} = 100\left(1 - \frac{EI\cos(\theta \pm \phi)}{EI\cos\theta}\right)$$

$$= 100\left(\frac{\cos\theta - \cos(\theta \pm \phi)}{\cos\theta}\right)$$

where θ is the angle between the line current and voltage and ϕ is the angle of error between the line-current flux and the voltage flux due to improper relation within the meter. This error is computed without reference to errors of calibration at full load. Full-load errors are independent of those just calculated and add to, or subtract from, them dependent upon their relative signs. The errors indicated, while computed for lagging power factor, are also applicable for leading power factor. The sign of the effect will change in going from a lagging power factor to a leading power factor. In other words, an improper lag adjustment which causes the meter to run slow on lagging power factor will cause it to run fast on leading power factor.

COMPENSATIONS

Although the three adjustments mentioned in previous paragraphs are the usual adjustments for a single-stator watthour meter, several other factors must be compensated for to make the meter the accurate instrument that it is. These "compensations" are built into the meter and provide corrections needed to make the meter register accurately under conditions of overload, temperature variation, frequency error, and voltage fluctuation.

Overload Compensation

As explained previously, the meter may be adjusted to record correctly at its nominal load. Unless it is compensated, however, it will not record correctly at loads as they increase up to the maximum load within its class range. Because electromagnetic steels are not perfect conductors of flux, the speed of rotation of the disk will tend to be proportionately less at higher loads. Also, a much greater effect at overloads is that due to damping caused by the interaction of the disk eddy currents with the fluxes that produce them. For example, the voltage coil

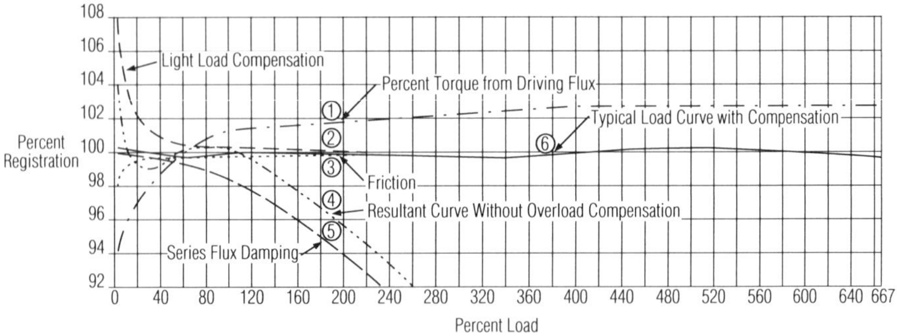

Figure 7-8. Factors of Accuracy.

produces eddy currents which interact with the current-coil flux to drive the disk, but the interaction of the voltage-coil eddy currents with the voltage-coil flux retards the disk. The voltage-coil flux is practically constant regardless of load, so its retarding effect can be calibrated out of the meter. Similarly, the fluxes produced by the current coil will act with the current-coil eddy currents to retard rotation of the disk. At rated load these self-damping effects are in the order of only 0.5 percent of the total damping. However, the retarding action increases as the square of the current flux. This is true because the retarding force is a function of the eddy currents multiplied by the flux, and in this case the eddy currents increase as the flux increases, so the retarding force increases as the flux multiplied by itself.

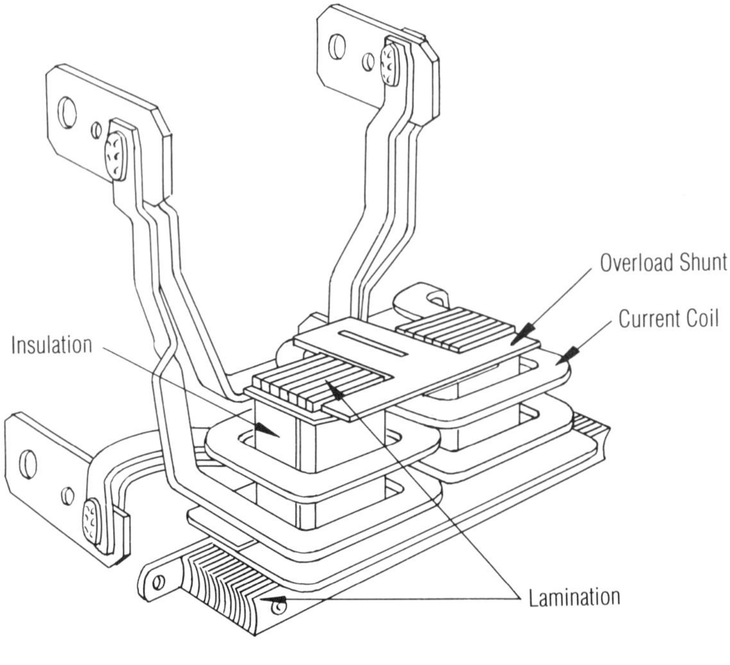

Figure 7-9. Simplified Diagram of Magnetic Circuit of Current Element for Overload Compensation.

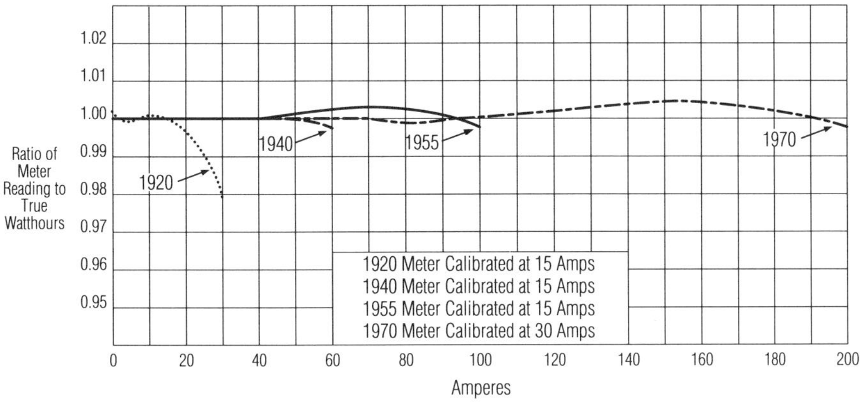

Figure 7-10. Heavy-Load Accuracy Curves.

Figure 7-8 shows the factors of accuracy for a meter with the typical load curve (6) of a model compensated meter. To negate the retarding or dropping accuracy shown as curve (4), which would result without overload compensation, a magnetic shunt is placed between (not touching) the poles of the current electromagnet and is held in place by non-magnetic spacers. (See Fig. 7-9.) This shunt has little effect below the point at which the accuracy curve of the meter would otherwise start to droop, but as the load increases the shunt approaches saturation causing the current flux which cuts the disk to increase at a greater ratio than the current. This causes an added increase in torque, which counteracts the droop in the accuracy curve up to the point at which the shunt is saturated. Beyond this point, which is usually beyond the maximum rated load of the meter, the accuracy curve drops very rapidly.

Figure 7-13 shows another diagram of the magnetic circuit for overload compensation on the current element. Other ways of minimizing the retarding effect are: (1) by proper proportioning of the voltage and current fluxes, so that the effective voltage-coil flux (about 4 percent of the total damping flux) is propor-

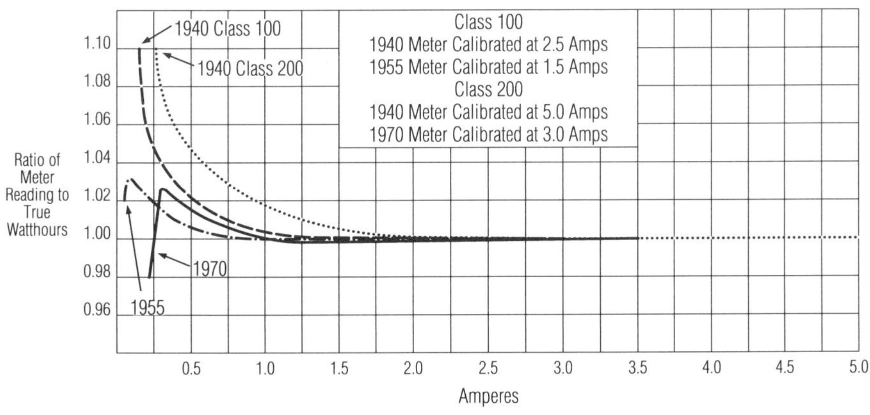

Figure 7-11. Light-Load Performance Curves.

tionately higher than the effective current-coil flux, (2) by use of stronger permanent magnets and lower disk speed, and (3) by design which gives the greatest driving torque while getting the least damping effect from the electromagnets. Present-day meters will accurately register loads up to 667 percent of the meter's nominal rating. Figure 7-10 shows comparisons of the accuracy of modern meters with that of those manufactured in 1920, 1940, and 1955.

At the same time that these improvements were being made, similar improvements were effected in light-load performance, as can be seen from the curves in Fig. 7-11.

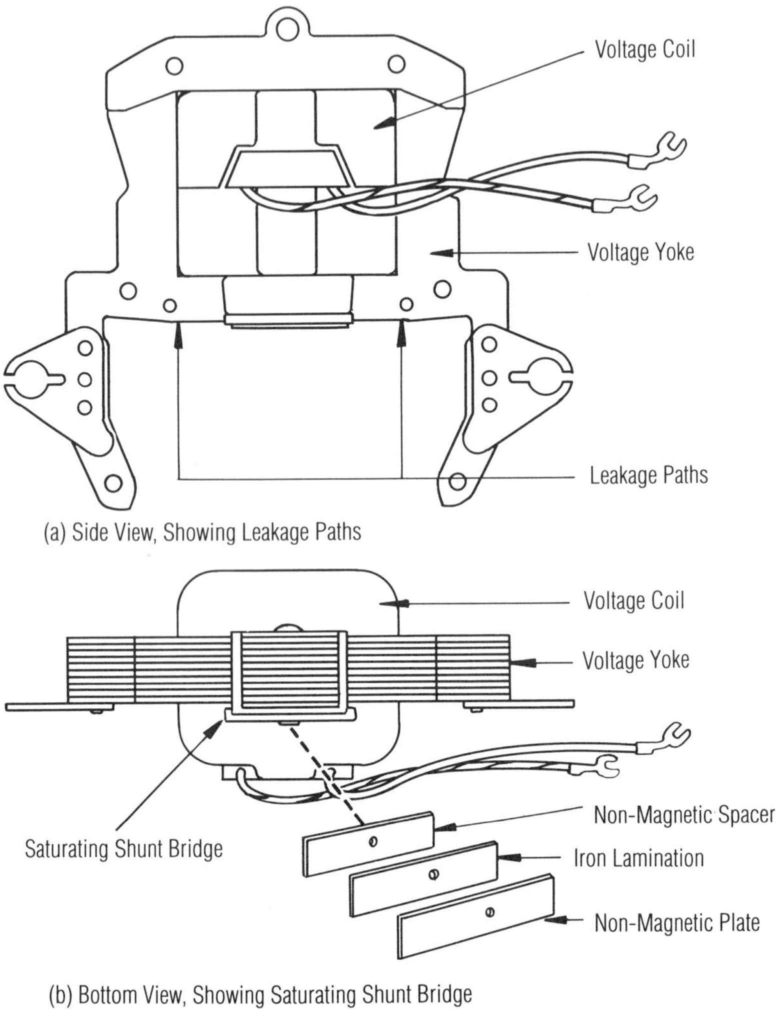

Figure 7-12. Simplified Diagram of Method of Compensation for Voltage Variation.

Voltage Compensation

Inaccuracies of registration in modern meters due to the usual range of voltage variations are very small. In a meter with no voltage compensation, errors resulting from voltage change are caused by:

1. The damping effect of the voltage flux,
2. Changes in the electromagnet characteristics due to changes in voltage, and
3. Changes in the effect of the light-load adjustment due to changes in voltage.

The damping effect of the voltage flux is similar to that of the current flux, with changes in effect being proportional to the square of the voltage.

The errors caused by electromagnet characteristics are due to the failure of the magnetic circuit to maintain straight-line properties under all conditions of flux density.

In an electromagnet the effective flux is not equal to the total flux. The ratio between the effective and the total flux determines many of the characteristics of the electromagnet. Improvements in the iron used have permitted a much closer approach to the desired straightline properties of the magnetic circuit. Finally, by use of saturable magnetic shunts similar to those used in the current magnetic circuit, voltage flux is controlled and the errors due to normal voltage variations are reduced to a negligible amount.

Since the light-load compensation is dependent only on voltage, any voltage change varies the magnitude of this compensation and tends to cause error. Increasing voltage increases light-load driving torque so that a meter tends to over-register at light load with overvoltage. Good meter design, which maintains a high ratio of driving torque to light-load compensating torque, reduces these errors to very small values.

Figure 7-12 shows one method of compensating for voltage variation.

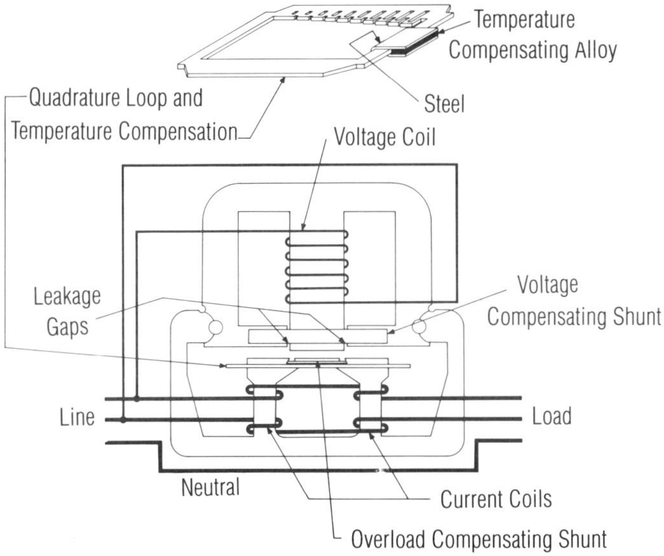

Figure 7-13. Overload, Voltage, and Class 2 Temperature Compensations.

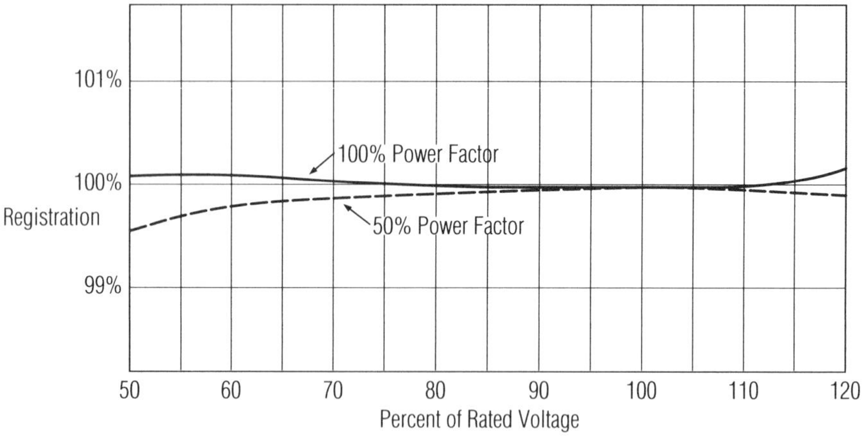

Figure 7-14. Voltage Characteristic Curves.

Figure 7-13 shows methods of compensating for overload, voltage variation, and temperature.

One of the improvements in meters of recent manufacture is the reduction in voltage errors to a degree that such a meter designed for use on 240 volts may, in most cases, be used on 120-volt services without appreciable error. Figure 7-14 shows a voltage characteristic curve for one of the modern meters.

Temperature Compensation

Watthour meters are subjected to wide variations in ambient temperature. Such temperature changes can cause large errors in metering accuracy unless the meter design provides the necessary compensation. Temperature changes can affect the strength of the retarding magnets, the resistance of the voltage and lag coils, the iron characteristics, the disk resistance, and other quantities which have a bearing on accuracy. Temperature errors are usually divided into two classes. Class 1 errors are those temperature errors which are independent of the load power factor, while Class 2 errors are those which are negligible at unity power factor but have large values at other than unity power factor.

Class 1 temperature errors are caused by a number of factors which produce a similar effect; namely, that the meter tends to run fast with increasing temperature. Since this is the effect caused by weakening the permanent magnet, the compensation for this class of error consists of placing a shunt between the poles of the permanent magnet to by-pass part of the flux from the disk. This shunt is made of a magnetic alloy which exhibits increasing reluctance with increasing temperature. With proper design, the shunt will by-pass less flux from the disk with increasing temperature so that the braking flux increases in the proper amount to maintain high accuracy at unity power factor over the entire temperature range.

Class 2 temperature errors, which increase rapidly with decreasing power factor, are due primarily to changes in the effective resistance of the voltage and lag circuits, which, in turn, cause a shift in the phase position of the total voltage

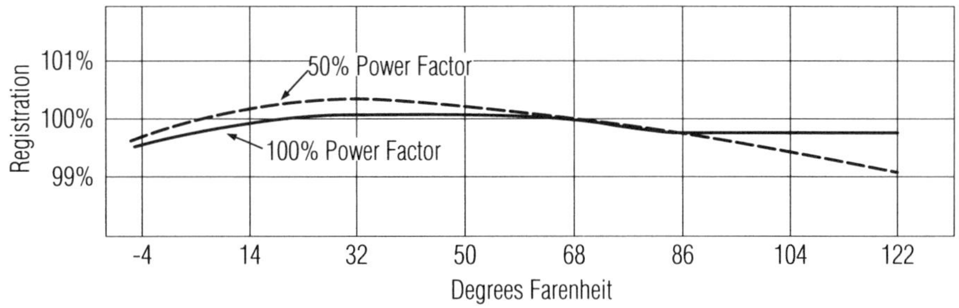

Figure 7-15. Temperature Characteristic Curves.

flux. Improved design has reduced these errors, and compensation in various forms has further minimized them. One compensation method consists of placing a small piece of material with negative permeability temperature characteristics around one end of a lag plate, or a small amount of the alloy in the magnetic circuit of a lag coil, to vary the reactance of the lag circuit in such a manner that the lag compensation remains correct with temperature change. Another method consists of over-lagging the voltage flux with a low-temperature-coefficient resistor in the lag circuit and adjusting the current flux with a lag circuit which contains a high-temperature-coefficient resistor. With this method, changes in one lag circuit due to temperature are, with proper design, counterbalanced by changes in the other lag circuit.

Some of the temperature effects tend to offset one another. An example of this is the change in disk resistance with temperature. An increase in disk resistance reduces electromagnet eddy-current flow which reduces driving torque. However, the same effect occurs with the eddy currents set up by the braking magnets, so braking torque decreases with driving torque and disk speed tends to remain constant.

Figure 7-13 shows the Class 2 temperature compensation of a three-wire meter and Fig. 7-15 shows the temperature characteristic curves of modern watthour meters, indicating the high degree of temperature compensation which has been secured by the methods previously outlined.

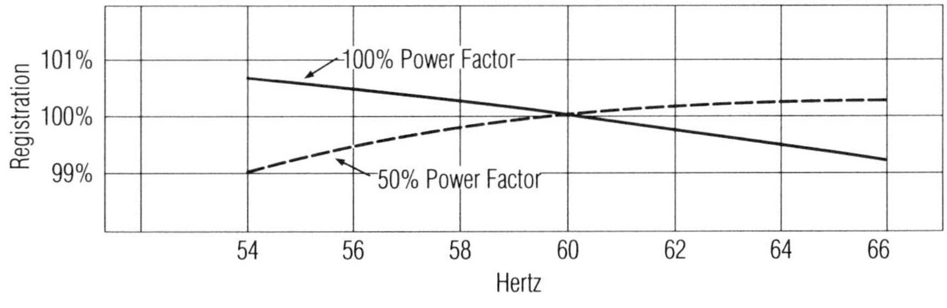

Figure 7-16. Frequency Curve of Modern Meter.

Anti-Creep Holes

Without anti-creep holes, the interaction of the voltage coil and the light-load adjustment might provide enough torque to cause the disk to rotate very slowly when the meter was energized. This "creep" would generally be in a forward direction, because the light-load adjustment is so designed that it helps overcome the effects of friction and compensates for imperfections of the electromagnet steels. In order to prevent the disk from rotating continuously, two diametrically opposed holes are cut in it. These holes add resistance to the flow of eddy currents caused by the voltage flux. Earnshaw's theorem explains that a conductor in a flux field tends to move to a position of least coupling between the conductor and the source of the flux field. Because of this, the disk will tend to stop at a position in which the anti-creep hole causes the greatest reduction in the eddy currents (sometimes moving backward a portion of a revolution in order to stop in this position). A laminated disk or one of varying thickness will also tend to stop in a position of least coupling.

Frequency Characteristics

Because of frequency stability of modern systems, variations in meter accuracy due to frequency variations are negligible. As frequency is increased, the reactance of the shunt coil increases and its exciting current decreases. The reactance of the eddy current paths in the disk is raised, thus limiting and shifting the phase of the eddy currents. Also, an increase in frequency raises the proportion of reactance to resistance in the shunt coil and the meter tends to become over-lagged. Any increase in reactance of the quadrature adjuster shifts its phase angle so that its action is more and more to reduce the flux, thus decreasing torque. Watthour meters are therefore slow on high frequencies, with the percent registration at 50 percent power factor lagging being higher than that for unity power factor. However, because of the stability of modern systems, specific frequency compensation is not required and in modern meters frequency variation errors are kept to a minimum by proper design.

Figure 7-16 shows the effect of frequency variations on modern meters.

Waveform

In determining the effects of harmonics on watthour meter performance the following facts must be borne in mind:

1. A harmonic is a current or a voltage of a frequency which is an integral multiple of the fundamental frequency. In a 60 hertz system, the third harmonic has a frequency of 180 hertz.
2. A distorted wave is a combination of fundamental and harmonic frequencies which, by analysis, may be broken down into such frequencies.
3. Currents and voltages of different frequencies do not interact to produce torque. A harmonic in the voltage wave will react only with the same harmonic in the current wave to produce torque.
4. To produce torque, two fluxes with time and space displacement are necessary.
5. A harmonic present only in the voltage circuit may have a small effect on meter performance. This is due to the torque component produced by the light-load adjustment.

6. Minor damping effects of harmonics in either voltage or current element are possible.

The magnetic shunt used for overload compensation can introduce harmonics in the current flux which are not necessarily present in the load current, particularly at high loads. To a lesser extent, this is also possible in the voltage flux, but, in general, unless extreme distortion of waveform exists, the errors due to harmonics will not degrade meter accuracy beyond normal commercial limits. However, when working with high accuracy watthour standards, the errors due to harmonics may be bothersome. In these cases it must be remembered that all meters, even when of the same manufacturer and type, do not exhibit identical reactions to the same degree of harmonics.

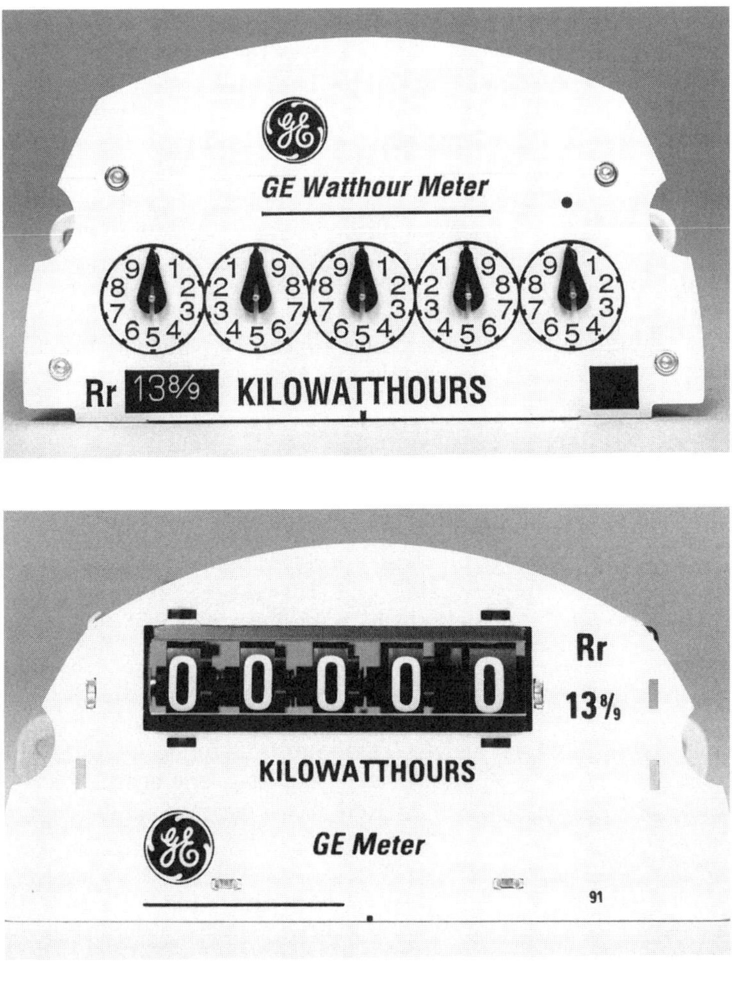

Figure 7-17. Clock-Type (top) and Cyclometer-Type Meter Registers.

Waveform distortion and resulting meter inaccuracies may be caused by over-excited distribution transformers and open-delta transformer banks. Some types of equipment, such as rectifiers and fluorescent lamps, may also cause distortion of the waveform. Welding equipment also causes poor waveform and presents a continuing metering problem, but in this instance other factors may have greater influence on meter errors.

In extreme cases of distortion, a separate analysis is necessary because each waveform has different characteristics. The distorted wave should be resolved into the fundamental and the various harmonic sine waves and calculations can then be made from this information.

METER REGISTERS

The third basic part mentioned in the beginning of this chapter is the register. The register is merely a means of recording revolutions of the rotor, which it does through gearing to the disk shaft. Either a clock (pointer-type) or a cyclometer-type register may be used.

Figure 7-17 shows a clock-type register and a cyclometer register. Both perform the same function, but the pointer-type has numbered dials on its face and the pointers turn to indicate a proportion of the number of revolutions the disk has made. In the cyclometer-type registers, numbers are printed on cylinders which turn to indicate a proportion of the number of revolutions of the disk. Since the purpose of the register is to show the number of kilowatt-hours used, the reading is proportional rather than direct. With the meter properly adjusted, the disk will revolve at a specified speed at full load. This speed and the rating of the meter determine the watthour constant, or K_h, which is the number of watthours represented by one revolution of the disk. The watthour constant K_h may be found by use of the formula:

$$K_h = \frac{\text{Rated Voltage} \times \text{Rated Current}}{\text{Full-Load RPM} \times 60}$$

The necessary gearing is provided so that the revolutions of the disk will move the first (or right-side) pointer or cylinder one full revolution (360 degrees) each time the rotor revolves the number of times equal to ten kilowatt-hours of usage. This is known as the gear ratio, or R_g. The register ratio, known as R_r is the number of revolutions of the wheel which meshes with the pinion or worm on the disk shaft for one revolution of the first dial pointer.

METER ROTOR BEARINGS

In order to support the shaft on which the rotor is mounted, bearings which will give a minimum amount of friction are used. Magnetic bearings are used in present-day meters. The weight of the rotor disk and shaft is in the order of 16 to 17 grams.

Modern meters have magnetic bearings consisting of two magnets which support the shaft and disk. The rotor is held up by mutual attraction when the bearing magnets are located at the top of the disk shaft or by repulsion of the magnets when they are located at the bottom of the shaft. One magnet is fastened to the meter frame and the other magnet is mounted on the disk shaft. Vertical alignment is provided by guide pins mounted on the meter frame at the top and bottom of the disk shaft, which has bushings mounted in each end. The only bearing pressures in this type of rotor support are slight side thrusts on the guide pins, since the shaft does not otherwise touch either the top or bottom supports, making the system subject to less wear. No part of this system requires lubrication. Additional advantages of this type of bearing system are reduced maintenance, less tilt error, and better ability to withstand rough handling. More details are available in the manufacturers' literature.

MECHANICAL CONSTRUCTION OF THE METER

The basic parts of the meter are assembled on a frame, mounted on a base, and then covered with a glass cover. The cover encloses the entire meter and is sealed to the base. The base and cover are so designed that it is almost impossible to tamper with the adjustments of the meter without leaving evidence.

Meters are maintained weathertight mainly by the design of the cover, base, and dust guard. The meter is allowed to breathe by providing an opening at the bottom of the base. This opening also allows any condensate that may form on the inside of the glass cover to drain out. The chief aid in allowing meters to operate outdoors and under varying humidity conditions is the self-heat generated in the stator which causes the cooler cover to act as the condenser under high-humidity operation. This is the reason meters should not be stored outdoors without being energized even though they are basically designed to be weathertight.

The materials and coatings used to prevent corrosion are generally the best materials economically available during that period of manufacture. Present-day meters are structurally almost completely made from high corrosion-resistant aluminum which has minimum contact with copper or brass materials. The iron laminations are protected by modern paints currently tested to be the best. A plastic base, glass cover, and stainless-steel cover ring complete the picture to give the meter its excellent corrosion resistance. In the application of corrosion-resistant finishes, consideration must also be given to the particular function of the part in question, such as exposure to the elements, wear resistance, and use as a current-carrying part.

One of the finishes applied to aluminum is anodizing. This finish is applied by use of an anodic process which passes a heavy current for a short period, converting the surface to aluminum oxide. This forms a very hard corrosion-resistant finish which is further improved by applying a sealer. This closes the pores in the oxide and prevents entrance of moisture. Another chromate-dip finish used is iridite. This applies the oxide coating and seals at the same time. It is used on cadmium-plated steel parts. Another finish that is applied to aluminum is alodine. This is a complex chromate gel which forms on the surface and seals the metal in one operation. When applied to aluminum, the parts take on an iridescent finish. It is used on parts such as the grid, register plates, and other parts not subject to wear or abrasion.

The copper and copper-bearing alloys that are current-carrying parts are tin-plated on contact surfaces, such as socket-meter bayonets. This not only gives superior protection against corrosion, but a superior contact results.

Brass screws are protected by a heavy nickel plating.

For the protection of ferrous metals, such as voltage and current electromagnet laminations, the part is, for some meters, immersed in a hot solution of phosphoric acid. This converts the surface to a hard iron phosphite which is further hardened by a sealer, usually paint. The voltage and current laminations are sealed by several coats of paint. Steel parts, such as register screws, are nickel-plated.

All the finishes described are constantly being improved with special emphasis on developing new agents to prevent corrosion.

Lightning and surge protection is provided by a combination of high insulation and surge levels built into the voltage coil and current coil and the provision of a ground pin in calibrated proximity to the current leads on the line side so that a lightning surge will jump the spark gap prior to entering the coils. The ground-pin gap is such as to cause a "spark over" at some voltage between 4,000 and 6,000 volts. The ground pin is attached to a strap in contact with the socket enclosure, which is grounded. The factory testing consists of hi-potting of both voltage coils and current coils at about 7,000 volts. The voltage coil, in addition, is exposed to a 10,000-volt surge of a 1.5 × 40 (crest in 1.5 microseconds, decay in 40 microseconds) wave shape to pick out any shorted turns in the windings. This combination has proved very successful in allowing meters to withstand repeated lightning surges and still allow continued accurate meter operation.

Present-day meters are built with permanent magnets that are practically unaffected by lightning surges.

TWO- OR THREE-WIRE METERS

Some single-stator meters are made so that by means of a very simple rearrangement of internal connections the meter can be converted from the connections used on 120-volt circuits to those needed for use on 240-volt circuits. Coil ends are brought to terminal boards and usually the connections can be changed with just a screwdriver. The basic theory, inherent accuracy, stability of calibration, insulation, and other desirable features of modern watthour meters are unaffected by the minor changes in internal connections that are made and this meter has the advantage of not becoming obsolete as the customer changes from 120-volt to 240-volt service.

The voltage coils are wound to give uniform flux distribution whether connected for 120-volt or 240-volt usage. A constant resistance-to-reactance ratio is maintained as the change is made; therefore a constant phase-angle relationship exists and no readjustment of the lag compensation is necessary.

The K_h constant is the same for either connection, since one current coil is used for 120-volt operation and two are used when the meter is connected for 240-volt operation. These meters are, of course, of standard dimensions so they are interchangeable with other single-phase, 120-volt or 240-volt standard-size meters. In general, changing internal connections has so little effect on calibration of the meter that it may be considered negligible. Details of internal connections may be found in Chapter 12, "Meter Wiring Diagrams."

THE THREE-WIRE METER

The three-wire meter is so designed that it has a voltage coil, connected across the two line wires, and two current coils, each connected in series with a line wire in such a way that the magnetic effects of currents flowing in the line wires are additive. The number of turns in each of the two current coils is one half as many as used in the current coil of a two-wire meter. According to Blondel's theorem, which is defined and discussed later (in the introduction to "Multi-Stator Meters," Part 2 of this chapter), two stators are required for accurate registration of energy flowing through a three-wire circuit. If, however, the voltages between each line wire and the neutral are single phase and exactly equal, the single-stator, three-wire meter is accurate. An unbalance in the voltage will cause accuracy proportional to one half the difference between voltages. Because modern systems are normally very closely balanced, any errors (usually considerably less than 0.2 percent) which do occur are negligible.

With the improved voltage compensation on modern meters, some utilities use the standard three-wire, 240-volt, single-stator meter on two-wire, 120-volt services in place of the standard two-wire meter or of the two- or three-wire convertible meters previously described. By connecting the two current coils in series, the meter K_h constant and registration are not changed on the two-wire service and the voltage compensation provides good performance at the 50 percent voltage operation.

GENERAL

Because of the care taken in their design and manufacture, and because of the long-wearing qualities of the materials used in them, modern watthour meters normally remain accurate for extended periods of time without periodic maintenance or testing. Probably no commodity available for general use today is so accurately measured as electricity.

PART 2—MULTI-STATOR METERS

This section on multi-stator watthour meters relies for its basic meter theory on single-stator meter data because a multi-stator meter is essentially a combination of single-stator meters on a common base. The differences are mainly in a few special features and in the various applications to polyphase power circuits.

BLONDEL'S THEOREM

The theory of polyphase metering was first set forth on a scientific basis in 1893 by Andre E. Blondel, engineer and mathematician. His theorem applies to the measurement of power in a polyphase system of any number of wires. The theorem is as follows:

> If energy be supplied to any system of conductors through N wires, the total power in the system is given by the algebraic sum of the readings of N wattmeters, so arranged that each of the N wires contains one current coil, the corresponding voltage coil being connected between that wire and some common point. If this common point is on one of the N wires, the measurement may be made by the use of N-1 wattmeters.

The receiving and generating circuits may be arranged in any desired manner and there are no restrictions as to balance among the voltages, currents, or power-factor values.

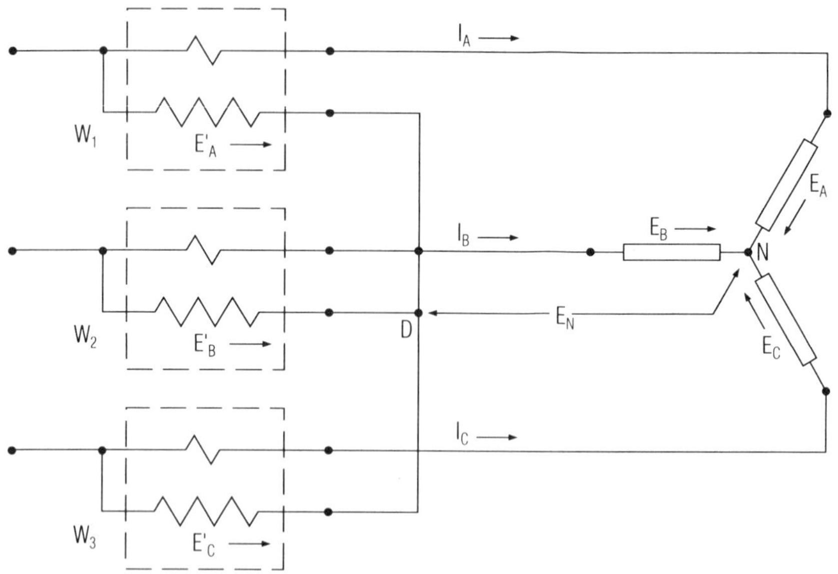

Figure 7-18. Diagram Used in Proof of Blondel's Theorem.

From this theorem it follows that basically a meter containing two stators is necessary for a three-wire, three-phase circuit, and a meter with three stators for a four-wire, three-phase circuit. Some deviations from this rule are commercially possible, but resultant metering accuracy, which may be worsened, is dependent upon circuit conditions which are not under the control of the meter technician. An example of such a deviation is the three-wire, single-stator meter previously described.

The circuit shown in Fig. 7-18 may be used to prove Blondel's theorem. Three watthour meters, or wattmeters, have their voltage coils connected to a common point D, which may differ in voltage from the neutral point N of the load, by an amount equal to E_N. The true instantaneous load power is:

$$\text{Watts}_{\text{Load}} = E_A I_A + E_B I_B + E_C I_C$$

Inspection of the circuit shows:

$$E_A = E'_A + E_N$$
$$E_B = E'_B + E_N$$
$$E_C = E'_C + E_N$$

Substituting in the equation for total load power:

$$\text{Watts}_{\text{Load}} = (E'_A + E_N) I_A + (E'_B + E_N) I_B + (E'_C + E_N) I_C$$
$$\text{Watts}_{\text{Load}} = E'_A I_A + E'_B I_B + E'_C I_C + E_N (I_A + I_B + I_C)$$

Since from Kirchhoff's law, $I_A + I_B + I_C = 0$, the last term in the preceding equation becomes zero, leaving

$$\text{Watts}_{\text{Load}} = E'_A I_A + E'_B I_B + E'_C I_C = W_1 + W_2 + W_3$$

Thus, the three watthour meters measure correctly the true load power. If, instead of connecting the three voltage coils at a common point removed from the supply system, the common point is placed on any one line, the voltage becomes zero on the meter connected in that line. If, for example, the common point is on line C, E'_C becomes zero and the preceding formula reduces to:

$$\text{Watts}_{\text{Load}} = E'_A I_A + E'_B I_B = W_1 + W_2$$

proving that one less metering unit than the number of lines will provide correct metering regardless of load conditions.

DEVELOPMENT OF THE POLYPHASE METER

The first American polyphase power systems were all two-phase. Knowledge of three-phase systems was limited at that time and their advantages were not fully understood. There was also some difference of opinion as to their merits compared with the two-phase system.

It was a simple matter to use separate single-stator watthour meters for metering the two-phase circuits and this was the general practice in America in the period of 1894-1898. Even the early three-phase systems were metered by combinations of two single-stator meters. The first true polyphase meters were produced in this country in 1898. Mainly, they consisted of the two-stator polyphase types for three-wire, three-phase and two-phase services.

The multi-stator polyphase meter is generally a combination of single-phase watthour-meter stators that drive a common shaft at a speed proportional to the total power in the circuit. The meter is essentially a multi-element motor, with a magnetic braking system, a register, a means for balancing the torques of all stators, and all the various adjustments and compensating devices found in single-stator meters. These components are assembled on a frame and mounted on a base which also contains the terminals. The base, the cover, and the terminals vary in their design according to the installation requirements of the meter.

The older types of these meters were multi-rotor meters with a rotor for each stator. One way of accomplishing this was to mount the stators side by side in a common case with the two rotor shafts recording on one dial through a differentially geared register. In such a meter, the stators do not need to be identical in current and voltage rating, but, if not identical, they must be so geared to the register that an equal number of watthours, as determined by the watthour constants of each stator, will produce the same registration on the dial. A second method of combining meter stators is to locate the stators one above the other in a common case, with the disks mounted on a common spindle to form a rotor which is geared to a single register. With this combination of stators, the stators do not need to be identical, but the watthour disk constant must be the same for each stator.

About 1935, the increased necessity for modifying the conventional types of polyphase meters, so as to reduce costs, weight, and space requirements resulted in a radical change in design whereby two stators were made to operate on one rotor disk. In 1939, one manufacturer went a step further and provided a meter with three stators operating on one rotor disk. These developments brought the size of polyphase meters down near that of the single-stator meter.

Most of the modern design features developed for single-stator meters are being applied to the multi-stator meter. All U.S. meter manufacturers supply self-contained capacities of Classes 100, 200, and 320, and Classes 10 and 20 for use with instrument transformers. Some manufacturers can furnish these meters with voltage-indicating lamps which are supplied at approximately 1.5 volts by a few separate turns on each voltage coil. One manufacturer supplies electromechanical indicators which are designed to be connected in series with each voltage coil. Magnetic-type bearings are being used on all meters of current manufacture.

Multi-stator meters are made in both "A" and "S" type bases. Type "A" (bottom-connected) are normally connected to the Line and Load by means of a test block. One disadvantage of the "A" base is that full Class 200 capacity is difficult to obtain due to the limitation placed on the size of the terminal chamber by standardization of meter base dimensions. The use of the Class 200 "S" (socket-type) meter increasing. Its advantages include the quick and easy insertion and removal of the meter in a compact meter socket and full Class 200 capacity. To test the meter, a socket test jack can be used or the meter can be removed for test. The socket can be furnished with a manual or an automatic bypass to allow removing or changing the meter without interrupting the customer's service.

Multi-stator Classes 10 and 20 meters are also furnished for switchboard use in semi-flush or surface-mounted cases, with or without drawout features. Drawout cases provide means to test the meter in place or to withdraw it safely from the case without danger of opening the current-transformer secondary circuits.

POLYPHASE METER CHARACTERISTICS AND COMPENSATIONS

Multi-stator polyphase meters have, in general, temperature, overload, voltage, and frequency characteristics similar to those of the single-stator meter and they are compensated in the same manner to improve these characteristics. Detailed explanations of these compensations have been given previously in this chapter.

The polyphase meter is basically a combination of two or more single-phase stators in one case, usually with a common moving element with such modifications as are necessary to balance torques and meet mechanical limitations. However, when single-phase stators are combined in the polyphase meter, the performance under unbalanced conditions does not always follow the independent single-phase characteristics.

The single-stator, two-wire meter is the most accurate form of the induction meter under all the conditions of the loads imposed upon it. To measure a four-wire service, three two-wire meters may be used with their registrations added to obtain total energy. From the point of view of accuracy, this is the ideal method of measurement. It is awkward, however, and introduces difficulties when measurement of demand is required. It must be understood that after compensation for interference and proper adjustments for balance, the accuracy of the multi-stator meter closely approaches that of the two-wire, single stator meter.

Driving and Damping Torques

In order to fully understand the theory of operation of a multi-stator watthour meter, it is necessary to analyze the driving and damping torques. Considering the single-stator, two-wire meter, the driving torque is directly proportional to voltage and load current and the cosine of the angle between them, except for a slight non-linearity due to the magnetizing characteristics of the steel. Damping torque should also theoretically be proportional to load, but this is not strictly true. The over-all damping flux has three separate components; namely, permanent-magnet damping, voltage-flux damping, and current-flux damping. The former is relatively constant and provides damping torque directly proportional to speed. Voltage-flux damping is also relatively constant since line voltage variations are normally small. Current-flux damping, however, varies with the square of the current. It therefore causes a definite divergence between torque and speed curves as the current load increases. This characteristic can be influenced in design by changing the ratio of the current or variable flux to the constant flux produced by the voltage coil and the permanent magnet. It cannot be eliminated.

In a typical modern single-stator meter at rated test-amperes and normal voltage, the damping torque from the permanent magnet is 96.7 percent of the total. The voltage flux furnishes 2.8 percent and the remaining 0.5 percent comes from the current flux. Since the current-damping component increases with the square of the current, the speed curve will obviously be below the torque curve as the load current increases. An uncompensated meter having 0.5 percent current damping at rated test amperes would be 2 percent slower at 200 percent test-amperes.

These underlying principles also apply to polyphase meters. However the combining of two or more stators to drive a common moving element introduces factors in performance that are not apparent from the performance of independent single-stator meters.

Individual-Stator Performance

Consider the simplest form of the polyphase meter—two stators driving a common moving element. The meter will perform as a single-phase meter if both elements are connected together on single-phase, that is, with the current coils connected in series and the voltage coils in parallel. With this connection, it is to all intents and purposes a single-phase meter and has all the characteristics of a single-phase meter. The fact that the two stators are coupled together on a single shaft makes no difference except to average the characteristics of the individual stators.

The same will be true, with the exception of interference errors, when the meter is connected in a polyphase circuit where the load is completely balanced. When the loads are not balanced, the polyphase meter no longer performs as a single-phase meter. If only one stator is loaded, the polyphase meter will tend to register fast. This is best illustrated by putting balanced loads on the two stators and checking the calibration. Then remove the load from one stator. The meter will be found to register fast all the way along the load curve as compared to the speed curve on combined stators. This is due to the variation of the over-all damping torque caused by the change in the current damping component.

Current Damping

Assume that a polyphase meter when operating at rated test-amperes has its damping components in the following typical relationships: 96.7 percent from a permanent magnet, 2.8 percent from voltage at rated voltage, and 0.5 percent from current at rated test-amperes.

When the load is taken off one stator, the driving torque drops 50 percent, but the total damping drops more than 50 percent since the current damping on the unloaded stator is eliminated, whereas it was a part of the total damping torque when both stators were loaded.

This characteristic changes with increasing load. For example, in a two-stator meter that runs 0.4 percent fast on a single stator at rated test-amperes as compared to the registration with balanced load on both stators at rated test-amperes, the difference in registration between single and combined stators may be as much as 3 percent at 300 percent rated test-amperes. This can be readily seen from the fact that the current-damping component from the stator that is not loaded would be nine times as great at 300 percent rated test-amperes as at 100 percent rated test-amperes. Therefore its elimination at 300 percent rated test-amperes takes away nine times 0.4 or about 3.6 percent from the total damping.

Unbalanced Loads

Due to the effect of current damping, a factor exists in the performance of polyphase meters that is not generally recognized. If the stators are loaded unequally, the registration will differ from that when the load is equally balanced or when the total load is carried by one stator, since the overload compensation is such as to cause the torque curve to go up in direct proportion to the increase in current damping on balanced loads. The departure from balanced-load performance, particularly on heavy loads, will be in proportion to the amount of overload compensation in the meter.

This can be understood if we take, for example, a 5-ampere, two-stator, polyphase meter and apply 5 amperes to each stator. The total current damping will be proportional to (5 amperes)^2 plus (5 amperes)^2 or 50 current-damping units. On the other hand, if 10 amperes is applied to one stator only, current damping will then be proportional to (10 amperes)^2 or 100 current-damping units. This is the same total energy and total flux, but they are divided differently in the stators and consequently produce different current damping. The current-squared damping law applies only to current flux produced in a single stator. It does not apply to the total currents in separate stators. In this case, the current damping of the single stator with 10 amperes is twice that of two elements with 5 amperes each.

Suppose the load is unbalanced so that there are 8 amperes on one stator and 2 amperes on the other. Then the total current damping is proportional to 8^2 plus 2^2 or 68 current-damping units as compared to 50 units with 5 amperes on each stator.

Interference Between Stators

Polyphase meters must have a high degree of independence between the stators. Lack of this independence is commonly known as "interference" and can be responsible for large errors in the various measurements of polyphase power.

Major interference errors are due to the mutual reaction in a meter disk between the eddy currents caused by current or voltage fluxes of one stator and any inter-linking fluxes that may be due to currents or voltages associated with one or more other stators. Specifically, these mutual reactions fall into three groups: voltage-voltage, current-current, and current-voltage or voltage-current. The following three paragraphs explain these reactions.

Voltage-Voltage Interference

The first, or "voltage-voltage," reaction in the disk is that due to the interlinkage of flux set up by the voltage coil of one stator with eddy currents caused by flux from the voltage coil of another stator. The magnitude of the so-called "interference torque" resulting from this reaction depends upon the relative position of the two voltage coils with respect to the center of the disk (for coils displaced exactly 180 degrees this torque is zero) and upon the phase angle between the two voltage fluxes. This torque could be very high unless these factors are thoroughly considered in proper design.

Current-Current Interference

The second, or "current-current," reaction is due to interlinkage of flux set up by the current coil of one stator with eddy currents caused by flux from the current coil of another stator. The magnitude of this second reaction again depends upon the relative position of the two stators (zero if at 180 degrees) and upon the magnitude of, and the phase angle between, the two current fluxes.

Current-Voltage or Voltage Current Interference

The third interference reaction, which may be described as "current-voltage" or "voltage-current," is due to the interlinkage of flux set up by the voltage or current circuits of one stator and the eddy currents caused respectively by the current or voltage fluxes from another stator. As before, the magnitude of this third type of reaction depends upon the relative geometrical position of the stators and the power factor of the circuit. The effect on the registration is a constant which is independent of the current load on the meter.

Interference Tests

Comparative tests to evaluate interference effects in a watthour meter have been established under ANSI-C12.

Interference tests are not part of the usual meter calibration procedure and, hence, are not performed in the meter shop. Since such tests are made to evaluate the manufacturer's design of a particular polyphase meter type they are usually performed in the meter laboratory.

The specialized interference tests require the use of a two-phase power source with two-stator meters and a three-phase source with three-stator meters. The test results do not give the specific interference errors which will be obtained in actual service, but if the results are within the established tolerances, assurance is obtained that the interference effects will not be excessive. Complete details of the tests may be obtained by reference to the previously mentioned national standards.

Design Considerations to Reduce Interference

Interference in a single-disk meter is reduced by proper design that includes control of the shape of the eddy current paths in the disk and the most favorable relative positions of the coils and stators. One of the common methods of reducing interference has been mentioned previously; namely, positioning two stators symmetrically about the disk shaft exactly 180 degrees from each other. This eliminates two of the three possible forms of interference.

Another method of reducing all types of interference is to laminate the disk. A number of separate laminations are used. Each lamination is slotted radially to form several sectors and the laminations are insulated electrically from each other. Because of the radial slots, the eddy currents in the disk are confined to the area around the stator which causes them and they cannot flow to a portion of the disk where they could react with fluxes from another stator to create interference torques. The lamination slots are usually staggered during manufacturing to provide sufficient mechanical strength for the disk and smoother driving torque during each disk revolution.

A third common method of reducing interference is to provide magnetic shielding around the voltage or current coil of each stator to keep the spread of flux to a minimum. Combinations of the preceding methods are also employed.

Meters with stators operating on separate disks or completely separate rotors are inherently free from the various effects listed before. However, proper design and spacing is still required to prevent voltage or current flux from one electromagnet reacting with eddy currents produced by flux from a second electromagnet.

MULTI-STATOR METER ADJUSTMENTS

The following is a description of the calibrating adjustments found in multi-stator meters. Details on the procedure involved in using these adjustments may be found in Chapter 14, "Watthour Meter Testing and Maintenance."

Most polyphase meters now in service contain two or three separate stators so mounted that their combined torque turns a single rotor shaft. As in single-phase meters, the adjustments provided for polyphase meters are the usual full-load, power factor, and light-load adjustments. In addition to these adjustments, polyphase meters have a fourth adjustment called "torque balance" designed to allow equalization of individual stator torques with equal applied wattage for accurate registration. Obviously, there is no requirement for torque balance in a single-phase meter.

Each stator in a multi-stator meter may contain a light-load adjustment, or a single light-load adjustment of sufficient range may be provided on one stator. All stators must contain power-factor compensation so that the phase relationships are correct in each stator. The power-factor compensation is adjustable on most meters, but some manufacturers make a fixed power-factor compensation at the factory which is not readily changed in the field. Only one full-load adjustment is provided on most modern polyphase meters, even though more than one braking magnet may be used. Torque-balance adjustments may be provided on all stators, or on only one stator in two-stator meters, or on two stators of a three-stator meter. In all cases, it is possible to equalize stator torques.

Torque-Balance Adjustment

For correct registration, the torque produced by each stator in a multi-stator meter must be the same when equal wattage is applied. A two-stator meter with one stator 5 percent fast and the other stator 5 percent slow would show good performance with both stators connected in series-parallel for calibration test on single-phase loading. However, if this meter were used to measure polyphase loads involving either low power factor or unbalance, the registration would be in serious error. To correct for this, each stator should be calibrated and adjusted separately to insure that each produces the same driving torque. The full-load adjustment cannot be used because it has an equal effect on the performance of all stators, so the torque-balance adjustment is provided for independently adjusting the torque of each stator.

Since the torque developed by a single stator is dependent upon the amount of flux (produced by the electromagnet) which passes through the disk, it follows that the torque for a given load can be varied by any method that will change the flux through the disk. A convenient way to effect this change is by providing a magnetic shunt in the air gap of the voltage-coil poles in the electromagnet. Moving this shunt into or out of the air gap bypasses a greater or a lesser portion of the voltage flux from the disk. This changes the disk-driving torque through a narrow range. The adjustment obtainable in this way is sufficient to allow the torques of the individual stators in a polyphase meter to be made equal to each other.

Two methods in general use for torque balancing are shown in Fig. 7-19 (a and b). The method shown in (a) uses two steel screws which can be turned into or out of the gaps situated in the voltage-coil iron just below the coil windings. The method shown in (b) uses a U-shaped soft iron wire which is inserted in the air gaps. This wire is attached to a yoke carried on threaded studs, which permits the magnetic shunt to be moved in and out of the air gap. After these adjustments have been set so that the torques of all stators are alike, the other meter adjustments can be made in much the same way as on single-phase meters.

Interdependence of Adjustments

Another characteristic of the polyphase meter is that any change in a full-load or light-load adjustment affects all stators alike. Of course, this does not apply to the power-factor or torque-balance adjustments.

The torques of the stators can be balanced at any unity-power-factor load value, but it is customary to make the balance adjustment at rated test-ampere load. The balance of the individual stator torques at other unity-power-factor load points will depend on how well the stator characteristics are matched. Any

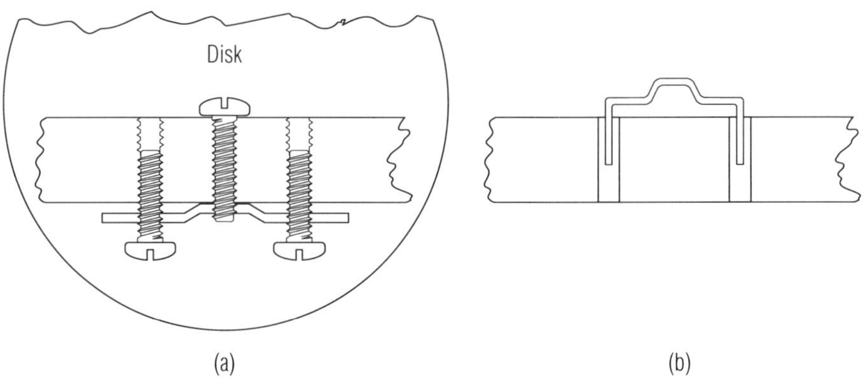

Figure 7-19. Methods of Shunting Voltage-Coil Air Gap for Torque-Balancing Adjustment in Multi-Stator Meters.

divergence that may exist cannot be corrected or minimized by the light-load adjustment or otherwise, except by attempting to select stators of the same characteristics. This is neither practical nor important.

In calibrating a polyphase meter at light load, it is proper to excite all voltage circuits at rated voltage. Under such condition, the overall accuracy is the same regardless of whether a single light-load adjuster is used for the complete calibration or whether, where more than one adjuster is provided, each is moved a corresponding amount. However, in the latter case, when a considerable amount of adjustment is necessary, it is the usual practice to move the adjusters of all stators about the same amount in order to assure sufficient range of adjustment and to avoid changes in torque balance at the 50 percent power factor test load.

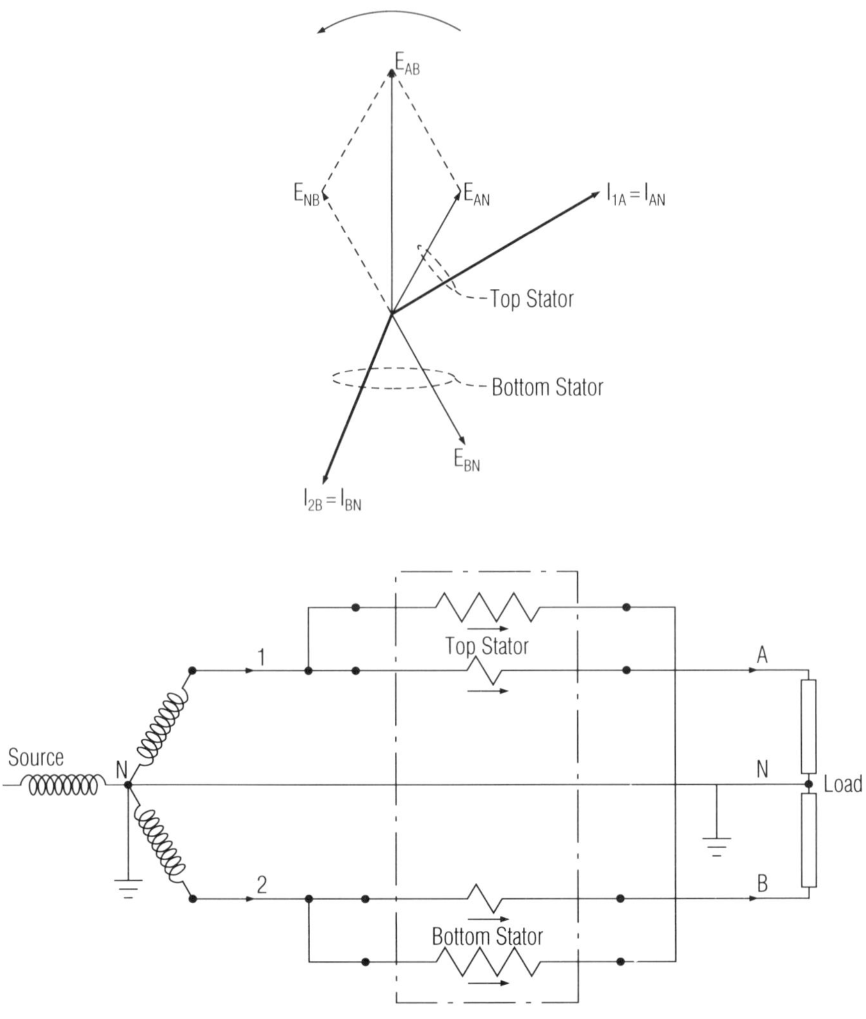

Figure 7-20. Two-Stator Meter on Three-Wire Network Service.

THREE-WIRE NETWORK SERVICE

Two-Stator Meter

Three-wire network service is obtained from two of the phase wires and neutral of a three-phase, four-wire wye system, as shown in Fig. 7-20. It is, in reality, two two-wire, single-phase circuits with a common return circuit and it has voltages which have a phase difference of 120 electrical degrees between them. The voltage is commonly 120/208 volts.

The normal method of metering a network service is with a two-stator meter connected as shown in Fig. 7-20. With this connection, which follows Blondel's theorem, each stator sees the voltage of one phase of the load. The phasors representing the load phase currents, I_{AN} and I_{BN}, are shown in the diagram lagging their respective phase voltages. The meter current coils carry the line currents, I_{1A} and I_{2B}, and inspection of the circuit shows that these currents are identical to the load phase currents. Hence, the meter correctly measures the total load power. Any loads connected line to line, between A and B in Fig. 7-20, will also be metered properly. With this type of meter, there are no metering errors with unbalanced load voltages or varying load currents and power factors.

Single-stator meters for measuring network loads have been developed and may be used with reasonable accuracy under particular load conditions. These meters are described under "Special Meters" in this chapter.

The conventional three-wire, single-stator, single-phase meter cannot be used for network metering. It will, of course, measure the 208-volt load correctly; but the two 120-volt loads are metered at 104 volts rather than at 120 volts and at a phase angle which is 30 degrees different from the actual. Therefore, for 120-volt balanced loads, meter registration will be close to 75 percent of the true value; but with unbalanced loads, the resulting meter error is a variable thereby rendering such metering useless.

THREE-WIRE, THREE-PHASE DELTA SERVICE

Two-Stator Meter

The three-wire, three-phase delta service is usually metered with a two-stator meter, in accordance with Blondel's theorem. The meter which is used has internal components identical to those used in network meters, but may differ slightly in base construction. Typical meter connections are shown in Fig. 7-21. In the top stator of the meter, the current coil carries the current in line 1A and the voltage coil has load voltage AB impressed on it. The bottom stator current coil carries line current 3C and its corresponding voltage coil has load voltage CB impressed. Line 2B is used as the common line for the voltage-coil connections.

The phasor diagram of Fig. 7-21 is drawn for balanced load conditions. The phasors representing the load phase currents I_{AB}, I_{BC}, and I_{CA} are shown in the diagram lagging their respective phase voltages by a small angle θ. By definition, this is the load power-factor angle. The meter current coils have line currents flowing through them, as previously stated, which differ from the phase currents. To determine line currents, Kirchhoff's current law is used at junction points A and C in the circuit diagram. Applying the law, the following two equations are obtained for the required line currents:

$$\dot{I}_{1A} = \dot{I}_{AB} - \dot{I}_{CA}$$
$$\dot{I}_{3C} = \dot{I}_{CA} - \dot{I}_{BC}$$

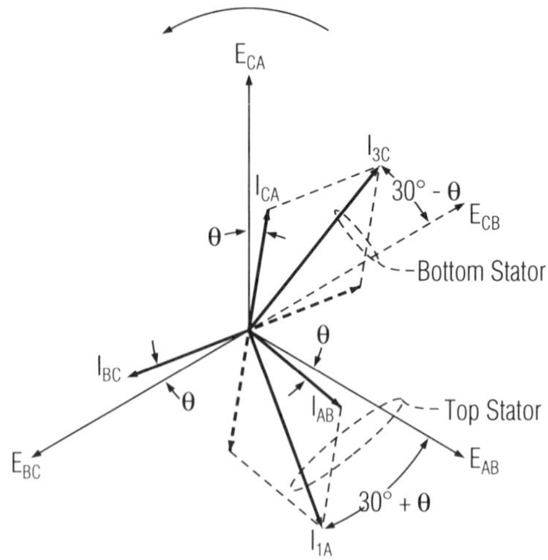

$$W = E_{AB} I_{1A} \cos(30° + \theta) + E_{CB} I_{3C} \cos(30° - \theta)$$

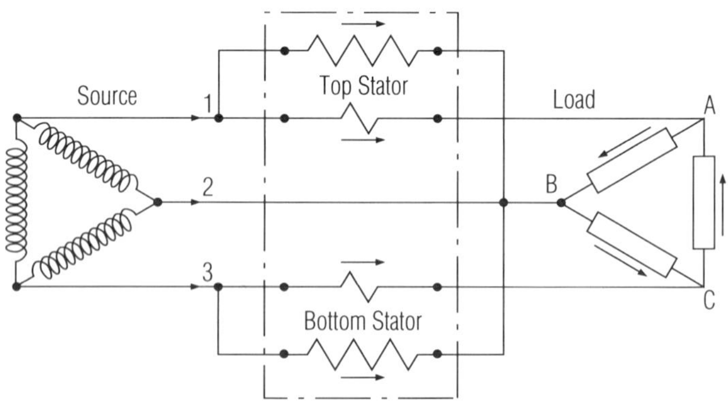

Figure 7-21. Two-Stator Meter on Three-Phase, Three-Wire Delta Service.

The operations indicated in these equations have been performed in the phasor diagrams to obtain I_{1A} and I_{3C} Examination of the phasor diagram shows that for balanced loads the magnitude of the line currents is equal to the magnitude of the phase currents times $\sqrt{3}$.

The top stator in Fig. 7-21 has voltage E_{AB} impressed and carries current I_{1A}. These two quantities have been circled in the phasor diagram and inspection of the diagram shows that for the general case the angle between them is equal to 30 degrees + θ. Therefore, the power measured by the top stator is $E_{AB}I_{1A}\cos(30° + \theta)$ for any balanced-load power factor. Similarly, the bottom stator uses voltage E_{CB} and current I_{3C}. These phasors have also been circled on the diagram and in this case the angle between them is 30 degrees – θ. The bottom stator power is then

$E_{CB}I_{3C}\cos(30° - \theta)$ for balanced loads. The sum of these two expressions is the total metered power.

Examination of the two expressions for power shows that even with unity-power-factor load the meter currents are not in phase with their respective voltages in the individual stators. With balanced-unity-power factor load, the current lags by 30 degrees in the top stator and leads by 30 degrees in the bottom stator. However, this is correct metering. To illustrate this more cleanly, consider an actual load of 15 amperes at unity power factor in each phase with a 240-volt delta supply. The total power in this load is:

$$3 \times E_{Phase} \times I_{Phase} \times \cos\theta = 3 \times 240 \times 15 \times 1 = 10{,}800 \text{ watts}$$

Each stator of the meter measures:

$$\text{Top Stator} = E_{AB}I_{1A}\cos(30° + \theta)$$

$$\text{Since } I_{1A} = \sqrt{3}\, I_{Phase}$$

$$\begin{aligned}\text{Top Stator} &= 240 \times \sqrt{3} \times 15 \times \cos(30° + 0°)\\ &= 240 \times \sqrt{3} \times 15 \times 0.866 = 5{,}400 \text{ watts}\end{aligned}$$

$$\text{Bottom Stator} = E_{CB}I_{3C}\cos(30° - \theta)$$

$$\text{Since } I_{3C} = \sqrt{3}\, I_{Phase}$$

$$\begin{aligned}\text{Bottom Stator} &= 240 \times \sqrt{3} \times 15 \times \cos(30° - 0°)\\ &= 240 \times \sqrt{3} \times 15 \times 0.866 = 5{,}400 \text{ watts}\end{aligned}$$

Total Meter Power = Top Stator + Bottom Stator = 5,400 + 5,400 = 10,800 watts = Total Load Power.

When the balanced-load power factor becomes lagging, the phase angles in the meter vary in accordance with the 30 degrees plus-or-minus-θ expressions. When the load power factor reaches 50 percent, the magnitude of θ is 60 degrees. The top stator phase angle becomes 30 degrees + θ = 90 degrees and, since the cosine of 90 degrees is zero, the torque from this stator becomes zero at this load power factor. To illustrate this with an example, assume the same load current and voltage used in the preceding example, but with 50 percent load power factor.

$$\text{Total Load Power} = 3 \times 240 \times 15 \times 0.5 = 5{,}400 \text{ watts}$$

$$\begin{aligned}\text{Top Stator} &= 240 \times \sqrt{3} \times 15 \times \cos(30° + 60°)\\ &= 240 \times \sqrt{3} \times 15 \times 0 = 0 \text{ watts}\end{aligned}$$

$$\begin{aligned}\text{Bottom Stator} &= 240 \times \sqrt{3} \times 15 \times \cos(30° - 60°)\\ &= 240 \times \sqrt{3} \times 15 \times 0.866 = 5{,}400 \text{ watts}\end{aligned}$$

Total Meter Power = 0 + 5,400 = 5,400 watts = Total Load Power.

With lagging load power factors below 50 percent, the top stator torque reverses direction and the resultant action of the two stators becomes a differential one, such that the disk rotation is that of the stronger stator. Since the bottom stator torque is always larger than that of the top stator, the meter disk continues to rotate in the forward direction but with proportionately slower speeds at power factors under 50 percent. Actually, on a balanced load, the two stators operate over the following ranges of power-factor angle when the system power factor varies from unity to zero: the leading stator from 30 degrees lead to 60 degrees lag; the lagging stator from 30 degrees lag to 120 degrees lag.

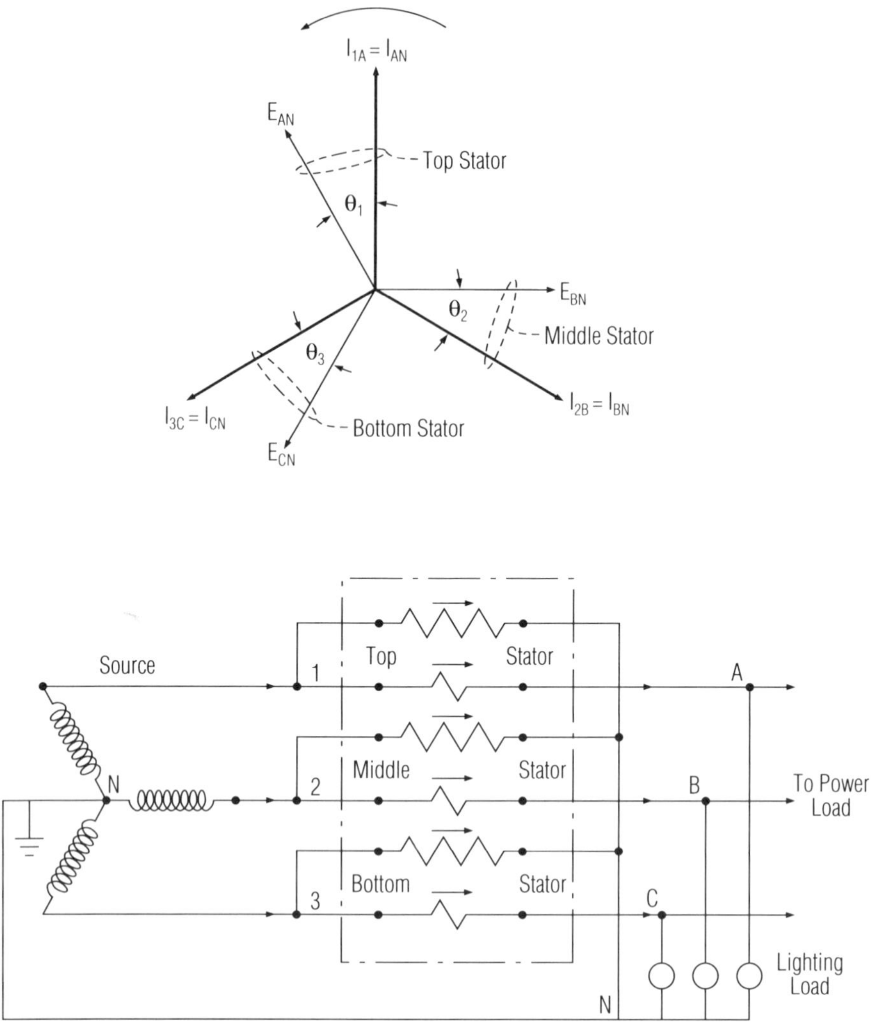

Figure 7-22. Three-Stator Meter on a Three-Phase, Four-Wire Wye Service.

FOUR-WIRE, THREE-PHASE WYE SERVICE

Three-Stator Meter

Figure 7-22 shows the usual meter connections for a three-stator meter on a four-wire wye service in accordance with Blondel's theorem. The neutral conductor is used for the common meter voltage connection.

The connection diagram shows three line-to-neutral loads and the phasor diagram shows the metering quantities. The diagram is drawn for balanced line-to-neutral loads, which have a lagging power-factor angle.

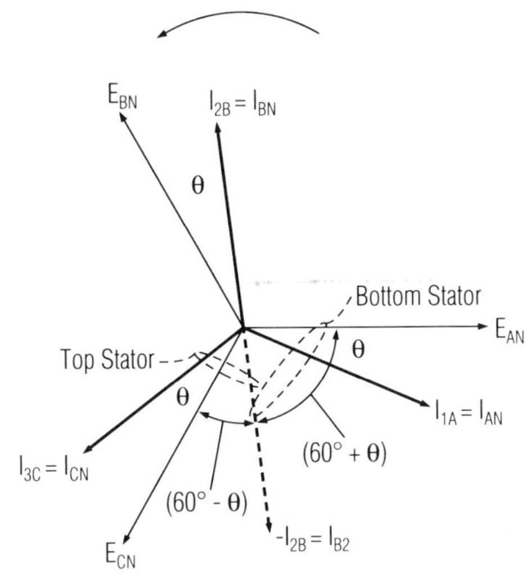

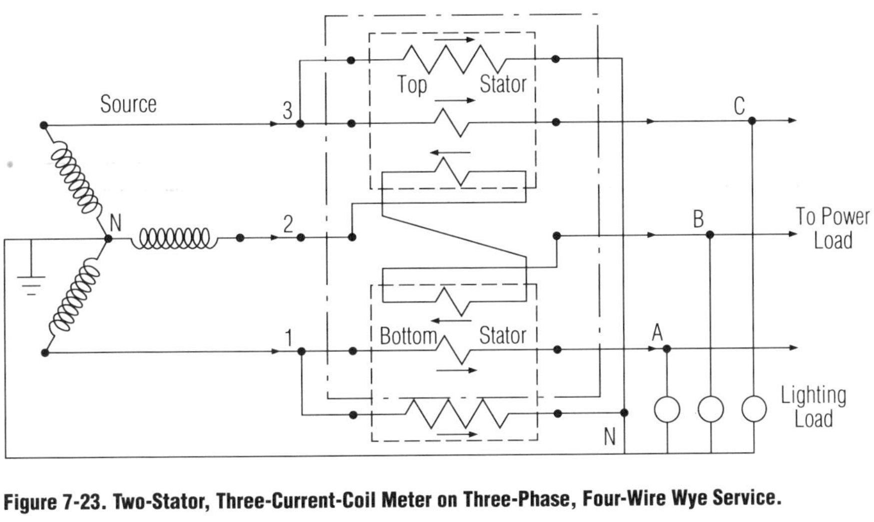

Figure 7-23. Two-Stator, Three-Current-Coil Meter on Three-Phase, Four-Wire Wye Service.

The expression for total meter power can be written as follows by inspection of the phasor diagram:

$$\text{Total Meter Power} = E_{AN}I_{AN}\cos\theta_1 + E_{BN}I_{BN}\cos\theta_2 + E_{CN}I_{CN}\cos\theta_3$$

which is the total power developed by the load.

If the loads are connected line-to-line, instead of line-to-neutral, the total load power will still be the same as the total meter power because it can be proven that any delta-connected load may be replaced by an equivalent wye-connected load.

Hence, there are no metering errors with unbalanced load voltages of varying load currents or power factors.

Two-Stator, Three-Current-Coil Meter

This meter employs two stators with two voltage coils and three current circuits. Two of the current circuits consist of separate current coils with the normal number of turns on each of the two stators. In addition, on each stator there is a second current coil with an equal number of turns. These two coils are connected in series internally to form the third meter current circuit which is commonly called the Z-coil circuit. Because this meter has three current circuits and only two voltage circuits, it is often called a 2 ½-stator meter, although this is technically incorrect.

Circuit connections are shown in Fig. 7-23 for the two-stator, three-current-coil meter on a four-wire wye service. Note that this connection does not fulfill Blondel's theorem, resulting in possible metering errors which will be discussed in following paragraphs. The two voltage coils are connected to measure the line-to-neutral voltages of the lines in which the single current coils are connected. The Z-coil is connected in the line which does not have a voltage coil associated with it. For correct metering the internal meter connections of this coil are reversed so that reverse direction is obtained in each stator as shown in the circuit diagram.

The Z-coil current reacts within the meter with the other two line-to-neutral voltages since it flows through both stators. This reaction is equivalent to the current acting with the phasor sum of the two line-to-neutral voltages. The assumption is made in this meter that the phasor sum of the two line-to-neutral voltages is exactly equal and opposite in phase to the third line-to-neutral voltage. When this condition exists, the metering is correct regardless of current or power-factor unbalance. The assumption, however, is correct only if the phase voltages are balanced. If the voltages are not balanced, metering errors are present, the magnitude of which depend upon the degree of voltage unbalance.

The phasor diagram of Fig. 7-23 shows the metering quantities. The diagram is drawn for balanced line-to-neutral loads which have a lagging power-factor angle θ. The Z-coil carries line current I_{2B}, but because its connections are reversed within the meter, the phasor current which reacts with the meter voltages is actually $-I_{2B}$, or I_{B2}. Thus, as shown in the phasor diagram, in the top stator, I_{B2} and I_{3C} react with E_{CN} while in the bottom stator, I_{B2} and I_{1A} react with E_{AN}. In both cases the angles between the voltage and current phasors are less than 90 degrees, so forward meter torque is developed. For the general case with a balanced load, the angle between I_{B2} and E_{AN} is 60 degrees + θ while the angle between I_{B2} and E_{CN} is 60 degrees – θ.

The expression for total meter power can be written as follows by inspection of the phasor diagram:

$$\text{Total Meter Power} = E_{AN}I_{1A}\cos\theta + E_{CN}I_{3C}\cos\theta + E_{AN}I_{B2}\cos(60° + \theta) + E_{CN}I_{B2}\cos(60° - \theta)$$

As an example, with a balanced load from each line to neutral of 15 amperes at 120 volts with a lagging power factor of 86.6 percent the true load power is:

$$\text{True Power} = 3\ E_{Phase}I_{Phase}\cos\theta = 3 \times 120 \times 15 \times 0.866 = 4676.4 \text{ watts}$$

For 86.6 percent power factor the phase angle θ is 30 degrees.

Metered Power = $120 \times 15 \times 0.866 + 120 \times 15 \times 0.866 + 120 \times 15 \times \cos(60° + 30°) + 120 \times 15\cos(60° - 30°)$

$$= 2 \times 120 \times 15 \times 0.866 + 120 \times 15\cos 90° + 120 \times 15\cos 30°$$
$$= 2 \times 120 \times 15 \times 0.866 + 120 \times 15 \times 0 + 120 \times 15 \times 0.866$$
$$= 3 \times 120 \times 15 \times 0.866$$
$$= 4676.4 \text{ watts}$$

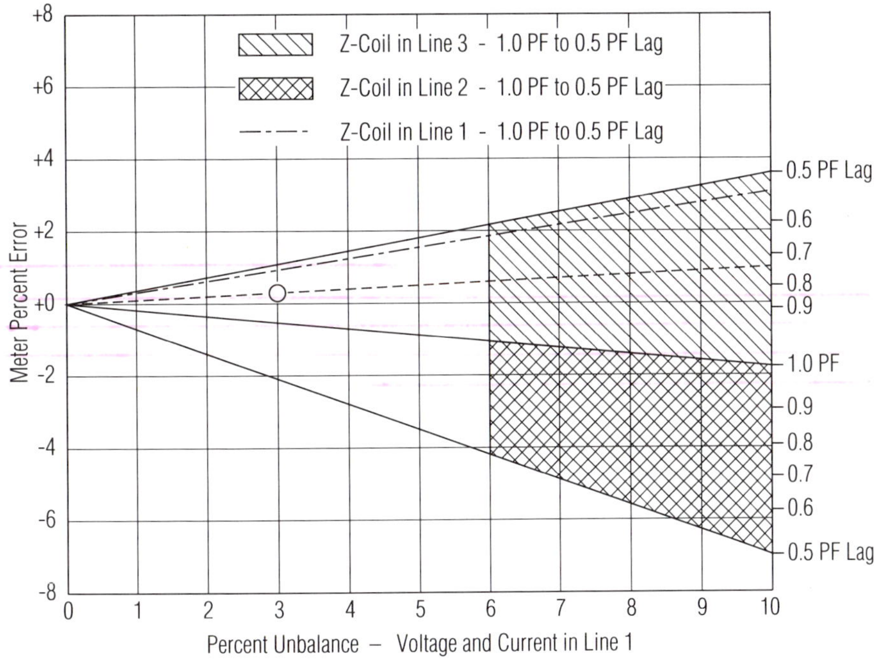

To use error chart, place a straight edge through the zero point on the Meter Percent Error scale and through the desired PF in the selected Z-coil location. Then select Percent Unbalance, and read percent error on the Meter Percent Error scale. For example, with the Z-coil in Line 3 and a 0.75 PF lag, a 3 percent voltage and current unbalance in Line 1 will result in a 0.35% error. See point O.

$$\text{Percent Unbalance} = \frac{\text{Maximum Deviation from Average}}{\text{Average}} \times 100$$

Figure 7-24. Error Curves for Equal Voltage and Current Unbalance in One Phase and for Three Possible Locations of Z Coil.

A similar proof of correct metering may be developed for a polyphase power load connected to lines A, B, and C.

As previously stated, a 2 1/2-stator meter is in error when the voltages are not balanced, either in magnitude or phase position. With unbalanced voltages, the amounts of any current unbalance and power-factor values also have a bearing on the amount of metering error as well as where the voltage unbalance occurs relative to the connection of the Z-coil. The curves of Fig. 7-24 are drawn for an assumed equal voltage and current unbalance in one load phase and for the three possible locations of the Z-coil. Using the curves for an assumed voltage and current unbalance of 2 percent in Phase 1, the following tabulation shows the variations in metering errors as the Z-coil is moved.

Meter Percent Error

Z-Coil in Line:	P.F. = 1.00	P.F. = 0.866 Lag
1	+ 0.66	+ 0.66
2	- 0.34	- 0.64
3	- 0.34	0.0

It must be remembered, however, that if the voltages remain balanced, the $2\,^1/_2$-stator meter will meter correctly with current and power factor unbalance.

While this method of metering does not follow Blondel's theorem and is not as accurate in cases of unbalanced voltages as the three-stator meter, it is acceptable for energy measurements by many users.

Two-Stator Meter Used with Three Current Transformers

This method of metering a four-wire wye service uses a conventional two-stator, two-current-coil meter with three current transformers in the circuit. The circuit connections are shown in Fig. 7-25.

The component currents in each current coil are indicated by the arrows on the circuit diagram of Fig. 7-25. Note that in both stators the third line transformer current, I_{2B}, is in opposition to the other line transformer current. The phasor diagram shows how these components add to produce the total current in each stator winding, I_X and I_Y.

Before the component currents are added, the phasor diagram for this connection is similar to that shown in Fig. 7-23 for the $2\,^1/_2$-stator meter, showing that this method is electrically equivalent to the $2\,^1/_2$-stator meter. The difference is that the currents are combined in the single-current-coil winding rather than their effects being combined in the form of flux in the core iron of the current coils. This method, naturally, also has the same accuracy limitations and errors as the $2\,^1/_2$-stator meter.

FOUR-WIRE, THREE-PHASE DELTA SERVICE

Three-Stator Meter

The four-wire delta service is used to supply both power and lighting loads from a delta source. The lighting supply is obtained by taking a fourth line from the center-tap of one of the transformers in the delta source. Correct metering according to Blondel's theorem requires a three-stator meter. Figure 7-26 shows the circuit and metering connections. Since the mid-tap neutral wire is usually grounded, this is the line used for the common meter voltage connection. The diagram shows the nominal voltage impressed on each voltage coil in the meter for a 240-volt delta source. The top stator voltage coil is commonly rated at 240 volts, although in service it operates at 208 volts. In order that the torque of each stator be equal for the same measured watts, it is necessary that the calibrating watts, or test constant, be the same for all stators. This necessitates that the current coils in the 120-volt stators have double the rating of the current coil in the 240-volt stator.

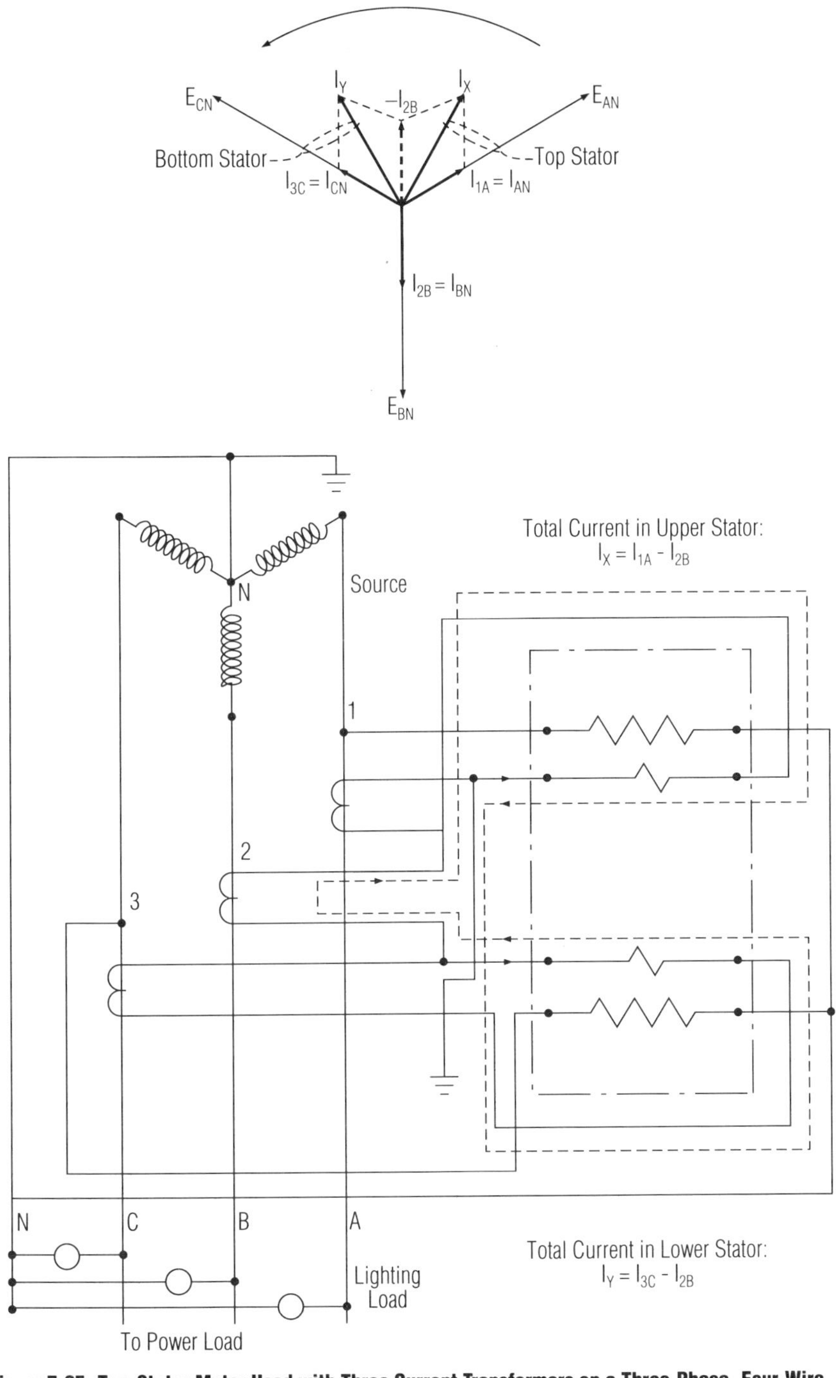

Figure 7-25. Two-Stator Meter Used with Three Current Transformers on a Three-Phase, Four-Wire Wye Service.

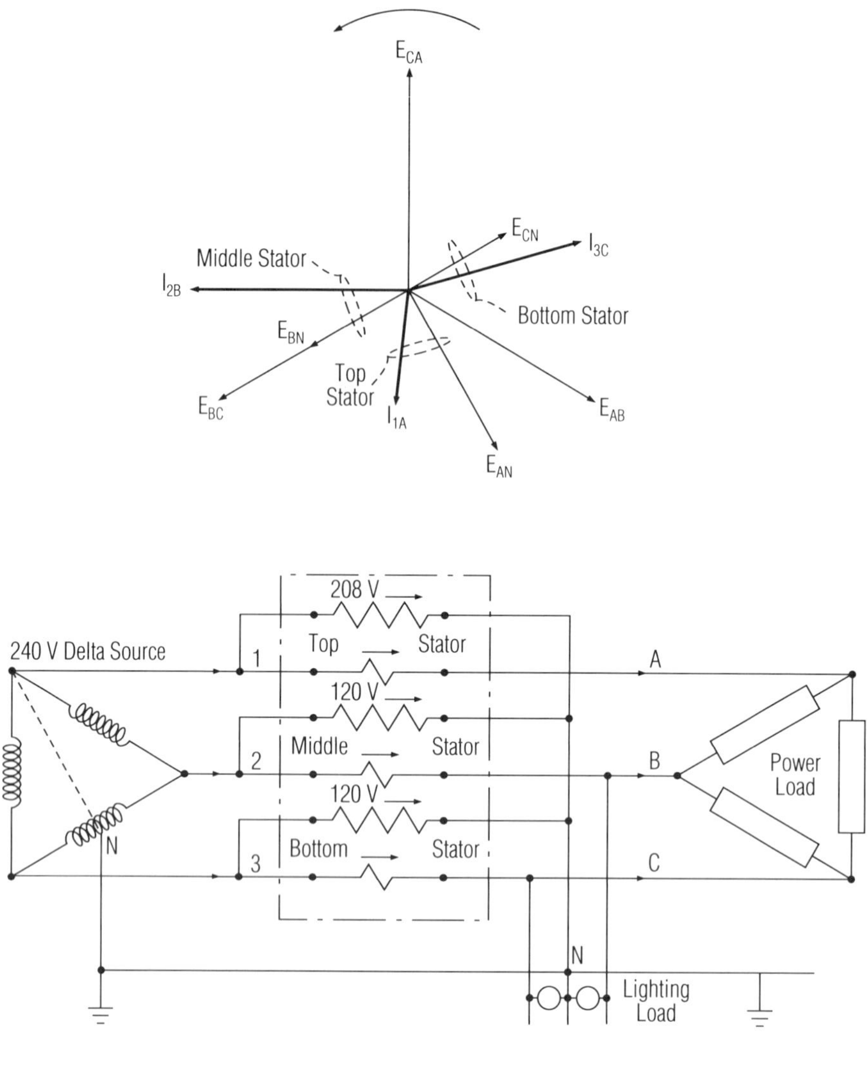

Figure 7-26. Three-Stator Meter on a Three-Phase, Four-Wire Delta Service.

The phasor diagram of Fig. 7-26 is drawn for a combined power and lighting load. To simplify the diagram, the current phasors for the individual loads are not shown. Since this method follows Blondel's theorem, it provides correct metering under any condition of voltage or load unbalance.

The advantages of using a three-stator meter for this application are (1) correct registration under all conditions of voltage, and (2) increased meter capacity in lighting phases, an important advantage when the load in these phases greatly exceeds that in the power phase. A disadvantage is that a special test procedure is required.

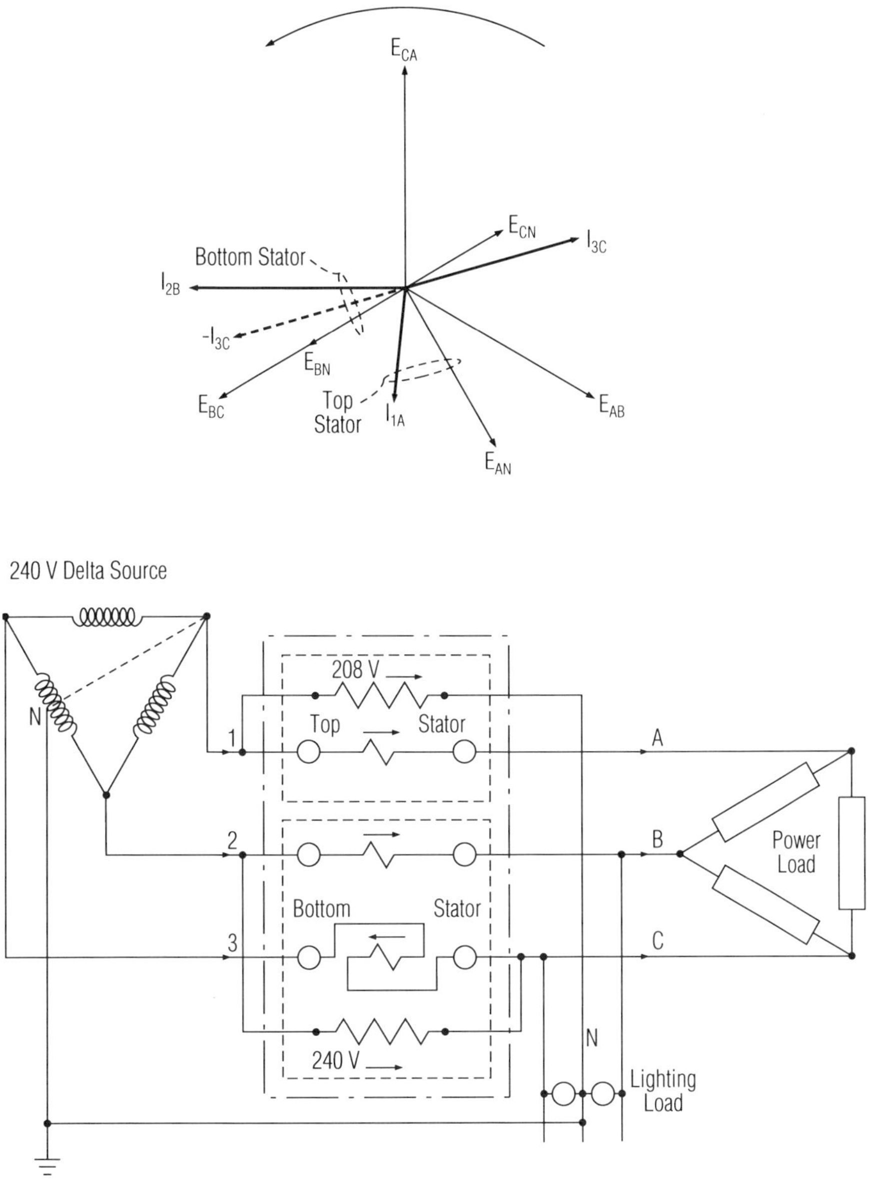

Figure 7-27. Two-Stator, Three-Current-Coil Meter on a Four-Wire Delta Service.

Two-Stator, Three-Current-Coil Meter

It is possible, as with the four-wire wye service, to meter a four-wire delta service with a two-stator, three-current-coil meter. However, it is a compromise with Blondel's theorem which again allows possible metering errors.

The meter connections are shown in Fig. 7-27. The bottom stator, which has two independent current coils, has its voltage coil connected across the two lines which supply the lighting load. The top stator is connected the same as in the three-stator meter of Fig. 7-26. It is obvious that both voltage coils now have the same voltage rating.

The phasor diagram of Fig. 7-27 is drawn for the identical load conditions used in Fig. 7-26 for the three-stator delta meter. However, the metering conditions here are different. The bottom stator now uses voltage E_{BC}. If the phasor current in the bottom coil, I_{3C}, were to act with this voltage, the developed torque would be backward since the angle between them is greater than 90 degrees. Therefore, to develop torque in the forward direction, the bottom current coil connections are reversed internally within the meter and the current phasor which produces torque is $-I_{3C}$ as shown on the phasor diagram. The bottom stator operates on the same principle previously described for the three-wire, single-phase meter with one current coil reversed.

Comparing this meter with the three-stator delta meter of Fig. 7-26, it can be seen that the middle and bottom current coils are now acting with a voltage of twice the rating that they do in the three-stator meter. To offset the voltage flux which is doubled, the current flux must be cut in half. Since the current is not changed, it is necessary to reduce the turns of each of these two coils by half. Again, the same principle applies as in the three-wire, single-phase meter.

One advantage of this meter over the three-stator meter in this application is that the same test procedure can be used as for the three-wire, three-phase meter. One disadvantage is the possible error under certain conditions of voltage and current in the three-wire lighting circuit. On small loads this may be of little importance as it exists wherever a standard single-stator, three-wire meter is used.

SPECIAL METERS

Single-Stator Meters for Three-Wire Network Service

Single-stator meters have been developed for use on three-wire network services. These meters do not conform to Blondel's theorem and are, therefore, subject to metering errors under certain conditions noted in the following paragraphs.

The schematic connections for the two types of meters now in use are shown in Fig. 7-28. Each meter has one voltage coil and two current coils. In one case, Fig. 7-28(a), the meter is designed to use line-to-line voltage (208 V) on the voltage coil and the other, Fig. 7-28(b), uses one line-to-neutral voltage (120 V) on its voltage coil. Obviously, any unbalance in line-to-neutral voltages will cause metering errors and, where unbalanced voltages exist, a two-stator meter should be used for accurate results.

The currents in the meter current coils are shifted in phase to provide correct metering. This is accomplished by impedance networks of resistors and inductors along with the current coils to split the total line currents and shift the phase position of the meter-current-coil current the desired amount. The number of turns and the impedance of the current coil may also be varied in design to obtain a usable meter. The current-impedance networks are shown in Fig. 7-28.

Phase sequence of voltages applied to these meters is extremely important since such meters can be designed to provide the correct phase shift of meter-coil current for only one phase sequence. Hence, if they are installed on the wrong

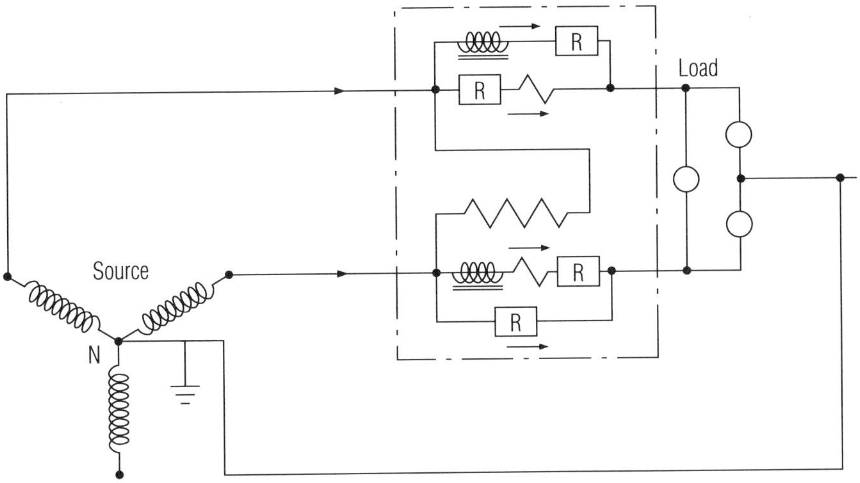

(a)

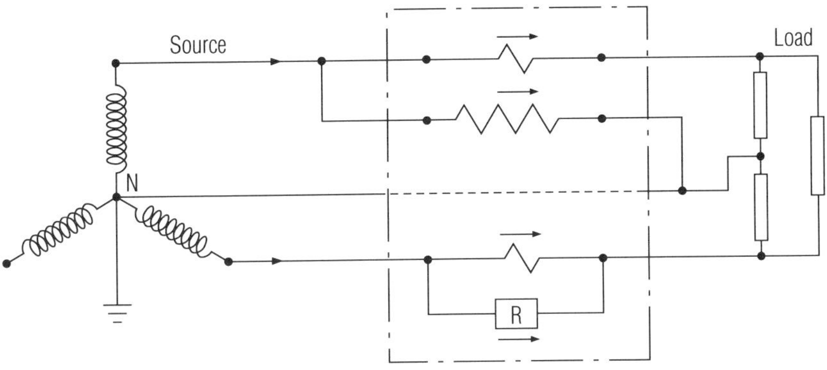

(b)

Figure 7-28. Single-Stator, Three-Wire Network Meters.

phase sequence their energy registration is useless. All meters of this type have a built-in phase-sequence indicator.

Universal Multi-Stator Meters

Universal multi-stator metering units permit the measurement of any conventional single-phase or polyphase service with a single type of meter. The meter is generally of the Class 10 or 20 socket type constructed with two split current coils and dual-range voltage coils. Associated instrument transformers are furnished for use within the meter-mounting enclosure, or external instrument transformers may be used. A specially designed terminal block in the meter-mounting enclosure provides independent connection of the meter. Correct measurement of any service depends on proper connection of the terminal block, making internal meter connection changes unnecessary. The universal unit is also available in self-contained ratings with provision for future use of instrument transformers.

MULTI-STATOR METER APPLICATIONS WITH INSTRUMENT TRANSFORMERS

The usual customer metering of polyphase services does not normally present major problems. However, with metering at the higher distribution voltages, either of distribution lines or customers at these voltages, instrument transformers are required and more difficult problems occasionally arise. In many cases, the actual circuit conditions are hard to determine, which in itself presents a metering problem. The question of whether or not a ground return path exists or can exist is also a problem, particularly with wye circuits.

Wye-Circuit Metering

Consider the circuit shown in Fig. 7-29(a). Here a four-wire wye circuit is derived from a delta-wye transformer bank. With this circuit there is no question that a three-stator meter, or its equivalent, is required to correctly meter any connected load. Three voltage transformers would be connected wye-wye as shown, with their primary neutral connected to the circuit neutral. When voltage transformers are connected in wye-wye there is a third-harmonic voltage generated in each primary winding. The neutral connection between voltage transformer neutral and system neutral provides a path for third-harmonic current flow, thereby keeping the third-harmonic voltages at a low value. If this path were not present, the transformer voltages would be highly distorted by the excessive third-harmonic voltages, causing meter errors. Also, serious hazards to voltage transformer insulation would exist because of large increases in exciting current due to harmonic voltages. The four-wire wye circuit may also be derived from a wye-wye distribution bank, wye auto-transformer, wye grounding bank, or zig-zag transformer connections. Regardless of its source, the metering connections previously discussed are correct.

The circuit of Fig. 7-29(b) is a delta-wye transformer bank with a three-wire secondary without ground. If it is known that there is no possibility of an actual or phantom ground in the secondary circuit, it may be metered as in any three-wire circuit by a two-stator meter. Two voltage transformers would be connected as shown using one of the lines for the common connection.

The three-wire wye secondary circuit with neutral grounded shown in Fig. 7-29(c), however, presents a major metering problem. Since the secondary circuit is three-wire, it is possible to use a two-stator meter. It is possible that this may not be correct metering, since loads may be connected from the unmetered line to ground and thereby fail to be metered. Also, loads connected from metered lines to ground will not be measured correctly. For correct metering under all load conditions a three-stator meter or equivalent must be used. In this case three voltage transformers connected wye-wye would be used as shown. To limit third-harmonic effects, the neutral of the voltage transformers must be connected

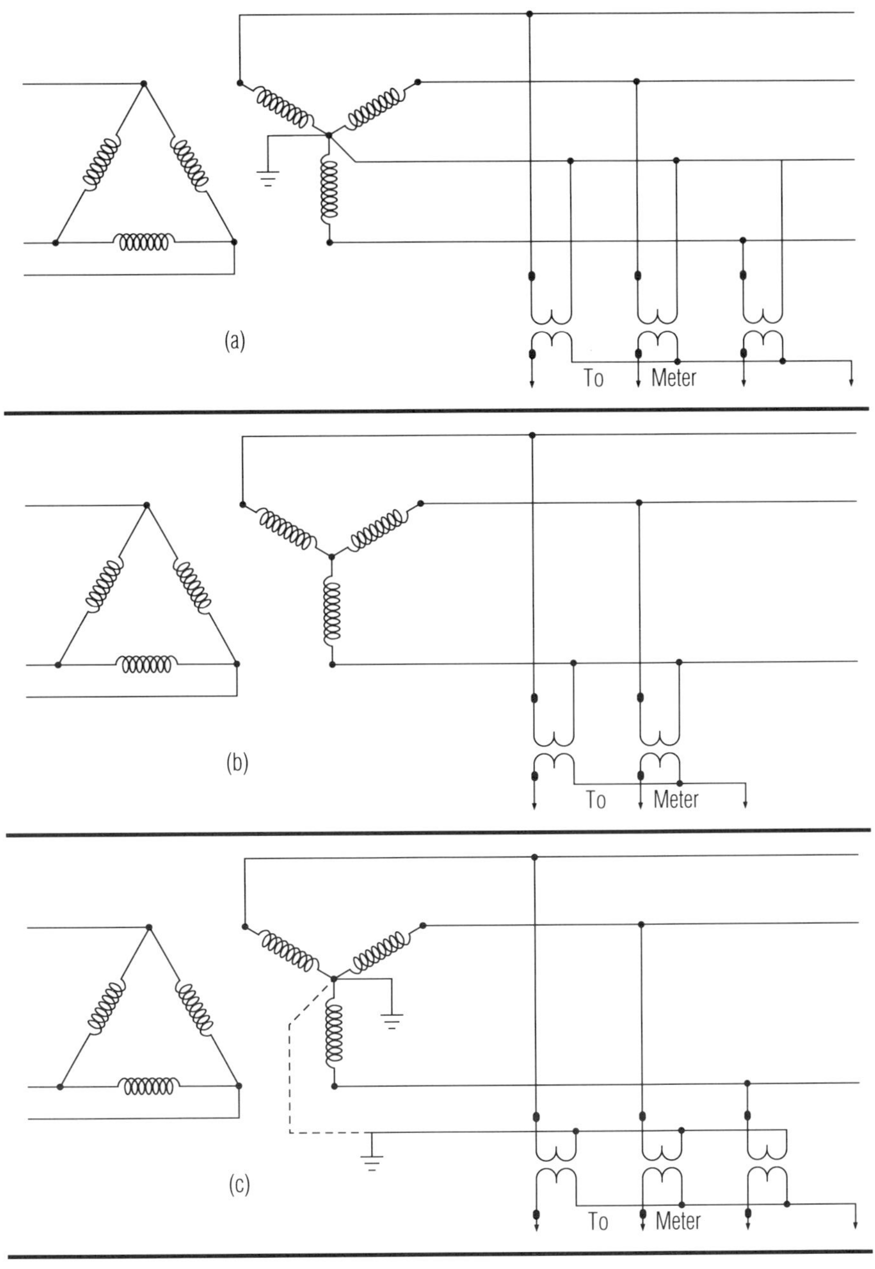

Figure 7-29. Wye-Circuit Voltage Transformer Connections.

to the neutral of the distribution bank by a low-impedance connection. If the meter is located at the same substation as the distribution bank, a ground connection to the station grounding grid may be sufficient or the two neutrals may be directly connected. If the metering location is at a considerable distance from the distribution bank, it may not be possible to establish a firm common ground. When the metering transformer neutral is left floating, the harmonic problem is again very serious. Harmonic voltages as large as 30 percent have been found in some instances. If the voltage transformer neutral is grounded at the metering location to an isolated ground, other problems exist. Differences in ground voltage under certain conditions can cause extremely hazardous conditions at the meter and a high-resistance ground may not eliminate the harmonic errors. Because of these conditions many companies require that a neutral conductor be run between the distribution and voltage transformers.

Many other problems can arise in wye-circuit metering, but they are too numerous to discuss here. In any such problem the presence of harmonics and their effects on metering should always be considered. Also, in metering transmission and distribution circuits a thorough understanding of circuit connections is necessary. For example, the three-wire wye connection with neutral ground is frequently encountered. With voltages in the order of 24,000 volts the designer is reasonably certain that customer loads will not be connected line to ground and two-stator metering will be correct. However, this may not always be the case. Further details on transformer connections will be found in Chapter 11, "Instrument Transformers."

PART 3—SOLID-STATE METERS

THE WATT/WATTHOUR TRANSDUCER

Solid-state metering was introduced to the electric utilities in the early 1970's in the form of a watt/watthour transducer. See Fig. 7-30. The advantages of solid-state electronic circuitry produced increased stability and accuracy surpassing the capabilities of the conventional electro-mechanical watthour meter, but at a significantly higher cost. Consequently, the watt/watthour transducer was most suitable to energy interchange billing and special applications where analog watt and digital watthour outputs were required.

The watt/watthour transducer provides an analog (watt) output signal in the form of a DC current and also a pulse (watthour) output from a Form C mercury-wetted relay or solid-state relay. The analog output may be used to drive a panel meter or strip-chart recorder or telemetered to a supervisory control system. The pulse output may be used to drive a totalizing register magnetic tape recorder, or a solid state recorder.

THE SOLID STATE WATTHOUR METER

Principle of Operation

A functional block diagram of an early watt/watthour transducer is shown in Fig. 7-31. The watt section is an electronic multiplier which uses the time-division-multiplier (TDM) principle to produce a pulse train which combines pulse-width and pulse-amplitude modulation. The pulse initiator section receives a DC current signal proportional to power from the watt section. Output pulses, propor-

Figure 7-30. Solid State Watt/Watthour Transducer.

tional to a convenient watthour-per-pulse rate, are fed from the KYZ output circuit to a register, tape recorder, electronic pulse counter, or other pulse-operated device. A complete description of the operation of the time-division-multiplier is included later in this chapter.

Solid State Watthour Meter

Compared with mechanical registers, solid state registers are more reliable and can be lower in cost. In addition, they can provide features not feasible with mechanical registers such as time-of-use measurements, switchable registers, and tamper detectors.

A typical meter consists of two sections: the multiplier and the register. In an electromechanical meter the multiplier consists of the voltage and current coils, and the meter disk; the register consists of the gears and dial indicators which count, store, and display the results of the multiplier. For clarity, in this section on solid state meters the following definitions apply: multiplier is a device which produces the product of a given voltage and current; register is a device which counts and displays the results of the multiplier; meter is an assembly which includes a multiplier and a register.

An electronic register is found on hybrid meters (meters with electromechanical multipliers and electronic registers), and on completely solid state meters. Most registers use a microprocessor which follows instructions stored as firmware to control the counting, storing, and displaying of data received from the multiplier.

All solid state meters must convert analog voltage and current signals into digital data. The digital data is sent to the register, either as serial data or parallel data. Serial data is a series of pulses where each pulse has a predetermined value, such as 0.6 watthours per pulse. Parallel data is typically in bytes and represents a new value.

To implement an electronic multiplier, meter manufacturers use one of these four approaches: time-division multiplication, Hall-effect technology,

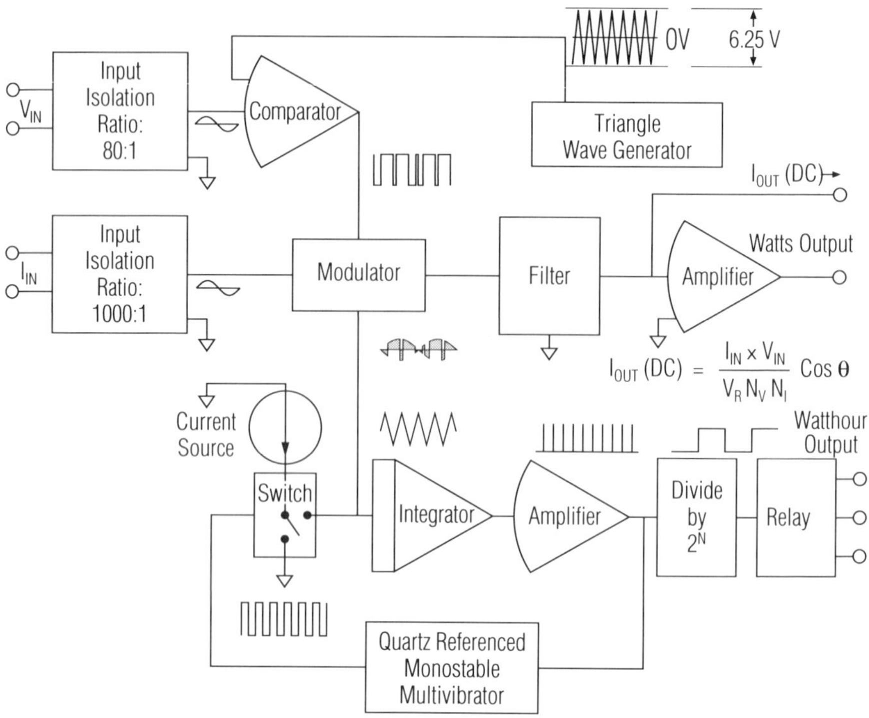

Figure 7-31. Functional Block Diagram Watt/Watthour Transducer.

transconductance amplifiers, or digital sampling techniques. Each method has advantages and disadvantages, and some manufacturers offer more than one type of electronic multiplier.

Characteristics common to all electronic multipliers are: the original input signals are scaled down to lower voltages to be compatible with solid state components, analog signals are converted to digital equivalents within the multiplier, and the phase angles between voltage and current are not measured directly.

Time-Division Multiplication

Time-division multiplication, also called mark-space-amplitude multiplication, computes power by using the common calculation of length-times-width to measure the area of a rectangle. The TDM multiplier develops a series of pulses where the width of each pulse is proportional to input voltage and the height of each pulse is proportional to input current. The area of each pulse then is proportional to power. Power can be integrated over time to develop an output signal for energy.

Figure 7-32(a) is a block diagram of a TDM multiplier, and waveforms within the multiplier.

The voltage and current reduction circuits typically are transformers, although a resistor-divider network could be used for voltage reduction. The voltage and current reduction blocks also provide protection from power transients.

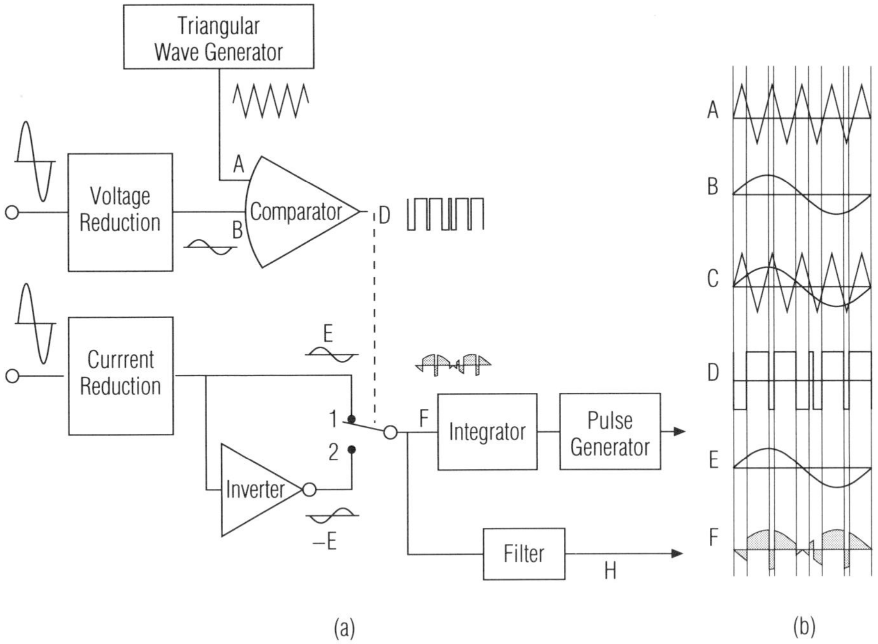

Figure 7-32. Time-Division Multiplication Waveforms.

The reduced voltage signal is compared with a triangular wave with a magnitude greater than the maximum of the input voltage signal. The frequency of the triangular wave varies with each manufacturer, with values typically between 800 Hz and 10 kHz. The comparator compares the input voltage with the reference voltage. For the positive half of the input voltage cycle, if the reference voltage is greater than the input voltage, the comparator output is negative; if the reference voltage is less than the input voltage, the comparator output is positive. For the negative half of the input voltage cycle, if the reference voltage is more negative than the input voltage, the comparator output is negative; if the reference voltage is less negative than the input voltage, the comparator output is positive. Figure 7-32(b), graph C, illustrates these comparisons. Graph D shows the output from the comparator, a signal with a fixed amplitude and a pulse width proportional to the input voltage.

The output from the comparator controls an electronic switch. When the comparator output is positive, the switch is set to input 1. When comparator output is negative, the switch is set to input 2.

Reduced current input is applied to switch input 1, and the inverse of that input is applied to switch input 2.

The output of the switch is shown on Fig. 7-32(b), graph F. The width of each shaded area is proportional to the width of the switch control signal, which is proportional to the input voltage. The height of each shaded area is proportional to the input current. The area of each shaded area is proportional to power for that period of time. While there appear to be positive as well as "negative" areas in the graph, the negative areas do not indicate power flow in the reverse direction.

The integrator and pulse generator convert power into energy measurements, and convert the analog information into digital data. The integrator sums individual areas, both positive and negative. When the accumulated total exceeds a predetermined value, the predetermined value is subtracted from the accumulated total and a pulse output is generated. The output of the pulse generator is sent to a register for storage and display.

By passing output signal F through an electronic filter, a signal proportional to instantaneous power can be produced. This waveform is shown as signal H, and can be transmitted to a SCADA (Supervisory Control and Data Acquisition) system for monitoring and control purposes.

A comparison of Fig. 7-31 with Fig. 7-32 shows that while several years have passed since the watt/watthour transducer of Fig. 7-31 was designed, the basic principles of TDM multiplication remain unchanged. The significant changes in the two designs result from the trend toward smaller, lower-cost components. In modern TDM multipliers, the design variations between manufacturers include: the frequency used by the triangular reference waveform; the rate at which integrated data is converted into pulses; the electronic components selected for the comparator, switch, and integrator; calibration and adjustments made available; method for scaling down voltage and current inputs; power supplies; and the technology used for the display.

Hall Effect Multiplication

The Hall effect can be explained as follows: When a current flows through a material which is in a magnetic field, a voltage appears across the material proportional to the product of the current and the strength of the magnetic field. This principle is illustrated in Fig. 7-33. In Fig. 7-33(a), there is no magnetic field and no voltage appears across the material. In Fig. 7-33(b), the magnetic field perpendicular to the path of the electrons displaces electrons toward the right side of the material. This produces a voltage difference side-to-side across the material. The voltage is proportional to the strength of the magnetic field and the amount of current flowing in the material.

Figure 7-34 illustrates application of the Hall effect to metering. A Hall effect device is inserted in a gap in a toroidal-shaped magnetic core. Current flows in the Hall effect device, based on line voltage across the device, after a reducing resistor. Current in a conductor looped around the magnetic core, creates a

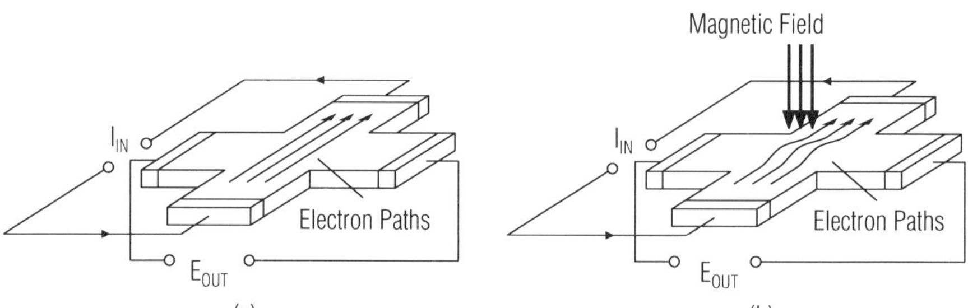

Figure 7-33. Hall Effect.

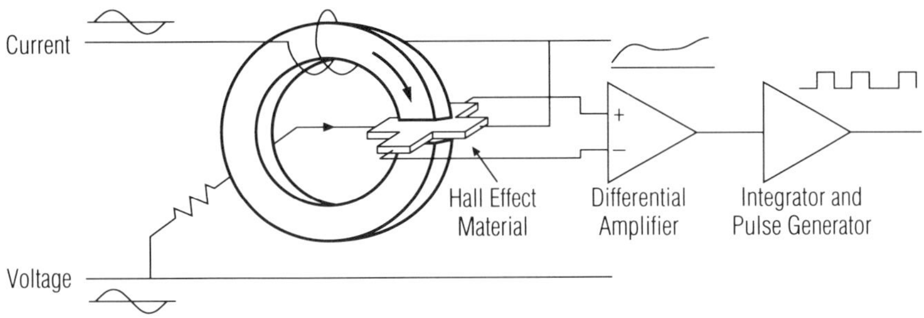

Figure 7-34. Hall Effect Applied to Metering.

magnetic field. The magnetic field flows around the core and through the Hall effect device, perpendicular to the flow of current.

The Hall effect voltage is sensed by a differential amplifier and supplied to an integrator and pulse generator. The integrator and pulse generator convert the power measured to an energy measurement, and convert analog information into digital data. The integrator calculates the area under the power curve, and stores accumulated data. When the accumulated total exceeds a predetermined value, the predetermined value is subtracted from the accumulated total, and a pulse output is sent to a register for counting, storing, and display of the measured data.

Transconductance Multiplier

The transconductance multiplier uses a differential amplifier, where bias current varies with an input signal. The circuit is illustrated in Fig. 7-35. In a metering application, input current in applied to the emitter of both transistors through resistor R_1. Input voltage is applied across the base of both transistors, causing

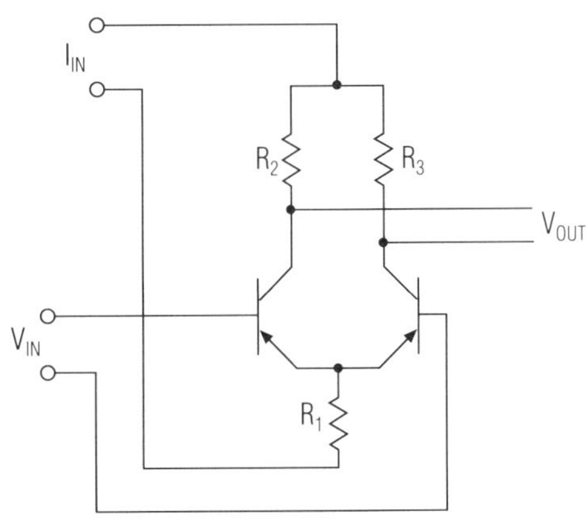

Figure 7-35. Transconductance Multiplier.

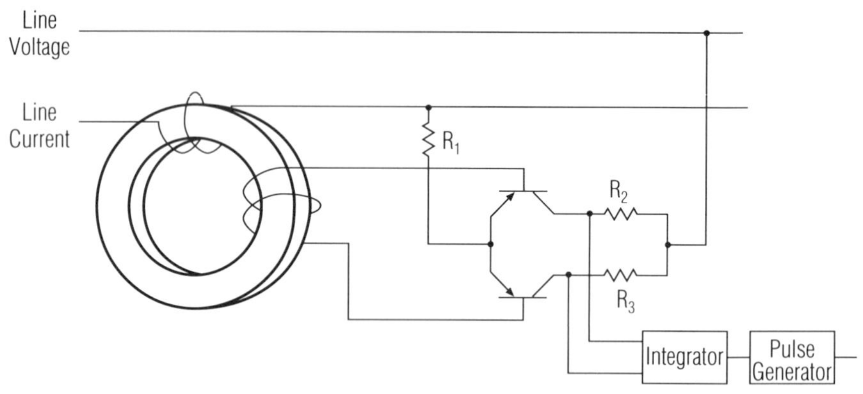

Figure 7-36. Transconductance Multiplier Applied to Metering.

one transistor to conduct more than the other transistor. The current flow difference causes different voltage drops across resistors R_2 and R_3. V_{OUT} is proportional to the bias current multiplied by the input voltage.

Figure 7-36 illustrates application of a transconductance amplifier for metering purposes. Bias current is developed from resistors R_1, R_2, and R_3 connected across the power line. Input voltage V_{IN} is developed from a secondary winding on a transformer (in this example a toroid core), where the primary winding carries line current. The output signal is fed to an integrator and pulse generator.

The integrator calculates the area beneath the power curve. When the sum of several areas exceeds a predetermined value, the predetermined value is subtracted from the integrated total and an output pulse generated. Output pulses are sent to a register for counting, storing, and display of the measured data.

Digital Multiplier

For all three multipliers described above, the analog voltage and current signals are multiplied together, then the results are converted to digital format. With the digital multiplier, the voltage and current analog inputs are immediately converted to digital equivalents, then multiplied using digital circuits.

An analog-to-digital converter measures the instantaneous value of the waveform, and converts each value to an equivalent digital word. An input sine wave for example, can be sampled several times within one cycle, and a digital equivalent of each instantaneous value can be stored in memory. Important specifications for analog-to-digital converters are: conversion time, the number of bits of resolution, and linearity.

Modern analog-to-digital conversion times are typically in the range of 20-50 microseconds. When the conversion time is short, changes to the input signal during the measuring "window" are small, increasing the accuracy of the measurement. When the time for each conversion is short, more samples can be taken during each cycle of the analog input, making the data collection process more accurate.

When sampling a voltage or current signal, the sampling rate determines the accuracy with which the signal is measured. The Nyquist theorem states: when

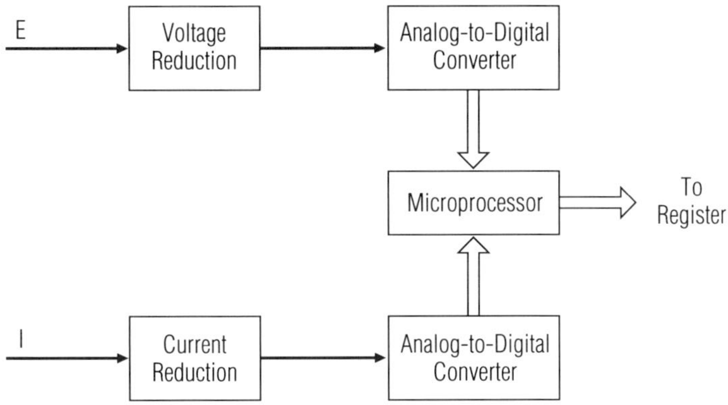

Figure 7-37. Digital Multiplier Block Diagram.

sampling an analog signal, to capture sufficient information about that signal, the sampling rate must be at least twice the highest frequency of interest in the analog signal. For example, if the highest frequency of interest in a signal is 60 hertz, the signal must be sampled at least 120 times per second to get a valid representation of that signal. If higher frequencies are of interest, a higher sampling rate must be used. For example, if measuring the seventh harmonic which has a frequency of 420 hertz is of interest, the signal must be sampled at least twice 420, or 840 times per second.

The sampling rate of solid state meters varies by model and by manufacturer. Typical rates are 4, 8, and 16 samples per 60 Hz cycle, rates equivalent to 240, 480, and 960 samples per second. Because two parameters, voltage and current, must both be measured, the digital multiplier must process information at two times these sampling rates. A three-phase meter taking two readings for each of three phases processes information at speeds up to 6,000 new readings each second.

The number of bits of resolution determines the granularity of the measurement. For example, if a 2-bit analog-to-digital converter is used to convert an analog signal which varies from 0 to +10 volts, the converter can output only 2^2, or 4 different digital values representing the input, or steps of 2.5 volts each. If a 12-bit analog-to-digital converter is used to convert a signal varying from 0 to +10 volts, the converter can output 2^{12}, or 4,096 different digital values, or steps of 2.44 millivolts each. In the second example, a change of only 2.44 millivolts at the analog input will cause a one-bit change in the output digital value. The resolution of analog-to-digital converters used in solid state meters varies by model and by manufacturer, depending on the accuracy of the application. Typical resolutions used are 4, 8, and 12 bits.

A block diagram of a digital multiplier in a power meter is shown in Fig. 7-37. This example shows two analog-to-digital converters although more could be used, especially in polyphase meters. The input voltage and current signals are scaled down by their respective reduction circuits, then applied to the analog-to-digital converters. The converter outputs are multiplied by the microprocessor, with the results stored in memory along with an accumulating total of the results.

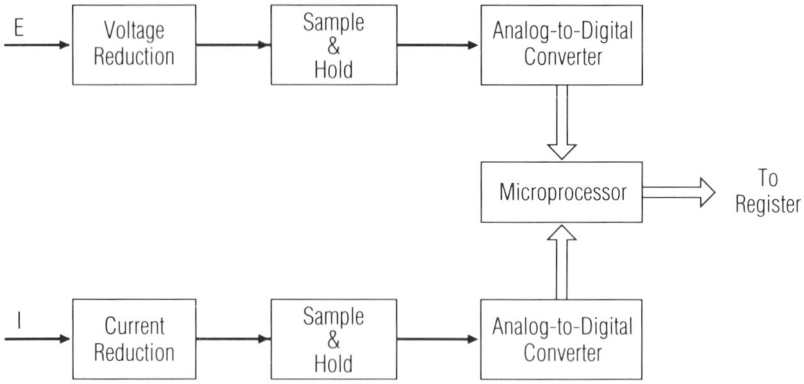

(a)

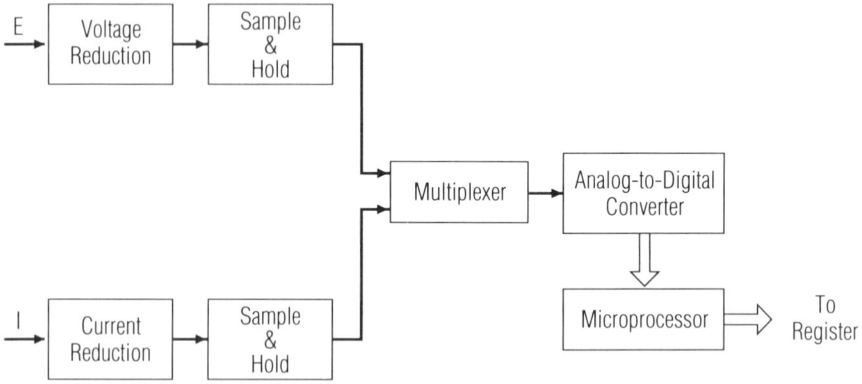

(b)

Figure 7-38. Variations of Digital Multiplier Designs.

When the accumulated total reaches a predetermined value, that predetermined value is subtracted from the total and one output pulse is generated. The output pulse indicates that one increment of power has been measured. The pulse is sent to an electronic register for storage and display.

Variations of digital multiplier designs are shown in Fig. 7-38. The sample-and-hold amplifiers shown in Fig. 7-38(a) allow the meter to measure voltage and current at the same instant, thus eliminating time skew in taking two readings. The multiplexer shown in Fig. 7-38(b) allows the meter to switch one analog input at a time to one analog-to-digital converter. This approach eliminates the cost of another analog-to-digital converter. Polyphase meters typically incorporate sample-and-hold amplifiers, multiplexers, or other techniques to trade off costs with accuracy requirements.

The digital multiplier calculates instantaneous power by multiplying digitized equivalents of the voltage and current signals, and can calculate other

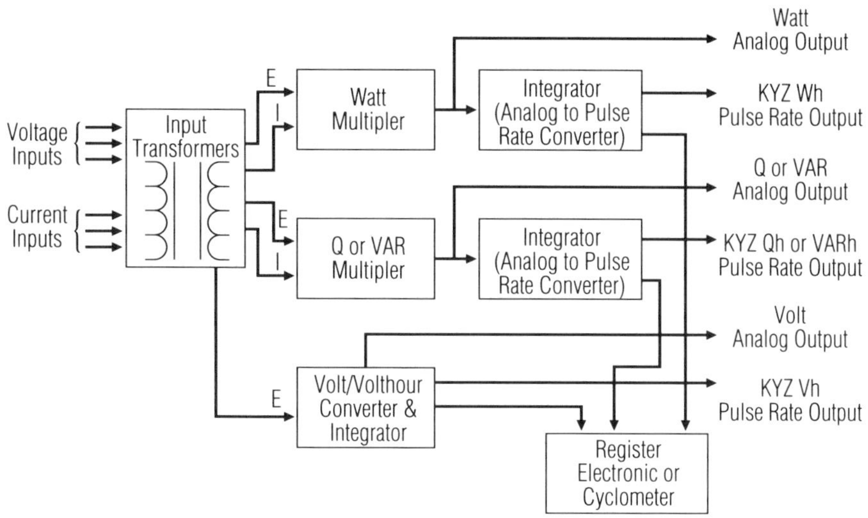

Figure 7-39. Solid State Three-Function Meter Block Diagram.

Figure 7-40. Electronic Multifunction Meter.

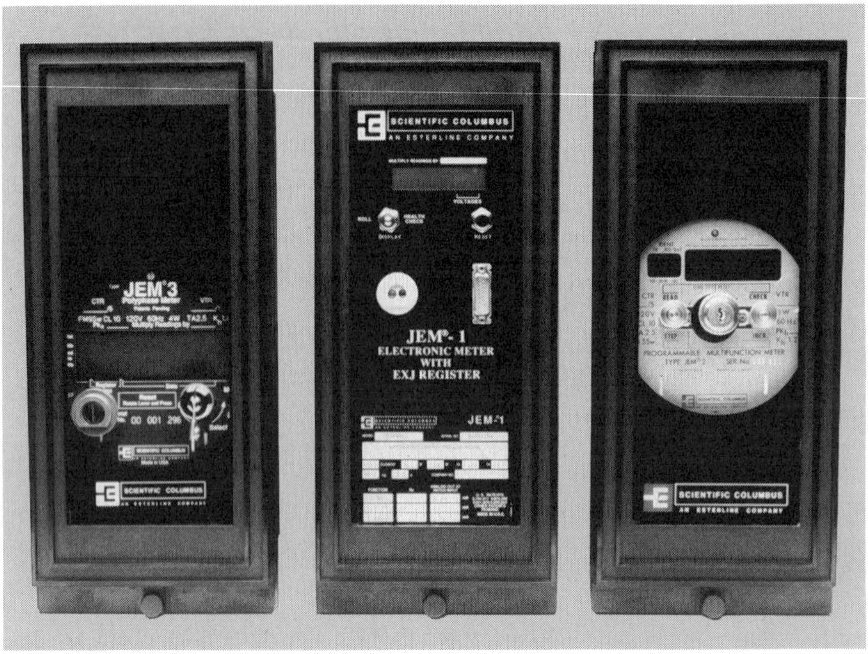

Figure 7-41. Electronic Polyphase Meter, Meter with Register, and Multifunction Meter.

values as well. By summing the products of a number of multiplications, then dividing by that same number, the result is average power consumed during the period. Increments of average power can be summed over time, to determine energy. To determine V_{RMS} and I_{RMS}, instantaneous values for voltage and current can be squared, their square roots computed, and the result is RMS values of voltage and current. If V_{RMS} and I_{RMS} are multiplied and integrated over a period of time, the result is arithmetic apparent power and energy (VA and VAhrs). To compute reactive power (VAR) it is necessary to multiply the current times the voltage, shifted 90 degrees in phase. In the digital meter this is accomplished by fixing the sampling rate for the digital multiplier at 4, or a multiple of 4, times the fundamental frequency of the voltage and current, so current and voltage are sampled every 90 degrees. To compute VAR, multiply the digitized values for current times corresponding digitized values for voltage which were sampled 90 degrees previously.

Figure 7-39 is a simplified block diagram of a typical three-function meter. Voltage and current inputs from instrument transformers are first reduced to low values by the input voltage and current transformers. These signals are then fed to appropriate multiplier or converter circuits where they are converted to DC signals and fed onto analog output circuits and integrator circuits as previously described. The pulse outputs of the integrators drive output KYZ relays and are also fed to the register section. In the register the pulses may be divided to a lower rate, if desired, for accumulating in the cyclometers or in a solid-state memory.

CHAPTER 8

DEMAND METERS

EXPLANATION OF TERM DEMAND

KILOWATT DEMAND is generally defined as the kilowatt load averaged over a specified interval of time. The meaning of demand can be understood from Fig. 8-1, in which a typical power curve is shown. In any one of the time intervals shown, the area under the dotted line labeled demand is exactly equal to the area under the power curve. Since energy is the product of power and time, either of these two areas represents the energy consumed in the demand interval. The equivalence of the two areas shows that the demand for the interval is that value of power which, if held constant over the interval, will account for the same consumption of energy as the real power. It is then the average of the real power over the demand interval.

The demand interval during which demand is measured may be any selected period but is usually 5, 10, 15, 30, 60, and in similar increments up to 720 minutes. The demand period is determined by the billing tariff for a given schedule.

Demand has been explained in terms of power (kilowatts) and usually this information has the greater usefulness. However, demand may be expressed in kilovars, kilovoltamperes, or other suitable units.

WHY DEMAND IS METERED

Two classes of expenses determine the total cost of generating, transmitting, and distributing electric energy. They are:

1. Capital investment items: depreciation, interest on notes, property taxes, and other annual expenses arising from the electric utility's capital investment in generating, transmitting, and distributing equipment, and in land and buildings.
2. Operation and maintenance items: fuel, payroll, renewal parts, workmen's compensation, rent for office space, and numerous other items contributing to the cost of operating, maintaining, and administering a power system.

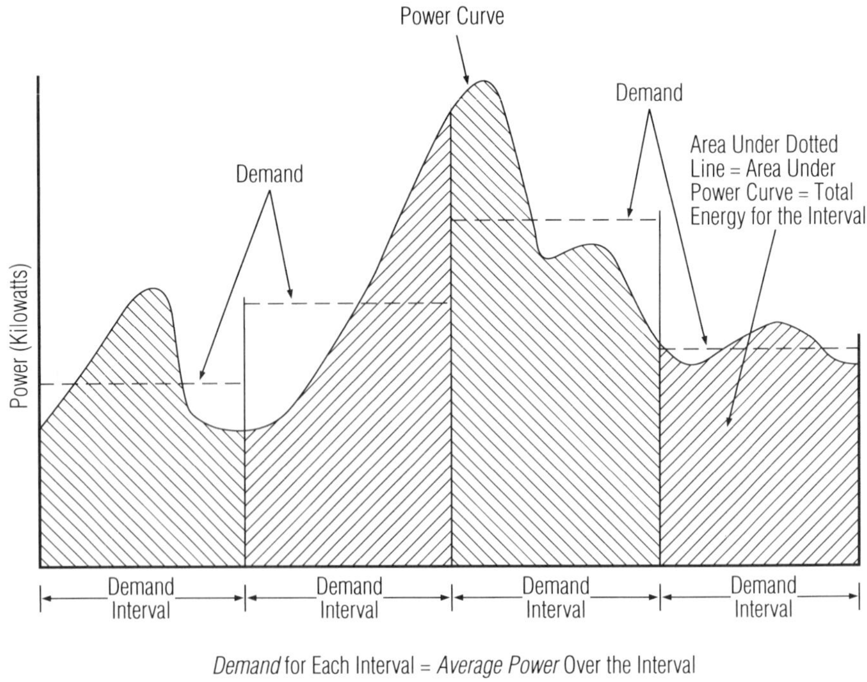

Figure 8-1. Power Curve Over Four Successive Demand Intervals.

In billing the individual consumer of electricity, the utility considers to what extent the total cost of supplying that consumer is determined by capital investment and to what extent it is determined by operation and maintenance expenses. Furnishing power to some consumers calls for a large capital investment by the utility. With other consumers, the cost may be due largely to operation and maintenance. The following two examples illustrate these two extremes of load:

1. In a certain plant, electricity is used largely to operate pumps which run at rated load night and day. The power consumed by the pump motors is low and the plant shares a utility-owned transformer with several other consumers. The amount of energy used each month is large because the pumps are running constantly. Therefore, the cost of supplying this consumer is largely determined by operating expenses, notably the cost of fuel. The capital investment items are relatively unimportant.
2. Another factory uses the same number of kilowatt-hours of energy per month as the pump plant of the previous example but consumes all of it in a single eight-hour shift each day of the month. The average power is therefore three times greater than for the pump plant and the rating (and size) of equipment installed by the utility to furnish the factory with energy must also be about three times higher. Costs rising from capital investment are a much greater factor in billing this consumer than in billing the operators of the pump plant.

Demand is an indication of the capacity of equipment required to furnish electricity to the individual consumer. Kilowatt-hours per month are no indication of the rating of equipment the utility must install to furnish a particular maximum power requirement during the month without overheating or otherwise straining its facilities. What is needed in this case is a measure of the maximum demand for power during the month. The demand meter answers this need.

MAXIMUM AVERAGE POWER

A commonly used type of demand meter is essentially a watthour meter with a timing element added. The meter sums up the kilowatt-hours of energy used in a specific time interval, usually 5, 10, 15, 30, or 60 minutes, and in similar increments up to 720 minutes. This demand meter thus indicates energy per time interval, or average power, which is expressed in kilowatts.

By means of a pointer-pusher mechanism or a recording device, a wattmeter could be made to preserve an indication of the maximum power delivered to a consumer over a month or some other period. Such a wattmeter would be simpler and cheaper to manufacture than a demand meter. It would not be an adequate substitute for a demand meter, however, because the wattmeter is sensitive to momentary changes in the real power. If the maximum power over a month were maintained for only a second, a wattmeter with a pointer-pusher device would preserve that value of maximum power for the meterman to read.

The capacity of most electrical equipment is limited by the amount of heating it can stand, and heating depends not only on the size of the load but on the length of time the load is maintained. A momentary overload such as the starting surge of a motor will not, in general, cause a temperature rise sufficient to break down insulation or otherwise damage the equipment. Therefore, the utility uses not a momentary value of maximum power, but maximum average power over an interval as a basis for billing.

GENERAL CLASSES

There are several types of demand meters:

- Integrating demand meters
- Thermal or lagged demand meters
- Pulse-operated demand registers
- Pulse-operated electronic demand meters (recorders)
- Solid state demand recorders
- Electronic time-of-use meters, with recorders under the meter cover
- Electronic time-of-use demand meters
- Electronic or mechanical pulse-accumulating demand recorders
- Electronic demand meters.

All of these meters have a common function, which is to measure power in such a way that the registered value is a measure of the load as it affects heating, and therefore the load-carrying capacity, of the electric equipment. Some demand meters have other functions which provide billing and load information for use in pricing.

Integrating Demand Meters

All integrating kilowatt demand meters in common use register the average power over demand intervals which follow each other consecutively and correspond to definite clock times. For example, if an integrating meter with a 15-minute interval is put into operation at 2:15 p.m. on a certain day, the first interval will be from 2:15 to 2:30 p.m., the next will follow immediately and will be from 2:30 to 2:45 p.m., and so on.

Integrating kilowatt demand meters are driven by watthour meters. The registering device turns an amount proportional to the watthours of energy in the interval. A timing mechanism returns the demand registering device to the zero point at the end of each interval. The final displacement of the registering device just before the timing mechanism returns it to zero is also proportional to the demand in the interval. This is true because the demand in the interval is equal to the energy consumed during the interval divided by the time, which is constant.

The integrating demand meters indicate only the maximum demand over a month or other period. The gears, shafting, and pointer-pusher of all integrating demand meters turn an amount proportional to the demand in every demand interval. In other words, a meter which indicates only maximum demand has a pointer-pusher and a pointer that indicates the maximum demand that has occurred during any interval since the pointer was reset to zero.

An integrating kilowatt demand meter is basically a watthour meter with added facilities for metering demand. The watthour meter is the driving element. The watt-hour and demand registering functions may be combined in a single device. Frequently, however, the demand register is physically separated from the driving element, sometimes by many miles.

Thermal or Lagged Demand Meters

In the thermal- or lagged-type demand meter, the pointer is made to move according to the temperature rise produced in elements of the meter by passage of currents. Unlike the integrating demand meter, the lagged meter responds to load changes in accordance with the laws of heating and cooling, as does electrical equipment in general. Because of the time lag, momentary overloading, instead of being averaged out, will have a minor effect on the lagged meter unless the overloading is held long enough or is severe enough to have some effect on the temperature of equipment. The demand interval for the lagged meter is defined as the time required for the temperature sensing elements to achieve 90 percent of full response when a steady load is applied. Like the integrating meter, the lagged meter is generally designed to register kilowatt demand.

The lagged type is essentially a kind of wattmeter designed to respond more slowly than an ordinary wattmeter. An important difference in these methods of metering demand is in the demand interval. In the integrating meter, one demand interval follows another with regularity, giving rise to the term "block interval." The thermal or lagged meter measures average load with an inherent time interval and a response curve which is based on the heating effect of the load rather than on counting disk revolutions during a mechanically timed interval.

These general classes of demand meters and their common principles and uses, are discussed in the following text.

WATTHOUR DEMAND METER

The watthour demand meter, as its name implies, combines in a single unit a watthour meter and demand meter. Such a meter may contain a watthour element combined with a mechanical, gear-driven, demand device or a watt-hour element

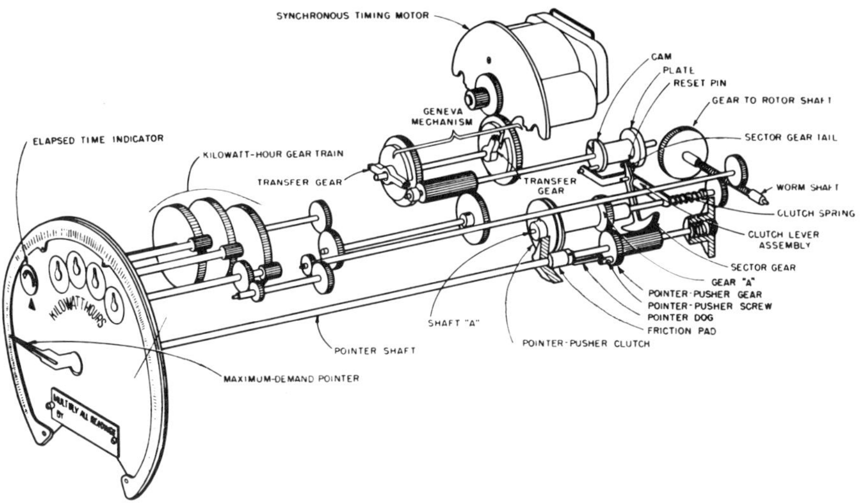

Figure 8-2. Gear Trains and Interval-Resetting Mechanism of an Indicating Watthour Demand Register.

and a thermal demand unit. The thermal and electronic demand meters will be discussed later in this chapter. Discussion here covers the method of mechanical demand measurement.

In a mechanical watthour demand meter, the watthour disk shaft drives two devices:

1. The gears and dial pointers through which the revolutions of the rotor are summed up in kilowatt-hours of energy.
2. The gears and shafting, which, working in conjunction with a timing motor or a clock, sum up the revolutions of the rotor during each demand interval in terms of kilowatts demand.

These two devices, which after their initial gearing to the disk shaft are independent of each other, comprise, in effect, two separate registers. They are commonly combined physically and referred to as the watthour demand register.

Three types of mechanical watthour demand meters are manufactured:

1. The *indicating type.* This type indicates only the maximum demand for each month or other period between resettings.
2. The *cumulative type.* This type also indicates the maximum demand during the period between resettings, but, in addition, by means of the resetting operation, the maximum demand for the period just ended is transmitted to dials and added to the total of previous maximums.
3. The *recording type.* By means of a pen moving across a chart, a record of demand for every demand interval is kept.

The gears, dials, and pointers by which the disk rotations of a watthour demand meter are translated into kilowatt-hours of energy are the same, in principle, as in the watthour meter register.

The demand pointer-pusher or recording device rotates a number of degrees proportional to energy utilization of each demand interval. Every 15 minutes, half-hour, or other demand interval, the timing mechanism performs two operations:

1. It releases a clutch, mechanically breaking the connection between the meter rotor and the pointer-pusher or recording mechanism.
2. It returns the pointer-pusher or recording mechanism to the zero point.

Then the clutch is re-engaged and the summing-up process begins again. The process of returning the pointer-pusher to zero at the end of each interval takes only a few seconds.

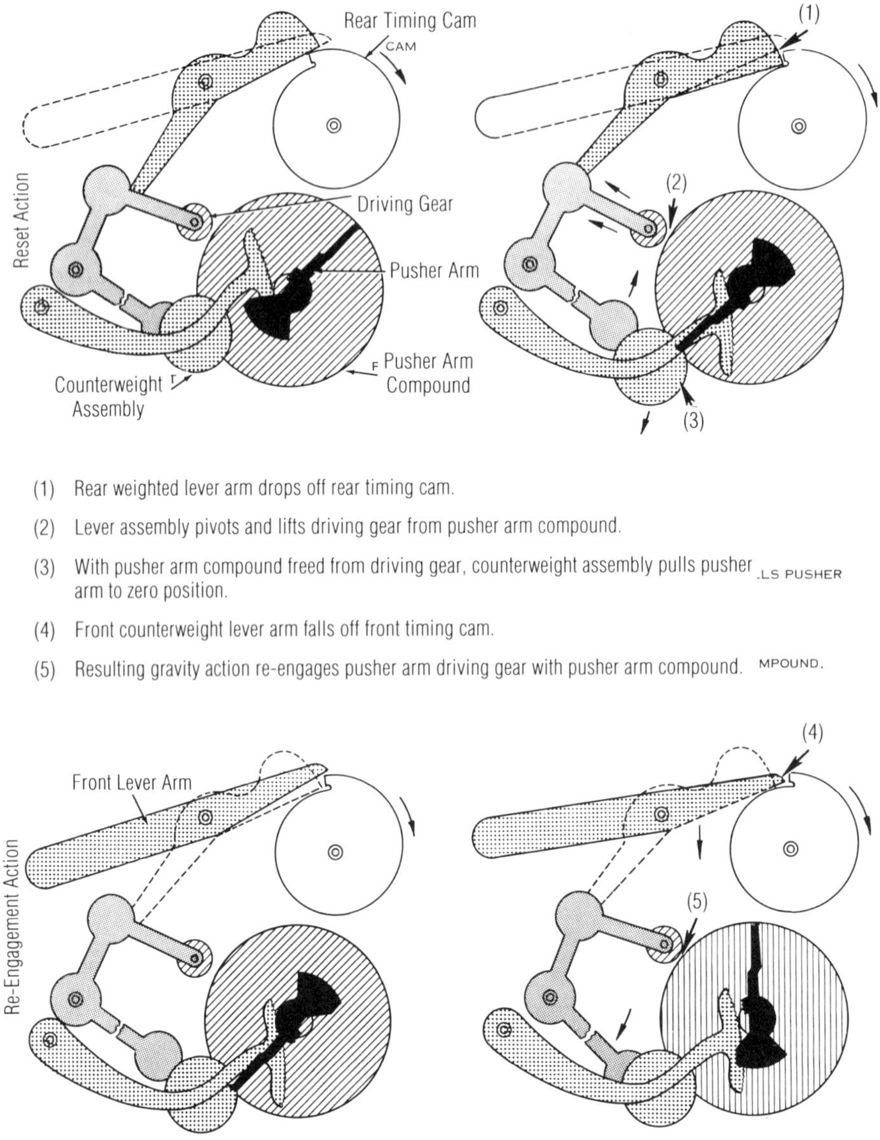

Figure 8-3. Gravity Reset for Indicating Demand Register.

The timing mechanism may be actuated by voltage from the metered circuit or by voltage from a separate circuit, or the mechanism may be a spring-driven device.

Indicating Type (Pointer Type)

The maximum demand is indicated on the graduated scale by the sweephand pointer. During each demand interval the demand pointer-pusher advances proportionally to the kilowatts demand. If the demand for a given interval is higher than any previous demand since the pointer was last reset, the pointer-pusher pushes the pointer upscale to indicate the new maximum demand. The pointer is held in this position by the friction pad.

At the end of each demand interval, the pointer-pusher is automatically returned to the zero point as follows (see Fig. 8-2): gear A, which meshes with and drives the pointer-pusher gear, is free to rotate on shaft A. Throughout each demand interval, however, the gear is driven by shaft A through the pointer-pusher clutch. At the end of the interval, the cam, driven by the synchronous timing motor through gears, transfer gears, and shafting, lifts the tail of the clutch lever, causing this lever to compress the clutch spring, thereby taking pressure off the clutch, disengaging it, and leaving the pointer-pusher assembly free to rotate. At this instant, the reset pin on the plate (which is also rotated by the synchronous motor) engages the tail of the sector gear and causes the gear to turn the pointer-pusher assembly backward to the zero point. Then the cam acts to close the clutch and registration is resumed.

An important variation of returning the pointer-pusher to zero is the gravity reset method. The energy for the reset operation is provided by a weight and pivot assembly geared to the pusher-arm shaft. The operation of the reset mechanism is triggered by dual timing cams. The drop-off of each of the cams is slightly displaced thus triggering the disengaging and resetting action. The resetting operation consists of the actions in Fig. 8-3.

Indicating Type (Dial)

The term dial type demand register is used to denote an indicating type demand register mechanism where the demand reading is in dial form. The dials return to zero when the monthly reset is performed. This type of demand display has better resolution than the pointer type due to a longer equivalent scale length.

Cumulative Type

The cumulative watthour demand meter goes one step further than the indicating watthour demand meter. A pointer, moved across a scale by a pointer-pusher mechanism, preserves the maximum demand until the meter is reset. The principle is the same as in the indicating meter, but the pointer and scale are much smaller. In addition, the meter preserves a running total of the maximum demands for consecutive months on small dials similar to watthour meter dials. The maximum demand for each month is added to the previous maximums on the dials when the meter is reset at the end of the month.

Except for the resetting device and the cumulative gear train, pointers, and dials, the cumulative demand register is the same in principle and operation as the indicating demand register. See Figs. 8-4 and 8-5.

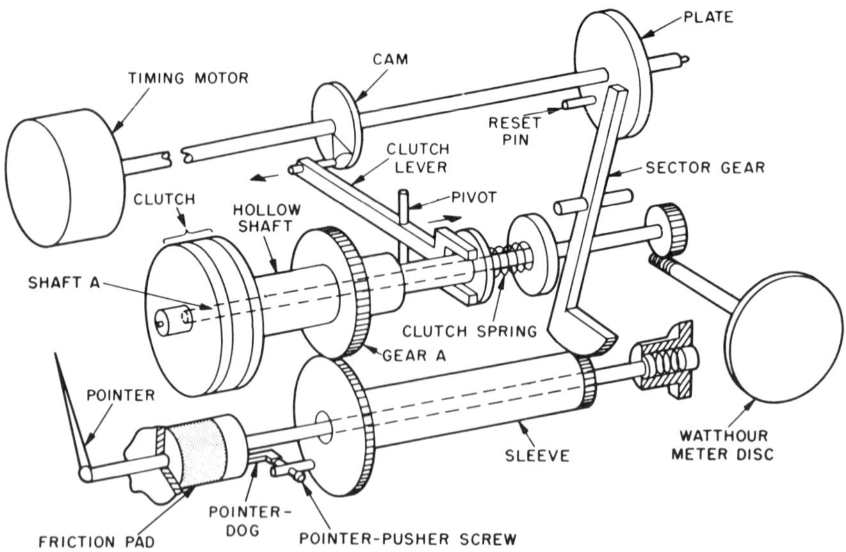

Figure 8-4. Simplified Schematic of the Interval-Resetting Mechanism of an Indicating Watthour Demand Meter.

Recording Type

Recording watthour demand meters are usually designed for use in polyphase circuits with instrument transformers. They keep a permanent record of the demand for every demand interval. Instead of driving a pointer-pusher mechanism, the meter rotor drives, through shafting and gearing, a pen which moves across a chart a distance proportional to the kilowatts demand. At the end of the interval, the pen is returned to zero.

The chart is moved a uniform amount each interval and the date and hours of the day are marked on the left side. Thus, the chart shows at a glance the demand which has occurred in each interval of each day.

Register Differences

There are two major differences among the various types of indicating demand registers: (1) the method of engagement between the demand pointer-pusher and the driving mechanism; (2) the method of resetting the pointer-pusher.

The method of engagement between the demand pointer-pusher and the driving mechanism may be by means of a flat disk clutch with felt facings or by means of a purely metal to metal contact, such as three fingers engaging the circumference of a disk. The clutch is released at the end of the demand interval, separating the pointer-pusher from the watthour meter driving gears so that the pointer-pusher can be returned to zero.

The return of the pointer-pusher to zero is done automatically in a minimum length of time by means of one of the following methods:

1. The demand interval timing motor.
2. A spring which is wound up during the demand interval.

3. A gravity reset mechanism which becomes operative as the clutch is released.

Re-engagement of the pointer-pusher mechanism occurs and the meter rotor begins its movement of the pointer-pusher upscale for the next demand interval.

Dual-range demand registers are also available. They provide a wider range of operational accuracy together with flexibility for growing loads. This is accomplished by a gear-shifting mechanism with a positive scale-plate interlock to eliminate the possibility of error.

A new form of block-interval demand register was announced in 1960. In this register the fixed scale, sweep-pointer type of demand indication was replaced by three small dials which show kilowatts demand in the same manner as the conventional kilowatt-hour dials. The register timing mechanism operates in the usual manner and resets the demand pointer driving and sensing pins to zero at the end of each demand interval. The driving and sensing pins do not change the demand reading unless a higher demand occurs, in which case the pins engage the demand pointers at the old value and drive them to the new value. The demand pointers are reset to zero by the meter reader operating the reset mechanism. This type of register provides a higher demand range, allowing greater flexibility in application and improved accuracy at less than full scale.

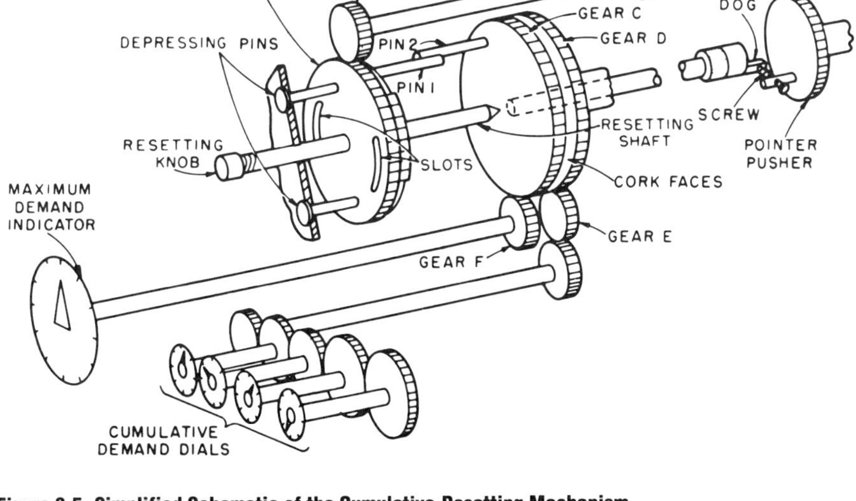

Figure 8-5. Simplified Schematic of the Cumulative-Resetting Mechanism of a Cumulative Watthour Demand Meter.

Register Application

The load range of watthour meters has been extended over the years until the 1961 design of self-contained meters is capable of carrying and measuring loads with currents up to 666 percent of its test-ampere value. Naturally, the full-scale values of demand registers had to keep pace with the meter capabilities. Use of a single, fixed scale for demand registers which had a 666 $^{2}/_{3}$ percent full-scale value would lead to serious errors in demand measurement. Nominal single-pointer demand register accuracy is 1 percent of full-scale value; so if the high scale were used on meters with lower capacity or on low-capacity services, the resulting values would produce low-scale readings with possible high errors. Consequently, a number of different full-scale demand values are available in this type of register, so that with proper register selection the demand readings will be above half-scale to give good accuracy.

Manufacturers normally supply registers with three overload capacities. These are capable of measuring demand with maximum load currents of 166 $^{2}/_{3}$ percent, 333 $^{1}/_{3}$ percent, and 666 $^{2}/_{3}$ percent of the meter rated test-ampere value. Some manufacturers apply class designations of 1, 2, and 6, respectively, to these register capacities. The full-scale demand values are available with the various register ratios normally used on a manufacturer's meters, thus allowing selection of a register with proper consideration being given to meter rating and service capacity. The various full-scale kilowatt values are obtained by changes in the gearing to the demand pointer-pusher with no change in register ratio, which is another indication of the independence between the kilowatt-hour gear train and the demand gear train in a demand register.

When using single-range demand registers, it is necessary to change the entire register to obtain a change in full-scale kilowatt value or class. The new dual-range registers are designed to allow operation in two different classes with a single unit. A gear shifting mechanism is provided which changes the full-scale demand value. When this shift is made, the demand scale or kW multiplier should be changed to correspond. The dual-range register may, therefore, be changed by a simple operation to perform in either of two different register classes without a change in register ratio or kWhr multiplier.

KILOVAR OR KILOVOLT-AMPERE DEMAND METER

It should be noted that the mechanical and the electronic type of demand meters can also be used to measure kilovar demand or kilovolt-ampere demand by connection of the watthour meter potential circuits to appropriate phase-changing voltage autotransformers.

In one type of recording kilovoltampere demand meter, a unique method is used to obtain the kilowatt demand and kilovolt-ampere demand, as well as instantaneous power factor and integrated kilovar-hours, kilovolt-ampere-hours, and kilowatt hours in the same instrument.

The meter consists of two watthour meter elements with a separate reactive compensator (auto-transformer) provided in order that one set of the watthour meter elements of the combination can be connected to measure kilovar-hours.

The values of kilovolt-ampere demand and instantaneous power factor are obtained by mechanical vector addition of active and reactive components of the power by mechanical solution of the equation $VA = \sqrt{W^2 + var^2}$.

This method of demand measurement is described in Chapter 9, "Kilovar and Kilovolt-Ampere Metering."

PULSE-OPERATED MECHANICAL DEMAND METER

Pulse-operated mechanical demand meters may be used in single circuits or in totalizing systems to determine the combined demand of several circuits. Distinctive features of pulse-operated demand meters are described in the manufacturers' section. The pulse operated demand meters may be located close to the meters which supply the pulses or they may be located some distance away with the pulses being transmitted over telemetering circuits.

The basic elements of a pulse-operated mechanical type demand meter are the demand-registering mechanism and the timing or interval-establishing mechanism. The demand-registering mechanism is energized through the remote actuating pulse generator. The same or, in some cases, a different source of voltage drives the timing mechanism and, in recording instruments, the chart drive mechanism when it is motor operated.

As the pulse initiator opens and closes, the operating current is intermittently interrupted, energizing and de-energizing the actuating assembly of the registering mechanism. This actuating assembly, for different types of meters, may consist of either an electromagnet, a solenoid, or a synchronous motor and contact arrangement. In either scheme the pulses received by the demand meter are converted to mechanical rotation, which advances the position of an indicating pointer, pen or stylus, or printing wheels, depending on the type of meter.

The timing mechanism, usually motor-driven, establishes the interval through which the registering device is allowed to advance. At the end of the predetermined demand interval, it effects a disengagement of the registering mechanism and returns the indicator pusher or recording device to a zero position ready to begin registration for the new demand interval.

Pulse Operated Electronic Demand Meters

These demand meters are similar in principle to mechanical demand meters, but having no moving parts they can accumulate pulses at a faster rate and provide better resolution and accuracy. Electronic demand meters are described later in this chapter.

Indicating Type

As in the direct-driven indicating meters, the pulse-operated sweephand indicating demand meter provides no record of demand when the demand pointer is manually reset at the end of the reading period.

Recording Type

Strip-chart and round-chart recorders produce a permanent demand record. Several styles of charts are available to suit particular applications. They may vary in scale marking, finish (for stylus or ink), duration, or other features.

Paper-Tape Type

The tape type of pulse-operated demand meter totals initiated pulses for each demand interval and may record these pulses in either of two forms, on paper tape as a printed number, or as a coded punching. Either form may be read manually or translated automatically to determine the demand values for each period.

THERMAL DEMAND METERS

Thermal demand meters are essentially ammeters or wattmeters of either indicating or recording type. Unlike the ordinary ammeter or wattmeter, the indication of which follows immediately any change of load, the thermal demand meter responds very slowly to load changes. See Fig. 8-6. Therefore, the indication of a thermal demand meter at any instant depends not only upon the load being measured at that instance but also on the previous values of the load. It represents, then, a continuous averaging of the load and so constitutes a measure of demand.

A block-interval-type demand meter, as well as a thermal demand meter, measures demand by averaging the load over a period of time. In block-interval demand, the average is a straight arithmetic average over a definite period of time with equal weight given to each value of load during that period. In thermal demand, the average is "logarithmic" and continuous, which means that the more recent the load the more heavily it is weighed in this average and that, as time passes, the importance of any instantaneous load value becomes less and less in its effect on the meter indication until finally it becomes negligible. While there may be some theoretical preference for thermal as compared to block demand due to the fact that it operates upon a heating curve similar to that of other equipment on the line, there is little difference in practice, since, in general, the measurement of demand by either method gives comparable response. It is true that on certain types of loads where severe peaks exist for short periods, considerable difference may result between the indication of block- and thermal-type meters; but, also, on such loads, considerable differences may result between the readings of two similar block demand meters due to peak splitting if they do not reset simultaneously .

Thermal watt demand meters are commonly combined with watthour meters to form a single measuring unit of energy and demand. In such combination meters, the potential to the thermal demand section may be supplied from separate small voltage transformers in the meter or from secondary potential windings on the potential coils of the watthour elements. Current to the thermal elements may be the entire load current or it may be reduced by small through-type current transformers which use the line conductors as single-turn primaries. Even though the watthour and demand elements may be electrically independent, their combination within the same meter housing provides savings in space requirements and installation costs.

Single- or dual-range thermal demand meters are presently available. The advantages of the dual-range feature are the same as previously described for the dual-range mechanical demand register.

The discussion which follows covers the theory of operation of only the single-range thermal demand meter.

Operation

The general principle of operation of a thermal demand meter may be explained by reference to Fig. 8-7. The bimetallic coils, each of which constitutes a ther-

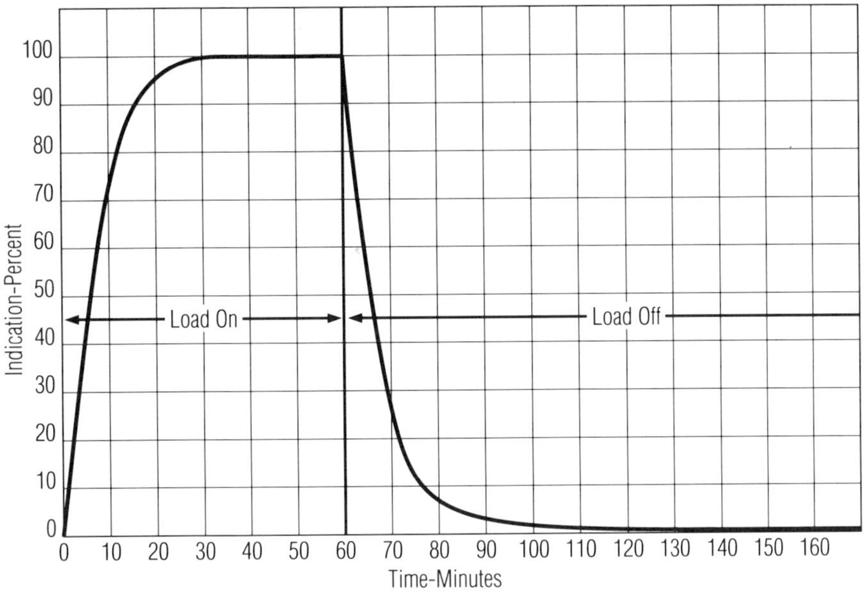

Figure 8-6. Time-Indication Curve of a Thermal Watt Demand Meter.

mometer, are connected to a common shaft in the opposing directions. The outer ends of these coils are fixed in relation to each other and to the meter frame. The shaft, supported between suitable bearings, carries a pointer. As long as the temperature of the two coils is the same, no motion of the pointer results even though this temperature changes, since the tendency of each of these coils to expand with rising temperature of one coil is higher than that of the other, there results a deflection of the pointer which is proportional to the temperature difference between the coils. The pair of coils thus constitutes a differential thermometer which measures a difference of temperature rather than an absolute temperature.

A small potential transformer in the meter has its primary connected across the line and its secondary line connected in series with two noninductive heaters. Each heater is associated with an enclosure and each enclosure contains one of the bimetallic strips. With potential only applied to the meter, a current $E/2R$ circulates through the heaters as shown in Fig. 8-7. This circulating current is directly proportional to the line voltage and passes through each one of the heaters. With only the potential circuit energized, heat is developed in the first heater at a rate:

$$W_1 = \left(\frac{E}{2R}\right)^2 \qquad R = \frac{E^2}{4R}$$

and in the second heater:

$$W_2 = \left(\frac{E}{2R}\right)^2 \qquad R = \frac{E^2}{4R}$$

Now analyze this same circuit with the current section energized with line current I. This line current enters the mid-tap of the voltage transformer secondary which is very carefully proportioned so as to cause the line current to divide into two equal parts. $I/2$ now passes through the heaters in parallel, then adds together to form again the line current I. Considering only the line current, heat is developed in the first heater at a rate:

$$W_1 = \left(\frac{I}{2}\right)^2 \qquad R = \frac{I^2R}{4}$$

and in the second heater:

$$W_2 = \left(\frac{I}{2}\right)^2 \qquad R = \frac{I^2R}{4}$$

Note that the currents $E/2R$ and $I/2$ add in one heater but subtract in the other. Heat is developed in the first heater at a rate:

$$W_1 = \left(\frac{E}{2R} + \frac{I}{2}\right)^2 \qquad R = \left(\frac{E^2}{4R^2} + \frac{EI}{2R} + \frac{I^2}{4}\right)R$$

and in the second heater:

$$W_2 = \left(\frac{E}{2R} - \frac{I}{2}\right)^2 \qquad R = \left(\frac{E^2}{4R^2} - \frac{EI}{2R} + \frac{I^2}{4}\right)R$$

subtracting:

$$W_1 - W_2 = 2\frac{EI}{2R}R = EI$$

If the temperature rise in each enclosure is proportional to the heat input, the temperature difference between the two enclosures will accordingly be proportional to the difference between these two values or simply EI. Since the values of current and voltage are taken as instantaneous values, the temperature difference is proportional to the power measured, regardless of power factor. If the current and voltage had been taken as effective, or root-mean-square, values,

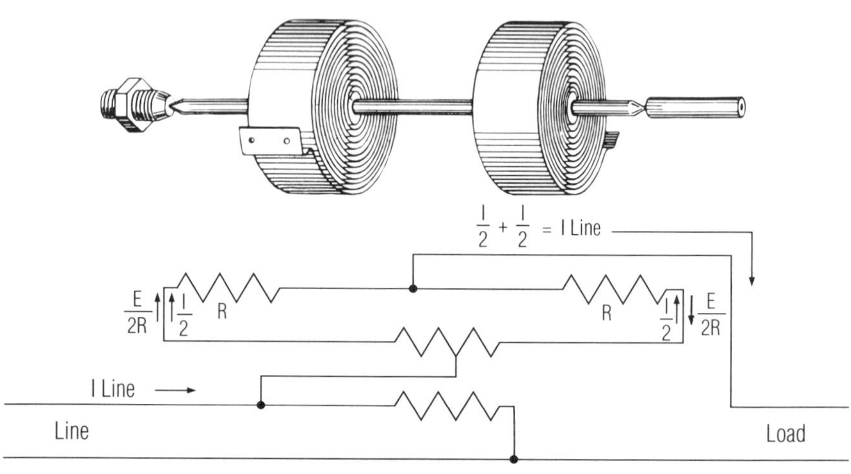

Figure 8-7. Two-Wire Watt Demand Meter with Current and Potential Circuits.

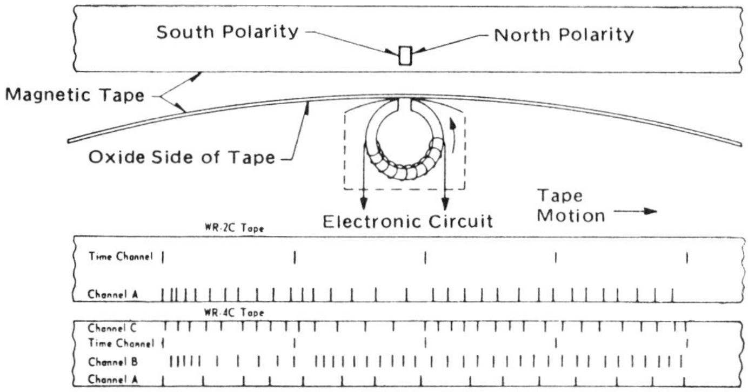

Figure 8-8. Recording Process, Principles of Operation.

then the addition and subtraction of the circulating current and the line current would have been phasorial, which would introduce a cosine term and would show that the temperature difference would be proportional to $EI\cos\theta$ or, again, the power measured by the meter.

It should be noted that in the wattmeter the deflection is proportional to the first power of $EI\cos\theta$ and the wattmeter scale is, accordingly, approximately linear.

To obtain different capacities of meters, the basic element may be shunted or supplied through a current transformer. (It may be noted that the current circuit is non-inductive, since the heaters are non-inductive and the current flows in opposite directions through the secondary of the voltage transformer.) While some demand meters carry line current, either shunted or unshunted through the heater elements, the use of the current transformer permits carrying the design of the meter to higher current ranges and has the advantage of increasing the flexibility of design and permitting operation of the heater itself at a lower insulation level, since it is completely isolated from line voltages. In addition, where both current and voltage transformers are employed, the resistance value of the heater circuits may be increased to take advantage of smaller connecting leads and, accordingly, lower thermal losses.

PULSE OPERATED ELECTRONIC DEMAND METERS

Magnetic Tape Pulse Recorder

In the magnetic tape pulse recorder, data pulses received from the initiating watthour meter are each individually recorded on one channel of the magnetic tape as they are received, while the tape slowly passes the recording head. Recorders are constructed with either one or three data channels. At the same time, a separate channel records an internally generated time pulse on the moving tape at the end of each demand interval. To determine the demand over a demand interval, it is necessary to add all the data pulses between time pulses. The process of reading the pulses from the tape is done at high speeds from 30 to 60 inches per second on special tape drive systems called translators. The translators accumulate the pulses for each interval and transfer the information to some suitable output mode such as computer compatible magnetic tape or punched cards. Modern magnetic tape pulse recorders, whether used for billing or load survey, generally employ a cartridge to facilitate the tape change procedure and to prevent damage to the tape. The typical tape cartridge contains a take-up reel, a supply reel, an automatic reel lock, a rubber idler gear, and a tape head pressure pad. The reels are set on bearings to reduce friction and the supply reel contains over 500 feet of quarter-inch instrumentation grade Mylar magnetic tape.

About 6 feet from the ends of the tape are reflective sections each about 6 inches long which serve to indicate to the translator that the end of the tape is approaching. When the cartridge is not in service, the spring loaded shaft supporting the reel lock on one end and the rubber idler on the other, holds the reel lock against both reels to prevent spillage of the tape during handling. When the cartridge is installed, the shaft is retracted releasing the pressure against the reels by exerting pressure on the rubber idler gear. Thus loaded, the rubber idler presses the tape against the recorder drive capstan which keeps the tape moving at the constant speed of a capstan.

The recorder drive mechanism provides a low torque drive to the takeup reel to prevent tape spillage after the tape passes the rubber idler and drive capstan combination. A slight drag is applied to the supply reel to keep the tape under tension and prevent bunching ahead of the idler and capstan. The pressure pad assures that a constant pressure is applied against the tape as it passes the cartridge head.

The tape, as it leaves the cartridge, passes over a magnet guide which magnetically orients the oxide to negative saturation and also erases any previous data on the tape supply. The primary purpose of the magnet guide is to bias the tape so that a high recorded signal level will be obtained on the time channel pulse that is recorded in a Return-to-Bias manner. Therefore, the recorder bias magnet should not be depended upon to erase previous data since record lengths may vary.

Figure 8-8 shows the principle of operation of the recording process. The recording head contains "C" shaped magnetic cores, the ends of which are separated by a small air gap or insulated space. A coil wound on each magnetic core is connected to an electronic circuit which either provides current of one polarity or the other through the winding or a capacitor discharge. The resultant current creates a flux at the air gap. The flux, in turn, seeks the path of least resistance, which is the oxide surface of the tape pressing against the air gap, and magnetizes it in a direction as determined by the polarity of the current in the winding.

SOLID STATE DEMAND RECORDERS

There is an increasing trend toward the use of solid-state recorders, away from magnetic tape and paper tape recorders which require considerably more maintenance and provide fewer functions.

Solid-state recorders use microelectronic circuits to count and store incoming pulses from the watthour meter. Low-power, large scale integrated (LSI) circuits provide today's solid-state recorders with new capabilities. Tamper detection, on-site data listing and display, and built-in modems which allow remote access to the demand recorder data via the telephone network, are a few of the functions that a solid-state recorder can provide in the same size package as a magnetic tape recorder.

Solid-state recorders have given rise to complete electronic metering systems which provide utilities with the ability to collect remote metering data which in turn improves utility cash flow and prevents revenue loss by the early detection of defective metering equipment.

Figure 8-9 is a block diagram of a typical solid-state demand recorder. The circuits and methods employed in this device which counts and stores pulses are representative of how solid-state technology is used in a demand recorder design, and are described further below.

Computer

The control of a solid state recorder is performed by the main computer which consists of a Central Processing Unit (CPU) with Read-Only Memory (ROM) and Random Access Memory (RAM).

The CPU fetches and executes program instructions stored in the ROM and RAM memory. Generally speaking, the stored program instructs the CPU to per-

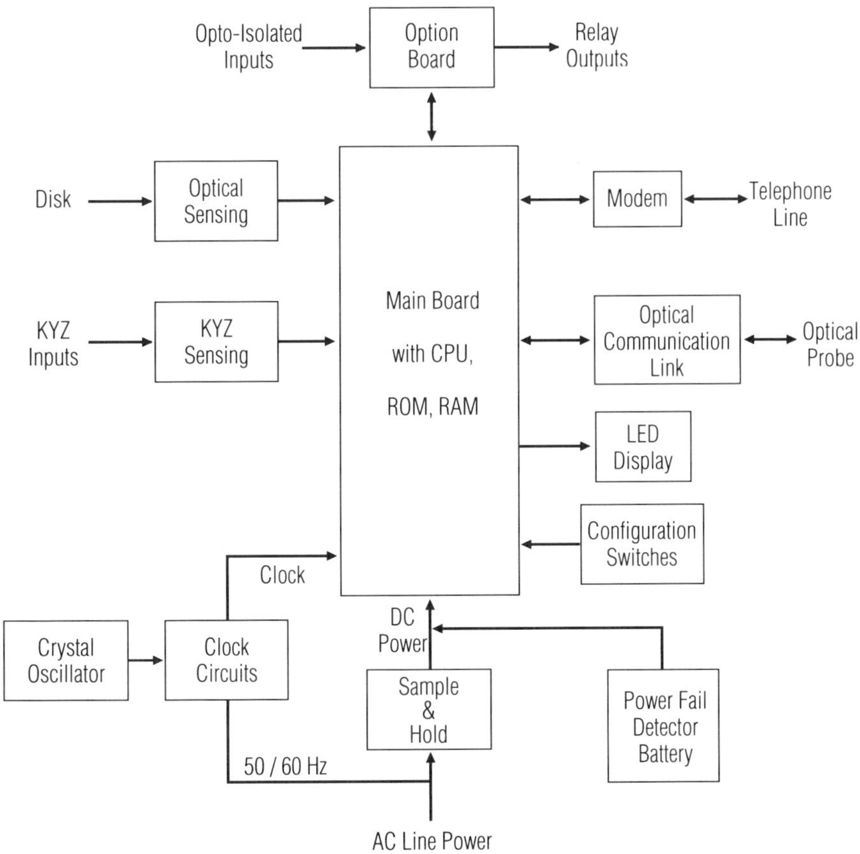

Figure 8-9. Solid State Demand Recorder, Block Diagram.

form two important tasks: maintain time, and count and store incoming energy pulses from the watthour meter pulse initiator.

Timekeeping

Solid state demand recorders can be programmed to accommodate a variety of demand interval periods. Intervals of 1, 5, 15, 30, and 60 minutes are commonly applied throughout the electric utility industry.

When the recorder is energized from system AC sources, it maintains time by tracking the line frequency of the utility system. If system voltage is lost or drops below a pre-determined value, a crystal oscillator powered by a battery takes over as the source for timing pulses.

Timekeeping, whether using the 60 hertz line frequency or the oscillator frequency, is generally performed by the CPU. When powered by the AC line, the CPU senses and counts each zero-crossing of voltage. Two zero-crossings indicate one cycle, and 3,600 cycles indicate an increment of one minute which can be the basis for demand interval timing and a "real time" clock. When powered by the internal battery, the CPU is programmed to count pulses from the crystal and to establish the time reference. In some designs, a special clock chip circuit develops the time reference.

Pulse Accumulation and Storage

The CPU counts pulses sent by the watthour meter pulse initiator, and stores the count in RAM. Solid-state demand recorders are available with a variety of optional input channels and data storage configurations. Commonly encountered recorder configurations provide 1, 2, 3, or 4 data input channels and 64 kilobytes of RAM memory for demand interval data storage. This amount of RAM is sufficient to store more than one month's worth of information.

Typically, each solid-state memory location corresponds with one demand interval. Unlike a magnetic tape recorder which stores each KY and KZ contact closure on magnetic tape, a solid-state recorder stores only the accumulated total of pulses received during each demand interval.

Power Supply

A power supply transforms system AC line voltages (for example: 120 VAC, 240 VAC, or 277 VAC) into the regulated low-power DC voltages required by the solid-state circuits. Supply voltages between 5 and 15 VDC and currents of less than one ampere are typical power requirements of solid state recorders.

Since the AC power to the recorder can sometimes be interrupted, recorder designs include a backup source of electrical power for use during power outages. Rechargeable nickel-cadmium batteries, non-rechargeable lithium cells, and lead acid cells are common battery technologies used in recorders.

Display

Most demand recorders are equipped with displays for on-site presentation of metering information. Liquid Crystal Display (LCD) technology is the most popular display technology. LCDs operate at very low power and are readable even in direct sunlight.

Output Circuits

Solid-state recorders have outputs for such functions as:

1. Load Control—Relay or solid-state outputs to control external electrical loads in accordance with a pre-programmed time schedule.
2. Demand Threshold Alert—Relay or solid-state outputs to indicate that current demand exceeded a programmed demand level.
3. Printer Output—Solid-state outputs to a printer for on-site printing of metering information.
4. End-of-Interval Indicator—Relay or solid-state output to indicate completion of a demand interval.
5. Pulse Input Indicator—Solid-state output, usually connected to an indicating lamp, to indicate that incoming pulses from a pulse initiator are being received by the recorder.

Self-Monitoring Capability

Many of today's solid-state recorders are equipped with self-monitoring diagnostic programs which continuously monitor critical operations. These diagnostic checks may test the RAM, ROM, program execution, and need for battery-backup. If the fault is serious, the self-monitoring program may trigger an orderly shutdown of recorder operations.

Communications

Information stored by solid-state recorders can be loaded automatically into portable terminals using an optical communication port, or sent by telephone lines to a centralized computer system.

Figure 8-10. Single Channel Recording Meter.

Physical Description

Solid-state recorders are generally available in two packaging configurations. Figure 8-10 shows a recorder which is integrally mounted within the watthour meter enclosure along with the pulse initiator. This configuration is generally used when a single channel of information is to be recorded, such as a residential load survey. This packaging style provides for simple, plug-in socket installation without the need for external wiring.

An alternative enclosure is shown in Fig. 8-11. This configuration is often used when multiple channels of information are to be recorded, such as revenue metering of a multi-unit industrial complex.

ELECTRONIC TIME-OF-USE (TOU) DEMAND METERS

Rising costs, environmental concerns and regulatory mandates have prompted many utilities to establish load management programs. A example of one such program is time-of-use metering.

Unlike a conventional tariff which charges the user for the quantity of energy used, a time-differentiated rate structure considers in addition *when* the energy is used. The user is charged a higher price for energy used during time periods when generation and delivery costs for the utility are higher.

Introducing the requirement to measure when energy is used increases the complexity of the meter. Mechanical gear trains controlled by electronic circuits were typical of the earliest TOU meters. These electromechanical meters have been replaced with completely electronic meters. See Fig. 8-12.

Figure 8-11. Solid State Demand Recorder.

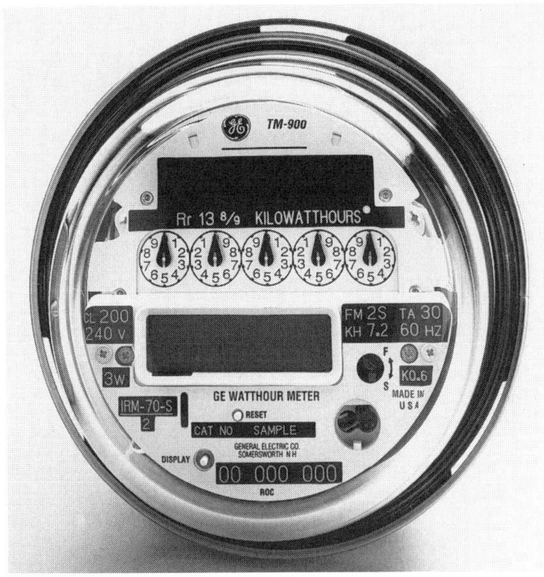

Figure 8-12. Time-of-Use Demand Meter.

TOU meters are designed with the flexibility to accommodate a variety of rate structures, which are defined by the utility. TOU meters are often equipped with programmable displays and rate schedules which accommodate energy (kilowatt-hours) and demand (kilowatt) registration for up to four daily rate periods. Some meter displays can be programmed to show the current rate of consumption or cumulative consumption for demand intervals from 1 to 240 minutes. For some meters, the measurement of rolling demand can be selected, and rolling demand will be calculated and displayed for programmed subintervals.

The indicated demand values can be reset (zeroed) by a mechanism mounted on the meter cover, or by an accessory device, such as a portable reader programmer.

ELECTRONIC DEMAND REGISTERS

Electronic demand registers were made practical with the development of non-volatile memory. This type memory can store billing data and programming constants, and retain this information during power interruptions without need for a battery. When the supply voltage drops below a minimum value, programming constants and billing data are transferred to the non-volatile memory from which it is recalled when power is restored.

The heart of the electronic register is a microcomputer which processes data. Figure 8-13 is a simplified block diagram of an electronic demand register showing the typical inputs to, and outputs from, the microcomputer.

Meter Disk Sensing

When used with an electromechanical meter, an electronic register must sense disk rotation. Most electronic registers use optical sensors which produce several

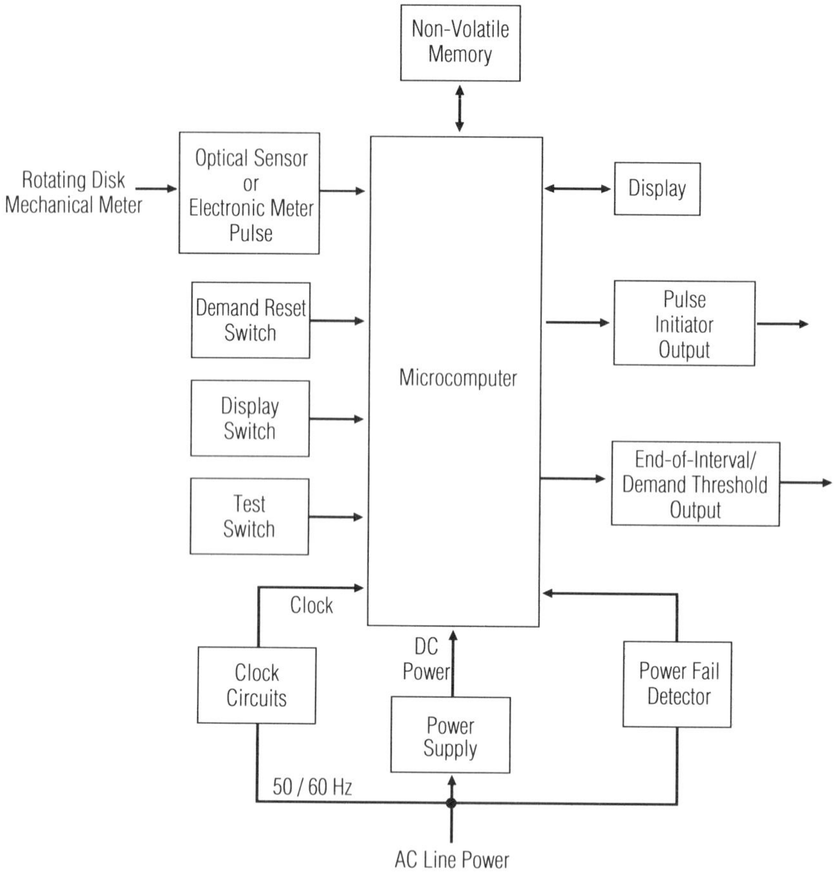

Figure 8-13. Electronic Demand Register, Block Diagram.

pulses for each disk revolution by sensing marks on the disk, or by detecting motion of a shutter mounted on the meter disk shaft.

When built into a totally electronic meter, the register receives pulses directly from the meter measurement circuits. Each pulse represents a value of energy consumed. Regardless of how pulses are generated, the microcomputer is programmed to convert each incoming pulse into an energy value.

Figure 8-14 illustrates a typical electromechanical meter. These meters are also available as models without dials, and without mechanical registers.

Digital (Pulse) Processing Microcomputer

Electronic demand registers process incoming pulses and perform the following functions:

1. Establish accurate and precise time intervals.
2. Count and accumulate pulses, convert pulse-counts into engineering units, and store and display data.

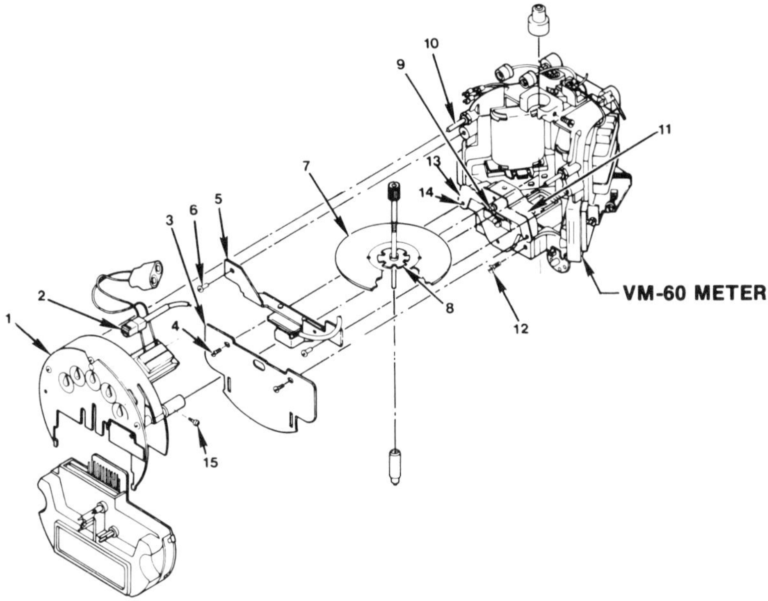

Figure 8-14. Electromechanical Meter with Electronic Demand Register.

3. Count and accumulate pulses for successive demand intervals, and detect and store the interval in which power consumption was maximum. At the end of each interval, reset the count to zero in preparation for the next interval.

Programming

Electronic registers may be pre-programmed with parameters such as:

1. The energy value of each input pulse.
2. Demand interval length (for fixed, or rolling calculations)
3. Type of demand display (maximum, or cumulative)
4. Display format (how many digits, and the decimal location)
5. Items to be displayed
6. The function of the output switch (for example: pulse initiator, end-of-interval, demand threshold alert).

Making changes in software allows one hardware design to serve a variety of applications. Program changes are analogous to a mechanical meter manufacturer selecting different gear ratios, dial scales, and timing motor speeds.

Rolling Demand-Capability

Mechanical registers use external pulse-operated demand recorders and translators to perform rolling-demand calculations. Electronic registers can include the ability to calculate rolling-demand, thus eliminating the need for separate pulse-operated recorders and subsequent translation of the data.

Rolling demand, also called "sliding window", is a process by which intervals are divided into a fixed number of subintervals. Instead of calculating demand only at the end of each interval, the calculation is performed at the end of each subinterval, and totaled and averaged for the interval. Greater accuracy results, and in some cases parameters such as maximum demand for an interval can

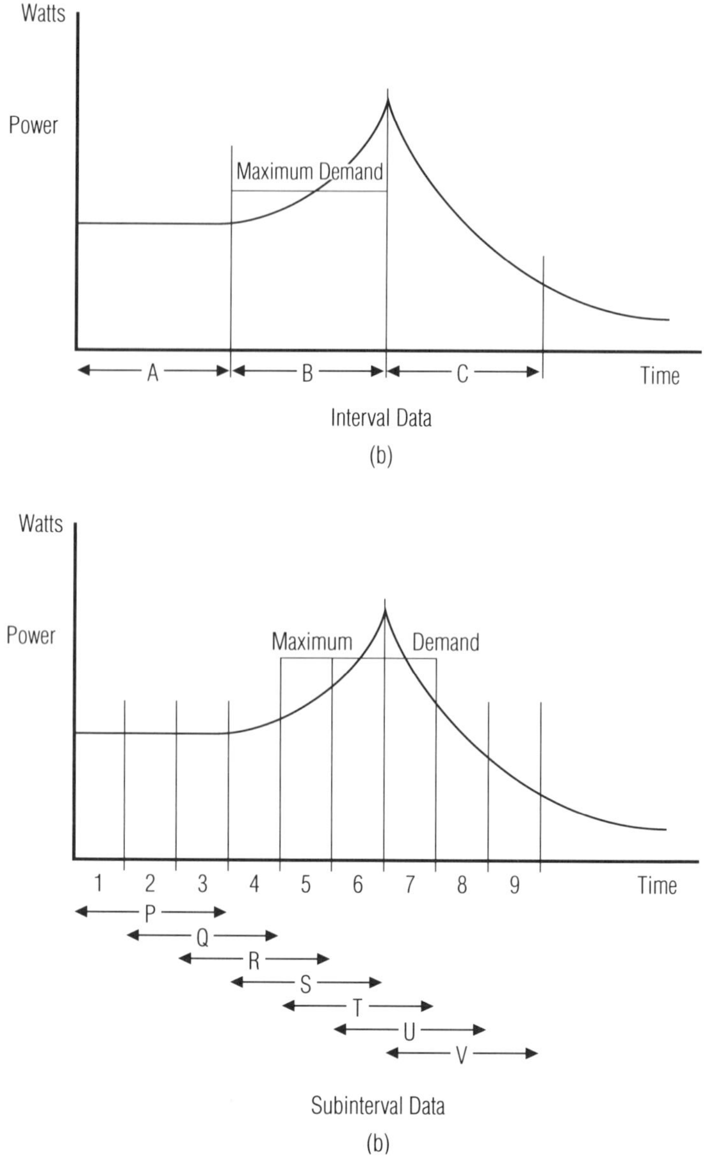

Figure 8-16. Fixed Demand, Rolling Demand Calculation.

change. For example, notice in the two curves shown in Fig. 8-15 that while the power consumption is the same, Maximum Demand is greater when subinterval data is the basis for the calculation. Notice also that the interval length does not change, and that subinterval demand calculations always use the most recent consecutive subintervals which make up the interval.

Display Modes

Electronic demand registers display several types of information. The normal display mode presents information needed on a regular basis such as data related to billing. An alternate mode can display information useful to field technicians or shop personnel, for verifying register program constants. A test mode provides information helpful during troubleshooting.

Output Signals

Most electronic demand registers are capable of providing output signals for communicating with external monitoring and control equipment. These output signals are generally a relay closure activated by the microprocessor. Outputs can include: pulse initiator output (KYZ), end-of-interval signal, and demand threshold alert.

Demand Accuracy

The accuracy of mechanical and thermal demand registers is typically expressed as a percent of full scale value. With these meters there is a fixed magnitude of error which can be present anywhere on the scale. The lower on the scale at which a reading is taken, the larger this fixed error is as a percent of actual reading. To minimize the effect of this error, meters are supplied with several full-scale values and, by proper register selection, readings can be taken above half-scale and provide reasonable accuracy. Manufacturers normally supply mechanical registers with three overload capacities designated Class 1, 2, and 6 to indicate register overload capacity.

Electronic demand registers are pulse-operated; the electronic counter receives a fixed number of pulses per disk revolution. The accuracy of electronic registers is typically expressed in terms of a pulse-count deviation over a given period of time. With electronic registers, accuracy is always within a few tenths of a percent, even at readings of less than 50% of "scale point." With electronic registers, it is not necessary to designate register classes, and readings are equally accurate throughout the scale range.

Demand Resolution

Many electronic registers provide resolution of one or more places to the right of the decimal point. This increased resolution makes it possible to capture readings closer to the actual demand. In most cases, the improvement in resolution yields a higher reading.

Improved Testability

Mechanical demand registers are tested by inputting enough meter disk revolutions, either directly or from a test board, to drive the demand pointer(s) to a reading above 25% of full scale, within the time window of the register's demand interval. If adjustment is required the test is repeated, requiring several demand intervals of time to complete the test procedure.

Electronic registers are not adjusted or calibrated. If a register is defective, the electronic module is simply replaced. Registers which incorporate a "Test Mode" further assist the test by injecting a few test pulses to quickly verify if the demand calculations are being performed correctly.

Electronic Demand for Multi-Function Meters.

Multi-function demand meters (registers) log and display demand and energy quantities for bi-directional, multi-function meters. Examples of the parameters displayed are kWh, kW, kVARh, kVARd, kQh, kQd, kV^2, power factor, average current and voltage per phase, and various maximum values with the times and dates they occurred.

Some multi-function registers have time-of-use functions with sufficient internal memory to perform interval recording. Some multi-function registers display energy and reactive power in four-quadrant values, such as forward and reverse kWh and kVARh.

Multi-function registers are similar to single-function registers, in that both monitor time, date, and other calendar data for a billing period.

CHAPTER 9

KILOVAR AND KILOVOLT-AMPERE METERING

FOREWORD

THE DESIGN OF an electric utility system is based on the total kilovolt-ampere (kVA) load to be served. KVA may be regarded as consisting of two components: kilowatts (kW) and kilovars (kVAR). Often revenue is derived solely from only one of these components, kilowatts. The ratio of kW to kVA is the power factor. It may also be defined as the ratio of power-producing current in a circuit to the total current in that circuit.

$$\frac{\text{kW}}{\text{kVA}} = \text{Power Factor} = \frac{\text{kW Current}}{\text{Total Current}}$$

A poor ratio of kW to kVA — low power factor — has a serious effect on the economic design and operating costs of a system. When power factor is low and rates are based only on kW, the utility is not being compensated for all the kVA it is required to generate, transmit, and distribute. To compensate for this, rate schedules have been established which take into consideration the power factor of the load being measured. These schedules take a variety of forms but in general they penalize poor power factor or reward good power factor.

The principal purpose of kVAR and kVA metering (or phase-displaced metering which is a general term describing this entire category) is in its application to the measurement of one or more of the quantities involved: power factor, average power factor, kVA, kVA-hours (kVAh), kilovars (kVAR), or kilovar-hours (kVARh). The method used depends upon the rate, the cost of the metering, and the accuracy required.

THEORY

The current required by induction motors, transformers, and other induction devices may be considered to be made up of two kinds of current: magnetizing current and power-producing current.

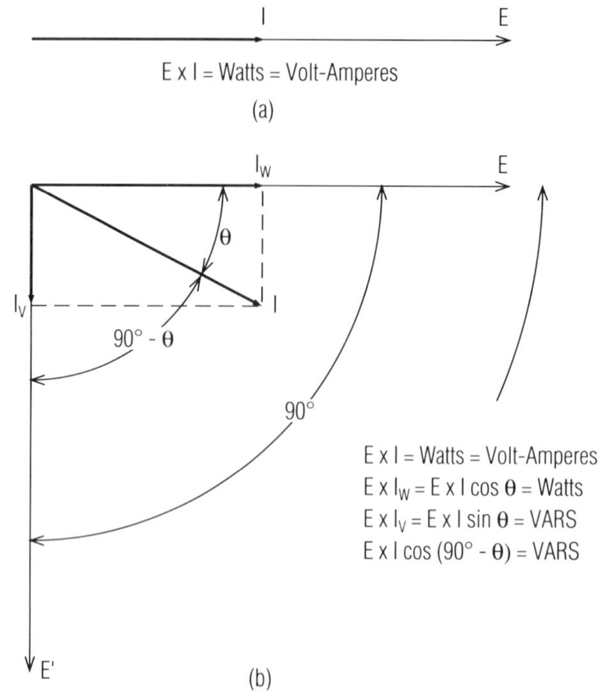

Figure 9-1. Phasor Relationships.

Power-producing current or working current is that current which is converted into useful work. The unit of measurement of the power produced is the watt or kilowatt.

Magnetizing current, which is also known as wattless, reactive, quadrature, or non-working current, is that current which is required to produce the magnetic fields necessary for the operation of induction devices. Without magnetizing current, energy could not flow through the core of a transformer or across the air gap of an induction motor. The unit of measurement of magnetizing volt-amperes is the var or kilovar. The word "var" is derived from "volt-amperes-reactive" and is equal to the voltage times the magnetizing current in amperes.

The total current is the current which would be read on an ammeter in the circuit. It is made up of both the magnetizing current and the power-producing current which add phasorially,

$$\text{Total Current} = \sqrt{(\text{kW Current})^2 + (\text{kVAR Current})^2}$$

Similarly,

$$\text{kVA or Apparent Power} = \sqrt{(\text{kW})^2 + (\text{kVAR})^2}$$

These relations are easily shown by triangles. See Fig. 4-22.

Phasor Relationships

In a circuit which contains only resistance, the current *I* is in phase with the voltage *E*. See Fig. 9-1(a). In this ideal case, watts equal volt-amperes.

When reactance (inductive or capacitive) is introduced into the circuit, the current is displaced or shifted out of phase with the voltage by an angle θ, depending on the relative amounts of resistance and reactance. Normally the reactance is inductive and the current I lags the voltage E. See Fig. 9-1(b). If the reactance should be capacitive, the current would lead the voltage.

The current I can be considered to be made up of two components: I_W which is in phase with E and which produces watts, and I_V which is displaced 90 degrees from E and which produces reactive volt-amperes (vars). By trigonometry:

$$I_W = I \cos \theta$$
$$\text{and } I_V = I \sin \theta$$

then:

$$E \times I \cos \theta = \text{watts}$$
$$\text{and } E \times I \sin \theta = \text{vars}$$

Again by trigonometry, the cosine varies from zero (0) for an angle of 90 degrees to one (1.0) for an angle of zero degrees. As the displacement angle becomes smaller the power factor improves. When the power factor is unity, watts and volt-amperes are equal to each other and reactive volt-amperes equal zero.

Phase Sequence

When phase-displaced metering is encountered, designers must have a definite knowledge of phase sequence. Watthour meters and wattmeters for measuring energy and power, respectively, can be correctly connected without consideration for phase sequence. However, when these meters are used to measure varhours or vars, the phase sequence must be known in order to make correct connections for forward rotation of the meter disk or upscale indication of the wattmeter.

Any phase-sequence identification, either letters or numbers, may be used. 1-2-3 has been established as standard and is therefore preferred. These phase

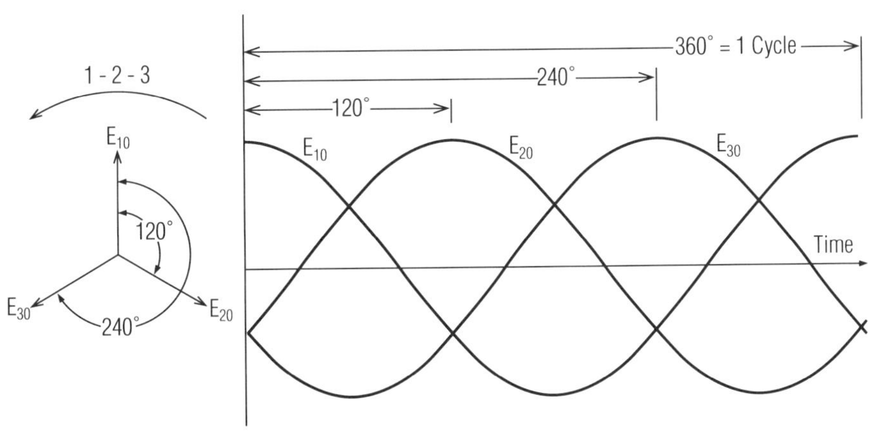

Figure 9-2. Phase Sequence and Phase Rotation.

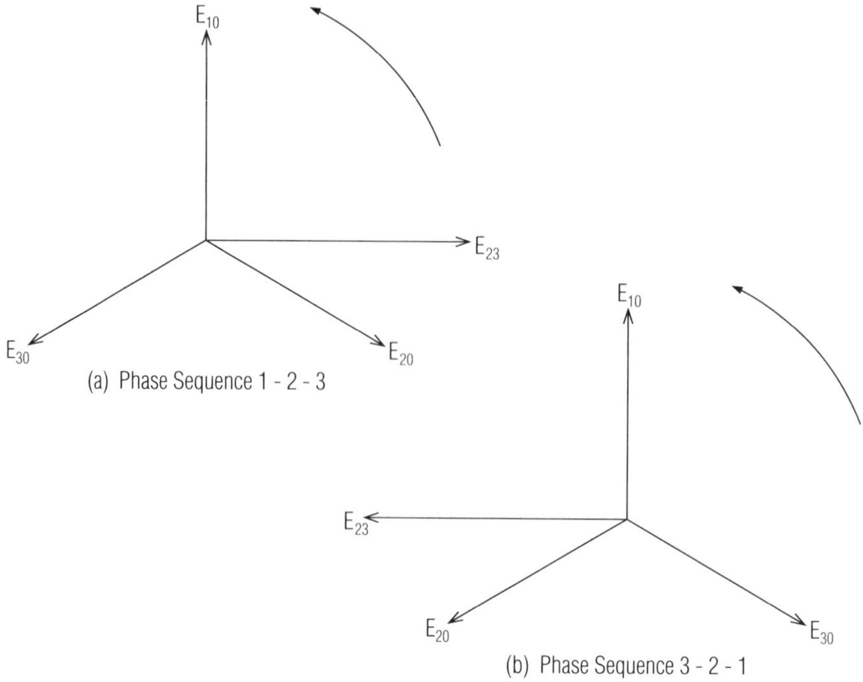

Figure 9-3. Importance of Phase Sequence.

identifications may not necessarily indicate the actual phases emanating from the generating station, but do indicate the sequence in which phase voltages reach their maximum values in respect to time. By common consent, counterclockwise phase rotation has been chosen for general use in phasor diagrams. In diagrams in which the curved arrow is omitted, counterclockwise phase rotation is always implied.

In Fig. 9-2, the sine wave E_{10} reaches its maximum value one-third of a cycle (or 120 degrees) before E_{20}. In turn, E_{20} reaches its maximum value one-third of a cycle before E_{30}. This phase sequence is E_{10}-E_{20}-E_{30} or, in conventional terms, 1-2-3. As the phasor diagram "rotates" counterclockwise, the phasors pass any point in the same sequence, E_{10}-E_{20}-E_{30}.

Most methods of phase-displaced metering use cross-phasing or autotransformers to obtain the desired phase shift. These methods require an integrated connection of all the phases. If phase sequence were not known, a connection intended to obtain a lagging voltage might result in a leading voltage which would result in incorrect metering. In Fig. 9-3(a), with supposed phase sequence 1-2-3, phasor E_{23} lags E_{10} by 90 degrees. In 9-3(b), with the opposite sequence, the same connection produces phasor E_{23} which leads E_{10} by 90 degrees.

There are various types of phase-sequence indicators available, from a lamp-reactor device to a small three-phase motor device. One type of phase-sequence indicator that operates similarly to a three-phase motor contains, as a rotor, a revolving disk connected so as to rotate in predetermined directions for 1-2-3 or 3-2-1 phase sequence.

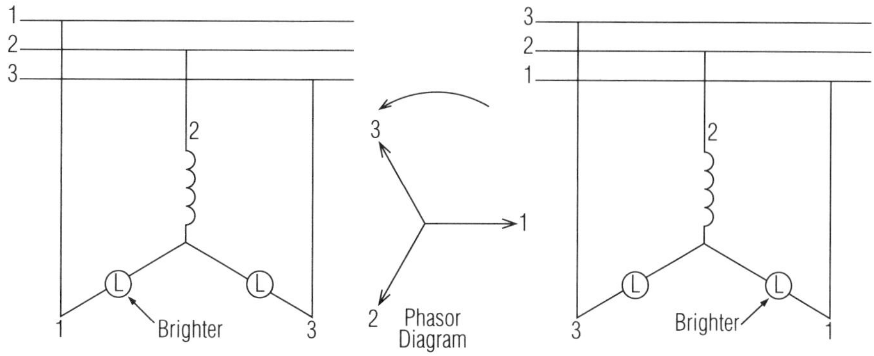

Figure 9-4. Use of Phase-Sequence Indicator.

A lamp-reactor, phase-sequence indicator can easily be constructed with two lamps and a watthour meter voltage coil (Fig. 9-4). If the common point of a wye-connected load is not grounded, it is free to shift from neutral. When the load consists of one reactor and two resistive components (such as lamps) the common point shifts toward the phase which lags the phase connected to the reactor. This results in less voltage on and, consequently, less light from the lamp connected to the lagging phase and more voltage on and brighter light from the lamp connected to the leading phase.

A phase-sequence indicator may be constructed for a service consisting of two phases and the neutral of a wye bank. This device uses two resistors, a capacitor, and a neon lamp, as shown in Fig. 9-5. The neon lamp lights only when the capacitor is connected to the lagging phase. This method also involves a shift of a common point. The neon lamp has such a high impedance that the point O (Fig. 9-5) can be said to be floating and so is free to shift so that either 120 volts or 0 volts exist between O and N, depending on the connection of the phase-sequence indicator.

Another method frequently used consists of a non-inductive resistor of 3,000-5,000 ohms, connected in series with the voltage circuit of a watthour meter (or a portable watthour meter standard), the current coil being connected to another phase. This method is the most convenient in routine testing of polyphase meters. If the meter is connected so that the phase in the current coil lags the phase in the voltage coil, introduction of the resistance will cause the meter to run considerably slower or reverse in direction. If the meter is connected with the current leading, the speed of rotation with resistance cut in will be increased but still forward in direction. A suitable non-inductive load may be used in the current circuit.

BASIC METHODS

It has been established that

$$W = E\,I\cos\theta$$

or watts equal volts × the working component of the current. This quantity is read by a wattmeter.

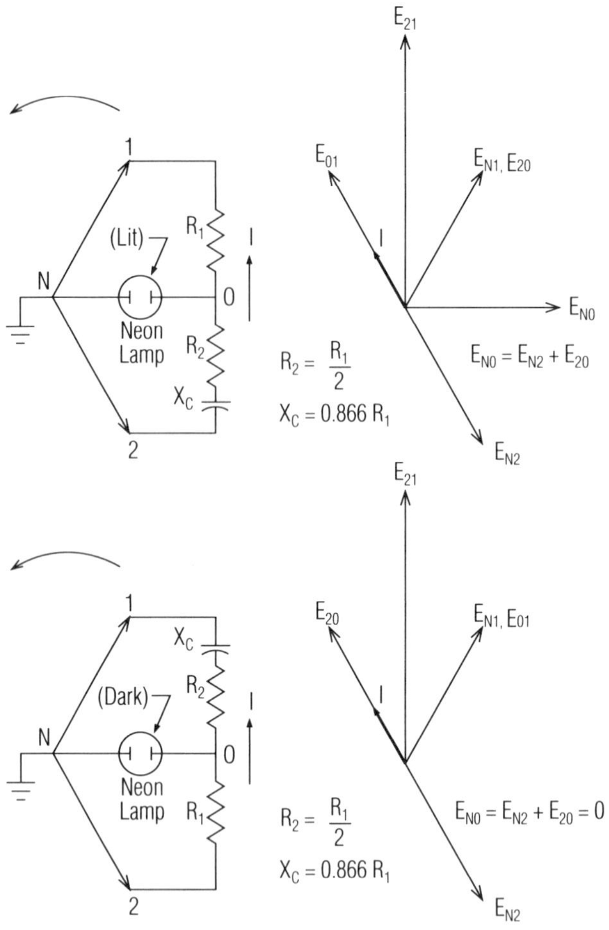

Figure 9-5. Use Of Special Phase-Sequence Indicator. Neon Lamp Lights Only When X_c is Connected to the Lagging Phase.

Similarly

$$\text{Vars} = E\,I \sin \theta$$

or vars equal volts × the magnetizing component of the current.

Since the magnetizing component of the current lags the working component by 90 degrees, this quantity could be read by a wattmeter if the voltage applied to the wattmeter could be displaced by 90 degrees to bring it in phase with the magnetizing current.

Volt-amperes are the product of volts and amperes without regard to the phase angle between these two quantities. Volt-amperes or kVA can be measured by a wattmeter in which the voltage is displaced by the amount of the phase angle between voltage and current. In this case power factor must be known and must be constant, although the meter will operate within commercial limits with limited variation of power factor.

Note that in both var metering and kVA metering a phase displacement is necessary. Hence the general term for metering which measures vars and kVA is "phase-displaced metering."

There are several basic approaches to phase-displaced metering. The method chosen depends upon the information desired, the meters and instruments available, and the degree of precision required from the measurements.

Some methods are applicable only to spot measurements, others can be used with integrating watthour meters for a continuous record.

Basic methods are:

1. Use of instruments (voltmeters, ammeters, wattmeters, etc.) to measure basic quantities.
2. 90-degree phase shift of voltage to measure vars.
3. Cross phasing.
4. Displaced voltage wattmeters. Shift voltage θ degrees (Fig. 9-1) to measure volt-amperes.
5. Use of special meters.

Combinations of two or more of the preceding methods may be used. To avoid any confusion in phase-displaced measurement, particularly where refinements are necessary, the following should be carefully observed:

The power factor of a circuit is the ratio between the active power (kW) and the apparent power (kVA). The power factor of a circuit is never greater than 1.0.

The power factor of a single-phase circuit is the cosine of the angle between the voltage and current and, in a wye system, of the angle between the respective phase voltages and currents.

The relationship is less apparent in a delta system. Power factor is the cosine of the angle between the respective line currents and imaginary wye voltages inscribed in the delta (see Fig. 9-6). The voltages which are available for measurement are the phase-to-phase delta voltages, which are 30 degrees out of

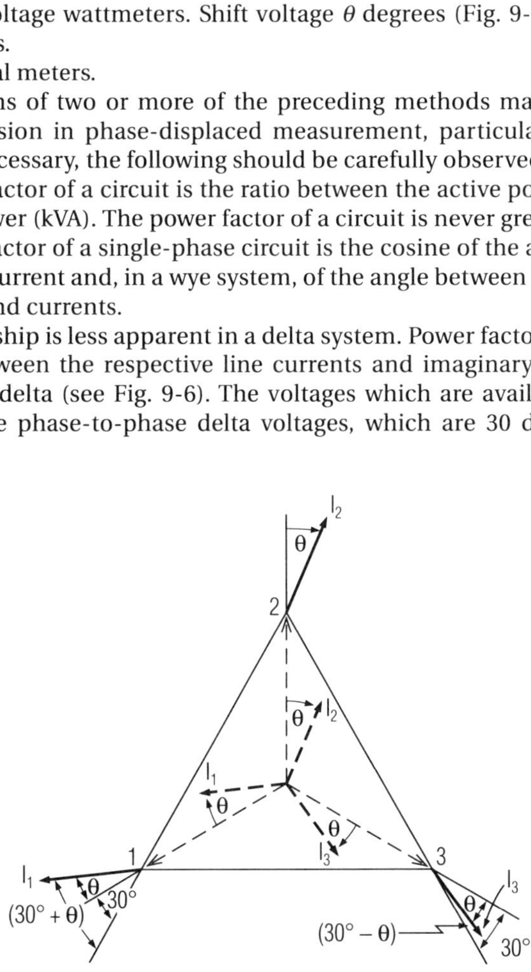

Figure 9-6. Power Factor Relationships in Delta Systems.

phase with the imaginary wye voltages. This introduces apparent phase angles of $(30 + \theta)$ and $(30 - \theta)$ degrees.

Referring again to Fig. 9-6, with both voltage coils connected to phase 2, phase 1 (leading phase 2) is associated with the $(30 + \theta)$ degrees angle and phase 3 (lagging phase 2) with the $(30 - \theta)$ degrees angle.

The assumption is made that all currents and voltages are perfect sine waves. Some instruments and methods will be in error if the sine wave is distorted.

Most methods of phase-displaced metering use cross-phasing or autotransformers to obtain the desired phase shift. These methods are subject to error if all phase voltages are not equal and at the proper phase angle displacement from each other.

In a balanced polyphase system, the currents and voltages are symmetrical. The term balanced polyphase system may be applied to a two-phase system as well as to a three-phase system.

The assumption is generally made that the circuit is inductive: line current lags applied voltage, as in Fig. 9-1. If the circuit is capacitive and line current leads applied voltage, then varmeters will indicate downscale and varhour meters will rotate backward. Most phase-displaced meters can be reconnected to reverse E' (Fig. 9-1) and give upscale indication or forward rotation on leading power factor. Varhour meters are usually equipped with detents to prevent occasional backward rotation in circuits in which the power factor varies and may sometimes go leading (capacitive).

Methods Using Instruments

In a two-wire, single-phase circuit, power factor, volt-amperes, watts, and vars can be measured by using a wattmeter, voltmeter, and ammeter. The power factor is equal to the power in watts indicated by the wattmeter divided by the product of the voltage across the voltage coil of the wattmeter and the current in the current coil of the wattmeter.

$$\text{Power Factor} = \frac{W}{E \times I} = \frac{\text{Watts}}{\text{Volts-Amperes}}$$

Vars may also be calculated from the preceding data:

$$\text{Vars} = \sqrt{(\text{Volt-Amp})^2 - \text{Watts}^2}$$

Variations of this method use a wattmeter and a varmeter or a power factor meter with a voltmeter and ammeter. This method may be used on balanced polyphase circuits or may be extended to unbalanced circuits by using more than one set of instruments.

In a balanced three-phase, three-wire circuit, the power factor may be readily determined from two wattmeter measurements or from measurements with two watthour meters. The equations for determining the power factor of a balanced threephase, three-wire circuit with two wattmeters are:

$$\text{Tan}\ \theta = \sqrt{3}\ \frac{W_2 - W_1}{W_2 + W_1}$$

$$\text{Power Factor} = \cos\theta$$

where θ is the angle of lag or lead of the current and W_1 and W_2 the readings of the wattmeters. The larger reading is W_1 and is always positive; the smaller reading W_2 may be either positive or negative.

Since the time in seconds for a specific number of revolutions of a watthour meter is inversely proportional to the load in watts, the following may be substituted for the preceding equation:

$$\text{Tan}\ \theta = \sqrt{3}\ \frac{S_1 - S_2}{S_1 + S_2}$$

where S_1 and S_2 are the times in seconds for a specific number of revolutions of watthour meters W_1 and W_2 respectively. W_1 and W_2 may be two single-phase meters or the two stators of a polyphase watthour meter. Table 9-1 furnishes a convenient means for determining the power factor corresponding to different values of the ratio

$$\frac{W_1}{W_2} \text{ or } \frac{S_2}{S_1}$$

Table 9-1. Value of Power Factor (Cosine θ) for $\frac{W_1}{W_2}$ or $\frac{S_2}{S_1}$.

For Positive Values of $\frac{W_1}{W_2}$ or $\frac{S_2}{S_1}$.

$\frac{W_1}{W_2}$	PF	$\frac{W_1}{W_2}$	PF	$\frac{W_1}{W_2}$	PF	$\frac{W_1}{W_2}$	PF	$\frac{W_1}{W_2}$	PF
0.848	0.99	0.543	0.89	0.381	0.79	0.246	0.69	0.117	0.59
0.790	0.98	0.525	0.88	0.367	0.78	0.233	0.68	0.104	0.58
0.747	0.97	0.507	0.87	0.353	0.77	0.220	0.67	0.092	0.57
0.712	0.96	0.490	0.86	0.339	0.76	0.207	0.66	0.079	0.56
0.681	0.95	0.473	0.85	0.325	0.75	0.194	0.65	0.066	0.55
0.654	0.94	0.457	0.84	0.312	0.74	0.181	0.64	0.053	0.54
0.628	0.93	0.441	0.83	0.298	0.73	0.168	0.63	0.040	0.53
0.605	0.92	0.426	0.82	0.285	0.72	0.156	0.62	0.026	0.52
0.583	0.91	0.410	0.81	0.272	0.71	0.143	0.61	0.013	0.51
0.563	0.90	0.396	0.80	0.259	0.70	0.130	0.60	0.000	0.50

For Negative Values of $\frac{W_1}{W_2}$ or $\frac{S_2}{S_1}$.

$\frac{W_1}{W_2}$	PF	$\frac{W_1}{W_2}$	PF	$\frac{W_1}{W_2}$	PF	$\frac{W_1}{W_2}$	PF	$\frac{W_1}{W_2}$	PF
0.013	0.49	0.154	0.39	0.312	0.29	0.498	0.19	0.729	0.09
0.027	0.48	0.169	0.38	0.329	0.28	0.519	0.18	0.756	0.08
0.040	0.47	0.183	0.37	0.346	0.27	0.540	0.17	0.783	0.07
0.054	0.46	0.199	0.36	0.364	0.26	0.562	0.16	0.811	0.06
0.068	0.45	0.214	0.35	0.382	0.25	0.584	0.15	0.840	0.05
0.082	0.44	0.230	0.34	0.400	0.24	0.607	0.14	0.870	0.04
0.096	0.43	0.246	0.33	0.419	0.23	0.630	0.13	0.901	0.03
0.110	0.42	0.262	0.32	0.438	0.22	0.654	0.12	0.933	0.02
0.124	0.41	0.278	0.31	0.458	0.21	0.678	0.11	0.966	0.01
0.139	0.40	0.295	0.30	0.478	0.20	0.703	0.10	1.000	0.00

Examples showing use of Table 9-1.

A polyphase meter has a watthour constant of 6. With the voltage circuit of only one stator energized, the meter makes ten revolutions in 52.7 seconds and with the voltage circuit of only the other stator energized it makes two revolutions in 30.9 seconds. To find the power factor:

$$S_2 \text{ for 10 revolutions} = 52.7 \text{ seconds}$$

$$S_1 \text{ for 10 revolutions} = 30.9 \times \frac{10}{2}$$

$$= 154.5 \text{ seconds,}$$

$$\frac{S_2}{S_1} = \frac{52.7}{154.5} = 0.341$$

From Table 9-1, cos θ (power factor) corresponding to $\frac{S_2}{S_1} = 0.341$ is 0.76.

From the equation:

$$\frac{K_h \times 3600 \times Rev}{t(S_1 \text{ or } S_2)} = W_1$$

the watts corresponding to $S_2 = 52.7$ and $S_1 = 154.5$ seconds are 4,100 and 1,400 respectively. $\frac{W_1}{W_2} = \frac{1400}{4100} = 0.341$, or cos $\theta = 0.76$, as before.

If the power factor had been such that the meter had run backward during the time, and $S_1 = 154.5$ seconds, then

$$\frac{S_2}{S_1} = \frac{52.7}{-154.5} = -0.341$$

Power factor corresponding to $\frac{S_2}{S_1} = -0.341$ is 0.27.

90-Degree Phase Shift

The most popular methods used in var metering involve quadrature voltages, i.e., a 90-degree phase shift of voltage is applied to a wattmeter or watthour meter. The 90-degree phase shift of voltage can be obtained in several ways.

Wattmeters and watthour meters are designed to indicate or to rotate at speeds proportional to the product of the voltage on the voltage coil, the current in the current coil, and the cosine of the angle between them:

$$W = EI \cos \theta$$

From Fig. 9-1, if the voltage E' were substituted for E in the wattmeter or watthour meter, it would indicate:

$$E'I \cos(90 - \theta) \text{ which equals } EI \sin \theta = \text{Vars.}$$

Simultaneous readings of watts and vars can be used to calculate power factor and volt-amperes.

Power Factor is equal to the cosine of the angle whose tangent is Vars/Watts, which in mathematical notation is shown as

$$\text{Power Factor} = \cos\left(\tan^{-1} \frac{\text{Vars}}{\text{Watts}}\right)$$

Table 9-2. Power Factors for $\frac{\text{Vars}}{\text{Watts}}$ or $\frac{\text{Varhours}}{\text{Watthours}}$.

PF	0.000	0.001	0.002	0.003	0.004	0.005	0.006	0.007	0.008	0.009
1.00	0.0									
0.99	0.1425	0.1351	0.1273	0.1189	0.1100	0.1004	0.0897	0.0776	0.0633	0.0448
0.98	0.2031	0.1978	0.1923	0.1868	0.1811	0.1752	0.1691	0.1628	0.1563	0.1496
0.97	0.2506	0.2462	0.2417	0.2372	0.2326	0.2279	0.2231	0.2183	0.2133	0.2082
0.96	0.2917	0.2878	0.2838	0.2799	0.2758	0.2718	0.2676	0.2635	0.2592	0.2550
0.95	0.3287	0.3251	0.3215	0.3179	0.3143	0.3106	0.3069	0.3031	0.2993	0.2955
0.94	0.3630	0.3596	0.3563	0.3529	0.3495	0.3461	0.3427	0.3392	0.3357	0.3322
0.93	0.3952	0.3921	0.3889	0.3857	0.3825	0.3793	0.3761	0.3728	0.3695	0.3663
0.92	0.4260	0.4230	0.4199	0.4169	0.4138	0.4108	0.4077	0.4046	0.4015	0.3984
0.91	0.4556	0.4527	0.4498	0.4468	0.4439	0.4409	0.4380	0.4350	0.4320	0.4290
0.90	0.4843	0.4815	0.4786	0.4758	0.4729	0.4701	0.4672	0.4643	0.4614	0.4585
0.89	0.5123	0.5095	0.5068	0.5040	0.5012	0.4984	0.4956	0.4928	0.4900	0.4872
0.88	0.5397	0.5370	0.5343	0.5316	0.5288	0.5261	0.5233	0.5206	0.5178	0.5151
0.87	0.5667	0.5640	0.5614	0.5587	0.5560	0.5533	0.5506	0.5479	0.5452	0.5425
0.86	0.5934	0.5907	0.5881	0.5854	0.5827	0.5801	0.5774	0.5747	0.5721	0.5694
0.85	0.6197	0.6171	0.6145	0.6119	0.6092	0.6066	0.6040	0.6013	0.5987	0.5960
0.84	0.6459	0.6433	0.6407	0.6381	0.6355	0.6329	0.6302	0.6276	0.6250	0.6224
0.83	0.6720	0.6694	0.6668	0.6642	0.6616	0.6590	0.6564	0.6538	0.6512	0.6485
0.82	0.6980	0.6954	0.6928	0.6902	0.6876	0.6850	0.6824	0.6798	0.6772	0.6746
0.81	0.7240	0.7214	0.7188	0.7162	0.7136	0.7110	0.7084	0.7058	0.7032	0.7006
0.80	0.7500	0.7474	0.7448	0.7422	0.7396	0.7370	0.7344	0.7318	0.7292	0.7266
0.79	0.7761	0.7735	0.7709	0.7682	0.7656	0.7630	0.7604	0.7578	0.7552	0.7526
0.78	0.8023	0.7997	0.7970	0.7944	0.7918	0.7892	0.7865	0.7839	0.7813	0.7787
0.77	0.8286	0.8260	0.8233	0.8207	0.8181	0.8154	0.8128	0.8102	0.8075	0.8049
0.76	0.8552	0.8525	0.8498	0.8472	0.8445	0.8419	0.8392	0.8366	0.8339	0.8313
0.75	0.8819	0.8792	0.8765	0.8739	0.8712	0.8685	0.8658	0.8632	0.8605	0.8578
0.74	0.9089	0.9062	0.9035	0.9008	0.8981	0.8954	0.8927	0.8900	0.8873	0.8846
0.73	0.9362	0.9335	0.9307	0.9280	0.9253	0.9225	0.9198	0.9171	0.9144	0.9116
0.72	0.9639	0.9611	0.9583	0.9555	0.9528	0.9500	0.9472	0.9445	0.9417	0.9390
0.71	0.9918	0.9890	0.9862	0.9834	0.9806	0.9778	0.9750	0.9722	0.9694	0.9666
0.70	1.0202	1.0173	1.0145	1.0116	1.0088	1.0060	1.0031	1.0003	0.9975	0.9947
0.69	1.0490	1.0461	1.0432	1.0403	1.0374	1.0345	1.0317	1.0288	1.0259	1.0231
0.68	1.0783	1.0753	1.0724	1.0694	1.0665	1.0636	1.0606	1.0577	1.0548	1.0519
0.67	1.1080	1.1050	1.1020	1.0990	1.0960	1.0931	1.0901	1.0871	1.0842	1.0812
0.66	1.1383	1.1352	1.1322	1.1291	1.1261	1.1231	1.1200	1.1170	1.1140	1.1110
0.65	1.1691	1.1660	1.1629	1.1598	1.1567	1.1536	1.1506	1.1475	1.1444	1.1413
0.64	1.2006	1.1974	1.1942	1.1911	1.1879	1.1848	1.1816	1.1785	1.1754	1.1722
0.63	1.2327	1.2294	1.2262	1.2230	1.2198	1.2166	1.2133	1.2101	1.2070	1.2038
0.62	1.2655	1.2622	1.2589	1.2556	1.2523	1.2490	1.2457	1.2425	1.2392	1.2359
0.61	1.2990	1.2956	1.2923	1.2889	1.2855	1.2822	1.2788	1.2755	1.2721	1.2688
0.60	1.3333	1.3299	1.3264	1.3230	1.3195	1.3161	1.3126	1.3092	1.3058	1.3024
0.59	1.3685	1.3649	1.3614	1.3578	1.3543	1.3508	1.3473	1.3438	1.3403	1.3368
0.58	1.4045	1.4009	1.3972	1.3936	1.3900	1.3864	1.3828	1.3792	1.3756	1.3720
0.57	1.4415	1.4377	1.4340	1.4303	1.4266	1.4229	1.4192	1.4155	1.4118	1.4082
0.56	1.4795	1.4756	1.4718	1.4680	1.4641	1.4603	1.4565	1.4528	1.4490	1.4452
0.55	1.5185	1.5145	1.5106	1.5067	1.5027	1.4988	1.4949	1.4910	1.4872	1.4833
0.54	1.5586	1.5546	1.5505	1.5465	1.5424	1.5384	1.5344	1.5304	1.5264	1.5224
0.53	1.6000	1.5958	1.5916	1.5875	1.5833	1.5792	1.5750	1.5709	1.5668	1.5627
0.52	1.6426	1.6383	1.6340	1.6297	1.6254	1.6211	1.6169	1.6126	1.6084	1.6042
0.51	1.6866	1.6822	1.6777	1.6733	1.6689	1.6644	1.6601	1.6557	1.6513	1.6470
0.50	1.7321	1.7274	1.7228	1.7183	1.7137	1.7091	1.7046	1.7001	1.6956	1.6911
0.49	1.7790	1.7743	1.7695	1.7648	1.7600	1.7553	1.7507	1.7460	1.7413	1.7367
0.48	1.8276	1.8227	1.8178	1.8129	1.8080	1.8031	1.7983	1.7934	1.7886	1.7838
0.47	1.8780	1.8729	1.8678	1.8627	1.8576	1.8526	1.8476	1.8426	1.8376	1.8326
0.46	1.9303	1.9249	1.9197	1.9144	1.9091	1.9039	1.8987	1.8935	1.8883	1.8832
0.45	1.9845	1.9790	1.9735	1.9680	1.9626	1.9571	1.9517	1.9463	1.9409	1.9356
0.44	2.0409	2.0352	2.0294	2.0238	2.0181	2.0124	2.0068	2.0012	1.9956	1.9900
0.43	2.0996	2.0936	2.0877	2.0817	2.0758	2.0700	2.0641	2.0583	2.0525	2.0467
0.42	2.1608	2.1545	2.1483	2.1422	2.1360	2.1299	2.1238	2.1177	2.1116	2.1056
0.41	2.2246	2.2181	2.2116	2.2052	2.1987	2.1923	2.1860	2.1796	2.1733	2.1670
0.40	2.2913	2.2845	2.2777	2.2710	2.2643	2.2576	2.2509	2.2443	2.2377	2.2311
0.39	2.3611	2.3539	2.3469	2.3398	2.3328	2.3258	2.3188	2.3119	2.3050	2.2981
0.38	2.4342	2.4267	2.4193	2.4119	2.4045	2.3972	2.3899	2.3826	2.3754	2.3682
0.37	2.5109	2.5031	2.4952	2.4875	2.4798	2.4721	2.4644	2.4568	2.4492	2.4417
0.36	2.5915	2.5833	2.5751	2.5669	2.5588	2.5507	2.5427	2.5347	2.5267	2.5188
0.35	2.6764	2.6677	2.6591	2.6505	2.6419	2.6334	2.6250	2.6165	2.6082	2.5998
0.34	2.7660	2.7568	2.7477	2.7386	2.7296	2.7206	2.7117	2.7028	2.6939	2.6852
0.33	2.8605	2.8508	2.8412	2.8316	2.8221	2.8126	2.8032	2.7938	2.7845	2.7752
0.32	2.9607	2.9504	2.9402	2.9300	2.9199	2.9099	2.8999	2.8900	2.8801	2.8703
0.31	3.0669	3.0560	3.0451	3.0344	3.0236	3.0130	3.0024	2.9919	2.9814	2.9710
0.30	3.1798	3.1682	3.1566	3.1452	3.1338	3.1225	3.1112	3.1000	3.0889	3.0779
0.29	3.3001	3.2877	3.2754	3.2632	3.2510	3.2390	3.2270	3.2151	3.2032	3.1915
0.28	3.4286	3.4153	3.4022	3.3891	3.3761	3.3633	3.3505	3.3377	3.3251	3.3126
0.27	3.5661	3.5520	3.5379	3.5239	3.5100	3.4962	3.4825	3.4688	3.4553	3.4419
0.26	3.7139	3.6986	3.6835	3.6684	3.6535	3.6387	3.6240	3.6094	3.5948	3.5804
0.25	3.8730	3.8565	3.8402	3.8240	3.8079	3.7919	3.7761	3.7604	3.7447	3.7293
0.24	4.0449	4.0271	4.0094	3.9919	3.9745	3.9572	3.9401	3.9231	3.9063	3.8896
0.23	4.2313	4.2119	4.1927	4.1737	4.1549	4.1362	4.1176	4.0992	4.0809	4.0628
0.22	4.4341	4.4130	4.3921	4.3714	4.3508	4.3305	4.3103	4.2903	4.2704	4.2508
0.21	4.6557	4.6326	4.6098	4.5871	4.5646	4.5424	4.5203	4.4985	4.4768	4.4554
0.20	4.8990	4.8736	4.8484	4.8235	4.7989	4.7744	4.7503	4.7263	4.7025	4.6790

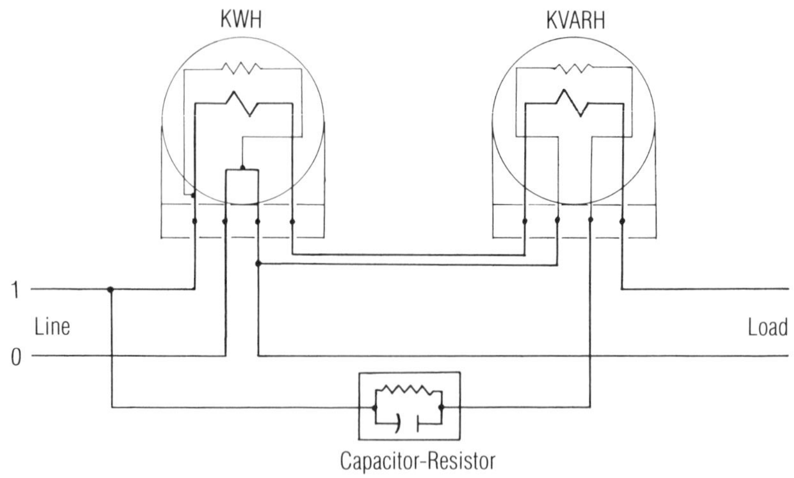

Figure 9-7. Measurement of Varhours in a Single-Phase Circuit.

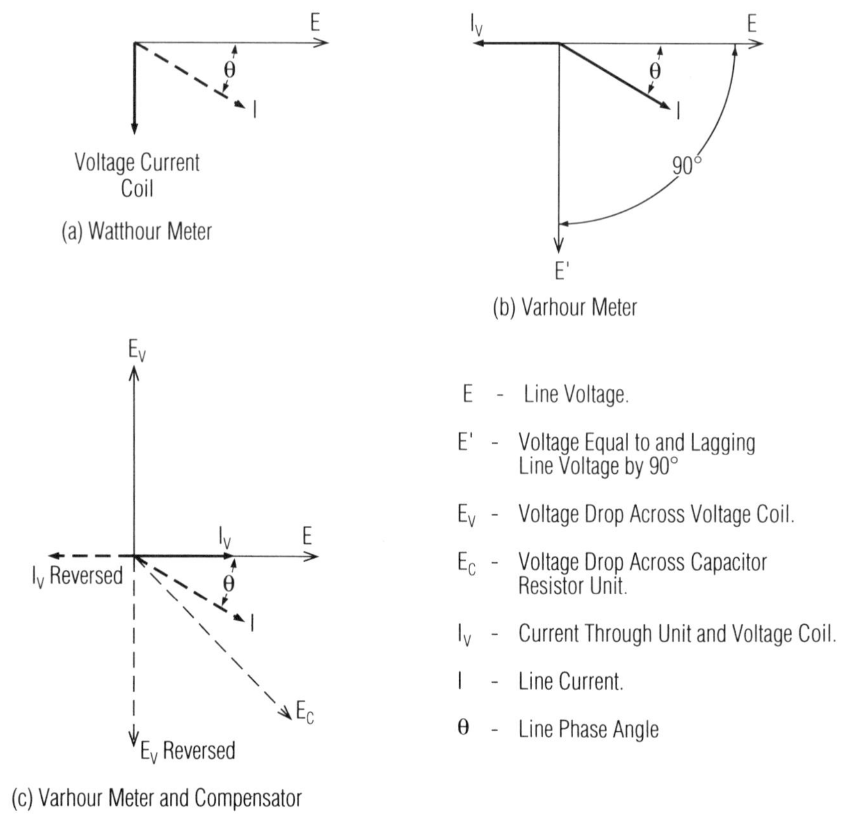

Figure 9-8. Phasor Diagram Showing Theory of Single-Phase Varhour Meter Phase-Shifting Unit.

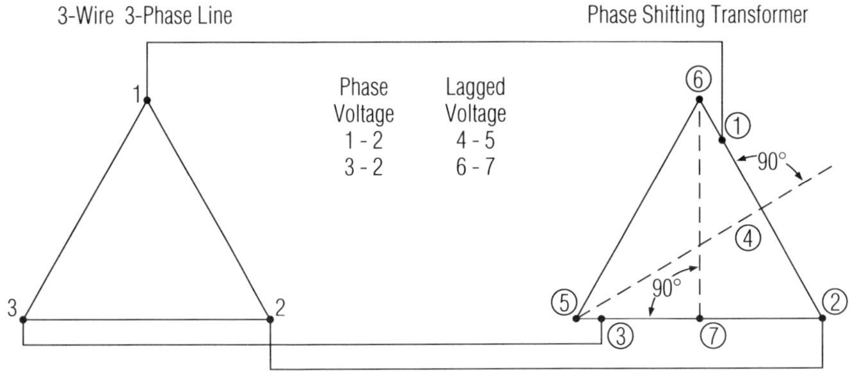

Figure 9-9. Elementary Diagram of Phase-Shifting Transformer for Three-Wire, Three-Phase Circuits.

or using watthour meters, Power Factor is equal to the cosine of the angle whose tangent is Varhours/Watthours, or in mathematical notation:

$$\text{Power Factor} = \cos\left(\tan^{-1}\frac{\text{Varhours}}{\text{Watthours}}\right)$$

A convenient method for solving these equations is provided in the Power Factor Chart, Table 9-2. The results of dividing vars by watts or varhours by watthours are tabulated. Perform the division, look up the answer in the table, and read the power factor to two decimal places in the column at the left. The third decimal place may be found at the head of the column in which the figure closest to the dividend appears. Additional accuracy may be obtained by interpolation. For example, if a wattmeter indicates 1,500 W and a varmeter 906 vars:

$$\frac{\text{Vars}}{\text{Watts}} = \frac{906}{1500} = 0.6040$$

Locate 0.6040 in Table 9-2. In the column to the left, read power factor of 0.85. The third number, 0.006, is directly above 0.6040. The power factor for these readings is 0.856.

Use of Capacitor-Resistor Unit

The measurement of varhours in a single-phase circuit is generally made by using a standard single-phase watthour meter which has a combination of resistance and capacitance in series with its voltage coil (Fig. 9-7).

The current in the voltage coil of a wattmeter or watthour meter lags the applied voltage by 90 degrees Fig. 9-8(a). If a voltage *E*' Fig. 9-8(b) lagging the phase voltage *E* by 90 degrees is applied to the voltage coil, then the current in the voltage coil will lag the phase voltage by nearly 180 degrees.

The circuit consisting of the capacitor-resistor unit and voltage coil in series with each other draws a current I_V Fig. 9-8(c) when connected across the phase voltage *E*. The voltage drop across the voltage coil (E_V) leads this current by 90 degrees and is equal in magnitude to phase voltage *E*. If the connections to the voltage coil are reversed, then E_V reversed equals *E*' and the coil voltage, current, and flux are correct for measuring vars.

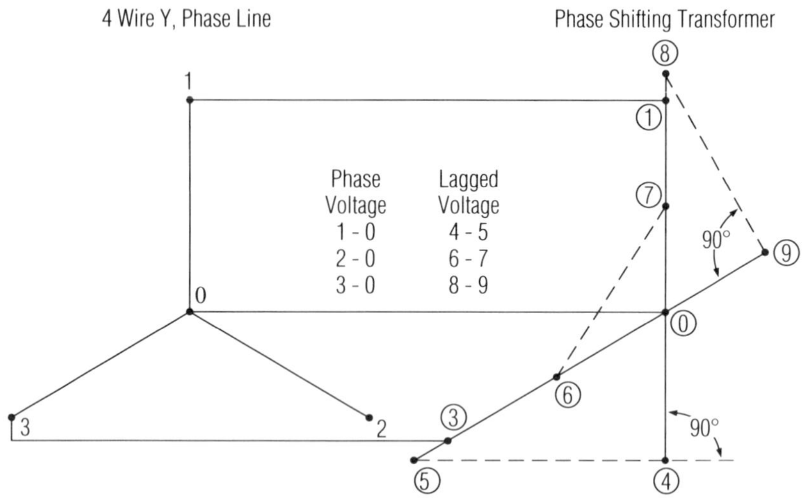

Figure 9-10. Diagram of Phase-Shifting Transformer for Four-Wire Wye, Three-Phase Circuits.

The normal and overload curves of this varhour meter will approximate those of a standard watthour meter. Similarly, the voltage curve will practically match the standard voltage curve. Frequency characteristics of the varhour meter as compared to a standard watthour meter will depend upon changes in the ratio of the capacitive reactance to the resistance and to the inductive reactance of the voltage coil when the frequency changes. Similarly, the overall accuracy of the varhour meter can be affected by changes in the capacitor caused by changes in temperature. The watts loss of the voltage circuit is greater because the current in the voltage coil is near unity power factor with respect to the line voltage.

Use of Autotransformers

The most common method used to obtain the desired phase shift in voltage is by a combination of autotransformers. The autotransformers not only shift the voltage of each phase the required number of degrees, but also supply this voltage at the same magnitude as the line voltage by tapping the windings at voltage points which, when added phasorially, result in the desired voltage (see Figs. 9-9 and 9-10). The autotransformer combination is known by a variety of names such as "phasing transformer," "reactive component compensator," "reactiformer," "voltage phasing transformer," and "phaseformer." Figs. 9-11 and 9-12 illustrate the application of the phaseshifting devices shown in Figs. 9-9 and 9-10.

These diagrams do not show all possible methods and connections, but are used to illustrate the principle. It is supposed that manufacturers will be able to supply connection diagrams of whatever variety of autotransformer is purchased.

With the addition of the proper autotransformer, a standard wattmeter or watthour meter becomes a varmeter or a varhour meter. Such a varmeter and a standard wattmeter can be used to derive volt-amperes and power factor by applying equations previously mentioned. When watthour meters are used on

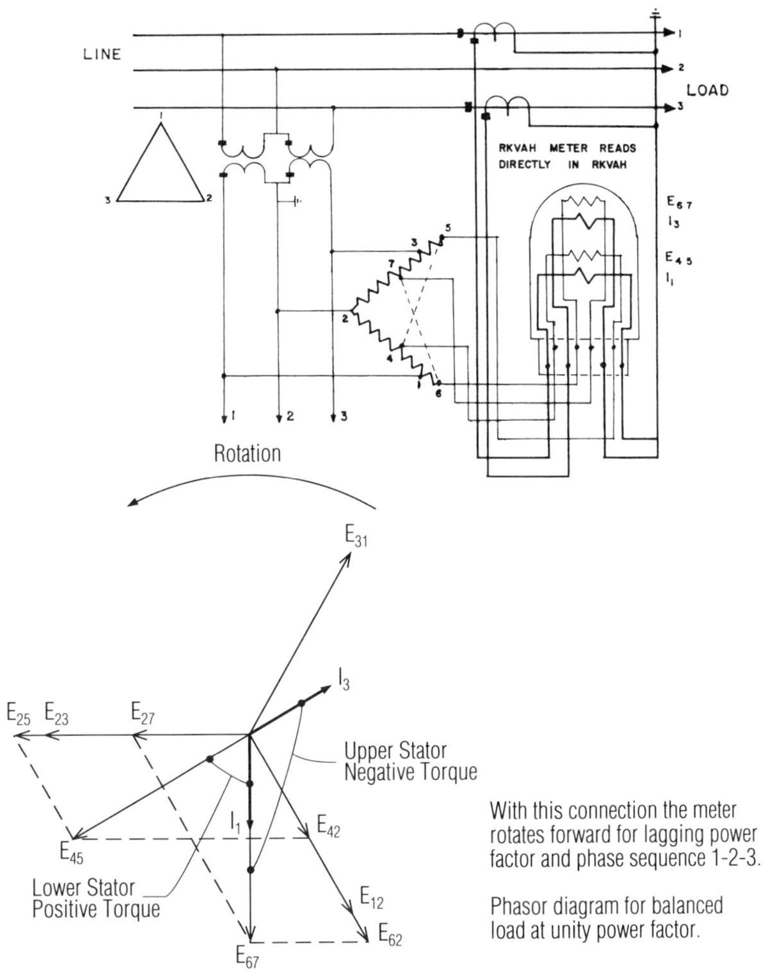

Figure 9-11. Connection of Three-Wire, Three-Phase, Two-Stator Varhour Meter with Autotransformers.

loads which vary constantly in magnitude and in power factor, the power factor value obtained,

$$\text{Power Factor} = \cos\left(\tan^{-1}\frac{\text{Varhours}}{\text{Watthours}}\right)$$

is an average power factor and does not necessarily represent the actual power factor at any one time. Similarly volt-ampere-hours calculated from these figures is also an average value and is not necessarily equal to the total volt-ampere-hours delivered (see Fig. 9-13).

Since phase-shifting autotransformers are essentially tapped, single-winding voltage transformers, the testing of them consists of verifying the various tap voltages in terms of the input voltage. The test should be performed with the burden, in volt-amperes, approximately equal to that which will be used in service.

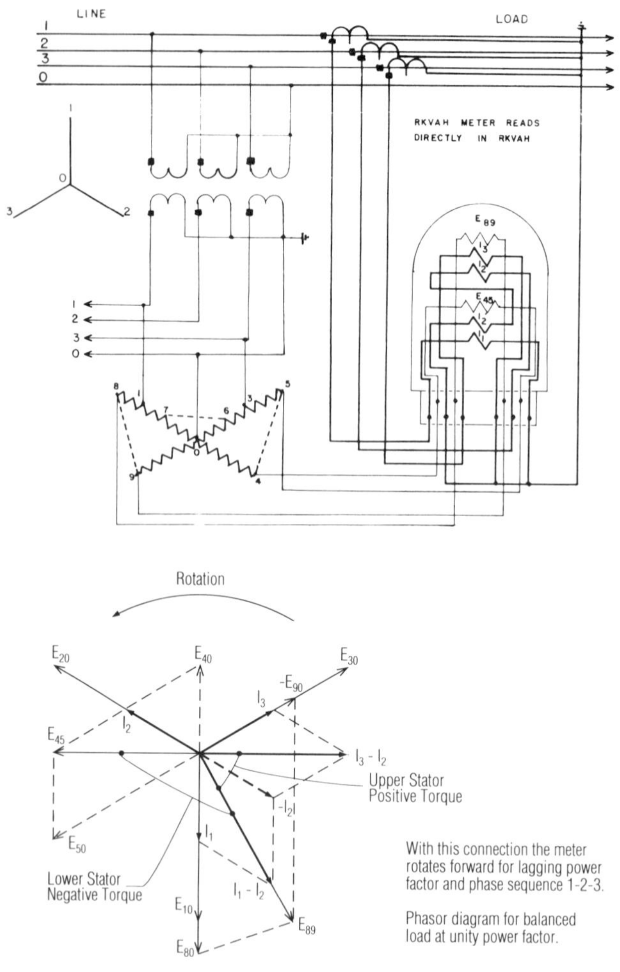

Figure 9-12. Connection of Four-Wire, Three-Phase Wye Varhour Meter with Autotransformer.

Because the values in the manufacturer's data table are usually expressed as "percent of applied voltage" the job may be simplified by using a voltmeter with a special scale or by energizing the transformer in multiples of 100 volts. Voltage readings may then be interpreted as a percent without lengthy calculations.

Cross Phasing

The inherent nature of a two-phase system readily provides the measurement of reactive volt-ampere-hours. The voltages of the two phases are normally displaced 90 degrees; therefore, by interchanging voltages the voltage from each phase is made to react with current from the other phase. In addition, the polarity of one of the voltage coils must be reversed in order to produce forward rotation on lagging power factor. See Fig. 9-14. This method is known as "cross phasing."

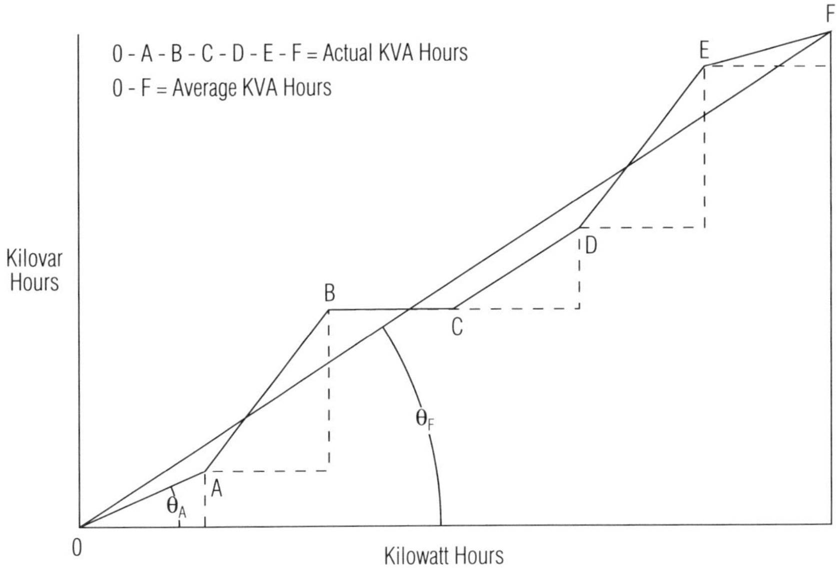

Figure 9-13. Instantaneous Power Factor (Cos θ_A) Does Not Necessarily Equal the Average Power Factor Calculated from kWh and RkVAh Readings (Cos θ_F).

A similar method of cross phasing may be applied to three-phase circuits. The current in each phase reacts with voltage between two line wires other than the one in which it is flowing.

In a three-wire, three-phase, delta meter, cross phasing is accomplished by interchanging voltage leads to the meter (Fig. 9-15). As in the two-phase meter, it

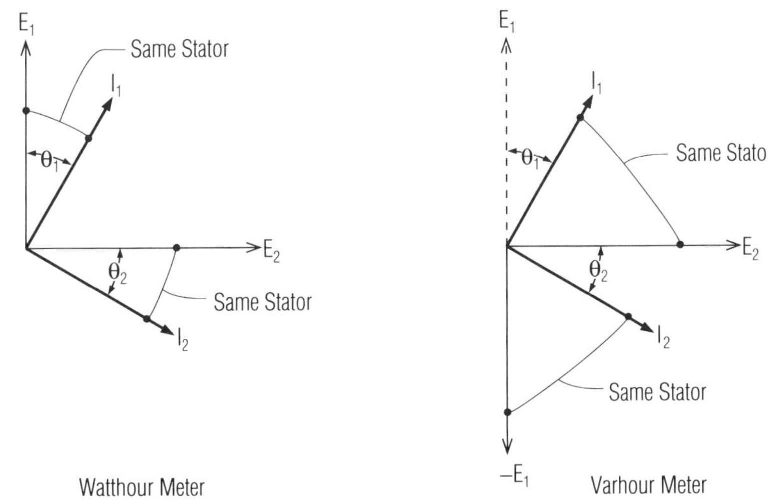

Figure 9-14. Phasor Diagram of Cross Phasing in a Two-Phase System.

is also necessary to reverse polarity of one stator's voltage with respect to its current. In Fig. 9-14 this has been done by reversing the voltage, although current leads may be reversed with the same result.

When the current in the phase leading the "common" phase is reversed, the varhour meter will rotate forward on lagging power factor and backward on leading power factor. When the other current is reversed the opposite is true.

It is necessary to multiply this meter's registration by 0.866 to read correct varhours.

With this connection, the varhour meter is actually metering only two phases of the three. The correcting multiplier is applied to correct for voltage magnitude and to include the third phase. Therefore, this method may only be used on a system with balanced voltages and currents.

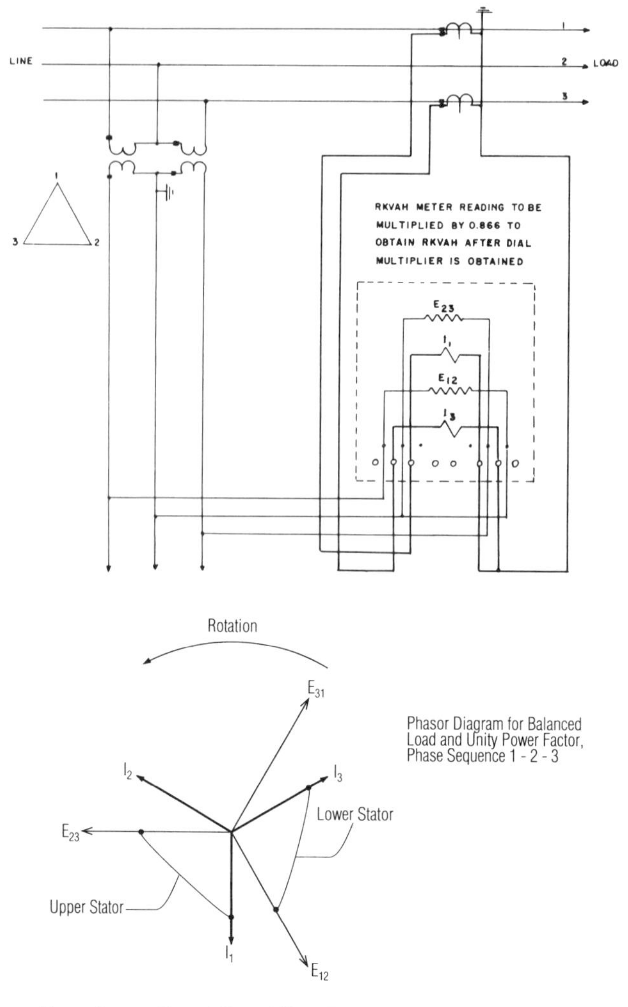

Figure 9-15. Cross-Phase Connection of Three-Wire, Three-Phase, Two-Stator Varhour Meter.

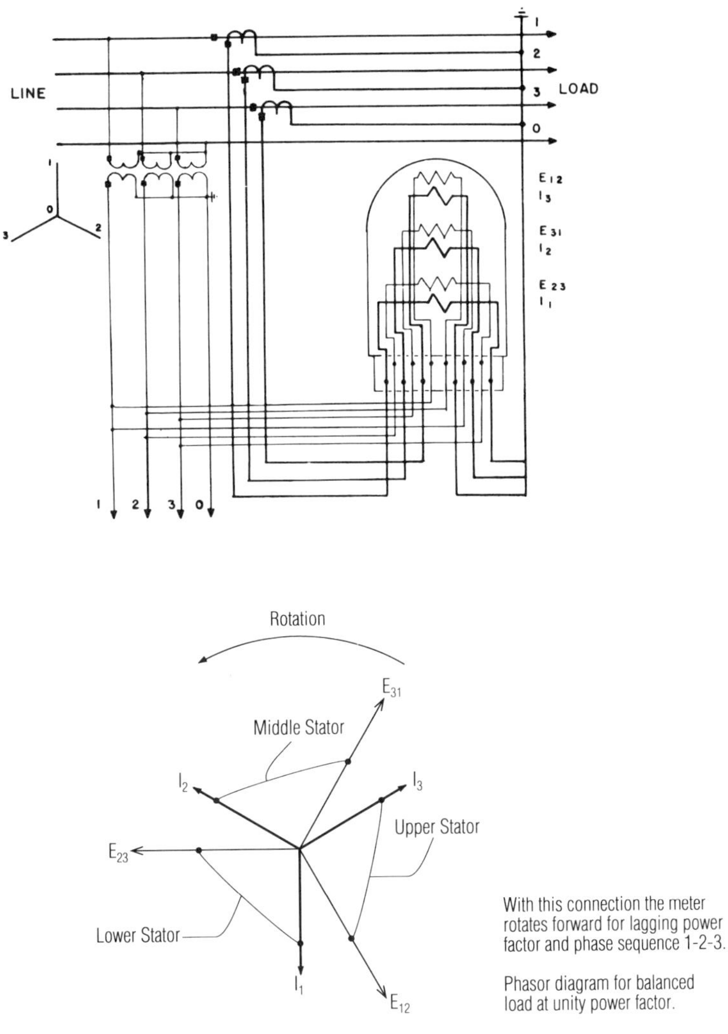

Figure 9-16. Cross-Phase Connection of Four-Wire, Three-Phase Wye, Three-Stator Varhour Meter.

In a wye system, since the voltage coils are impressed with line-to-line voltage which is $\sqrt{3}$ or 1.732 times as great as the required voltage, the registration is correspondingly high. A correction factor of 0.577 must be applied for correct registration (Fig. 9-16). A $2^1/_2$-stator meter may be used in the same manner and a two-stator meter may be used if current-transformer secondaries are connected in delta. The 0.577 correction factor also applies in these cases.

Displaced-Voltage Meters

The displaced-voltage method of measuring kilovolt-ampere hours consists of using a watthour meter in which the voltage has been shifted to an angle corre-

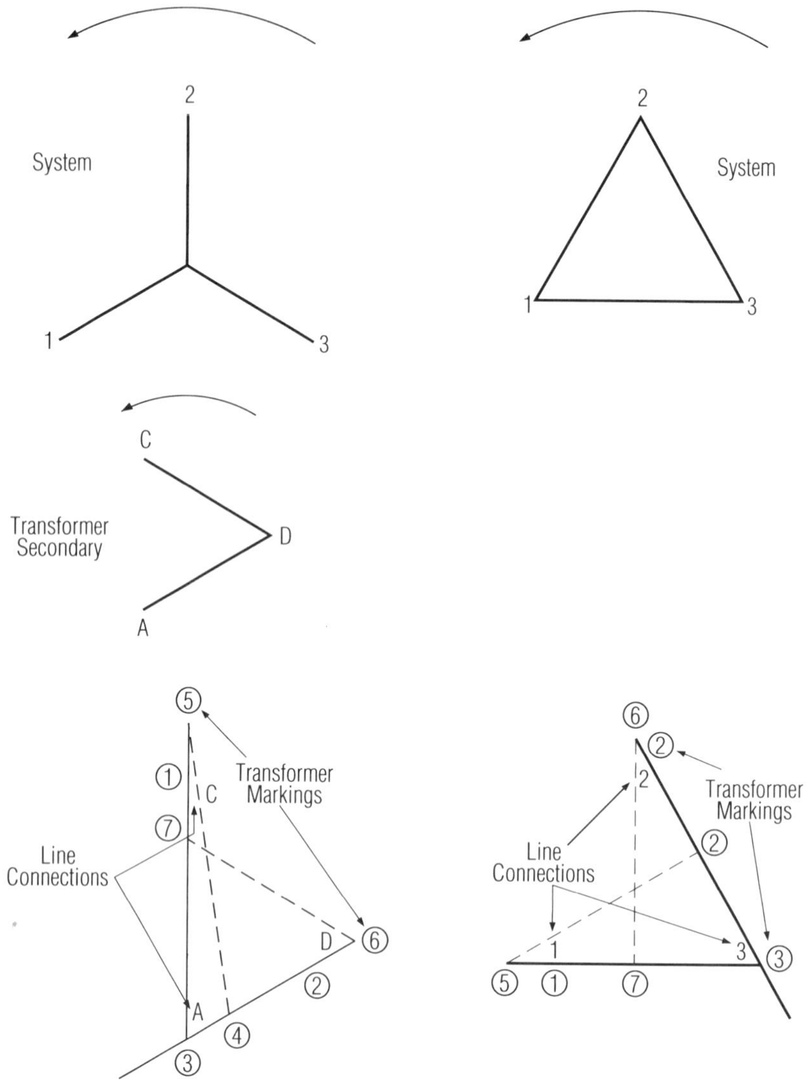

Figure 9-17. Phasor Diagrams of Autotransformer for RkVAh and kVAh Measurements on Three-Phase, Three-Wire and Three-Phase, Four-Wire Wye Circuits. Phasor Relations Shown for Only One Power Factor Tap (0.82). For Use with Two-Stator Meter.

sponding to the load power factor so as to be in phase with the line current yet equal in magnitude to the original circuit voltage. Thus, the phase angle between voltage and current is practically eliminated, the power factor in the meter is unity, and the registration of the meter will be proportional to kilovolt-ampere hours over a limited range of power factor. This voltage displacement may be accomplished either by the use of autotransformers (see Fig. 9-17) or by setting

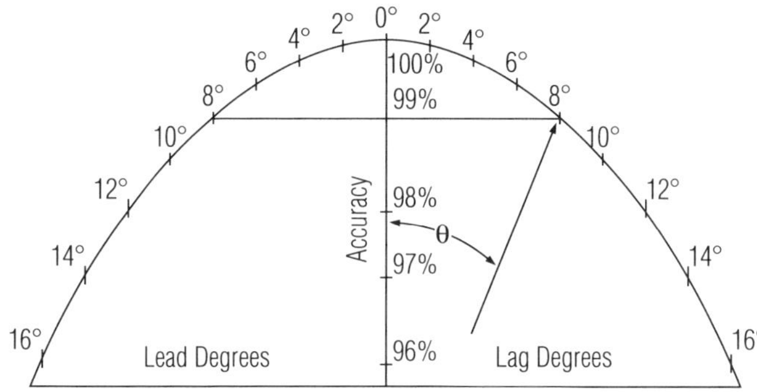

Figure 9-18. Accuracy of a Watthour Meter When Measuring Volt-Ampere Hours Near 1.00 Power Factor.

the lag adjustment in the meter to provide the desired phase shift instead of adjusting it to lag voltage flux by 90 degrees as is usually the case. The registration will be true volt-ampere hours only at an assumed power factor. However, by connecting the meter to an approximate predetermined power-factor tap of a phase-shifting autotransformer, the accuracy will be satisfactory for most commercial measurements over an appreciable range of power factor (Fig. 9-18). A periodic check of the power factor of a circuit must be made at reasonable intervals to see that the power factor is still within the range of taps to which the meter is connected. Calibrating the meter 1 percent fast as a watthour meter greatly extends the range of power factor over which the meter is within ±1 percent. On this basis the power-factor ranges for the listed mean power-factor settings are shown in Table 9-3.

Table 9-3. Range of Measurement of kVA Meter by the Displaced-Voltage Method.

Mean P.F.	P.F. Range
0.98	1.00 - 0.98
0.92	0.98 - 0.82
0.87	0.95 - 0.75
0.82	0.92 - 0.69
0.80	0.90 - 0.66
0.73	0.85 - 0.58
0.71	0.83 - 0.55
0.50	0.66 - 0.32

At 1.0 power factor, kilowatts and kilovolt-amperes are the same. From this zero phase-angle condition the power-factor angle may change as much as 8 degrees lag or lead before the cosine of the angle, which is the power factor, changes as much as 1 percent (Fig. 9-18). The deviation may go to 11.5 degrees lead or lag without the cosine changing more than 2 percent.

But if the initial condition be one where the power factor is, for example, 0.707 lagging, that is, the current lagging the voltage by 45 degrees, then a shift of the phase angle from 45 degrees lag to 56 degrees (an 11 degree shift) changes the power factor from 0.707 at 45 degrees to the cosine of 56 degrees which is 0.56, a power factor change of 15 percent.

Lagging the voltage 45 degrees by means of an autotransformer brings the meter voltage and meter current into phase with each other when the load power factor is 0.707 lagging. Now a shift in the load phase angle of 11 degrees from 45 to 56 degrees produces a shift in the meter from 0 to 11 degrees, a 2 percent change in power factor. This meter will now register within 2 percent of 100 percent at load phase angles of from 34 to 56 degrees. If the meter is adjusted 1 percent fast, it will register within ±1 percent of 100 percent over a range of ±11.5 degrees from the assumed phase angle (total range 23 degrees). Figure 9-19 shows typical accuracy curves.

A capacitor-resistance unit, such as is used in varmeters, may be used instead of an autotransformer to obtain the desired phase shift of voltage. The unit is not designed to shift voltage 90 degrees, but an amount corresponding to the assumed power factor so that the meter reads in volt-amperes rather than vars.

Special Meters

A varmeter is a meter with built-in means of obtaining the 90-degree phase shift. The meter may incorporate an internal capacitor-resistance unit or autotransformer. It is installed in the same manner as a wattmeter.

A volt-ampere meter is similar to a varmeter but with a built-in phase shift corresponding to the assumed power factor, as described previously.

One special meter includes a watthour meter, a varhour meter using autotransformers, and a mechanical ball mechanism which adds vectorially the rotations of the two meters to give a reading in kilovolt-ampere hours. This meter gives actual kilovolt-ampere hours which may be interpreted on a register which may be a graphic, indicating, or cumulative demand register. Such a meter will yield readings of kilowatt-hours, kilovar-hours, kilovolt-ampere hours, kilowatt demand, kilovolt-ampere demand, and an indication of power factor at the moment the meter is being observed.

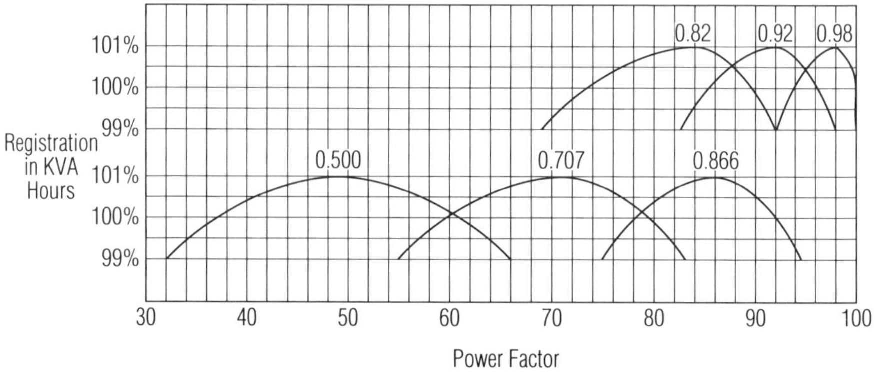

Figure 9-19. Accuracy of a Volt-Ampere Hour Meter Adjusted One Percent Fast for Typical Power Factor Ranges.

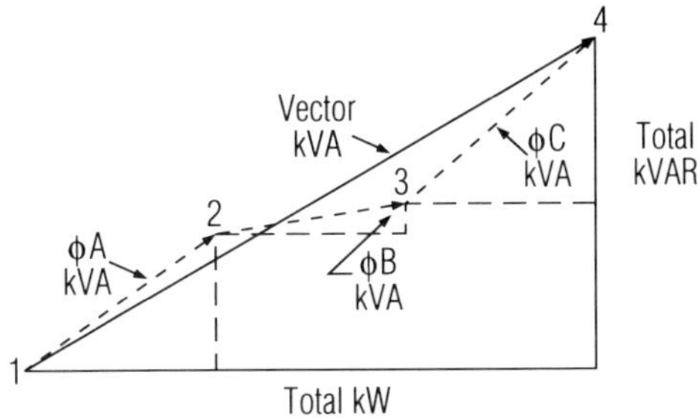

Figure 9-20. Comparison of Vector and Arithmetic Summing.

Some special varhour meters are in production. These meters are used on specified phase voltages in conjunction with a watthour meter. No autotransformer or other external means of phase shifting is necessary. Their principle of design is that the quadrature voltage necessary for metering vars can be obtained for each phase by correct phasor addition of the other two phase voltages and by applying constant correction factors. This is done by building a multistator meter and passing a given phase current first through one stator with one voltage then through the second stator with a second voltage. The resulting torque is proportional to the product of the current, the vector sum of the voltages, and the power factor of the angle between them. This torque is equal to vars. Correction constants are applied in the design of the voltage coils of the meters to avoid a special multiplier as is needed in cross-phased meters.

Detailed information on the design and operation of these meters may be obtained from the manufacturers.

Solid-State Kilovolt-Ampere Metering

An electronic meter is available with kVA demand, kVAh, and kVA analog output capabilities. It features true rms response (VA = rms current × rms voltage) and arithmetic summing of the rms volt-amperes of each element.

True rms response is important when accurate volt-ampere metering of loads with distorted waveforms is needed. Arithmetic summing results in more accurate metering than vector summing when voltage, power factor, or load unbalance exists. Vector summing is required to obtain volt-amperes when separate kWh and reactive meters are used to feed magnetic tape recorders or solid-state data collecting equipment. Under all voltage and load conditions, the arithmetic sum will always be equal to or greater than the vector sum and will more accurately meter the true volt-ampere load.

Referring to Fig. 9-20, vector kVA is the straight line from 1 to 4. Arithmetic kVA is the sum of line segments 1 to 2, plus 2 to 3, plus 3 to 4. By inspection, it is obvious that the vector kVA sum is less than the sum of the individual kVA's of the individual phases.

Phase sequence can be important, depending on the type multiplier used and whether the apparent power desired is arithmetic power or vector power. With arithmetic power, calculated from V_{RMS} and I_{RMS}, phase sequence is not important. With vector apparent power, calculated from real and reactive powers, the method which is used to obtain the 90° phase shift for reactive power, can be

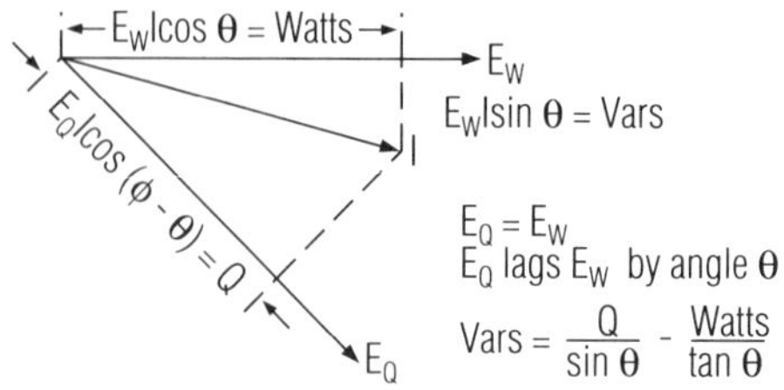

Figure 9-21. Vector Relationships for Q-hour Metering.

phase-sequence sensitive. Consult the manufacturers for details on the use of specific meters.

Q-Metering

The Q-hour meter is a conventional watthour meter in every detail. But unlike the varhour meter discussed previously, a single Q-hour meter has the capability of measuring both leading and lagging vars over some definable range of power factor. It was noted previously that a displacement of the watthour meter voltage by 90° produced vars. By displacing the voltages, any angle, other than 0° or 90°, the torque on the watthour meter will not be proportional to watts or to vars, but is proportional to a "quantity" called Q.

From Fig. 9-21 the following relationships exist:

$$\text{Watts} = E_W I \cos\theta$$
$$\text{Vars} = E_W I \sin\theta$$
$$Q = E_Q I \cos(\phi - \theta)$$
$$|E_Q| = |E_W|$$

Expanding the equation:

$$Q = E_Q I \cos(\phi - \theta)$$
$$Q = E_Q I \cos\phi \cos\theta + E_Q I \sin\phi \sin\theta$$

Since $|E_Q| = |E_W|$

$$Q = (E_W I \cos\theta) \cos\phi + (E_W I \sin\theta) \sin\phi$$
$$Q = (\text{watts}) \cos\phi + (\text{vars}) \sin\phi$$

After rearranging terms:

$$(\text{vars}) \sin\phi = Q - (\text{watts}) \cos\phi$$

or,

$$\text{Vars} = Q/\sin\phi - (\text{watts}) \cos\phi/\sin\phi$$

But,

$$\cos\phi/\sin\phi = 1/\tan\phi$$

Therefore,

$$\text{Vars} = Q/\text{Sin}\,\phi - \text{Watts}/\tan\phi$$

This is the general expression for the relationship of watts, vars and Q for any lagging angle ϕ.

Although any angle of lag is theoretically obtainable by using a phase-shifting transformer, it would be desirable to eliminate the need for another piece of equipment. Fortunately, an appropriate angle of lag of 60° is readily available from both a three-wire, three-phase and a four-wire wye, three-phase circuit by the simple expedient of cross-phasing. A 60° angle of lag will result in the Q-hour

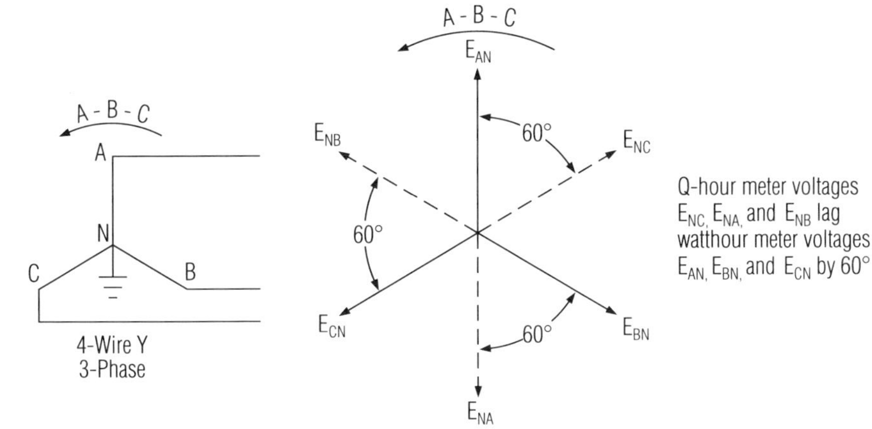

Figure 9-22. Phase Displacement of 60° from Three-Phase, Four-Wire, Wye System.

meter having forward torque for any power factor angle between 150° (-86.7% pf) lagging and 30° (86.7% pf) leading. Figures 9-22 and 9-23 illustrate how a 60° phase displacement of voltages may be obtained from both four-wire wye, three-phase and three-wire, three-phase circuits by cross-phasing.

Using an angle of lag of 60°, the general expression for vars previously developed may be written in a form without reference to trigonometric functions.

Substituting values for sin 60° and tan 60° into the general expression,

$$\text{Vars} = Q/\sin\phi - \text{Watts}/\tan\phi$$

will give

$$\text{Vars} = Q/\sqrt{3/2} - \text{Watts}/\sqrt{3},$$

which reduces to

$$\text{Vars} = \frac{2Q - \text{Watts}}{\sqrt{3}}$$

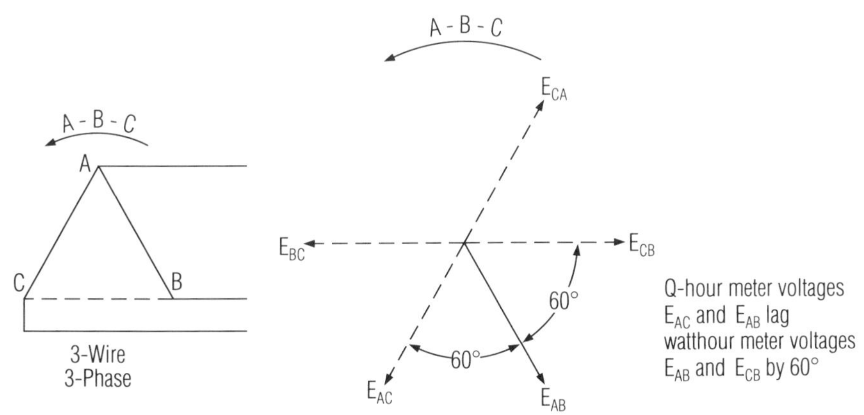

Figure 9-23. Phase Displacement of 60° from Three-Phase, Three-Wire System.

See Fig. 9-24 for the phasor relationships that exist when the Q-meter voltage has been lagged 60° from the wattmeter voltage.

It was noted previously that the Q-hour meter measures both lagging varhours and leading varhours over a specified range, the limits of which are determined by the degrees of lag of the Q-hour meter voltages. Figure 9-25 illustrates the useful range of the Q-hour meter using a 60° lag of voltages. It may be noted that forward torque exists on the Q-hour meter from a 90° lagging pf angle to a 30° leading angle. Relative measurements (torques) for the watthour meter and corresponding power factor angles are also shown. Note that the watthour meter also has forward torque over pf range equal to that of the Q-hour meter. If the Q-hour meter is to provide measurement for both leading and lagging power factor, how is the distinction between leading and lagging varhours to be made when the meter gives positive (forward) readings for both? Note from Fig. 9-25 that the calculated vars are positive between 0° power factor and 90° lagging power factor angle. The relationships may be used to determine whether vars are leading or lagging:

If (2Q – watts) is positive, pf is lagging

If (2Q – watts) is zero,pf is 1.0

If (2Q – watts) is negative, pf is leading

Other relationships are:

If $Q/W > 0.5$, pf is lagging

If $Q/W = 0.5$, pf is 1.0

If $Q/W < 0.5$, pf is leading

When power factors are leading more than 30° (86.7% pf), the 60° lagged Q-Hour meter cannot be used since it reverses at that point. In such cases, either a Q-hour meter lagged by some appropriate angle less than 60° or separate leading and lagging reactive meters are required. Also, at locations where power flow may be in either direction, conventional reactive meters rather than Q-meters should be used.

Solid-State "Q" Metering

The solid-state Q meter is based on principles described in Chapter 7 for solid-state multipliers. For time-division multipliers, Hall-effect multipliers, and multipliers using transconductance amplifiers, the 60° voltage phase shift is

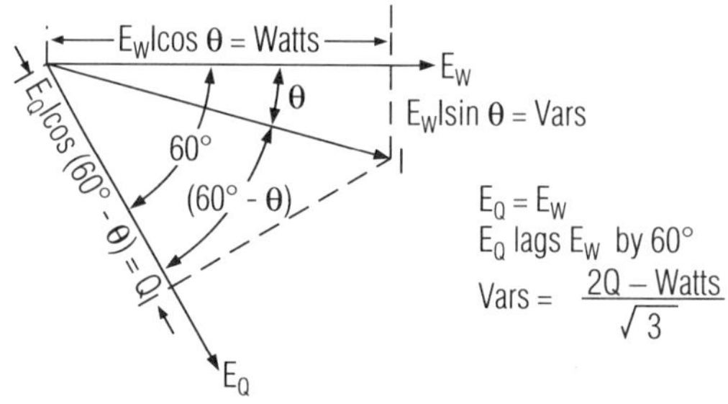

Figure 9-24. Q-meter Voltage Lagged 60° from Wattmeter Voltage.

accomplished by internal cross-phasing of the connections. For digital multipliers using microprocessors, a digital phase shifter accomplishes the 60° phase shift, or Q is calculated from

$$Q = \frac{\text{Real Power}}{2} + \frac{\sqrt{3} \times \text{Reactive Power}}{2}$$

pf Angle	Power Factor	Q	Watts	Vars
90° lag	0.000	$\frac{3}{2}$	0	$\sqrt{3}$
60° lag	0.500	$\sqrt{3}$	$\frac{\sqrt{3}}{2}$	$\frac{3}{2}$
30° lag	0.867	$\frac{3}{2}$	$\frac{3}{2}$	$\frac{\sqrt{3}}{2}$
0° lag	1.000	$\frac{\sqrt{3}}{2}$	$\sqrt{3}$	0
30° lead	0.867	0	$\frac{3}{2}$	$-\frac{\sqrt{3}}{2}$

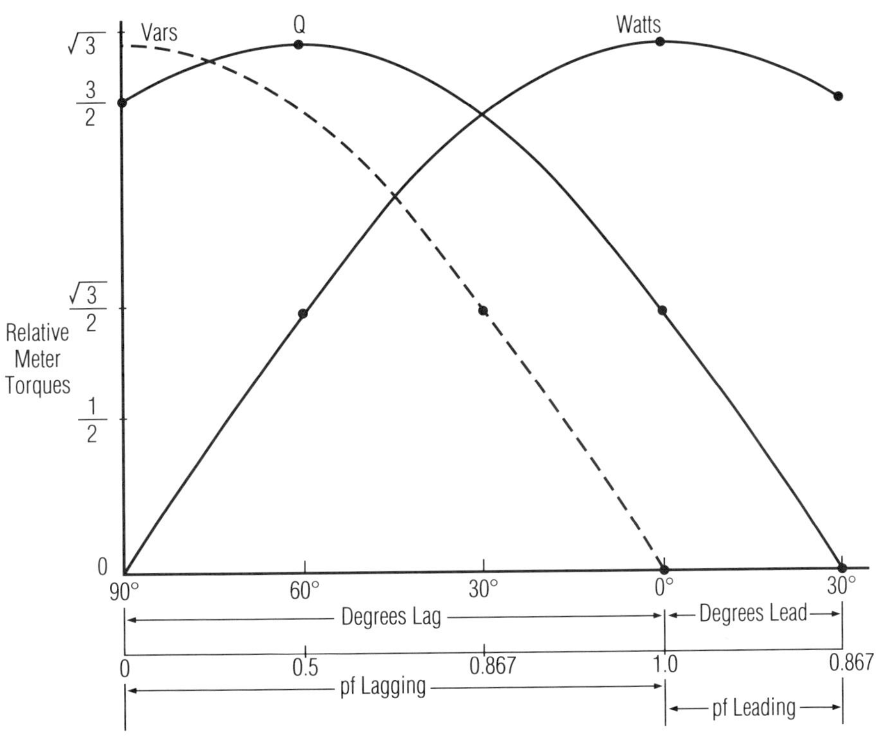

Figure 9-25. Useful Range of Q-hour Meter with 60° Lag of Voltages.

For multipliers not using microprocessors, the 60° phase shift is accomplished through internal cross-phasing of the connections. Phase sequence must be taken into consideration when the multiplier chosen produces the 60° phase shift by internal cross-phasing of connections.

SOLID-STATE METERING

Solid-State Multiple-Function Metering

Electronic multiple-function meters can measure active quantities such as watt demand and watthours, reactive quantities such as var demand and var-hours, and apparent quantities such as VA-demand and VA-hours. These measurements can be for uni-directional or for bi-directional applications.

Figure 9-26 shows a diagram which illustrates relationships between active power (watts), reactive power (vars), and vector apparent power (volt-amperes). The familiar Power Triangle has been incorporated into Fig. 9-26 in each of the four quadrants.

Quadrant	Power Factor	Watts	Vars
I	Lag	Delivered (+)	Delivered (+)
II	Lead	Received (-)	Delivered (+)
III	Lag	Received (-)	Received (-)
IV	Lead	Delivered (+)	Received (-)

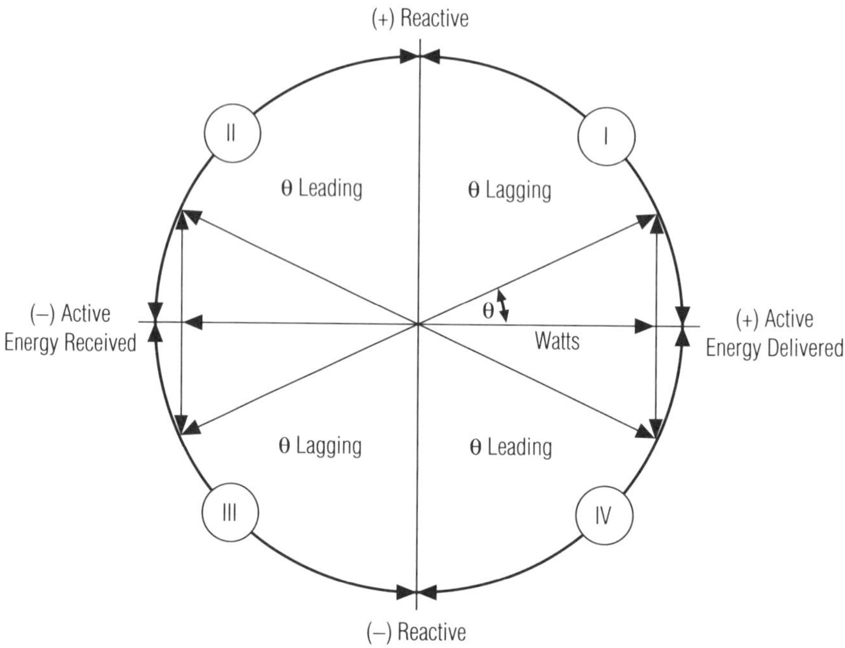

Figure 9-26. Relationships Between Watts, Vars, and Volt-Amperes.

Energy sold by the source may be defined as delivered. Energy purchased may be defined as received.

For any given voltage and current, VA is constant and an unsigned, absolute quantity. For any given phase angle θ

$$\text{Watts} = EI \cos \theta$$
$$\text{Vars} = EI \sin \theta$$

The signs (plus, minus) and directions (delivered, received) are for conventionally polarized instrument transformer connections, with respect to the power source.

The following quantities, or combinations of quantities, can be measured by electronic multiple-function meters:

kWh, kW demand delivered	Quadrants I and IV
kWh, kW demand received	Quadrants II and III
kVAh, kVA demand delivered	Quadrants I and IV
kVAh, kVA demand received	Quadrants II and III
kVARh, kVAR demand active quantities delivered, p.f. lagging	Quadrant I
kVARh, kVAR demand active quantities received, p.f. leading	Quadrant II
kVARh, kVAR demand active quantities received, p.f. lagging	Quadrant III
kVARh, kVAR demand active quantities delivered, p.f. leading	Quadrant IV

The following combinations of reactive quantities are available on some meters as displayed quantities or as non-displayed outputs. Consult the specifications of individual meter manufacturers.

kVARh and kVAR demand delivered	Quadrant I plus II
kVARh and kVAR demand received	Quadrant III plus IV

These two combined quadrant var quantities are equivalent to the performance of a rotating detented var meter.

kVARh and kVAR demand delivered, net	Quadrant I minus IV
kVARh and kVAR demand received, net	Quadrant III minus II

These two combined quadrant var quantities are equivalent to the performance of a rotating un-detented var meter.

Figure 9-27 illustrates the relationships between active power (watts), reactive power (vars), and the quantity called Q. Notice that Q is delivered (is positive) when the phase angle is between 0° and 90°, or between 330° and 360°. Q is received (is negative) when the phase angle is between 150° and 270°. Q is not defined when the phase angle is between 90° and 150°, or between 270° and 330°.

The capabilities of multiple-function meters are accomplished using either of two designs. In one design, illustrated in Fig. 9-28 and Fig. 9-29, a separate

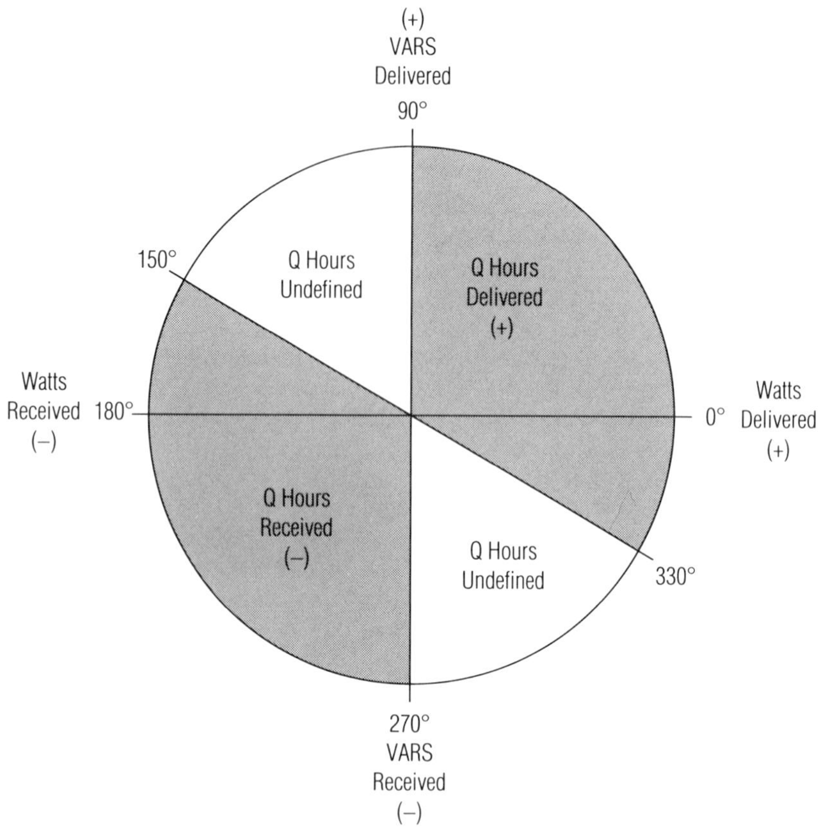

Figure 9-27. Relationships Between Watts, Vars, and Q.

multiplier is used for each function. Figure 9-28 shows a design based on a time-division multiplier with separate multipliers for real power, reactive or Q power, and volt or volt-hour measurements. Each multiplier provides data to a microprocessor-based electronic register for counting, storing, and displaying measured data. Figure 9-29 shows another three-function meter, with dedicated multipliers supplying data to a single register. In these designs the multiplier is not microprocessor-based and each multiplier requires a separate calibration and adjustment procedure.

A second method uses a single microprocessor as the multiplying element for all functions and is illustrated in Fig. 9-30. The microprocessor provides the electronic register with real, reactive, and apparent powers; V_{RMS} and I_{RMS} values; Q power; and power factor. The electronic register counts, stores, and displays the data.

Solid-State Reactive Metering

The solid-state reactive meter uses principles described in Chapter 7 for solid-state multipliers. For time-division multipliers, Hall-effect multipliers, and mul-

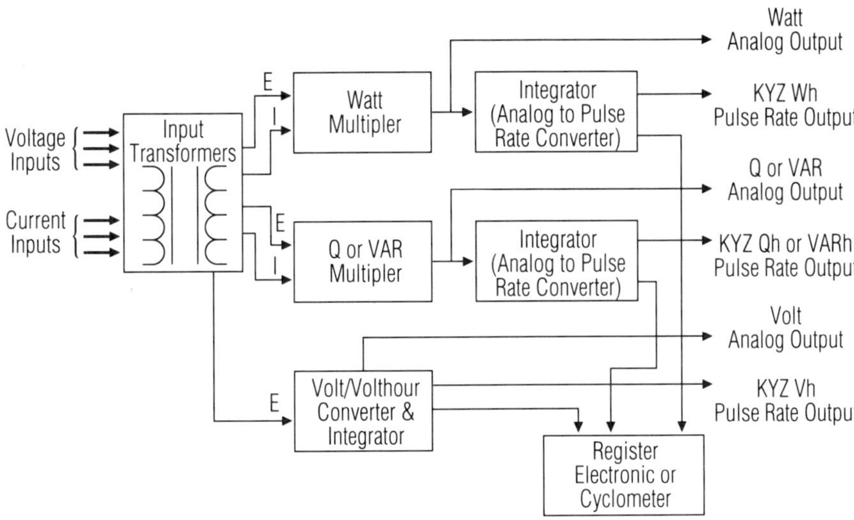

Figure 9-28. Solid-State Three-Function Meter with Multiple Outputs.

tipliers using transconductance amplifiers, the 60° voltage phase shift is accomplished by internal or external phase shifting or through cross-phasing connections. For digital multipliers using microprocessors, a digital phase shifter accomplishes the 90° phase shift, and reactive power can be calculated from

$$\text{Reactive Power} = V \times I \times \cos(\theta + 90°) \text{ or}$$
$$= V \times I \times \sin\theta.$$

For multipliers not using microprocessors, the 90° phase shift is accomplished through an external phase-shifting transformer, or cross-phasing the connections. Depending on the method used to produce the phase shift, phase sequence may need to be taken into consideration.

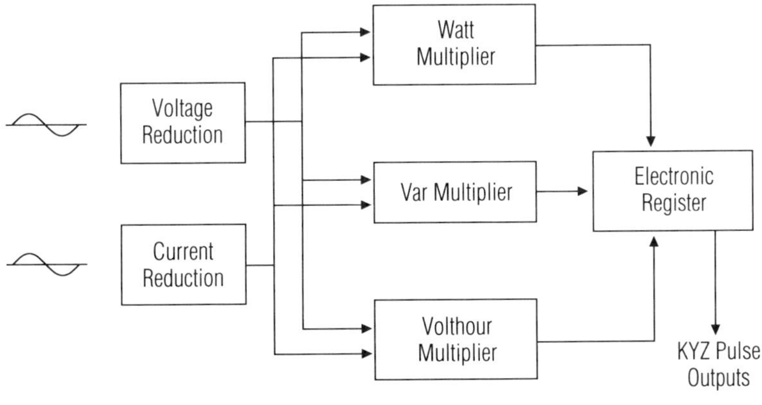

Figure 9-29. Solid-State Three-Function Meter with Single Outputs.

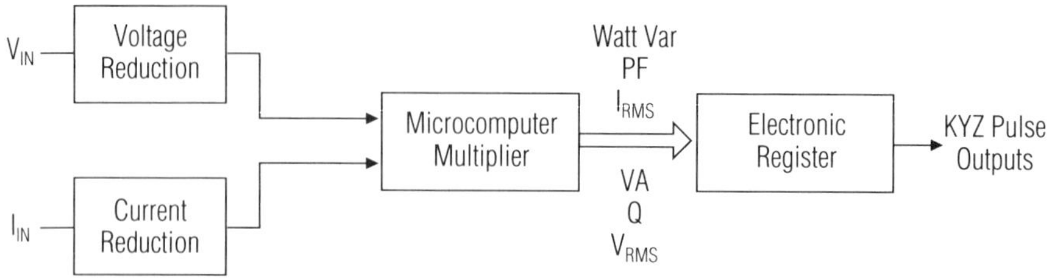

Figure 9-30. Solid-State Meter with Central Microprocessor.

CHAPTER 10

SPECIAL METERING

COMPENSATING METERING FOR TRANSFORMER AND LINE LOSSES

INTRODUCTION

SOMETIMES IT IS IMPRACTICAL, either for physical or cost reasons, to install metering equipment at the contractual billing point. As a solution, the metering equipment can be installed at a location which is more economical or more accessible, and then compensate the meter for losses which occur between the metering point and the billing point.

For example, instead of measuring power consumption on the high-voltage side of a power transformer with high-voltage meters, it is less costly to use lower-voltage meters and measure the power on the low-voltage side of the customer's transformer. In some cases, supply power can be measured more accurately at the lower usage voltage, rather than at the supply voltage. Compensating metering is needed if the billing point is located at an inaccessible point, such as a remote boundary between two utilities or the mid-span of a river crossing. For these conditions, where transformer losses or line losses between the metering equipment and the billing point are significant, compensating metering for the losses should be considered. The decision to use compensating metering can be influenced by local regulations and practices, the layout of the substation, and the ability to obtain contractual agreements with the customers.

Because compensation is based on mathematical formulas, errors are introduced when the formulas do not accurately describe actual physical conditions. For example, the formulas may assume temperature and frequency are constants and always 75°F and 60 Hz, when actually both parameters are variables. Unbalanced voltages and currents can effect the accuracy of loss measurements. In general, transformer losses represent less than 2% of the capacity of a transformer bank, and errors due to loss compensation have a negligible effect on the measurements of total kilowatt-hours and total revenue.

OBJECTIVE

The objective of compensating metering is to determine unmetered losses which occur between the billing and metering points, and then either record the losses on a loss meter or combine the losses with the metered portion of the load on a single meter. The common practice is to combine the losses with the metered portion, which duplicates values which the meter would have recorded had it been located at the billing point.

Energy dissipated between the billing and metering points cannot be measured directly. The losses are calculated indirectly using transformer theory, circuit theory, and currents and voltages at the meter test switch or meter socket.

Commercially available compensating meters operate with formulas approved by meter engineers and regulatory agencies. These formulas add (or subtract) simulated losses to the metered load and record compensated meter readings, or uncompensated readings with simulated losses directed to a separate loss meter.

TRANSFORMER LOSSES

Losses in the transformer are caused by hysteresis, eddy currents, and load currents. Hysteresis losses are derived from energy expended as the magnetic field within the transformer continually changes intensity and direction. Eddy current losses are caused by energy expended by current, induced by the magnetic field, circulating within the transformer core. Eddy current losses are minimized by building the transformer core from electrically-resistive steel formed into thin, insulated laminations. Load losses, or I^2R losses, are caused by current passing through the transformer windings and the resistance of those windings.

Transformer losses are either no-load losses, also called core losses or iron losses, or load losses, also called copper losses.

LINE LOSS COMPENSATIONS

Lines are considered to be resistances and have I^2R losses. The lengths, spacings, and configurations of lines are usually such that inductive and capacitive effects can be ignored. Bus losses are treated the same as line losses. If line and bus losses are to be compensated, they are included as part of the transformer load losses.

TRANSFORMER LOSS TESTS

Power transformers are tested for losses by the manufacturer prior to shipping. Characteristics of individual transformers vary and, for accurate compensation, it is important to obtain actual test results for the each power transformer involved. *Real* losses are given in the test results. *Reactive* losses are generally not given but can be calculated based on *real* losses, magnetizing current, and transformer impedance.

An open circuit test at rated voltage measures no-load (core) losses at rated voltage.

A short circuit test at base current measures load (copper) losses at rated load. Base current is the current at rated kVA and rated voltage.

Magnetizing current and transformer impedance are determined by other tests and are included with the transformer test results.

TRANSFORMER LOSS MODEL

The transformer loss model for loss compensation is based on mathematical relationships with metered voltage and current as the variables.

No-load (core) loss watts are proportional to V^2
Load (copper) loss watts are proportional to I^2

No-load (core) loss vars are proportional to V^4
Load (copper) loss vars are proportional to I^2

The first pair of relationships (no-load loss watts and load loss watts) for rated current and voltage are given in the manufacturer test results. The second pair of relationships (no-load loss vars and load loss vars) must be derived from the respective loss kVAs and the power factor angles of the losses.

No-load loss kVA equals rated transformer kVA times per-unit magnetizing current. Load loss kVA equals rated transformer kVA times per-unit impedance.

The power factor angle for no-load loss is the angle whose cosine equals the ratio of no-load loss watts to no-load kVA. The power factor angle for load loss is the angle whose cosine equals the ratio of load loss watts to kVA.

No-load loss vars and load loss vars are the products of their respective kVAs times the sines of their respective power factor angles.

The power triangle may also be used to determine vars. Vars equal the square root of kVA^2 minus watts^2.

BIDIRECTIONAL ENERGY FLOW

Usually if energy is delivered by a high-voltage line to a customer, it passes through the customer's transformer for use at a lower voltage. Compensation for losses between the high-voltage billing point and the lower-voltage metering point is added to meter percent-registration.

If energy flow is from a low-voltage circuit to a high-voltage circuit, compensation is subtracted from the meter percent-registration.

Normally transformer loss compensators are wired as additive but they can be ordered wired as subtractive. A pair of compensators, one additive and one subtractive, and a pair of detented watthour meters can meter and compensate bidirectional energy flow.

Bidirectional energy flow and compensation can be accomplished with one solid-state meter with a compensation option. Watthours as well as varhours can be metered and compensated by one solid-state meter.

METER LOCATION

The transformer-loss meter, or any meter compensated for transformer losses, should be connected to the low-voltage side of the circuit ahead of the disconnection device, so that the core losses may be registered even though the load is disconnected. For most types of construction the metering point is the point where the voltage circuit (or metering voltage transformer) is connected to the low-voltage bus.

Conductor copper losses on the low-voltage side between the billing point and the metering point should be included in the registration. These losses may be negligible, but in some cases it is desirable to include conductor losses in calibrating calculations if these losses assume appreciable proportions. Conductor copper losses are added to the copper losses of the transformer, since both vary with the square of the current.

TRANSFORMERS WITH TAPS

Usually transformers are provided with taps to permit adjustment of utilization voltages. The loss data for transformers supplied by the manufacturers is generally based on the rated voltage. Iron losses in watts at rated voltage are the same as those existing when connection is made to a tap and its rated voltage applied. For copper losses of transformers, it is sufficiently close to consider that the losses are divided equally between the high-voltage and the low-voltage sides and that the site of the conductor is the same throughout each of the windings. Taps on the high-voltage side are the most common in use. When metering on the low-voltage side, if copper loss is given for the rated voltage V_R and tap voltage V_T is used, the multiplying factor M_T to be applied to the copper loss at rated voltage will be:

$$M_T = \frac{V_R}{2V_T + 0.5}$$

For taps on the low-voltage side with metering also on the low-voltage side:

$$M_T = \frac{V_T}{2V_R} + \frac{1}{2}\left(\frac{V_T}{V_R}\right)^2$$

Multipliers calculated for most common taps are shown in Fig. 10-1. Where taps are used in both windings, both multipliers are required. Where taps might be changed rather frequently, or for use with automatic tap changers, the best performance is obtained by basing the adjustment on the median tap.

Percent Tap	Tap on High-Voltage Winding	Tap on Low-Voltage Winding
86.6	1.077	0.808
90.0	1.056	0.855
92.5	1.041	0.890
95.0	1.026	0.926
97.5	1.013	0.963
100.0*	1.000*	1.000*
102.5	0.988	1.038
105.0	0.976	1.076
107.5	0.965	1.115
110.0	0.955	1.155

* Tap on which copper loss data is based.

Figure 10-1. Copper-Loss Multipliers for Common Transformer Taps with Low-Voltage Metering.

TRANSFORMER CONNECTIONS

Consideration must be given to the relationship existing between the load current through the power transformer winding and the current through the metering current transformer. The connections of the power transformers may affect the copper-loss calculations for determination of the meter calibrations. This will apply to the calibration of both the transformer-loss meter and the transformer-loss compensator.

Open-Delta Connections

For open-delta connections, full-load losses occur when the transformers are supplying 86.6 percent of the sum of the kVA ratings of the two transformers. Therefore, if an open delta bank were used, the value of the full-load line current would be multiplied by 0.866 to determine the current in the transformer windings at full-load transformer losses. This will apply to both the loss-meter and the loss-compensator calculations.

Scott Connections

In Scott connections used for three-phase to two-phase transformation, the teaser transformer is connected to the 86.6 percent tap. The copper-loss multiplier for the teaser transformer is 1.077 or 0.808 as shown in Fig. 10-1, depending on whether the metering is on the two-phase or three-phase side. For the main transformer, the copper-loss multiplier is 0.875 for metering on the three-phase side and 1.167 for metering on the two-phase side. This applies to both the loss-meter and loss-compensator calculations.

Three-Phase, Four-Wire Delta Connections

Three-phase, four-wire delta connections with a two-stator meter involve a special feature. When final tests of the meters are made with stators in series, the percent loss to which the transformer-loss compensator is adjusted applies to the stator connected to the two-wire current transformer. The stator connected to the three-wire current transformer measures the vector sum of the two currents displaced by 120 degrees. For this stator the percentage copper loss is multiplied by 1.155. For operation in service the loss increment then will be divided equally between the two stators. Similarly, with the transformer-loss meter the speed of the stator connected to the three-wire current transformer should be increased by the relation:

$$1 \text{ to } \frac{5^2}{4.33^2} = 1 \text{ to } 1.333 \text{ or } 33.3\ \%$$

LOSS COMPENSATION METHODS

Transformer-Loss Meters

The basic measurement requirements for transformer losses call for a meter having one voltage-squared stator and one or more current-squared stators, depending on the number of metering current circuits. All stators are combined on the same shaft which drives a register of proper ratio to record the losses in kilowatt-hours or kilovar-hours. The E^2 stator consists of a standard watthour meter voltage coil and a low-current (possibly 50 mA) winding, which is connected in series with an adjustable resistor that serves as a core-loss adjustment. Registration is proportional to E^2. An I^2 stator consists of a standard current coil and a low-voltage voltage coil connected across the current coil and a series resistor in the current circuit. Registration, therefore, is proportional to I^2.

Standard transformer-loss meters equipped with two or three stators are supplied by the manufacturer and are designed for 3.6 seconds per revolution of

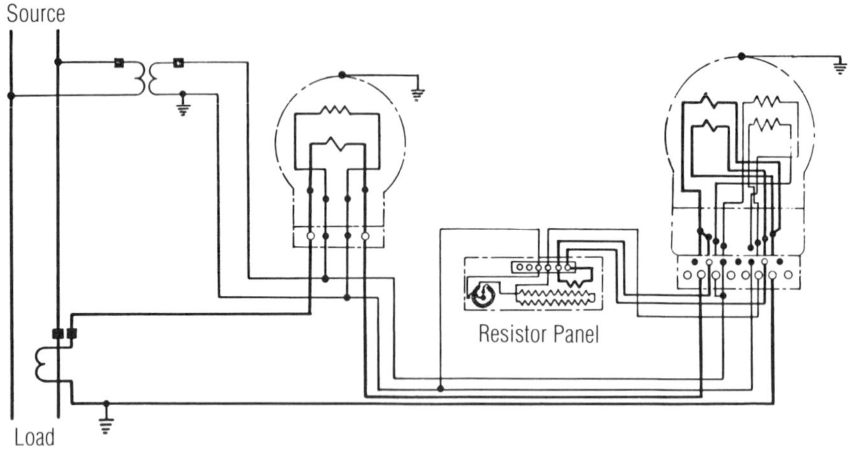

Figure 10-2. Principle of Loss Meter.

the disk with 5 amperes flowing through each of the I^2 stators. Four-stator meters with 5 amperes flowing through each of the three I^2 stators are designed for 2.4 seconds per revolution of the disk. These values are for basic timing only; the meters are to be adjusted for each specific application. Calibration of heavy load is accomplished by adjusting the permanent magnets and light load by the light-load adjustment associated with the voltage coil of the E^2 stator. The watthour

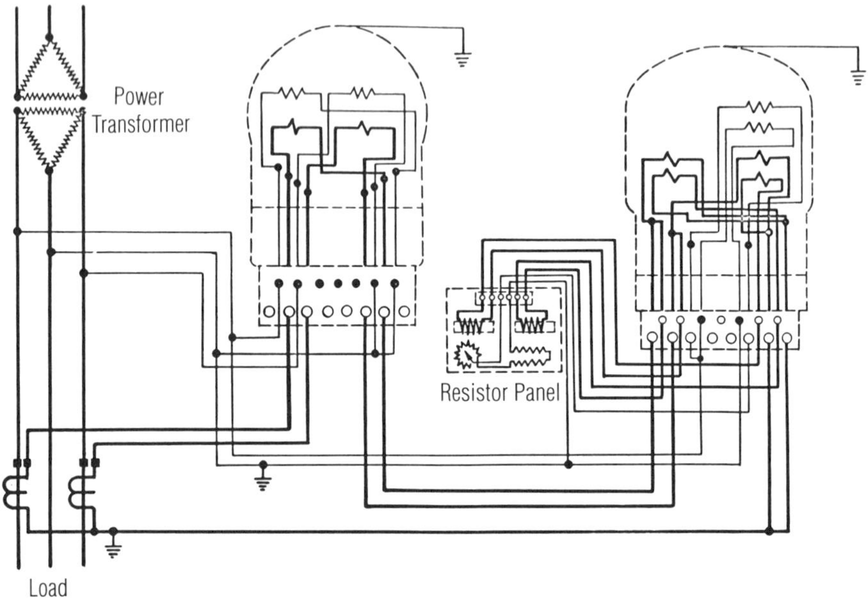

Figure 10-3. Connections for a Three-Phase, Three-Wire Loss Meter Installation.

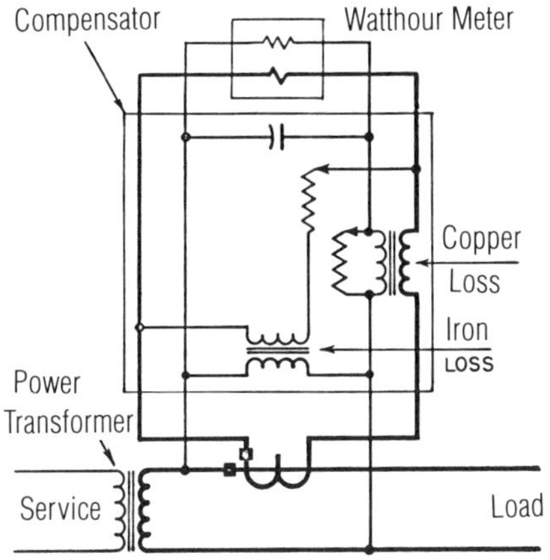

Figure 10-4. Principle of Transformer-Loss Compensator.

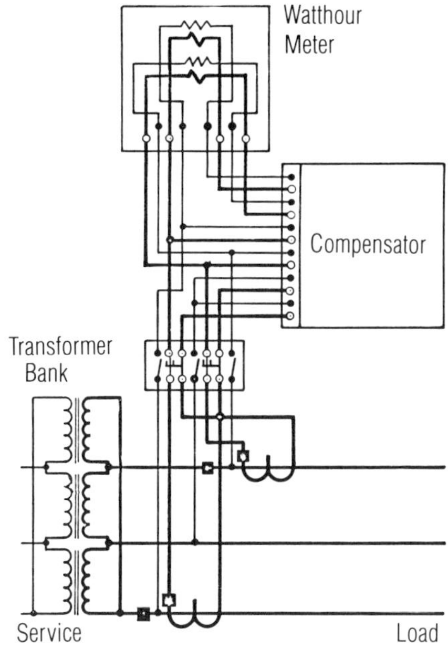

Figure 10-5. Connections of Meter and Transformer-Loss Compensator on a Three-Phase, Three-Wire Circuit.

constant (K_H) and the register ratio (R_R) are selected so that the loss registration is in kilowatt-hours and the desired register constant (K_R) obtained. If a demand register is used, the loss increment of maximum demand will be in kilowatts. Since the maximum loss demand will be coincident with the maximum load demand except under very rare load conditions, the loss demand may be added to the load demand for billing purposes. Figure 10-2 shows the basic principles employed in the transformer-loss meter. Figure 10-3 shows the connections of a typical three-phase, three-wire, transformer-loss meter installation.

In all of the following descriptions of methods to be used for recording losses, it must be remembered that any change in the original set of conditions will have an effect on the results obtained. Should losses be calculated for paralleled lines, paralleled transformers, or for individual transformers, any changes in the connections thereof or the methods of operation will call for a new set of calculations. The compensating meter has been generally superseded by the transformer-loss compensator, because of the greater simplicity of the latter for the combined load-plus-loss measurement in a single meter. However, it remains as a useful instrument in those cases where separate loss measurements are required.

Transformer-Loss Compensator

In compensating for losses using the transformer-loss compensator, losses are added into the registration of the standard watthour meter that is used to measure the customer's load on the low-voltage side of the transformer bank. The transformer-loss compensator is connected into the current and the voltage circuits

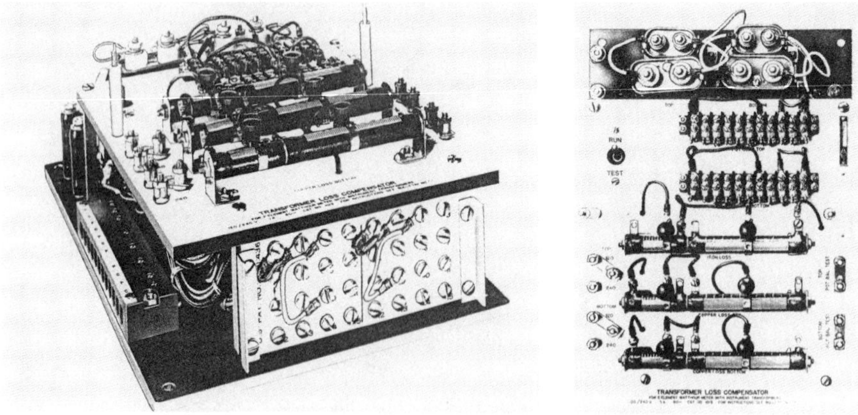

Figure 10-6. Front and Bottom Views of a Two-Element Transformer-Loss Compensator.

and, when properly calibrated, the losses will be included in the watthour meter registration.

For iron-loss compensation a 115/230:3 volt transformer, with its primary connected to the metering voltage supply and the 3 volt secondary connected in series with an adjustable resistor, provides a current which is passed through the current coil of the meter and is equivalent to the iron loss. As the iron-loss current so produced is proportional to the voltage, the iron-loss increment as measured by the meter is proportional to the square of the voltage.

To include the copper-loss increment, a small current transformer has its primary connected in series with the current coil of the watthour meter and an adjustable resistor connected to its secondary. The connections are such that the drop across the resistor is added to the voltage applied to the voltage coil of the meter. As the copper-loss component so produced is proportional to the current, the copper-loss increment as registered by the meter is proportional to the square of the current. Compensation for the flow of voltage-circuit current through the copper-loss element is provided. Figure 10-4 shows the principle of operation. Figure 10-5 shows the connection of a transformer-loss compensator for a typical three-wire, three-phase circuit. Three-element compensators suitable for connection to three-stator meters differ only in the addition of the third element. Figure 10-6 illustrates a two-element transformer-loss compensator.

In applying this method, the losses are determined in percent of the load at the light-, heavy-, and inductive load test points of the meter. The meter normally used to measure the customer's load remains a standard measuring instrument and the compensator is calibrated by tests on the meter with and without the compensator. A switch is provided on the compensator for this purpose.

Resistor Method

The resistor method of loss compensation uses standard metering equipment which has been specially adapted to measure losses. No-load losses are compensated by a resistor which is energized by the meter voltage and has its watts-loss measured by one stator of the watthour meter. Compensation for load losses is accomplished by setting the meter full-load adjustment and light-load adjustment to include the percent losses in meter percent registrations.

The resistor can be in the meter or in the meter enclosure. A resistor located in the enclosure is easier to install, replace, and calibrate than a resistor in the meter.

Solid-State Compensating Meter

A complete compensating meter combines transformer loss compensation with meter functions within the meter package. Modern compensating meters are solid-state meters with several optional compensation provisions.

Although solid-state meters compensate using the same mathematics as electromechanical meters combined with transformer loss compensators, they differ markedly in their method of operation. Instead of using resistances, tapped transformers, and capacitors to modify currents and voltages, solid-state compensating meters convert the current and voltage to digital inputs for microprocessor circuits, which then perform the mathematical operations to compute the losses and add them to the kilowatt-hour and kilovar-hour registers.

LOSS CALCULATIONS

Determination of Losses

For both methods of metering losses the following information is requested from the manufacturer of the transformers: kVA rating, iron loss at rated voltage, copper loss at 75°C at rated full-load current, rated voltage, and voltage taps provided in the high-voltage and/or low-voltage side. If power-factor tests are to be made on the low-voltage side or if varhour losses are to be compensated for, the following additional information is required: percent exciting current at rated voltage and percent impedance at 75°C. From a field inspection of the installation, data are obtained relative to transformer connections; taps in use; and the length, size, and material of conductors from the low-voltage terminals of the transformers to the metering point.

Application

The first step in applying loss compensation on any specific service, as pointed out before, is the collecting from the manufacturer of certain data concerning the individual transformers and the securing of certain field data. It is recommended that forms be provided with which the average meter employee can become familiar. By filling in the data secured from these various sources, the procedures can more easily be followed. The application of the transformer-loss meter or transformer-loss compensator may be better understood by the following steps applied to a specific installation.

Typical Example

A 9,999 kVA transformer bank, consisting of three 115,000:2,520 volt, 3,333 kVA transformers, is connected delta-delta. Metering at 2,520 volt will be by a two-stator watthour meter connected to two 3,000:5 ampere current transformers and two 2,400:120 volt voltage transformers. It is desired to compensate the metering to record, in effect, the load on the 115,000 volt side of the power transformer using (a) a transformer-loss meter and (b) a transformer-loss compensator. A field inspection indicates that the connection from the low-voltage

LOSS COMPENSATING METER INSTALLATION
FIELD DATA

Customer's Name or Station JOHN DOE MFG. CO

Address 1000 MARKET STREET

Transformer Data (From Name Plate)

Make	Type	Serial No.	KVA Rating	Rated Voltage	Tap Voltage In Use	% Imp. 75° C.
WESTG	SL-OA	6530499	3,333	115,000/2,520	115,000/2520	8.16
WESTG	SL-OA	6530500	3,333	115,000/2,520	115,000/2520	8.03
WESTG	SL-OA	6530501	3,333	115,000/2,520	115,000/2520	8.12

Available Taps: H. V. 115,000; 112,125; 109,250; 106,375; 103,500 **L. V.** 2,520

Connections: Hi V: Delta ✗ **Wye** **Wire** 3 **Low V: Delta** ✗ **Wye** **Wire** 3

Current Transformer Ratio 3,000:5 **Voltage Transformer Ratio** 2,400:120

Secondary conductor data: Show size, length, and kind of secondary conductor from power transformer terminals to potential transformer connections. Make sketch below showing power transformer connections, secondary conductors, and ground connections.

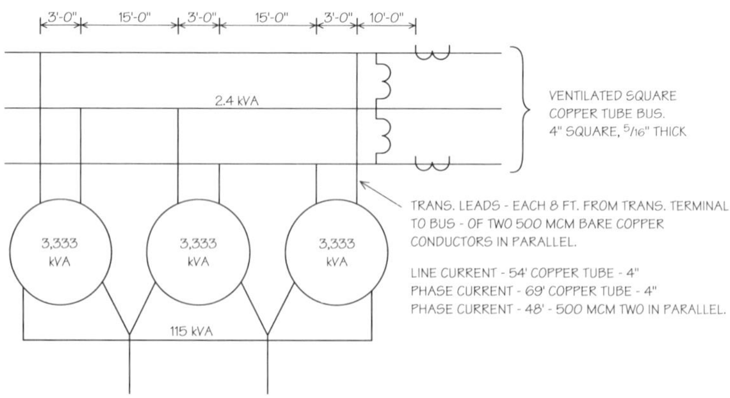

By ____________

Date ____________

Figure 10-7. Field Sheet for Obtaining Loss Data on Power Transformers.

terminals of the transformer to the metering point (where voltage transformers are connected) consists of the following conductors:

Carrying Phase Current: 96 ft. of 500 kcmil copper conductor and 69 ft. of square copper ventilated tubing, 4 in. × 4 in. × 5/16 in. thick.

Carrying Line Current: 54 ft. of 4 in. × 4 in. × 5/16 in. square ventilated copper tubing.

LOSS CALCULATIONS (WATTS)

Customer's Name or Station JOHN DOE MFG. CO

Address 1000 MARKET STREET

Power Transformer Data (Furnished By: WESTINGHOUSE ELECTRIC CORP. **)**

Make	Type	Serial No.	Rated KVA	Rated Voltage	Tap Voltage In Use	Fe Loss At Rated Voltage	Cu Loss At Rated Voltage
WESTG	SL-0A	6530499	3,333	115,000/2,520	115,000/2,520	9,650	18,935
WESTG	SL-0A	6530500	3,333	115,000/2,520	115,000/2,520	9,690	18,400
WESTG	SL-0A	6530501	3,333	115,000/2,520	115,000/2,520	9,340	18,692

Connections: PRIMARY Delta ✗ **Wye** **Wire** 3 **SECONDARY Delta** ✗ **Wye** **Wire** 3

Instrument Transformer Ratio: CT 3,000:5 **VT** 2,400:120

Fe Loss Calculations:

Total Fe losses at power transformer at rated voltage:

9,650 + 9,690 + 9,340 = 28,680 WATTS

Total Fe losses at 120 volts (voltage transformer secondary):

$$\left(\frac{2{,}400}{2{,}520}\right)^2 \times 28{,}680 = 26{,}013.6 \text{ WATTS}$$

Cu Loss Calculations:

Transformer coil current (secondary) at full load:

$$\frac{3{,}333{,}000}{2{,}520} = 1{,}322.6 \text{ AMPS}$$

Line current at transformer bank (secondary) at full load:

$$1{,}322.6 \times \sqrt{3} = 2{,}290.9 \text{ AMPS}$$

Total transformer Cu losses at rated voltage and kVA:

18,935 + 18,400 + 18,962 = 56,027 WATTS

Total transformer Cu losses at voltage tap other than rated:

____

Transformer Cu losses at current transformer full load:

$$\left(\frac{3{,}000}{2{,}290.9}\right)^2 \times 56{,}027 = 96{,}078.8 \text{ WATTS}$$

Secondary conductor losses at current transformer full load:

$$\text{LINE CURRENT} = \frac{3{,}000^2 \times 54 \times .00275}{1{,}000} = 1{,}336.5 \text{ WATTS}$$

$$\text{PHASE CURRENT} = \frac{1{,}732^2 \times 69 \times .00275}{1{,}000} = 569.2 \text{ WATTS}$$

$$\text{PHASE CURRENT} = \frac{1{,}732^2 \times 48 \times .01099}{1{,}000} = 1{,}582.5 \text{ WATTS}$$

Total Watts Cu loss at current transformer at full load:

96,078.8 + 1,336.5 + 569.2 + 1,582.5 = 99,567.0 WATTS

Calculated By: ________ **Date:** ________ **Checked By:** ________ **Date:** ________

Figure 10-8. Calculations for Transformer Watt Losses.

Figure 10-7 shows a suggested field-data form on which the information obtained in the field has been entered. This data serves as a basis for requesting loss information from the manufacturer and for loss calculations.

Calculation Methods

Figures 10-8 and 10-9 show suggested forms for making loss calculations. The

LOSS CALCULATIONS (VARS)

Customer's Name or Station JOHN DOE MFG. CO

Address 1000 MARKET STREET

Power Transformer Data (Furnished By: WESTINGHOUSE ELECTRIC CORP. **)**

Make	WESTG	WESTG	WESTG	
Type	SL-OA	SL-OA	SL-OA	
Serial Number	6530499	6530500	6530501	
Rated kVA	3,333	3,333	3,333	
Rated Voltage	115,000/2,520	115,000/2,520	115,000/2,520	
Tap Voltage in Use	115,000/2,520	115,000/2,520	115,000/2,520	
Fe Loss Rated Voltage	9,650	9,690	9,340	
Cu Loss at Full Load	18,935	18,400	18,692	
% Imp. at 75°	8.16	8.03	8.12	
% Exciting Current	1.00	1.06	0.91	

Instrument Transformer Ratio: CT 3,000:5 **VT** 2,400:120

Core Loss

$$VA = 3{,}333{,}000 \times .01 = 33{,}330$$
$$\text{Cos } \theta = \frac{9{,}650}{33{,}330} = .2895$$
$$\text{Sin } \theta = .9571$$
$$\text{VARs} = 33{,}330 \times .9571 = 31{,}900$$

$$VA = 3{,}333{,}000 \times .0106 = 35{,}329$$
$$\text{Cos } \theta = \frac{9{,}690}{35{,}329} = .2742$$
$$\text{Sin } \theta = .9617$$
$$\text{VARs} = 35{,}329 \times .9617 = 33{,}975$$

$$VA = 3{,}333{,}000 \times .0091 = 30{,}330$$
$$\text{Cos } \theta = \frac{9{,}340}{30{,}330} = .3079$$
$$\text{Sin } \theta = .9514$$
$$\text{VARs} = 30{,}330 \times .9514 = 28{,}855$$

TOTAL VARs = 94,730

$$\text{AT RATED VOLTAGE} = \left(\frac{2400}{2520}\right)^4 \times 94{,}730 = 77{,}934.2 \text{ VARs}$$

Copper Loss

$$VA = 3{,}333{,}000 \times .0816 = 271{,}972$$
$$\text{Cos } \theta = \frac{18{,}935}{271{,}972} = .06962$$
$$\text{Sin } \theta = .9975$$
$$\text{VARs} = 271{,}972 \times .9975 = 271{,}292$$

$$VA = 3{,}333{,}000 \times .0803 = 267{,}639$$
$$\text{Cos } \theta = \frac{18{,}400}{267{,}639} = .06874$$
$$\text{Sin } \theta = .9976$$
$$\text{VARs} = 267{,}639 \times .9976 = 266{,}996$$

$$VA = 3{,}333{,}000 \times .0812 = 270{,}639$$
$$\text{Cos } \theta = \frac{18{,}692}{270{,}639} = .06906$$
$$\text{Sin } \theta = .9976$$
$$\text{VARs} = 270{,}639 \times .9976 = 269{,}989$$

TOTAL VARs = 808,277

$$\text{AT CURRENT TRANSF. FULL LOAD} = \left(\frac{3000}{2290.9}\right)^4 \times 808{,}277 = 1{,}386{,}086 \text{ VARs}$$

Calculated By: ______ **Date:** ______ **Checked By:** ______ **Date:** ______

Figure 10-9. Calculations for Transformer Var Losses.

loss data received from the manufacturer and the secondary conductor data obtained from the field have been entered and the iron and copper losses have been calculated. Total iron loss and total copper loss results will be used later for determining calibrations for transformer-loss meters and transformer-loss compensators. Thus far, the procedure is identical for both transformer-loss meters

LOSS METER CALIBRATIONS

Customer's Name or Station JOHN DOE MFG. CO

Address 1000 MARKET STREET

Loss Meter Calibration Data:

Co. No. 68480 **Serial No.** 27891397 **Make** G. E.

Type V-21-A **Amps.** 5 **Volts** 120 **Phase** 3 **Wire** 3 **Stators** 2

K_h 100 R_r 100 R_g 10,000 **Reg. Type** M-30 K_r 100 **Demand K** 100

Connections: PRIMARY Delta ✗ **Wye** **Wire** 3 **SECONDARY:** ✗ **Wye** **Wire** 3

Cu Element Calibration:

Cu loss at full load of current transformer 99,567 **(watts).**

Seconds / disk revolution with 5 amperes through meter coils connected series 3.6156 **seconds.**

Fe Element Calibration:

Fe loss at 120 volts 26,013.6 **(watts).**

Seconds / disk revolution 13.839 **seconds.**

The above information from calculations below:

K_h **(Basic):** $\frac{99{,}567 \times 3.6}{3{,}600} = 99.57$

Selected — K_h 100 R_r 100 K_r 100 **Demand K** 100

Cu Loss — Sec / disk revolution at 5 amperes:

$$\frac{100 \times 3{,}600}{99{,}567} = 3.6156 \text{ seconds}$$

Fe Loss — Sec / disk revolutuion at 120 volts:

$$\frac{100 \times 3{,}600}{26{,}013.6} = 13.839 \text{ seconds}$$

Calculated By: ________ **Date:** ________ **Checked By:** ________ **Date:** ________

Figure 10-10. Calculations for Determining Loss Meter Calibrations.

and transformer-loss compensators. Throughout the calculations an empirical assumption is made that within small variations of the rated voltage, iron loss varies as the square of the voltages and the vars loss varies approximately as the fourth power of the voltages.

Loss Compensation Formulas

Calculations should be referred either to the high voltage circuit or to the low voltage circuit. Once the choice is made, the convention should be followed for all subsequent calculations to avoid confusion. The formulas which follow all refer to the high voltage circuit.

Note: VTR = Voltage Transformer Ratio
CTR = Current Transformer Ratio

LWFe = Watt losses due to iron (core-loss watts)

$$\text{LWFe} = \text{No-load Test Watts} \times \frac{(\text{Meter Voltage} \times \text{VTR})^2}{(\text{Rated Transformer Volts})^2}$$

LWCu = Watt losses due to copper

$$\text{LWCu} = \text{Load Test Watts} \times \frac{(\text{Meter Test Current} \times \text{CTR})^2}{(\text{Transformer Test Amps})^2}$$

LVFe = Var losses due to iron (core-loss vars)

$$\text{LVFe} = \text{No-load Vars} \times \frac{(\text{Meter Voltage} \times \text{VTR})^4}{(\text{Rated Transformer Volts})^4}$$

LVCu = Var losses due to copper

$$\text{LVCu} = \text{Load Vars} \times \frac{(\text{Meter Test Current} \times \text{CTR})^2}{(\text{Transformer Test Amps})^2}$$

$$\text{No-load Vars} = \text{Transformer Test kVA} \times 1{,}000 \times \sin\theta\,\text{Fe}$$

Where θ Fe is the angle whose cosine is $\dfrac{\text{No-Load Test Watts}}{\text{Transformer Test kVA} \times 1000}$

$$\text{Load Vars} = \text{Transformer Test kVA} \times 1{,}000 \times \sin\theta\,\text{Cu}$$

Where θ Cu is the angle whose cosine is $\dfrac{\text{Load Test Watts}}{\text{Transformer Test kVA} \times 1000}$

%LWFe = Percent Iron Watt Loss (Core Loss)

$$\%\text{LWFe} = \frac{\text{LWFe} \times 100}{\text{Meter Test Current} \times \text{Meter Voltage} \times \text{CTR} \times \text{VTR}}$$

%LWCu = Percent Copper Watt Loss (Load Loss)

$$\%\text{LWCu} = \frac{\text{LWCu} \times 100}{\text{Meter Test Current} \times \text{Meter Voltage} \times \text{CTR} \times \text{VTR}}$$

%LVCu = Percent Copper Var Loss (Load Vars)

$$\%\text{LVCu} = \frac{\text{LVCu} \times 100}{\text{Meter Test Current} \times \text{Meter Voltage} \times \text{CTR} \times \text{VTR}}$$

%LVFe = Percent Iron Var Loss (Core Vars)

$$\%\text{LVFe} = \frac{\text{LVFe} \times 100}{\text{Meter Test Current} \times \text{Meter Voltage} \times \text{CTR} \times \text{VTR}}$$

Meter test current is test current with metered current elements (coils) in series, and with unity power factor. Meter test voltage is test voltage with meter voltage elements (coils) in parallel, and with unity power factor.

$$
\begin{aligned}
\text{Meter test current} &= \text{Meter TA} \times \text{K} \\
\text{Meter test current} &= \text{5.0 amperes (transformer rated)} \times \text{K} \\
\text{Meter test current} &= \text{50\% Class (JEMTECH)} \times \text{K} \\
\text{Where K} &= \text{3 (3-element meter)} \\
\text{K} &= \text{2 (2-element meter)} \\
\text{K} &= \text{4 (2-1/2 element meter)}
\end{aligned}
$$

$$\text{Transformer Test Amps} = \frac{\text{Transformer Test kVA}}{\sqrt{3} \times \text{Transformer Rated kV}}$$

Calculate %LWFe, %LWCu, %LVFe, and %LVCu for full load, single phase, and with unity power factor.

If the test voltage remains constant:

%LWCu is proportional to test load (test amps)
%LWCu is inversely proportional to power factor

%LWFe is inversely proportional to test load (test amps)
%LWFe is inversely proportional to power factor

%LVCu is proportional to test load (test amps)
%LVCu is inversely proportional to power factor

%LVFe is inversely proportional to test load (test amps)
%LVFe is inversely proportional to power factor

These relationships permit light-load test and 50%-power-factor-load test values to be easily derived from full-load test values.

Percent Watthour Losses

Percent watthour losses, calculated at full-load meter test, can be used to determine the percent watthour losses at light load and at 50% lag load. For example, the percent *copper* loss for the meter light-load (10%) test is one-tenth the *copper* loss for the meter full-load (100%) test. The percent *iron* loss at meter light-load test is 10 times the *copper* loss at meter full-load. The percent *copper* loss at a meter lag load (50% power factor) is twice the *copper* loss of the meter full-load test. The percent *iron* loss at meter lag load is twice the *iron* loss of the meter full-load test.

The percent losses are added to the percent registration of the uncompensated meter to obtain the calibration percent registrations of the compensated meter.

TRANSFORMER-LOSS METER

Calibrations

Figure 10-10 is a suggested form that might be used for transformer-loss meters. This form lists the calculations necessary for calibration of the transformer-loss meter to be used with the typical installation that has been selected for this example. The final calibrating data is listed in seconds per revolution at 120 V for the iron-loss element and at 5 amperes for the copper-loss element. These quantities are selected in order that testing values for all loss installations will be uniform; also, because 120 V is the normal rated secondary voltage of most voltage transformers and 5 amperes is the secondary current of the current transformer operating at its full rating. The calculations for var losses are needed only if power-factor determinations or varhour measurements are to be made on the low-voltage side. If var losses are required, the form shown in Fig. 10-10 would be used in the same manner as described for watt losses.

Meter Test

Transformer-loss meters usually are calibrated and tested using an ammeter, a voltmeter, a stop watch, and a variable supply of voltage and current, and applying the revolutions per second as determined from calculations. If the number of meter installations warrants, a portable standard transformer-loss meter may be used. This consists of an E^2 and an I^2 element which can be operated separately—the E^2 element to operate at 3.6 seconds per revolution for 120 V and the I^2 element at 3.6 seconds per revolution for 5 amperes. Service meters can be tested by comparison with this test meter using the ratio of the meter to the standard revolutions. In such cases, variations in current or voltage during the test would be compensated for automatically.

Initial adjustment of a transformer-loss meter consists of balancing the individual stator torques. For this test, voltage only is applied to the E^2 stator. Five amperes of current is passed through each I^2 stator individually and the separate torques adjusted by the torque balance adjustments until they are equal. An arbitrary current of 45 to 60 mA is then passed through the current coil of the E^2 stator and the torque adjusted to equal that produced by each of the I^2 stators.

Full-load calibration is accomplished by applying rated voltage only to the E^2 stator and passing 5 amperes through the I^2 stators in series. The meter speed is adjusted using the braking magnets until the calculated copper-loss speed is obtained. Following the full-load test, with voltage only still applied to the E^2 stator, the current in the I^2 stators is reduced to an arbitrary light-load value and the meter speed is adjusted to the correct value using the light-load adjustment on the E^2 stator. If a value of $1\frac{1}{2}$ amperes is selected for the light-load test, the correct meter speed will be 9 percent of the value at full load, since speed varies with the square of the current in the I^2 stators, and the time per revolution will be $11\frac{1}{9}$ times that at full load.

The final calibration test consists of removing all current from the I^2 stators and connecting the external resistors to supply current to the E^2 stator. The E^2 stator current is then adjusted by the resistors to produce the calculated speed value for transformer-core-loss measurement at rated voltage.

TRANSFORMER-LOSS COMPENSATOR

Calibrations

Should a transformer-loss compensator be used, Fig. 10-11 shows a suggested form that may be used for calculations and calibration data. The losses at light-load, full-load, and inductive-load test points are calculated in terms of the watt loads at which the meter is tested. For a polyphase meter at the full-load test point, this is the product of the number of meter stators and the volt-ampere rating of each stator. The percent loss at any other loads may be calculated by proportion. While copper losses vary as the square of the load current, percent copper losses vary directly with the percent load on the meter. Similarly, percent iron losses vary inversely with the percent load on the meter. It is therefore simple to determine a percent loss for any load. For meter test loads at power factors other than 1, it is necessary only to divide the percent iron and percent copper losses at power factor 1.0 by the desired power factor. Thus, calibrations for percentage losses need be made only for the full-load test point. For the other loads, they are determined by proportion as shown in Fig. 10-11. As in the case of the compensating meter installation, the calculation of var losses is made only if power-factor determinations or varhour measurement on the low-voltage side are to be compensated to the high-voltage side. The var-loss calculations are performed in a similar manner as watt-loss calculations shown in Fig. 10-11.

Test

A transformer-loss compensator and a watthour meter are connected for test in the same manner as normally applies to a watthour meter. With the compensator test switch in the test position, thus disconnecting the compensator from the watthour meter, the watthour meter is tested in the usual way. When all of the tests on the watthour meter are completed, the compensator test switch is opened, which will place the copper-loss elements in the circuit. Current is applied to each stator separately with the appropriate copper-loss resistor in the circuit in each case. In Fig. 10-11 the percent copper loss at full load was calculated as 0.691 percent. With 5 amperes in each stator, one at a time, the performance is adjusted by the compensator loss resistors to give single-stator performance values 0.7 percent faster than the values obtained in the final meter calibration. (For three-phase, four-wire delta meters, see comments earlier in this chapter under "Transformer Connections.") After this check, the compensator test switch is closed in the normal position, which includes the iron-loss element in the circuit, and meter test connections are made for the series test. With light-load current, the iron-loss element is adjusted to the calculated light-load setting point. Since the copper-loss units are also in the circuit, the value for this test shown in Fig. 10-11 is the total final percent registration of 101.9 percent. A final check should then be made with the compensator at all loads. It must be remembered that inductive-load tests are for check purposes only, since it is not possible to make any compensator adjustments under this condition. If the compensator adjustments are correct under the other test conditions, the meter performance with inductive load should agree with the calculated value. In making inductive-load tests it must be remembered that any deviation in power factor from the nominal value of 0.5 lagging will result in differences from the desired performance. For inductive load tests, therefore, it is important that the test results be compared with the desired values for the power factor at which the test is made.

Tests in service follow the same general principles. A test on the meter with and without the compensator determines the accuracy of either the meter or the compensator. The compensator's performance is indicated by the difference between the results of the test with and without the compensator.

If the compensator requires adjustment, errors on the heavy-load test are corrected with the copper-loss adjustment, those at light load with the iron-loss adjustment. If inductive-load tests are desired in the field, it is important to establish and correct for the true power factor of the test load. The phase angle of

DETERMINATION OF METER CALIBRATIONS AND COMPENSATOR SETTINGS

KWH Meter:

Co. No. 67614 Ser. No. 19022727 Make WESTG Type R1-2

Amps 5 Volts 120 Phase 3 Wire 3

K_h 2/3 R_r 3,000 R_g 15,000 K_r 12,000 Demand K 12,000

Current Transformer Ratio 3,000:5 Voltage Transformer Ratio 2,400:120

Demand Element:

Co. No. ______ Ser. No. ______ Make ______ Type ______ Volts ______

Demand Interval 30 Minutes Demand Constant 12,000

Loss Compensator:

Co. No. 164 Ser. No. 15409 Make E. S. Type 1014 Volts 120

Phase 3 Wire 3 Delta or Wye D Elements 2

DETERMINATION OF METER CALIBRATIONS:
Full load meter rating for series test:

2 EI Cos θ = 2 × 2400 × 3000 = 14,400,000 WATTS

Percent Iron Loss: $\frac{26{,}013.6 \times 100}{14{,}400{,}000} = .180\%$

Percent Copper Loss: $\frac{99{,}567 \times 100}{14{,}400{,}000} = .691\%$

Percent Meter Calibration with Compensator:

Load	Stator	Meter Calibration	% Fe Loss	% Cu Loss	% Total Loss	Final % Registration
Light	Series	100.0	1.80	.0691	1.869	101.9
	Top or Right					
	Mid or Split					
	Bot or Left					
Full	Series	100.0	.180	.691	.871	100.9
	Top or Right					
	Mid or Split					
	Bot or Left					
Inductive	Series	100.0	.360	1.382	1.742	101.7
	Top or Right					
	Mid or Split					
	Bot or Left					

Calculated By: ______ Date: ______ Checked By: ______ Date: ______

Figure 10-11. Calculations for Determining Transformer Compensator Calibration.

the loading devices and possible variations in the three-phase line voltages all have an effect on the true power factor of the test load.

RESISTOR METHOD

There are other methods which are being used to measure transformer losses that employ standard metering equipment especially adapted to measure losses.

One such method uses a standard watthour meter connected to the secondary side of the transformer bank in the usual manner.

Copper-loss compensation is effected by applying corrections to the observed meter registration at heavy load and at light load in a manner similar to the application of instrument transformer corrections. In general, the loss is added to the load kilowatt-hour registration by applying a negative correction equal in value to

$$100\left[\frac{\text{kW bank losses}}{\text{kW load}}\right]$$

so that the observed meter registration is correct at the test load employed. These corrections are computed from the known copper loss of the transformer bank at rated load and take into account the power factor of the customer's load.

Since copper losses vary as the square of the load current, the required adjustment at any load point, I, is equal to I^2/I, or is in direct proportion to the load current I. For example, the adjustment at 10 percent meter load will be one-tenth of that required at 100 percent load. In this manner, the performance curve of the watthour meter, so adjusted, will closely follow the copper-loss curve of the power transformer throughout its load range.

Core-loss compensation is obtained by adding a single-phase load to the watthour meter in the form of a fixed resistor mounted within the meter. The resistor is connected so that it is energized by the meter voltage and its watts loss is measured by one stator of the watthour meter. The actual watts load added to the meter is equal to the bank core loss divided by the product of the instrument transformer ratios used.

Since the resistance is fixed, the watts loss varies as the square of the voltage and the measurement of this watts loss is equivalent to direct measurement of the core loss of the power-transformer bank which also varies approximately with the square of the voltage.

Formula for Calculating Resistance

The value of the resistance to be used for this resistor is determined by the following equation:

$$R = \frac{E_{mo}^2 \times N_v \times N_c}{1000L_i \dfrac{E_{mo}^2}{E_m^2}} = \frac{E_m^2 \times N_v \times N_c}{L_i \times 1000}$$

Where

R = Resistance in ohms

E_m = Calculated Meter Voltage

$= \dfrac{\text{Rated secondary voltage of transformer bank}}{N_v}$

E_{mo} = Meter operating voltage

N_v = Ratio of instrument voltage transformers

N_c = Ratio of instrument current transformers

L_i = kW core loss of transformer bank at rated voltage

= kW core loss at calculated meter voltage E_m

Formula for Computing Compensating Corrections

The percentage registration for a compensated meter should be as follows:

$$\text{Percentage Registration} = 100\left[\frac{\text{kW load} + \text{kW bank losses}}{\text{kW load}}\right] \tag{1}$$

$$\text{Percentage Registration} = 100\left[\frac{\text{kW load}}{\text{kW load}}\right] + 100\left[\frac{\text{kW bank losses}}{\text{kW load}}\right] \tag{2}$$

The first component, $100\left[\frac{\text{kW load}}{\text{kW load}}\right]$, represents the meter registration due to the load.

The second component, $100\left[\frac{\text{kW bank losses}}{\text{kW load}}\right]$, represents the meter registration due to the losses in the transformer bank.

As it is desirable that an overall correction be applied which includes the compensation required for both the core loss and the copper loss, the watthour meter is tested with the compensating-resistor current added to the test-load current in the meter but not in the portable standard watthour meter. The following formulas are derived on this basis.

A. Copper-Loss Correction

From formula (2) above, the correction for copper loss may be stated as:

$$\text{Copper-loss correction} = -100\left[\frac{\text{kW bank copper loss}}{\text{kW load}}\right]\% \tag{3}$$

Let L_c = kW copper loss of transformer bank at rated kVA load
$\cos\theta$ = Average power factor of customer's load
T = Transformer bank kVA rating at full load
A = kW load in percent of rated kW load of transformer bank

Then formula (3) becomes:

$$\text{Copper-loss correction at rated load of bank} = -100\,\frac{L_c}{T\cos\theta}\,\% \tag{4}$$

Since copper losses vary as the square of the load current, the kW = 100 copper loss at any load A is equal to

$\left[\left(\frac{A}{100}\right)^2 L_c\right]$ and since the kW load at A is equal to $\left(\frac{A}{100}\right)$T $\cos\theta$, then

Copper-loss correction at any load *A* is:

$$A = -\left[\frac{100\left(\frac{A}{100}\right)^2 L_c}{\left(\frac{A}{100}\right)T\cos\theta}\right] = -100\left[\frac{\left(\frac{A}{100}\right)L_c}{T\cos\theta}\right]\% \tag{5}$$

To convert load A to test load

Let S = Test load in percent of meter rating

$$F = \frac{\text{Meter capacity [See note in Section C, below]}}{\text{Rated kVA of transformer bank} \times 1{,}000}$$

whence $A = SF$. Substituting SF for A, formula (5) may be expressed as:

$$\text{Copper-loss correction at test load } S = -100\left[\frac{\left(\frac{SF}{100}\right)L_c}{T\cos\theta}\right]\% \tag{6}$$

B. Core-Loss Correction

From Formula (2), the correction for core loss may be stated as:

$$\text{Core-loss correction} = -100\left[\frac{\text{kW bank core loss}}{\text{kW load}}\right]\% \tag{7}$$

Let L_i = kW core loss of transformer bank at rated secondary voltage. Then core-loss correction at rated kVA load is

$$= -100\left[\frac{L_i}{T\cos\theta}\right]\% \tag{8}$$

Since core losses do not change with load current, the kW core loss as

$$= -100\left[\frac{L_i}{\left(\frac{A}{100}\right)T\cos\theta}\right]\% \tag{9}$$

Since $A = SF$, then core-loss correction at test load S

$$= -100\left[\frac{L_i}{\left(\frac{SF}{100}\right)T\cos\theta}\right]\% \tag{10}$$

Since the kW core loss of the transformer bank is given at rated voltage and since at the time of the meter test the operating voltage of the bank may differ from the rated voltage, the value of core loss for the operating voltage is determined as follows:

Let E = Rated secondary voltage of the transformer bank

E_o = Operating voltage of the transformer bank

$$\text{Then kW core loss at } E_o = \left(\frac{E_o}{E}\right)^2 \times L_i \tag{11}$$

Since voltage measurements are taken at the meter terminals, let:

$$E_m = \text{Calculated meter voltage} = \frac{\text{Rated secondary voltage of transformer bank}}{\text{Voltage transformer ratio}} = \frac{E}{N_v}$$

$$E_{mo} = \text{Meter operating voltage at time of test} = \frac{E_o}{N_v}$$

From which $E = E_m \times N_v$ and $E_o = E_{mo} \times N_v$

Formula (1) then becomes:

$$\text{kW core loss at } E_o = \left(\frac{E_{mo} \times N_v}{E_m \times N_v}\right)^2 L_i = \left(\frac{E_{mo}}{E_m}\right)^2 L_i \tag{12}$$

Formula (10) may be restated:

Core-loss correction at test load S and any meter voltage E_{mo} =

$$-100\left[\frac{L_i\left(\frac{E_{mo}}{E_m}\right)^2}{\left(\frac{SF}{100}\right) T \cos\theta}\right]\% \tag{13}$$

C. Combined Copper- and Core-Loss Correction

Combining formulas (6) and (13), the combined copper-loss and core-loss correction becomes:

Compensated metering correction at test load S and meter operating voltage can be expressed as:

$$E_{mo} = -\frac{100}{T\cos\theta} \times \left[\frac{SF\ L_c}{100} + \frac{100L_i\left(\frac{E_{mo}}{E_m}\right)^2}{SF}\right]\% \tag{14}$$

Where

L_c = kW copper loss of transformer bank at rated kVA load
L_i = kW core loss of transformer bank at rated voltage
T = Transformer bank kVA rating at full load
$\cos\theta$ = Average power factor of customer's load
E_{mo} = Meter operating voltage at time of test
E_m = Calculated meter voltage

$$= \frac{\text{Rated secondary voltage of transforemer bank}}{\text{Voltage transformer ratio}}$$

S = Test load in percent of meter kVA rating
= 100 percent for heavy-load test = 10 percent for light-load test

$$F = \frac{\text{Meter capacity (See Note below)}}{\text{Rated kVA of transformer bank} \times 1{,}000}$$

Note: Meter capacity is defined as (meter primary amperes) × (meter primary volts) × (number of stators), except for three-phase, three-wire, 2-stator meters, for which meter capacity is (meter primary amperes) × (meter primary volts) × $\sqrt{3}$.

An example of the data necessary in calculating settings for resistor-type compensated meters follows.

Transformer Bank Rating (T)	1,500 kVA
Rated Primary Voltage of Transformer Bank	26,400 volts
Rated Secondary Voltage of Transformer Bank (E)	2,400 volts
Service Characteristic of Transformer Bank Secondary	3-phase, 3-wire, delta
Ratio of Metering Voltage Transformers (N_v)	2,400:120 volts

Ratio of Metering Current Transformers (N_c)	400:5 amperes
Calculated Meter Voltage ($E_m = E/N_v$)	120 volts
Core Loss of Transformer Bank (L_i at E)	3.680 kW
Copper Loss of Transformer Bank (L_c at T)	13.130 kW
Customer's Average Power Factor ($\cos\theta$)	1.0

$$\text{Ratio Factor } (F) = \frac{\text{Primary Meter Volts} \times \text{Primary Meter Amps} \times \sqrt{3}}{T \times 1{,}000} = 1.11$$

Using this data, compensating resistance can be computed:

$$\text{Compensating Resistance} = R = \frac{E_m^{\,2} \times N_v \times N_c}{L_i \times 1000} = 6{,}261 \text{ ohms}$$

Compensated meter corrections, for assumed values of E_{mo} (meter operating voltage at time of test) and with S (test load, in percent) assumed to be 100% and then 10%, can be computed and are shown in the table below.

$$\text{Compensated Metering Correction} = -\frac{100}{T\cos\theta}\left[\frac{SF \times L_c}{100} + \frac{100\,L_i}{SF} \times \left(\frac{E_{mo}}{E_m}\right)^2\right]\%$$

Compensated Meter Correction at

E_{mo}(Volts)	S = 100%	S = 10%
110	–1.2%	–2.0%
112	–1.2%	–2.0%
114	–1.2%	–2.1%
116	–1.2%	–2.2%
118	–1.2%	–2.2%
120	–1.2%	–2.3%
122	–1.2%	–2.4%
124	–1.2%	–2.5%
126	–1.2%	–2.5%

Figure 10-12 gives the connections for the testing of a watthour meter with its compensating resistor.

Occasionally situations arise where it is necessary to subtract the transformer losses from the kilowatt-hour registration. To accomplish this, the fixed resistor is connected so that its watts loss is subtractive, the sign before the formula is reversed, and the meter speed adjusted with a positive correction applied.

SOLID-STATE COMPENSATING METERS

Derivation of Coefficients

Solid-state meters with transformer loss compensation (TLC) options are sometimes programmed by the manufacturer using loss data provided by the meter purchaser, and are sometimes programmed by the user. Consult with the manufacturer of each particular model of meter, to determine if that model is programmed at the factory, or by the user.

The meter performs internal calculations to generate compensation values, and adds those values to measured watts and vars. When the meter is programmed, coefficients based on TLC percentages and other meter-specific information are

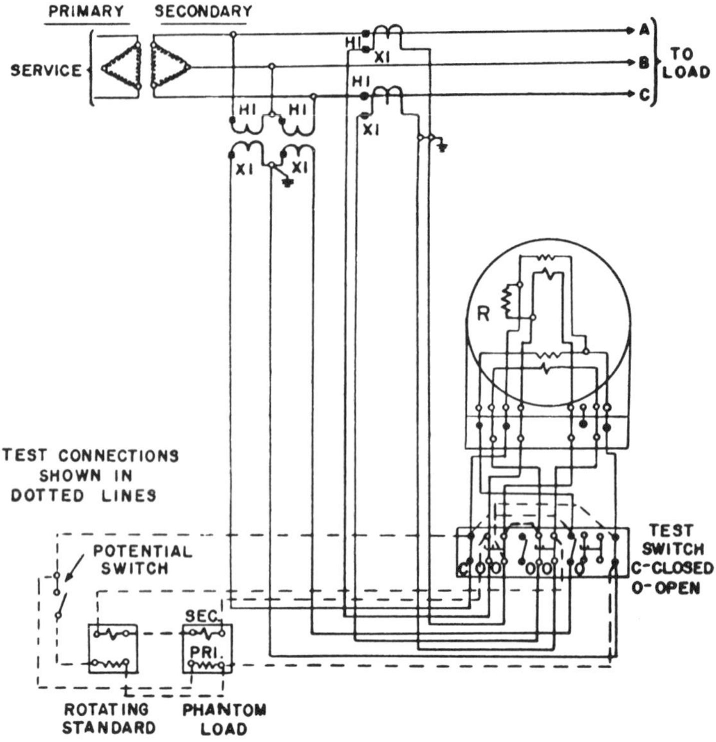

Figure 10-12. Test Connections for Three-Phase, Three-Wire, Two-Stator Watthour Meter with Compensating Resistor.

stored in the meter's "configuration PROM". The coefficients then can be accessed each time a loss calculation is performed.

Transformer loss compensation provides compensation for all metered and derived quantities. It causes the display registers, memory, and KYZ output quantities to be compensated.

SUMMARY

The methods described in this section are useful as compared with metering on the high-voltage side:

1. When the metering cost is appreciably lower than for metering on the high-voltage side.
2. For exposed locations on the system, where high-voltage instrument transformer equipment may be expected to be troublesome because of lightning or other disturbances.
3. When the limited available space makes the installation of high-voltage metering equipment difficult, hence more expensive.
4. When a customer with a rate for low-voltage service is changed to a high-voltage service rate.

Generally speaking, metering on the high-voltage side should be preferred:

1. Where the cost of high-voltage metering is lower than for methods with loss compensation.
2. Where multiple low-voltage metering installations are necessary in place of one metering installation on the high-voltage side.
3. Primary metering is necessary where a part of the load is used or distributed at the supply voltage.

Comparing the compensating meter with the transformer-loss compensator, the advantages of the compensating meter are:

1. Iron losses are registered at no load, regardless of how small they are.
2. The load measurement on the low-voltage side by the conventional watthour meter is available as a separate quantity from the losses.

The disadvantages of the compensating meter are:

1. Two meters are used, whereas primary metering requires only one. This complicates the determination of maximum demand.
2. Special test equipment, not ordinarily carried by meter testers, is required.

TOTALIZATION

Introduction

Totalization is the combining the metering of two or more electrical circuits. Revenue metering of a customer who is served by two or more lines and is billed on coincident demand must be totalized. Totalization is done in real time, addition of the individual meter readings does not suffice.

Often totalization is not necessary, but is used to simplify meter reading and the subsequent accounting procedures. Monthly kilowatt-hour usage, for example, can be the addition of the individual meter readings or one totalized meter reading.

Totalization is the algebraic addition of like electrical quantities, done on a real time basis. The addition can be as simple as combining the kilowatt-hour usage of two circuits, or as complicated as solving an algebraic algorithm to determine the net usage for an installation with multiple internal and external sources and loads.

Several methods to totalize the metering of multiple circuits are available.

Parallel Current-Transformer Secondaries

Totalization before measurement, such as paralleling the secondaries of the current transformers in two or more circuits having a common voltage source, is shown in Fig. 10-13. Here the secondaries of the current transformers on Line 1 of circuits A and B have been connected in parallel at the coil of the meter. A similar arrangement would be used on Line 3. Precautions must be taken because of the difficulties arising from the increased effects of burden and the flow of exciting current from one transformer to the other during unbalanced load conditions. The most important of these precautions are:

1. All of the transformers must have the same nominal ratio regardless of the ratings of the circuits in which they are connected.
2. All transformers which have their secondaries paralleled must be connected in the same phase of the primary circuits.
3. The secondaries must be paralleled at the meter and not at the current transformers.

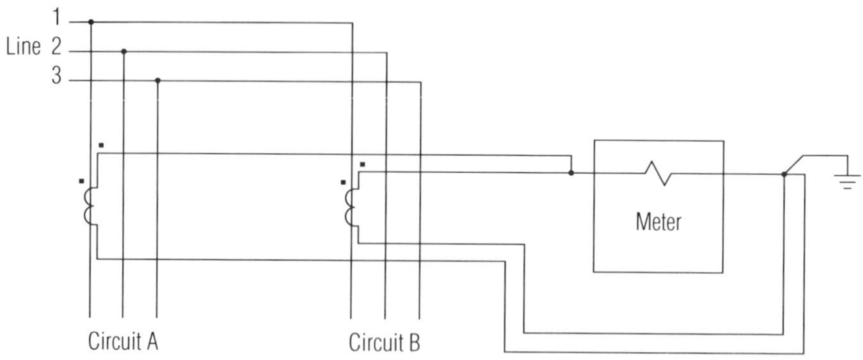

Figure 10-13. Simplified Connection Diagram for Parallel-Connected, Current-Transformer Secondaries.

4. There should be only one ground on the secondaries of all transformers. This should be at their common point at the meter.
5. Modern current transformers with low exciting currents and, therefore, little shunting effect when one or more current transformers are "floating" at no load should be used. Three or more "floating" current transformers might have an effect that should be investigated.
6. The secondary circuits must be so designed that the maximum possible burden on any transformer will not exceed its rating. The burden should be kept as low as possible as its effects are increased in direct proportion to the square of the total secondary current.
7. A common voltage must be available for the meter. This condition is met if the circuits share a common bus that is normally operated with closed bus ties.
8. A common voltage must be available for the meter.
9. Burdens and accuracies must be carefully calculated.
10. If adjustments are made at the meter to compensate for ratio and phase angle errors, the ratio and phase angle error corrections used must represent the entire combination of transformers connected as a unit.
11. The watthour meter must be of sufficient current capacity to carry, without overload errors, the combined currents from all the transformers to which it is connected.
12. Low-voltage, low-burden-capability current transformers are not suited to this application since the burden imposed on parallel secondaries may be very high.
13. Meter voltage usually is equipped with a throw-over relay to avoid loss of meter voltage in the event the normal supply is de-energized.

Parallel Current-Transformer Primaries

Under certain conditions, particularly with window-type transformers, primaries may be paralleled. With the proper precautions, acceptable commercial metering may be obtained with this method. Without proper consideration of all factors involved, the errors may be excessive, particularly at low current values.

Multistator Meters

Mechanical totalization is accomplished by combining, on one watthour meter shaft, the number of stators necessary to properly meter the circuits involved.

Mechanical totalizing is in common use for the totalization of power and lighting circuits. A three-stator meter for example, is available for the totalization of one three-wire, three-phase power circuit plus one single-phase lighting circuit, either three- or two-wire.

Mechanical totalization can be extended to include other combinations within the scope of multistator watthour meters. In determining the suitability of such meters, consideration must be given not only to the number of stators required for the totalization, but also to the practical space limitation for the required number of meter terminals.

Although it is not necessary to have circuits of identical rating for this method of totalization, the circuits must be such as to give the same watthour constant to each meter stator. For instance, if one circuit has voltage transformers of 24,000:120 and current transformers of 100:5, while the second circuit has voltage transformers of 2,400:120 and current transformers of 1,000:5, the transformer multiplying factors are the same for both circuits, 24,000:120 × 100:5 = 2,400:120 × 1,000:5, and totalization through a totalizing meter is feasible.

PULSE TOTALIZATION

Pulse totalization is the algebraic addition of pulses from two or more pulse generating meters.

PULSE INITIATORS

A pulse initiator is the mechanism, either mechanical or electrical, within the meter which generates a pulse for each discrete amount of a metered quantity.

The pulse initiator can be a small switch or pair of switches mounted within the meter. It may also be an electronic device. The pulse initiator may consist of a light-sensitive device positioned so as to receive reflections directly from the bottom of the disk or from reflective vanes on a shaft driven from the disk. Whether the pulse initiator assembly is commonly geared to the disk shaft or receives reflections from the disk itself, the number of pulses is directly proportional to the number of disk revolutions.

The output pulses are normally Form C closures, commonly known as KYZ or double-throw, single-pole, or their electrical equivalent. Each closure is one pulse.

In some applications, Form A closures are used. These can be obtained from a Form C by using only KY or only KZ. Obtaining a Form A from a Form C doubles the value of each pulse.

Since the pulses are directly proportional to disk revolutions, a definite watthour value can be assigned to each pulse. If the ratio between pulses and disk revolutions is 1:1, the watthour value of a pulse must be, and is, equal to the primary watthour constant of the meter. Thus, in a receiving instrument which counts the number of pulses, this count, multiplied by the primary watthour constant, gives the value of the energy measured. It is important to note that in certain mechanical pulse receivers one pulse is a latching and the next an advancing pulse. There may be confusion regarding this type of device since only half the pulses are recorded. Regardless of the action of the receiver, this chapter considers one closure of the pulse initiator as one pulse transmitted.

The pulses are initiated by disk revolutions of a watthour meter. Hence, the value of the pulse is in terms of energy and not of power. Before any time factor is introduced the pulse has a value in watthours rather than watts. In a magnetic tape recorder the time channel, in effect, divides the quantity received by time and permits a reading in watts over the time period. This does not change the character of the value of the pulse as received. In this type of device the interval pulses, when multiplied by the appropriate constant, yield demands in kilowatts; the total, when multiplied by the appropriate constant, yields kilowatt-hours.

There are certain requirements in the design and operation of pulse initiators that should be kept in mind when employing this type of telemetering. In any pulse system the initiator is the only source of information transmitted. Therefore, particularly when used for billing, the performance of the mechanism generating the pulses must match the accuracy of the requirement. A pulse initiator, free from faults, has the accuracy of the watthour meter in which it is installed. The final answer given by the receiver cannot be more accurate than the initiating pulse.

To achieve the desired accuracy, good design of the pulse initiator, proper application, and high order of maintenance are necessary. The good design must be provided by the manufacturers. Proper application means matching the capabilities of the pulse initiator, the communication channel, and the receiver.

It must be remembered that as the value in kilowatt-hours of each pulse is increased, the possible "dollar" error in the demand determination by a miscount of even one pulse becomes correspondingly greater. The lower limit of pulse value and correspondingly greater rate of sending pulses is determined by pulse initiator design and receiver capabilities.

Characteristics of Pulse Initiators

There are two basic output circuits of pulse initiators used for telemetering of kilowatt-hours, the 2-wire and the more common 3-wire.

The pulse initiator output circuit may be energized with AC or DC. Direct current, in conjunction with polarized relays, or electronic type demand devices, is used where transmission over telephone lines is necessary. The polarized relay permits true 3-wire operation over a 2-wire circuit. It is also used on long circuits to avoid the attenuation due to the capacity effect of some communication cables.

Electronic pulse initiators of the mercury-wetted-relay or transistor-switch types provide quick-make, quick-break action in delivering pulses to the transmitting circuit and obtain all of the advantages of this construction with none of the disadvantages associated with the mechanical types.

The output pulse circuits of pulse initiators are usually low power circuits. Most receivers, totalizers, and pulse equipment designed for meter pulse applications are also low power devices and therefore require no intermediary device. However, the installation must have interposing relays, impedance correcting circuits, or a combination thereof to protect the pulse initiator where there are long runs of connecting wire, multiple devices activated by one initiator, or devices having inductive or capacitive input characteristics.

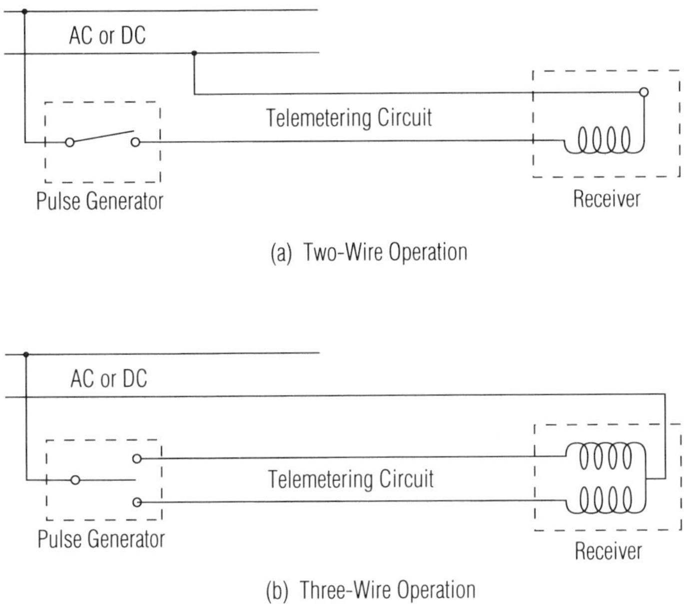

Figure 10-14. Simplified Diagrams Illustrating Basic 2-Wire and 3-Wire Pulse Circuits.

Types of Pulse Initiators

Mechanical pulse initiators with cam-and-leaf construction depend upon the meter disk for their driving force.

Electronic pulse initiators may have a shuttered disk, an output shaft with reflective vanes or reflective spots on the rotor of the meter that work in conjunction with a light source to furnish an input signal to an amplifier whose output is connected to the pulse transmitting circuit. Electronic pulse initiators can be designed that permit a much greater number of output pulses per meter disk revolution than cam-and-leaf devices.

Pulse initiators of meters with electronic registers are sometimes integral with these registers. The registers are driven, not by direct gearing as with a mechanical device, but by light or infrared reflections off the meter disk. The pulse initiator supplies output pulses which are an integral number of the pulses driving the register. Normally the register input is 12 pulses per revolution. Output pulses can therefore be 12, 6, 4, 3, 2, or 1 pulse per revolution. The reciprocals 1/12, 1/6, 1/4, 1/3, 1/2, and 1/1 are known as the M_p or R/I of the initiator. The choice of M_p or R/I is made when programming the register.

Some meters with electronic registers have no provision for supplying output pulses. These meters must be equipped with a separate pulse initiator. Having a separate initiator has the advantage of providing a redundant revolution-counting circuit which can serve as a check against the meter registration.

Electronic meters which do not have a rotating disk often generate pulses to indicate a quantity of energy which is identical to that quantity metered by one revolution of the disk of an equivalent mechanical meter. This equivalent revolution pulse permits testing the meter using conventional test procedures. Often, the output pulse is programmable in terms of M_p or R/I. The output pulse is also programmable in terms of K_e, where K_e is a discrete amount of the metered quantity per pulse, for example, kilowatt-hours/pulse.

Maintenance of Pulse Initiators

Proper maintenance of mechanical pulse initiators requires correct adjustment of contacts for maximum tension and, hence, minimum resistance, when closed; adequate clearance with contacts open; and low friction loading on the meter rotor.

If the contact points become discolored or pitted, the points should be dressed with a burnishing tool and then with paper until they are bright but not necessarily flat. If employed with discretion, fine crocus cloth is sometimes useful. The use of a file on these points is bad practice and should be avoided. If contacts require filing, they should be replaced because their service life will be limited. Small pits in the contact points will not impair operation as long as the points are clean.

The same mesh conditions required for mechanical pulse initiators must be obtained for electronic pulse initiators; however, friction due to mechanical make and break of contacts has been eliminated in solid-state pulse initiators.

Kilowatt-hour and Kilowatt Constants

Meter pulse initiators normally have a three-wire pulse output. The wires are designated K, Y, and Z. The pulse output is a series of alternating KY and KZ closures. Contact closure between K and Y is a pulse, as is closure between K and Z. Most initiators have break-before-make closures. The output can be thought of as a single-pole double-throw switch. In relay terms, the output is a dry, Form C relay. It is useful to know the mechanical terminology, even though solid-state switches are increasingly used in new installations.

Each pulse represents a distinct quantity of kilowatt-hour energy. The kilowatt-hour value of each pulse is K_e.

Meter K_d is programmable in most solid-state meters and electronic registers which have KYZ pulse outputs. It is derivable for pulse initiators on electromechanical meters using constants listed in manufacturers' literature as P/DR, Pulses/Disk Revolution, R/I, Revolutions/Impulse, or M_p, Revolutions per pulse. All of these constants are either equivalents or reciprocals — the particular constant depending on the preference of the meter manufacturer. Depending on the model and type, these constants are either programmable or must be specified when purchased. Meter K_e equals K_h (watthours per revolution) × R/I (revolutions per pulse) divided by 1,000.

Overall K_e, that is the effective K_e of pulses delivered to an end device by the output of the pulse totalizer or relay, may be different from meter K_e. If the totalizer or any interposed relay has an input/output ratio other than 1:1, the output K_e will equal the meter K_e times the product of the input/output ratios.

If a two-wire device is operated by a three-wire pulse system, the value of each two-wire pulse will be double the three-wire value.

For a particular demand interval, each pulse represents a distinct quantity of kilowatt demand. The demand inteŕval is programmed into the pulse receiver, a demand indicator, pulse recorder, translation system, and other devices. The value of the demand pulse, K_d, is equal to the overall K_e divided by the demand interval expressed in hours.

Meter pulse systems determine demand by counting the total quantity of pulses accumulated over a complete demand interval. Devices that operate on pulse rate, or duration, or the times between pulses might be erroneous, especially when pulse totalization is being performed.

The following gives the nomenclature, equations, and method of calculating the application of pulse initiators.

Required data:

Maximum kW demand expected, for example, kilowatts.
Demand interval, in hours.
Pulse receiving capacity of demand meter per interval.

Nomenclature:

kWh = kW × demand interval in hours.
Pulse = The closing and opening of the circuit of a two-wire pulse system. The alternate closing and opening of one side and then the other of a three-wire system is equal to two pulses.
K_e = kWh/pulse, i.e. the energy.
M_p = Meter disk revolutions per pulse.
T_i = Demand interval in hours.
R_p = Ratio of input pulses to output pulses for totalizing relay(s).
N_p = Number of pulses required to advance receiver.
K_d = Kilowatts per incoming pulse at receiver = $\dfrac{K_e \times R_p}{T_i}$
K_h = Secondary watthour constant.

Kilowatts divided by the pulse receiving capacity of the demand meter gives a possible value of K_d. Obviously no demand meter would be intentionally run to exactly full scale at the demand peak. Usually a choice of one-half to three-quarters of the maximum pulse capacity per interval of the receiver is reasonable and will allow for load growth. Also, the choice should be such as to obtain convenient values of K_e and K_d.

If revolutions per pulse is too small a fraction, the value of K_e must be increased and a lower scale reading on the demand meter accepted.

kW Constants

Where K_d is the kilowatt value of the incoming pulse and where T_i = the incoming demand interval in hours,

$$\frac{K_e\,(\text{final})}{T_i}$$

Kilowatt dial, chart, or tape multiplier = Kd × kW ratio where the kW ratio equals the number of incoming pulses to give a reading of 1 on the demand dial, chart, or tape.

Available R/I, M_p, and other ratios are generally in the range of 1:10 to 10:1. Indicators which operate on reflections off the meter disk are normally 2, 4, 6, or

8 pulses per revolution. Questions concerning the availability of specific ratios should be referred to the meter manufacturer.

TOTALIZING RELAYS

Where the totalization of more than two circuits is required, an intermediate totalizing relay is generally necessary. This relay must be capable of adding pulses and, when required, subtracting other pulses from the positive sum and retransmitting the algebraic sum to a receiving device. When a totalizing relay with an input-to-output ratio other than 1 to1 is used, it must be considered in adjusting the pulse value of the meter.

If, for example, a relay is used with a 4 to 1 ratio, it is necessary to furnish four pulses to the relay for every one that is retransmitted to the receiver. Pulse values for the meter must have 1 to 4 ratio to the values for the receiver. For example, if the receiver pulse value was 38.4 kWh, it would be 38.4/4 = 9.6 kWh at the meter.

Totalizing relays serve to combine pulses produced by two or more meters and to retransmit the total over a single channel. Electronic types eliminate most of the problems of maintenance associated with the older electromechanical mechanisms but in no way relieve the situation of limiting pulse rates to prevent "overrun" of the relays. Attention to pulse rates (pulses per minute) is especially important if several relays are operated in cascade to accommodate a large number of meters in a single totalizing network.

The maximum pulse rate at which any relay can be operated is often limited by the receiver. Electronic data logging receivers are capable of accepting very high pulse rates with very short time duration per pulse, but many electromechanical receivers require relatively low rates and relatively long pulses.

In electronic totalizing relays, all input circuits of a relay must be interrogated (scanned) in turn and all pulses present must be outputted to the receiver before any channel can be again interrogated. A limiting condition is pulses on all channels simultaneously (burst condition).

For a multi-channel relay the maximum input rate must be no greater than output rate divided by the number of channels. For example, a seven-channel relay with a relay ratio of 1 to 1 and an output rate of 56 pulses per minute has an input rate of 8 pulses per minute per channel (56/7=8). In order to provide some safety margin, a rate of 7 pulses per minute is published.

If the relay ratio is other than one to one, this factor must be considered. The formula used is output rate times the ratio factor divided by the number of channels ($56 \times 4/7 = 32$ for a four-to-one relay).

It must be kept in mind that the higher the output rate of the relay, the shorter will be the duration of the pulses. This too can be a limiting factor for some receivers.

A general discussion of the electromagnetic type totalizing relays can be found in the Eighth edition of this *Handbook* in Chapter 10. For specific operating characteristics of modern totalizing relays, the manufacturer's literature should be consulted.

MULTICHANNEL PULSE RECORDERS

Many applications which previously required a totalizing relay and a single channel pulse counter can now be performed by a multichannel pulse recorder used in conjunction with a "smart" translation device. Normally, the recorder stores pulse counts from one to four meters. Periodically, the recorder is interrogated and the information is transferred to a computer which performs the translation. The computer is programmed to combine those channels which are to be totalized, and to produce reports for revenue billing, load survey, market research, and other applications.

A typical recorder stores 35 days of data in 15 minute increments. The time base is usually programmed to be that of the demand interval. A smaller increment would fill the recorder memory in proportionally less time, perhaps 11 or 12 days. Similarly, a larger increment would permit more days of data to be stored; if the increment were one hour, 140 days of data could be stored.

One useful memory design is the wrap-around type. Wrap-around memory overwrites the most current data over the oldest data. Another useful property is that the data is not erased when the recorder is interrogated. These properties permit redundancy in data storage. The recorder always retains the most recent 35 days of data. The database at the translation site has all data up through the last interrogation. The field data can be reread if the communications channel introduced errors or if data stored at the central computer is corrupted.

Data transmission can be by telephone, radio transmissions, cable TV lines, power line carrier, microwave, or other communication medium with the ability to transmit data accurately. In the event the communication system fails, most solid-state recorders can be interrogated by a portable reader coupled to the recorder via an optical port.

Each time there is communication between an individual recorder and the central computer, the central computer runs a time check and corrects the individual unit to the system time. Each unit keeps time using the 60 hertz power system as reference. In addition, each unit has a backup source for keeping time during power outages.

When the communication system is telephone, radio transmissions, or other medium which is constantly available, individual units can call or be called at frequent intervals, such as once a day. By frequent interrogation, problems can be discovered and solved promptly. Additionally, the units can be self-diagnostic, including tamper detectors which initiate calls to report alarms.

PULSE ACCESSORIES

Auxiliary Relays

In some cases of totalization it may be found necessary to operate more than one device from the same watthour meter pulse initiator or to operate an AC device and a DC circuit from the same pulse source. Auxiliary relays may be used for this purpose. In general, such relays have a single 3-wire input and two or three similar isolated 3-wire output circuits. Relays are available to convert 2-wire pulses to 3-wire and 3-wire pulses to 2-wire.

Another auxiliary relay often used for this and other purposes is the polarized relay. A polarized relay is a direct current operated relay. It permits the use of the 2-wire circuit to transmit 3-wire pulses. This is done by reversing the polarity of the direct current applied to the relay coil. It is equivalent to using a positive pulse for one side of the 3-wire circuit and a negative pulse for the other.

Polarized relays are found to be necessary in the totalization and telemetering of pulses over some distance. The features of these relays are the low current at which they operate, 1 to 20 mA DC; the positive action upon polarity change, preventing stray currents from causing incorrect operation; and the fact that the contacts for retransmitting consist of two sets which can be paralleled or used to drive separate circuits. The polarized relay maintains a closed transmitting contact until the opposite polarity is received. Polarized relays require only two wires between source of pulses and the relay, an obvious advantage for distance metering.

PULSE-COUNTING DEMAND METERS

Pulse-counting demand meters, as their name implies, are the receiving devices which count the pulses transmitted by the telemetering system. By counting for a time interval, such as 15 minutes, and then resetting, a block-interval demand measurement is obtained.

The final read-out may be a demand register, a round chart, a strip chart, a printed figure, a punched tape, or a magnetic tape. Limitations in regard to the rate at which pulses can be received, and also in regard to the total number of pulses accepted in any demand interval, are characteristic of all pulse-counting demand meters. Manufacturers' publications should be consulted for details of individual demand meters.

NOTES ON PULSE TOTALIZATION

When selecting or designing a pulse-totalization system there are a number of practical considerations that must not be ignored.

The pulse-receiving capability of the totalizing recorder is a case in point. Here are four factors to be considered:

1. Total number of pulses that can be recorded during any one demand interval.
2. Ability to record clusters of pulses at a much higher rate than capability over entire demand interval. Cascaded intermediate relays may result in transmission of two to eight almost simultaneous pulses. If the receiver fails under these conditions, the intermediate relay circuits must be redesigned to space pulses.
3. Sensitivity to pulse duration must be considered. Where pulse duration may be less than one-quarter of a second it is obviously futile to employ a receiver requiring pulses of a minimum duration of three-quarters of a second.
4. Pulse-receiving mechanisms which fail to respond to less than perfect contact closures will require excessive maintenance. Receivers should tolerate some variation in pulse current as well as in pulse duration.

In any complex pulse-totalization system, pulse values must be established for both transmitting and receiving instruments. If it is desired to retain the same pulse value at initiator and at receiver, it may be necessary to cascade totalizing relays rather than combine all pulses in one relay. For example, certain types of six- or eight- circuit totalizing relays cannot be operated with any degree of reliability with a 1:1 ratio of incoming to outgoing pulses. A 2:1 ratio doubles the value of the pulse. It is often possible to employ, in such a case, two 4-element relays with 1:1 pulse ratios, the outputs of which are combined on a 2-element relay again with a 1:1 pulse ratio. In this manner, pulse values at the receiver can be kept at the same values as when initiated.

METERING TIME-CONTROLLED LOADS

Certain loads that lend themselves readily to control as to time of usage are sometimes served by utilities under a special rate schedule. Usually a watthour meter in conjunction with a time switch is employed in order that the load may be disconnected during the time of system peak demands. In other cases, registration of kilowatt-hours during such off-peak periods may be segregated on a two-rate register and include not only the registration of the energy consumed by a particular device but also the general service load. The domestic water heater is probably the most common load of this type and the following forms of water heater metering are sometimes used to fit the various rate schedules that have been devised by the utilities.

Water Heater Loads

Perhaps the most frequently used method in the past has employed a regular watthour meter for the house load and a separate meter and time switch combination for the water heater load. The time switch is operated by a synchronous motor and its contacts open the water heater circuit during the on-peak periods specified within the special rate schedules provided. Combination single-phase watthour meters and time switches are available included within a single case. The combination watthour meter and the time switch may be equipped either with the conventional four- or five-dial register or it may be equipped with a two-rate register having two sets of dials so that the off-peak and on-peak energy can be indicated on separate dials. Such devices have a number of varied applications. In some instances, it may be desirable to register the on-peak load of the entire service on one set of dials or, if the rate schedule so provides, the house load might be recorded during on-peak periods with the water heater disconnected.

The connection of the combination meter and time switch for control of off-peak water heater load with a separate meter for the house load is illustrated in Fig. 10-15. This combination is applicable where the water heater load is supplied at a completely different rate than the house load.

Figure 10-16 shows a combination meter and time switch with a single rate register and double-pole contacts controlling the off-peak water heater load. The meter registers both the house load and the water heater load, the latter being disconnected during predetermined on-peak periods. This arrangement is suitable when the regular domestic rate applies also to the water heater load or where a fixed block of kilowatt-hours during the billing period is assigned to the water heater load by the tariff.

Figure 10-17 shows a combination meter and time switch with a two-rate register and double-pole contacts for disconnecting the water heater during on-peak periods. This arrangement might be applied where the house load used during off-peak periods is charged for at the same rate as the water heater load. The connections are the same as shown in Fig. 10-16.

There is also available a combination meter and time device with a two-rate register but without contacts for the water heater load. The water heater remains

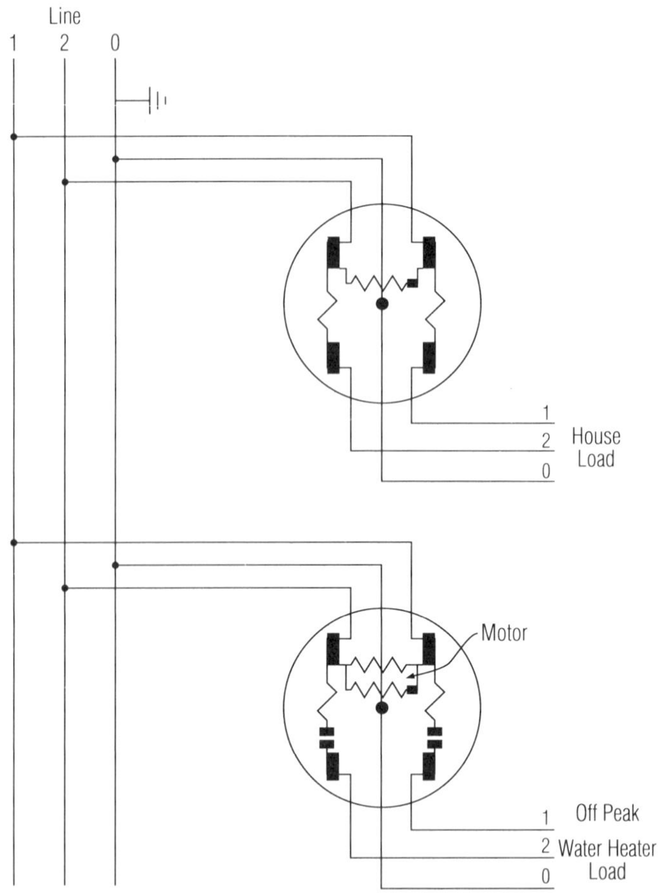

Figure 10-15. Connection for Combination Watthour Meter and Time Switch and Separate Meter for Residential Load.

connected to the service at all times. The time device serves to control only the two sets of register dials. In this case, the water heater load may be billed at the regular domestic rate during peak periods and all loads billed at a different rate during off-peak periods or a specified number of kilowatt-hours during the billing period may be assigned by the tariff to the water heater load.

A number of the arrangements provided by two-rate registers and time switch combinations may also be obtained by separate meters and time switches.

The use of time switches and similar methods of control have been replaced in some measure by a so called fixed block or floating block in the rate schedule. Under this method, the water heater load is supplied at a lower rate than the domestic load and a fixed number or a variable number of kilowatt-hours are assigned to the water heater during the billing period. This method requires only one single-dial meter, without a time switch, for the entire water heater and domestic load.

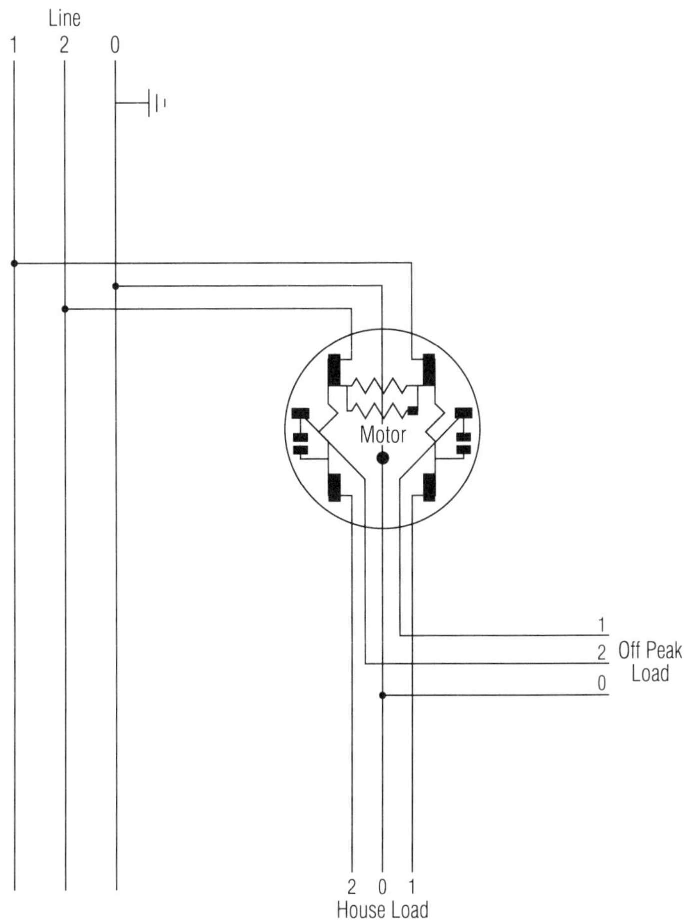

Figure 10-16. Connection for Combination Watthour Meter and Time Switch.

ELECTRONIC REGISTERS

In the late 1970's and early 1980's, manufacturers began taking advantage of the developments in miniaturization of electronic components and introduced standard watthour meters with sophisticated registers which could record the customer's usage on several registers apportioned as to time of day, and, where applicable, day of the week. One such demand meter, shown in Fig. 10-18, records total kilowatt-hours on a continuously cumulative basis, similar to standard kilowatt-hour meters. In addition, the register can record energy and maximum demand for each of the three daily time periods.

The electronic register shown in Fig. 10-18 allows program selection of up to three possible demand presentations. The three demand presentations are cumulative, continuously cumulative, and indicating. The register may be programmed to display any or all of the three demand formats. Demand is calculated as kilowatt-hours divided by the demand interval, which is program selectable.

Figure 10-17. Combination Time Switch and Two-Rate Register.

Figure 10-18. Demand Meter with Cover Removed.

Intervals of 15, 30, 60, 120, 240 minutes, or the entire TOU may be selected. In addition, rolling demand measurements can be program selected, with rolling demand calculations occurring at programmed subintervals. The register display verifies demand reset by a continuous "all eights" display. The "all eights" display can also be used to verify that the display segments are operating properly. Many solid-state registers have timed load control outputs which can operate a contactor. The contactor, sized to make and break the water heater load, controls the appliance.

KILOWATT-HOUR MEASUREMENTS ABOVE PREDETERMINED DEMAND LEVELS

For load studies, rate studies, or other special applications, a differential register is available which will register the total kilowatt-hours on a service as well as the number of kilowatt-hours consumed during a period when the demand is in excess of a predetermined kilowatt demand level. This is accomplished with a differential gear, one side of which is driven at a speed proportional to a given kilowatt-hour load by a synchronous motor and the other side from the watthour meter disk. A ratchet prevents reverse rotation when the speed of the meter-driven gear is less than that of the motor-driven gearing. The result is a reading on one set of dials of the total kilowatt-hours and, on another set of dials, the kilowatt-hours used during the period when the demand is greater than the preset level. The application of this device is illustrated for an actual load curve in Fig. 10-19. The shaded area indicated above the horizontal line represents the energy used at a higher demand level.

There is also available another special meter to record the so-called excess energy above a predetermined set point or, in effect, a predetermined demand point. Such a meter consists of a watthour meter having an extra stator rated at 120 V and 50 mA and connected for reverse torque. There is also provided an external adjustable resistor panel and a standard voltage transformer to supply

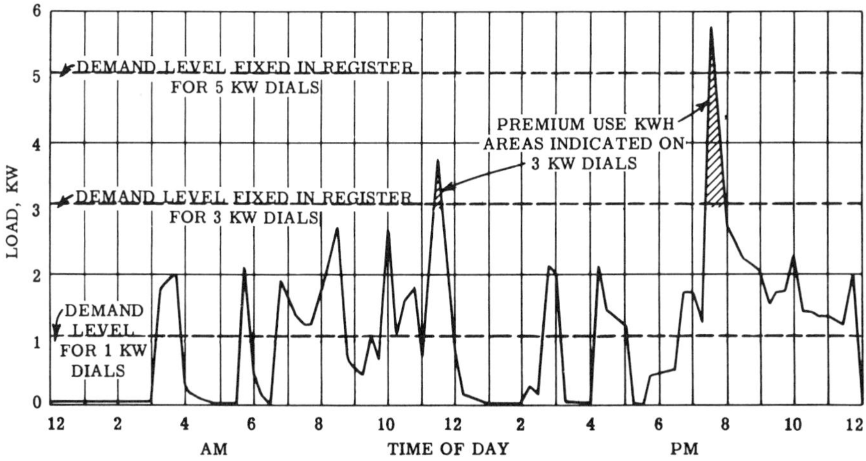

Figure 10-19. Load Curve Showing Measurement by Differential Register.

all of the power for the negative torque stator. The meter torque is negative up to the point of excess so no registration results until the excess set point is passed. Below and at the excess point the meter will not register and a detent will prevent reverse rotation of the shaft. When the excess point is passed, the meter shaft rotates and the registration of kilowatt-hours above the preset limits will occur. This meter may have certain applications for specially designed rate schedules. It has some slight advantage over the register employing the differential gearing because it can be more easily calibrated for a change in the excess point without requiring mechanical gear changes.

LOAD-STUDY METERS

Load study implies load profile. Most major meter manufacturers offer solid-state registers with load profile capabilities. The register typically stores 35 days of load profile data with a 15-minute interval and is programmable to other intervals. Depending on the software, translation of the data is usually done on a mainframe computer or a personal computer. Various software packages are available. The reports are typically graphical plots of 15-minute average demands (or the demands averaged over the programmed interval), and tabular presentations of kilowatt demand, kilowatt hours, and meter revolutions.

Temporary studies of installations, that is installations where the study is for a short time, are often read with a hand-held device to avoid the costs of installing a remote communications capability. Load studies of longer duration are usually linked for remote access via a communications network such as the telephone network or packet radio.

CHAPTER 11

INSTRUMENT TRANSFORMERS

REASONS FOR USING INSTRUMENT TRANSFORMERS

IT WOULD BE DIFFICULT AND IMPRACTICAL to build self-contained meters to measure the energy in high-voltage or high-current circuits. To provide adequate insulation and current-carrying capacity the physical size of the meters would have to be enormously increased. Such meters would be costly to build and would expose the meter technician to the hazards of high voltage. The use of instrument transformers makes the construction of such high-voltage or high-current meters unnecessary.

Instrument transformers are used for the following reasons:

1. To insulate, and thereby isolate, the meters from the high-voltage circuits.
2. To reduce the primary voltages and currents to usable sizes and to standard values that are easily metered with meters having a common secondary rating.
3. To provide a means of combining secondary voltages or currents phasorially to simplify metering.

The instrument transformers deliver accurately known fractions of the primary voltages and currents to the meters. With proper register ratios and dial multipliers the readings of the meters can be made to indicate the primary kilowatt-hours.

DEFINITIONS

Definitions which relate mainly to instrument transformers are listed below. Other general definitions are included in Chapter 2 of this Handbook and in ANSI Std 100-1988, *Standard Dictionary of Electrical and Electronic Terms.*

Bar-Type Current Transformer—A transformer with a fixed and straight single primary turn which passes through the magnetic circuit. The primary and secondary(s) are insulated from each other and from the core(s) and are assembled as an integrated structure.

Burden of an Instrument Transformer—The active and reactive power consumed by the load on the secondary winding. Burden is expressed either as impedance with the effective resistance and reactive components, or as volt-amperes and power factor at a specified current, voltage, and frequency.

Bushing-Type Current Transformer—A transformer with an annular core, no primary winding, and a secondary winding insulated from and permanently assembled on the core which can fit over a secondary bushing in a pad-mount transformer.

Grounded-Neutral Terminal-Type Voltage Transformer—A transformer with the neutral line of the primary permanently connected to the case or mounting base.

Indoor Transformer—A transformer which must be protected from weather.

Insulated-Neutral Terminal-Type Voltage Transformer—A transformer with the neutral line of the primary insulated from the case or base, and connected to a terminal with insulation for a lower voltage than required by the line terminal. The neutral line may be connected to the case or mounting base in a manner which allows temporary disconnection for dielectric testing.

Leakage Flux—Magnetic flux, produced by current in a transformer winding, which flows outside the windings.

Low Remanence Current Transformer—A transformer with a remanence less than 10% of maximum flux.

Multiple-Secondary Current Transformer—A transformer with one primary and three or more secondaries, each on separate magnetic circuits.

Multi-Ratio Current Transformer—A transformer with three or more ratios obtained by taps on the secondary winding.

Rated Current—The current in the primary upon which the performance specifications are based.

Rated Secondary Current—Rated current divided by the marked ratio.

Rated Secondary Voltage—Rated voltage divided by the marked ratio.

Rated Voltage—The primary voltage upon which the performance specifications of a voltage transformer are based.

Series-Parallel Primary Current Transformer—A transformer with two insulated primaries which can be connected in series or in parallel, and provides different rated currents.

Tapped-Secondary Current (Voltage) Transformer—A transformer with two ratios, obtained by a tap on the secondary winding.

Wound-Type Current Transformer—A transformer with a primary with one or more turns mechanically encircling the core. The primary and secondary windings are insulated from each other and from the core, and are all assembled as an integrated structure.

BASIC THEORY OF INSTRUMENT TRANSFORMERS

The Voltage Transformer

The Ideal Voltage Transformer Connection Diagram

Figure 11-1 shows the connection diagram for an ideal voltage transformer. Note that the primary winding is connected across the high-voltage line and the secondary winding is connected to the voltage coil of the meter. When 2,400 volts are applied to the primary of this voltage transformer, 120 volts are developed in the secondary by transformer action. This secondary voltage is applied to the voltage coil of the meter. Since there is no direct connection between the primary and secondary windings, the insulation between these windings isolates the meter from the primary voltage. Note that one end of the secondary winding is connected to ground. This provides protection from static charges and insulation failure.

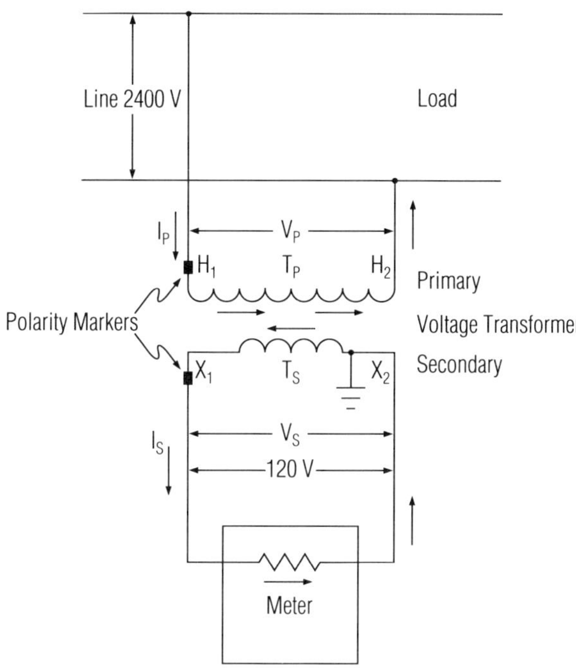

Marked Primary Voltage = 2400 V
Marked Secondary Voltage = 120 V
Marked Ratio = 2400:120 = 20:1

$$\text{Turn Ratio} = \frac{T_P}{T_S}$$

$$\text{True Ratio} = \frac{V_P}{V_S}$$

Figure 11-1. The Ideal Voltage Transformer.

Polarity

In Fig. 11-1, the polarity markers are used to show the instantaneous direction of current flow in the primary and secondary windings of the voltage transformer. They are so placed that when the primary current I_P is flowing into the marked primary terminal H_1, the secondary current I_S is at the same instant flowing out of the marked secondary terminal X_1. These markings enable the secondaries of the voltage and current transformers to be connected to the meter with the proper phase relationships. For example, in the case of a single-stator meter installed with a voltage and a current transformer, reversal of the secondaries from either transformer would cause the meter to run backward .

Secondary Burden

In Fig. 11-1, the voltage coil of the meter draws a small current from the secondary winding. It is therefore a burden on the secondary winding. The ANSI Standard C57.13 defines the burden of an instrument transformer as follows:

> "That property of the circuit connected to the secondary winding that determines the active and reactive power at its secondary terminals."

The burden on a voltage transformer is usually expressed as the total volt-amperes and power factor of the secondary devices and leads, at a specified voltage and frequency (normally 120 volts below 25 KV and 115 volts above 25 KV, and 60 Hz).

Marked Ratio, Turn Ratio, True Ratio

The marked ratio of a voltage transformer is the ratio of primary voltage to secondary voltage as given on the rating plate.

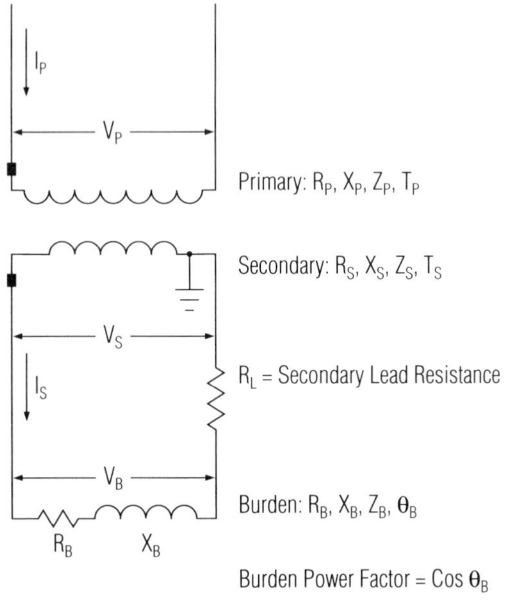

Figure 11-2. The Actual Voltage Transformer with Burden and Lead Resistance.

The turn ratio of a voltage transformer is the ratio of the number of turns in the primary winding to that in the secondary winding.

The true ratio of a voltage transformer is the ratio of the rms primary voltage to the rms secondary voltage under specified conditions.

In an ideal voltage transformer, the marked ratio, the turn ratio, and the true ratio would always be equal and the reversed secondary voltage would always be in phase with the impressed primary voltage. It must be strongly emphasized that this ideal voltage transformer does not exist. It has been assumed that the ideal voltage transformer is 100 percent efficient, has no losses, and requires no magnetizing current. This assumption is not true for any actual voltage transformer.

The concept of the ideal voltage transformer is, however, a useful fiction. Modern voltage transformers, when supplying burdens which do not exceed their accuracy ratings, approach very closely to the fictional ideal. Most metering installations involving instrument transformers are set up on this ideal basis and in most cases no corrections need be applied. Thus, in the example shown in Fig. 11-1, it would normally be assumed that the meter voltage coil is always supplied with 1/20th of the primary voltage. If this assumption is to be valid, the limitations of actual voltage transformers must be clearly understood and care taken to see that they are used within these limitations.

The Actual Voltage Transformer—The Phasor Diagram (Fig. 11-3)

In the ideal voltage transformer the secondary voltage is directly proportional to the ratio of turns and opposite in phase to the impressed primary voltage. In an actual transformer an exact proportionality and phase relation are not possible because:

a. The exciting current that is necessary to magnetize the magnetic core causes an impedance drop in the primary winding, and
b. The load current that is drawn by the burden causes an impedance drop in both the primary and secondary windings.

Both of these produce an overall voltage drop in the transformer and introduce errors in both ratio and phase angle. The net result is that the secondary voltage is slightly different from that which the ratio of turns would indicate and there is a slight shift in the phase relationship. This results in the introduction of ratio and phase angle errors as compared to the performance of the ideal voltage transformer.

Figures 11-2 and 11-3 are the schematic and phasor diagrams of an actual voltage transformer. The phasor diagram (Fig. 11-3) is drawn for a transformer having a one-to-one turn ratio and the voltage-drop and loss phasors have been greatly exaggerated so that they can be clearly separated on the diagram. These are normal conventions used when drawing phasor diagrams for transformers and do not invalidate any of the results to be derived.

The operation of the voltage transformer may be explained briefly by means of the phasor diagram (Fig. 11-3) as follows:

The flux ϕ in the core induces a voltage, E_S in the secondary winding lagging the flux by 90 degrees. A voltage equal to nE_S (where n is the turn ratio) is also induced in the primary winding lagging ϕ by 90 degrees. To overcome this induced voltage, a voltage $E_P = -nE_S$ must be supplied in the primary. Thus, E_P must lead E_S by 180 degrees and therefore leads the flux ϕ by 90 degrees.

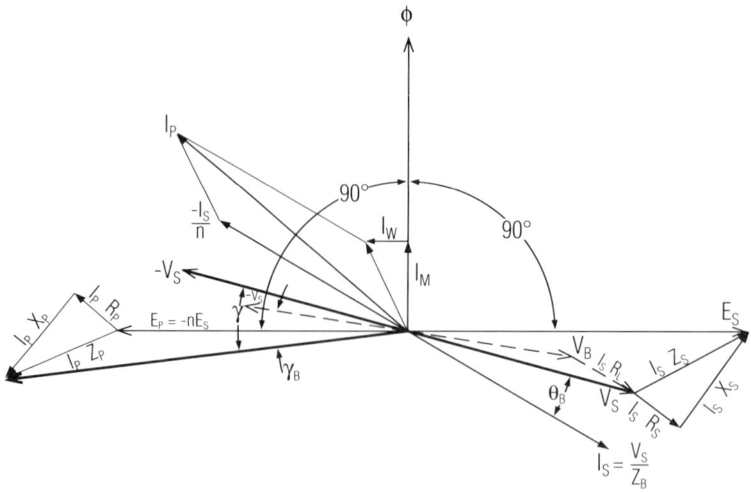

ϕ = Flux in Core
I_M = Magnetizing Current
I_W = Core Loss Current
I_E = Exciting Current
V_P = Impressed Primary (Terminal) Voltage
E_P = Voltage Required to Overcome Induced Primary Voltage
I_P = Primary Current
R_P = Resistance of Primary
X_P = Reactance of Primary
Z_P = Impedance of Primary
E_S = Induced Secondary Voltage
V_S = Secondary Terminal Voltage
I_S = Secondary Current
R_S = Resistance of Secondary
X_S = Reactance of Secondary
Z_S = Impedance of Secondary
R_L = Resistance of Secondary Leads
V_B = Voltage at Terminals of Burden
R_B = Resistance of Burden
X_B = Reactance of Burden
Z_B = Impedance of Burden
θ_B = Phase Angle of Burden
$\cos \theta_B$ = Power Factor of Burden
γ_B = Apparent Phase Angle of Voltage Transformer at Burden Terminals
T_P = Turns on Primary
T_S = Turns on Secondary
N = Turns Ratio = T_P / T_S
V_P / V_S = True Ratio of Voltage Transformer
V_P / V_B = Apparent Ratio of Voltage Transformer at Burden Terminals
γ = Phase Angle of Voltage Transformer

Figure 11-3. Phasor Diagram of Voltage Transformer.

The secondary current I_S is determined by the secondary terminal voltage V_S and the impedance of the burden Z_B. Theoretically, the true burden "seen" by a voltage transformer includes the leads R_L in series with the connected instruments. In practice, the effect of the leads on the total burden is very small and is neglected. I_S is equal to $\frac{V_S}{Z_B}$ and lags V_S by a phase angle θ_B where $\cos \theta_B$ is the power factor of the burden. (This burden power factor should not be confused with the power factor of the load being supplied by the primary circuit.)

The voltage drop in the secondary winding is equal to $I_S Z_S$ where Z_S is the impedance of this winding. This drop is the phasor sum of two components $I_S R_S$ and $I_S X_S$, where R_S and X_S are the resistance and reactance of the secondary winding. The voltage drop $I_S R_S$ must be in phase with I_S, and the voltage drop $I_S X_S$ must lead I_S by 90 degrees. The induced secondary voltage E_S is equal to the phasor sum of $V_S + I_S Z_S$ and V_S is the phasor difference $E_S - I_S Z_S$.

I_M is the magnetizing current required to supply the flux ϕ and is in phase with the flux. I_W is the current required to supply the hysteresis and eddy current losses in the core and leads I_M by 90 degrees. The phasor sum of $I_M + I_W$ is the exciting current I_E. This would be the total primary current if there were no burden on the secondary.

When a burden is connected to the secondary, the primary must also supply the reflected secondary current $-\frac{I_S}{n}$. The total primary current I_P is therefore the phasor sum of I_E and $-\frac{I_S}{n}$.

The voltage drop in the primary winding is equal to $I_P Z_P$, where Z_P is the impedance of the primary winding. This drop is the phasor sum of two components $I_P R_P$ and $I_P X_P$, where R_P and X_P are the resistance and the reactance of the primary winding. The voltage drop $I_P R_P$ must be in phase with I_P and the drop $I_P X_P$ must lead I_P by 90 degrees. The primary terminal voltage V_P is equal to the phasor sum of $E_P + I_P Z_P$.

The phasor $-V_S$ is obtained by reversing the secondary voltage phasor V_S. In practice, this simply amounts to reversing the connections to the secondary terminals. This reversal is automatically taken care of by the polarity markings, and, if these are followed, the terminal voltage from the marked to the unmarked secondary lead will be $-V_S$.

In Fig. 11-3, note that the reversed secondary voltage phasor $-V_S$ is not equal in magnitude to the impressed primary voltage V_P and that $-V_S$ is out of phase with V_P by the angle gamma (γ). In an ideal voltage transformer of one-to-one ratio, $-V_S$ would be equal to, and in phase with, V_P. In the actual voltage transformer this difference represents errors in both ratio and phase angle.

True Ratio and Ratio Correction Factor (RCF)

The true ratio of a voltage transformer is the ratio of the rms primary voltage to the rms secondary voltage under specified conditions.

In the phasor diagram, Fig. 11-3, the true ratio is $\frac{V_P}{V_S}$. It is apparent that this is not equal to the one-to-one turn ratio $\frac{T_P}{T_S}$ upon which this diagram was based. V_S in this case is smaller in magnitude than V_P as a result of the voltage drops in the transformer.

The turn ratio of a voltage transformer $\frac{T_P}{T_S}$ is built in at the time of construction and the marked ratio is indicated on the nameplate by the manufacturer. These ratios are fixed and permanent values for a given transformer.

The true ratio of a voltage transformer is not a single fixed value since it depends upon the specified conditions of use. These conditions are secondary burden, primary voltage, frequency, and waveform. Under ordinary conditions primary voltage, frequency, and waveform are practically constant so that the true ratio is primarily dependent upon the secondary burden and the characteristics of the particular voltage transformer.

The true ratio of a voltage transformer cannot be marked on the nameplate since it is not a constant value, but a variable which is affected by external conditions. The true ratio is determined by test for the specified conditions under which the voltage transformer is to be used. For most practical applications, where no corrections are to be applied, the true ratio is considered to be equal to the marked ratio under specified ANSI standard accuracy tolerances and burdens.

Thus it might be found that the true ratio of a voltage transformer having a marked ratio of 20:1 was 20.034:1 under the specified conditions. However, the true ratio is not usually written in this way because this form is difficult to evaluate and inconvenient to use. The figure 20.034 may be broken into two factors and written 20×1.0017. Note that 20 is the marked or nominal ratio of the voltage transformer which is multiplied by the factor 1.0017.

This factor, by which the marked ratio must be multiplied to obtain the true ratio, is called the ratio correction factor, or RCF. True Ratio = Marked Ratio × Ratio Correction Factor.

$$\text{RCF} = \frac{\text{True Ratio}}{\text{Marked Ratio}}$$

Phase Angle

It is apparent from Fig. 11-3 that the reversed secondary voltage $-V_S$ is not in phase with the impressed primary voltage V_P. The angle gamma (γ) between these two phasors is known as the phase angle of the voltage transformer and is usually expressed in minutes of arc. (Sixty minutes of arc is equal to one degree.)

In the ideal voltage transformer the secondary voltage V_S would be exactly 180 degrees out of phase with the impressed primary voltage V_P. The polarity markings automatically correct for this 180-degree reversal. The reversed secondary voltage $-V_S$ would therefore be in phase with the impressed primary V_P and the phase angle γ would be zero.

In the actual voltage transformer, the phase angle γ represents a phase shift between the primary and secondary voltages in addition to the normal 180-degree shift. The 180-degree shift is corrected by the reversal that occurs when the polarity markings are followed, but the phase angle γ remains. This uncorrected phase shift can cause errors in measurements where exact phase relations must be maintained.

ANSI C57.13 defines phase angle of an instrument transformer as the phase displacement, in minutes, between the primary and secondary values. The phase angle of a voltage transformer is designated by the Greek letter gamma (γ) and is positive when the secondary voltage from the identified to the unidentified terminal leads the corresponding primary voltage.

The phase angle of a voltage transformer is not a single fixed value, but varies with burden, primary voltage, frequency, and waveform. It results from the voltage drops within the transformer as shown in Fig. 11-3. Under ordinary conditions, where voltage, frequency, and waveform are practically constant, the phase angle is primarily dependent upon the secondary burden and the characteristics of the particular voltage transformer.

Effects of Secondary Burden on Ratio and Phase Angle

It is apparent from Fig. 11-3 that any change in the secondary current I_S will change the relative magnitudes and phase relations of the primary terminal voltage V_P and the secondary terminal voltage V_S. Since the secondary current I_S is a function of the burden impedance Z_B, the true ratio $\frac{V_P}{V_S}$ and the phase angle γ are affected by any change in burden. Figure 11-5 shows the metering accuracy curve of a voltage transformer referred to connected burden.

Effects of Primary Voltage on Ratio and Phase Angle

A change in primary voltage causes a nearly proportional increase or decrease in all of the other voltages and currents shown in the phasor diagram, Fig. 11-3. If this proportionality were exact, no change in true ratio or phase angle would result from a change in voltage. However, the exciting current I_E is not strictly proportional to the primary voltage V_P, but varies according to the saturation curve of the magnetic core as shown in Fig. 11-4. Note that the change in exciting current for the normal operating range of 90 to 110 percent of rated primary voltage is very nearly linear. Above 110 percent rated voltage, the core is rapidly approaching saturation and the exciting current I_E increases more rapidly than the primary voltage V_P. This could result in a change in true ratio and phase angle with voltage if the transformer is operated at more than 110 percent of its rated voltage. The exact point of satuıation depends upon the particular design. Some voltage transformers may show greater changes with voltage than others.

In the normal operating range, and even much below this range, the change of true ratio and phase angle with voltage is very small with modern, well-designed voltage transformers.

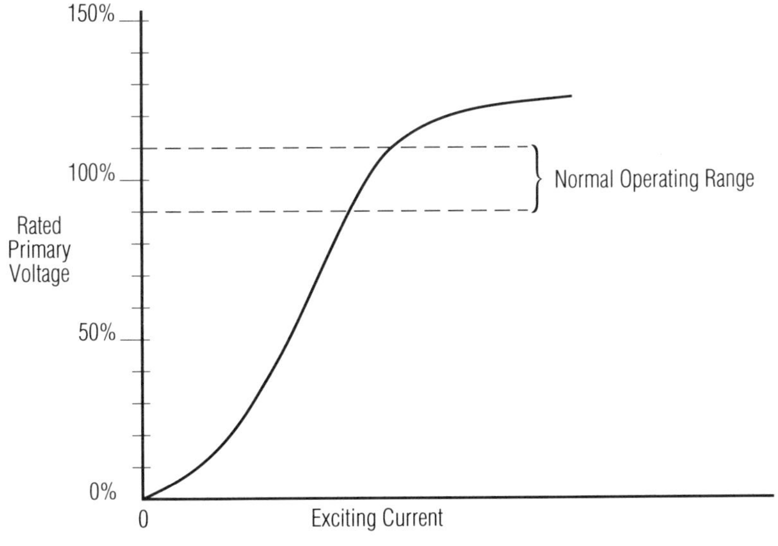

Figure 11-4. Typical Saturation Curve for a Voltage Transformer.

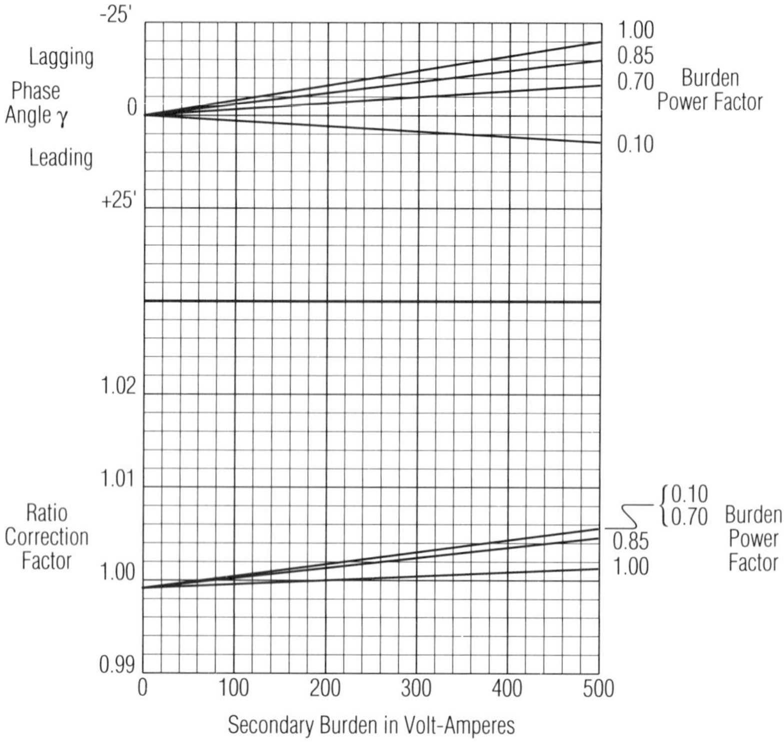

Figure 11-5. Characteristic Ratio and Phase-Angle Curves for a Voltage Transformer at 60 Hertz, 120 Volts.

Effects of Frequency on Ratio and Phase Angle

A change in frequency changes the impedance of the voltage transformer and the burden. Increasing frequency increases the reactance of the transformer (X_P and X_S) and would increase the voltage drops $I_P X_P$ and $I_S X_S$ were it not for the fact that the secondary current I_S would decrease because of an increase in the burden reactance X_B. (See Fig. 11-3.) These two effects tend to cancel each other to some extent, but depend upon the ratio of resistance to reactance in the transformer and the burden. In addition, the exciting current I_E decreases rapidly at higher frequencies and increases at lower frequencies. At lower frequencies, the core will saturate at voltages below the normal rating and large changes in ratio and phase angle could occur.

Thus a small increase in frequency may have little effect, whereas a small decrease may result in appreciable change in true ratio and phase angle. A drastic decrease in frequency results in excessive exciting current and overheating of the voltage transformer.

Voltage transformers are normally designed for a single frequency, though they can be designed to work satisfactorily for a small range of frequencies, such as 50-60 hertz. In utility work this frequency is usually 60 hertz. Since power system frequency is closely regulated, the problem of varying frequency does not normally arise.

Effects of Waveform on Ratio and Phase Angle

Since any distorted waveform of the impressed primary voltage may be considered equivalent to a mixture of a sinusoidal voltage at the fundamental frequency and sinusoidal voltages at higher harmonic frequencies, waveform distortion would also have an effect on the true ratio and phase angle.

If the burden is a magnetic core device requiring a large exciting current, this may result in a waveform distortion in the secondary current I_S. However, this error is included if the transformer is tested with this burden. In testing voltage transformers, care must be used to avoid overloading the primary voltage supply which could produce a distorted primary voltage waveform.

Effects of Temperature on Ratio and Phase Angle

A change in temperature changes the resistance of the primary and secondary windings of the voltage transformer. This results in only slight changes of ratio and phase angle as the voltage drops in the transformers are primarily reactive and the secondary current is determined by the impedance of the burden. The change in accuracy is usually less than 0.1 percent for a 55° C change in temperature.

Effects of Secondary Lead Resistance on the Ratio and Phase Angle as Seen by the Meter

The true ratio and phase angle of a voltage transformer are defined in terms of the terminal voltages V_P and V_S. The true secondary burden is defined in terms of the impedance connected to the secondary terminals and therefore includes the secondary lead resistance R_L as shown in Fig. 11-2. The resistance of the secondary leads R_L is small compared to the impedance of the burden Z_B, so that ordinarily the lead resistance does not change the secondary burden sufficiently to make any appreciable difference in the ratio and phase angle at the voltage transformer terminals.

However, the meter is not connected directly to the secondary terminals, but at the end of the secondary leads. The voltage at the meter terminals is the burden voltage V_B as shown in Fig. 11-2 and not the secondary terminal voltage V_S. V_B differs from V_S by the phasor drop $I_S R_L$ that occurs in the leads. (See Figs. 11-3 and 11-6.)

This voltage drop is in phase with the secondary current I_S and therefore causes the burden voltage V_B to be slightly different in magnitude and slightly shifted in phase relation with respect to the secondary terminal voltage V_S.

The effect of this line drop in terms of ratio correction factor and phase angle may be calculated as shown in Fig. 11-6. Figures for a typical example have also been shown to illustrate the use of these equations. In this example, the ratio correction factor 1.0009 and the phase angle + 1.7 minutes due to the secondary lead resistance were small and could be ignored in all but the most exacting applications. If a greater lead resistance or a heavier secondary burden had been assumed, then this effect would be much greater. For example, if the lead resistance R_L was increased to one ohm and the secondary current I_S to one ampere at 0.866 burden power factor (θ_B) then the ratio correction factor and phase angle due to the leads would rise to 1.0073 and + 14.3 minutes. Such an error could hardly be ignored.

It should be emphasized that the effect of the secondary lead resistance, in causing a change in apparent ratio and phase angle at the meter terminals, is a straight lead-drop problem and is not due to the voltage transformer in any way. The effect would be exactly the same if an ideal voltage transformer were used.

In spite of the fact that this lead-drop effect is not due to the voltage transformer, it is sometimes convenient to include this drop during the test of a voltage transformer by determining the apparent ratio and phase angle between the primary terminals of the transformer and the terminals of the burden of the end of the actual or simulated secondary leads. This apparent ratio $\left(\frac{V_P}{V_B}\right)$ and apparent phase angle (γ_B) are indicated by the dashed-line phasors V_B, $-V_B$, and $I_S R_L$ on the phasor diagram, Fig. 11-3. This is the total RCF and phase angle that must be used to correct the readings of the meter, as both the transformer and lead-drop errors are included. In making acceptance tests to determine if the

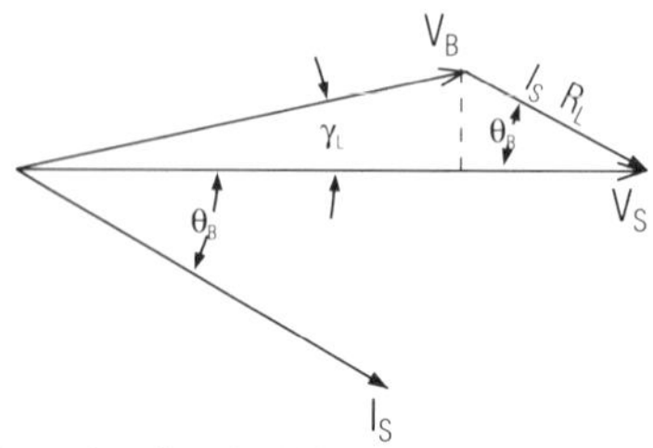

V_S = Secondary Terminal Voltage (Example: 120 Volts)
V_B = Voltage at Terminals of Burden (Example, Calculated: 119.896 Volts)
I_S = Secondary Current (Example: 0.24 Amperes)
θ_B = Phase Angle of Burden (Example: 30°)
$\cos \theta_B$ = Power Factor of Burden (Example: .866)
R_L = Resistance of Secondary Leads (Example: 0.50 Ohms)

RCF_L = Ratio Correction Factor (Caused by Secondary Leads Only)

Example:

$$RCF_L = \frac{V_S}{V_B} \approx \frac{V_S}{V_S - I_S R_L \cos \theta_B} = \frac{120}{120 - (0.24)(0.5)(.866)} = 1.0009$$

γ_L = Phase Angle in Minutes (Caused by Secondary Lead Resistance Only)

Example:

$$\gamma_{L\,(\text{min.})} \approx \frac{I_S R_L \sin \theta_B}{V_S} 3438 = \frac{(0.24)(0.5)(0.5)(3438)}{120} = +1.7 \text{ Minutes}$$

Note: The constant 3438 is the number of minutes in a radian.
The above formulas are approximations based on small angles.

Figure 11-6. Phasor Diagram and Calculation of the Ratio Correction Factor and Phase Angle Due to the Secondary Lead Resistance Only (Applies to Voltage Transformer Secondary Leads).

transformers meet specifications, the tests must be made at the transformer secondary terminals as the lead drop is not caused by the transformer.

In actual practice the ratio and phase angle errors due to secondary lead drop are usually limited to small values by strict limitations of allowable lead resistance and secondary burden. This lead drop is troublesome only in exceptional cases where long leads and heavy burdens are required. In case of doubt, a calculation, using the formulas given in Fig. 11-6, will quickly indicate the magnitude of the error involved.

Effects of Common Secondary Leads on Ratio and Phase Angle as Seen by the Meter

In a polyphase circuit where two or three voltage transformers are used, it is the normal practice to use one wire as the common neutral secondary lead for all of the voltage transformers.

This fact must be taken into account when measuring or calculating the effect of the lead drop on the ratio and phase angle at the meter. If three voltage transformers are connected in wye as shown in Fig. 11-20, the neutral secondary lead carries no current with a balanced burden. If two voltage transformers are connected in open delta as shown in Fig. 11-18, the neutral secondary carries $\sqrt{3}$ times the current of the other leads for a balanced burden on phases 1-2 and 2-3.

If, because of long secondary leads or heavy burdens, the lead-drop effect causes significant error, then the use of a common secondary lead increases the difficulties of determining this effect by test or calculation. Calculations must be made phasorially, taking into account the magnitude and phase relation of the current in each secondary lead.

Polyphase Burdens

When the secondaries of two or three voltage transformers are used to supply interconnected polyphase burdens, it becomes difficult to determine the actual burden in each transformer. Calculations of burden must be made phasorially and become exceedingly complex where several polyphase and single-phase burdens are involved. Such calculations can be avoided by testing at the burden under actual or simulated three-phase conditions. This is required only in the most exacting applications where corrections based on the actual burden must be applied. In most cases, burdens are kept within the ratings of the voltage transformers and no corrections are applied.

Methods of Compensating Voltage Transformers to Reduce Ratio and Phase Angle Errors

Voltage transformers are designed to have low exciting current and low internal impedance. This reduces the ratio and phase angle errors. In addition, the turn ratio may be made slightly different than the marked ratio. This is done to compensate the transformer for minimum error at a specific burden instead of at zero burden. If the transformer is used with a burden approximating the design burden, the errors may be greatly reduced.

Permanence of Accuracy

The accuracy of a voltage transformer does not change appreciably with age. It may change due to mechanical damage or to electrical stresses beyond those for which the transformer was designed.

The Current Transformer

The Ideal Current Transformer Connection Diagram

Figure 11-7 shows the connection diagram for an ideal current transformer. Note that the primary winding is connected in series with one of the high-voltage leads carrying the primary current, and the secondary winding is connected to the current coil of the meter. When 600 amperes flow through the primary winding of this current transformer, 5 amperes are developed in the secondary winding by transformer action. This secondary current is passed through the current coil of the meter. Since there is no direct connection between the primary and secondary windings, the insulation between these windings isolates the meter from the voltage of the primary. Note that one end of the secondary winding is connected to ground. This provides protection from static charges and insulation failure.

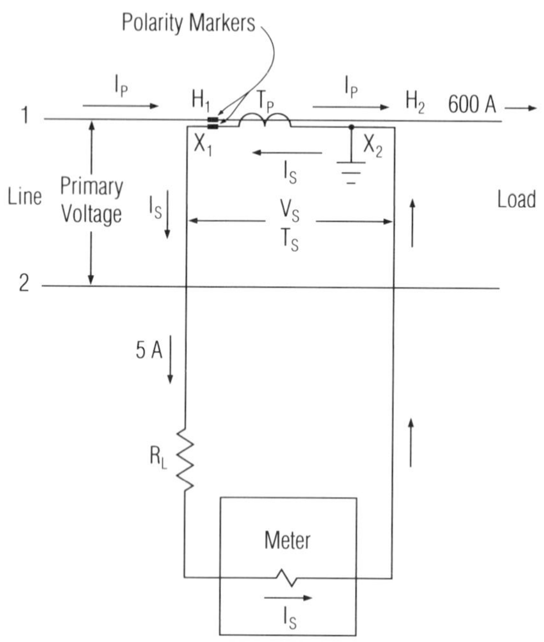

Marked Primary Current = 600 A

Marked Secondary Current = 5 A

Marked Ratio = 120:1

Turn Ratio $= \frac{T_S}{T_P}$

True Ratio $= \frac{I_P}{I_S}$

Figure 11-7. The Ideal Current Transformer.

Polarity

In Fig. 11-7 the polarity markers are used to show the instantaneous direction of current flow in the primary and secondary windings of the current transformer. They are so placed that when the primary current I_P is flowing into the marked primary terminal H_1, the secondary current I_S is at the same instant flowing out of the marked secondary terminal X_1. These markings enable the secondaries of the current and voltage transformers to be connected to the meter with the proper phase relationships. For example, in the case of a single-stator meter installed with a current and a voltage transformer, reversal of the secondaries from either transformer would cause the meter to run backward.

Secondary Burden

In Fig. 11-7 the impedance of the current coil of the meter and the resistance of the secondary leads causes a small voltage drop across the secondary terminals of the current transformer when the secondary current I_S is flowing. The current transformer must develop a small terminal voltage V_S to overcome this voltage drop in order to maintain the secondary current. The impedance of the meter and resistance of the secondary leads is therefore a burden on the secondary winding.

This burden may be expressed as the total volt-amperes and power factor of the secondary devices and leads at a specified current and frequency (normally, 5 amperes and 60 hertz). It is often more convenient to express current-transformer burdens in terms of their total resistance in ohms and inductance in millihenries, or as total ohms impedance at a given power factor and frequency.

While the basic definition of burden for a current transformer and voltage transformer is the same in terms of active and reactive power supplied by the instrument transformer, the effect of burden impedance is the reverse in the two cases. Zero burden on a voltage transformer is an open-circuit or infinite impedance, while zero burden on a current transformer is a short-circuit or zero impedance.

The impedance of the current coil of the meter in Fig. 11-7 is very low so that the current transformer is operated with what amounts to a short circuit on its secondary winding. This is the normal condition of operation for a current transformer.

Marked Ratio, Turn Ratio, True Ratio

The marked ratio of a current transformer is the ratio of primary current to secondary current as given on the rating plate.

The turn ratio of a current transformer is the ratio of the number of turns in the secondary winding to the number of turns in the primary winding. (Note: This is just the opposite of a voltage transformer. A voltage transformer that steps down the voltage has more turns on the primary than the secondary. A current transformer that steps down the current has more turns on the secondary than on the primary.)

The true ratio of a current transformer is the ratio of rms primary current to the rms secondary current under specified conditions.

In an ideal current transformer, the marked ratio, the turn ratio, and the true ratio would always be equal and the reversed secondary current would always be in phase with the impressed primary current. It must be strongly emphasized that this ideal current transformer does not exist.

The concept of the ideal current transformer is, however, a useful fiction. Modern current transformers, when supplying burdens which do not exceed their accuracy ratings, approach very closely to this fictional ideal. Most metering installations involving instrument transformers are set up on this ideal basis and, in most cases, no corrections need be applied. Thus, in the example shown in Fig. 11-7, it would normally be assumed that the meter current coil is always supplied with 1/120 of the primary current. If this assumption is to be valid, the limitations of actual current transformers must be clearly understood and care taken to see that they are used within these limitations.

The Actual Current Transformer—The Phasor Diagram (Fig. 11-9)

In the ideal current transformer the secondary current is inversely proportional to the ratio of turns and opposite in phase to the impressed primary current. In reality, however, an exact inverse proportionality and phase relation are not possible because part of the primary current must be used to excite the core. The exciting current may be subtracted phasorially from the primary current to find the amount remaining to supply secondary current. Therefore, the secondary current will be slightly different from the value that the ratio of turns would indicate and there is a slight shift in the phase relationship. This results in the introduction of ratio and phase angle errors as compared to the performance of the "ideal" current transformer.

Figures 11-8 and 11-9 are the schematic and phasor diagrams of an actual current transformer. The phasor diagram, Fig. 11-9, is drawn for a transformer having a one-to-one turn ratio and the voltage drop and loss phasors have been greatly exaggerated so that they can be clearly separated on the diagram.

Basically, the phasor diagram for a current transformer is similar to that for the voltage transformer. However, in the current transformer the important phasors are the primary and secondary current, rather than the voltages.

The operation of the current transformer may be explained briefly by means of the phasor diagram (Fig. 11-9) as follows:

The flux ϕ in the core induces a voltage E_S in the secondary winding lagging the flux by 90 degrees. A voltage equal to $\frac{E_S}{n}$, where n is the turn ratio $\left(\frac{T_S}{T_P}\right)$, is also induced in the primary winding lagging ϕ by 90 degrees. To overcome this in-

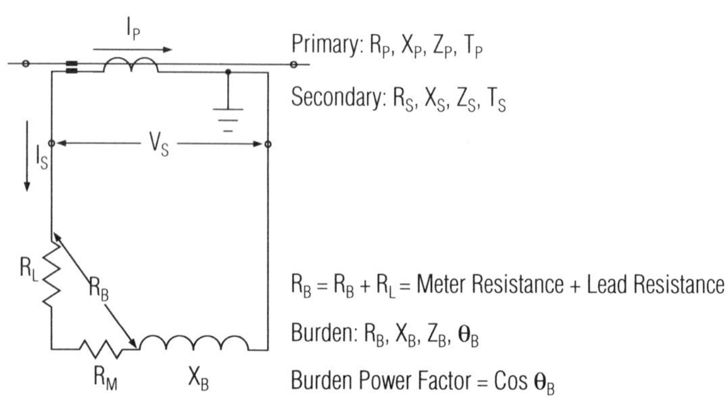

Figure 11-8. The Actual Current Transformer with Burden.

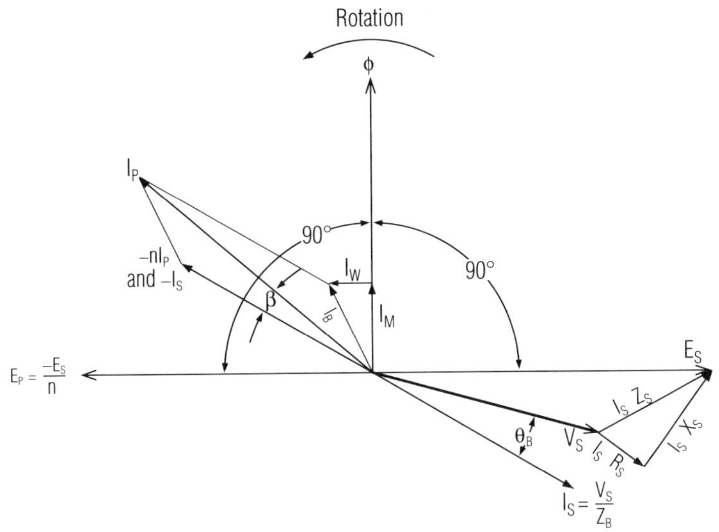

ϕ	=	Flux in Core	R_B	=	Resistance of Burden
I_M	=	Magnetizing Current	X_B	=	Reactance of Burden
I_W	=	Core Loss Current	Z_B	=	Impedance of Burden
I_E	=	Exciting Current	θ_B	=	Phase Angle of Burden
E_P	=	Voltage Required to Overcome Induced Primary Voltage	$\cos \theta_B$	=	Power Factor of Burden
I_P	=	Primary Current	T_P	=	Turns on Primary
E_S	=	Induced Secondary Voltage	T_S	=	Turns on Secondary
V_S	=	Secondary Terminal Voltage	n	=	Turns Ratio = T_S/T_P
I_S	=	Secondary Current	I_P/I_S	=	True Ratio of Current Transformer
R_S	=	Resistance of Secondary	β	=	Phase Angle of Current Transformer
X_S	=	Reactance of Secondary			

Figure 11-9. Phasor Diagram of Current Transformer.

duced voltage, a voltage $E_P = -\frac{E_S}{n}$ must be supplied in the primary. Thus, E_P must lead E_S by 180 degrees and therefore leads the flux by 90 degrees.

The secondary current I_S is determined by the secondary terminal voltage V_S and the impedance of the burden Z_B. I_S is equal to $\frac{V_S}{Z_B}$ and lags V_S by a phase angle θ_B, where $\cos \theta_B$ is the power factor of the burden. (This burden power factor should not be confused with the power factor of the load being supplied by the primary circuit.) The burden impedance Z_B is made up of the burden resistance R_B and the burden reactance X_B. (See Fig. 11-8.) Note particularly that the burden resistance R_B is equal to the sum of the meter resistance R_M and the secondary lead resistance R_L. Since the total impedance of current transformer burdens is very low, usually less than one ohm, the lead resistance R_L is an appreciable part of the burden and cannot be neglected. In some cases the resistance of the secondary leads may constitute the greater part of the burden impedance.

The voltage drop in the secondary winding is equal to $I_S Z_S$, where Z_S is the impedance of this winding. This drop is the phasor sum of two components $I_S R_S$ and $I_S X_S$, where R_S and X_S are the resistance and reactance of the secondary winding. The voltage drop $I_S R_S$ must be in phase with I_S, and the voltage drop $I_S X_S$ must lead I_S by 90 degrees. The induced secondary voltage E_S is equal to the phasor sum of $V_S + I_S Z_S$ and V_S is the phasor difference $E_S - I_S Z_S$.

I_M is the magnetizing current required to supply the flux ϕ and is in phase with the flux. I_W is the current required to supply the hysteresis and eddy current losses in the core, and leads I_M by 90 degrees. The phasor sum of $I_M + I_W$ is the exciting current I_E.

The primary must supply the reflected secondary current $-nI_S$. The total primary current I_P is therefore the phasor sum of I_E and $-nI_S$.

With a low-impedance burden connected to the secondary winding, the impedance of the primary winding is extremely low, since the reflected impedance of the secondary is approximately proportional to the square of the turn ratio, and the primary winding of a step-down current transformer has fewer turns than the secondary.

The primary current in the current transformer is determined by the load on the primary circuit of the installation, The voltage drop in the primary winding is therefore very small, even with full-rated current in the primary line, because of the low impedance of this winding. The induced secondary voltage E_S and the secondary terminal voltage V_S are both small because the transformer is essentially short circuited by the low-impedance burden. Therefore, the voltage E_P required to overcome the voltage $\frac{E_S}{n}$ induced in the primary is also very small.

Since the true ratio of a current transformer is $\frac{I_P}{I_S}$, it is not ordinarily necessary to consider the primary voltage or the voltage drops in the primary, since they do not affect the value of either the primary or secondary currents.

The phasor $-I_S$ is obtained by reversing the secondary current phasor I_S. In Fig. 11-9, which is for a one-to-one transformer, $-I_S$ is coincident with $-nI_S$. This reversal is automatically taken care of if the polarity markings are followed.

In Fig. 11-9, note that the reversed secondary current phasor $-I_S$ is not equal in magnitude to the impressed primary current phasor I_P and that $-I_S$ is out of phase with I_P by the angle beta, β. In an ideal current transformer of one-to-one ratio, $-I_S$ would be equal to and in phase with I_P. In the actual current transformer, this difference represents errors in both ratio and phase angle.

True Ratio and Ratio Correction Factor (RCF)

The true ratio of a current transformer is the ratio of the rms primary current to the rms secondary current under specified conditions.

In the phasor diagram, Fig. 11-9, the true ratio is $\frac{I_P}{I_S}$. It is apparent that this is not equal to the one-to-one turn ratio $\left(\frac{T_S}{T_P}\right)$ upon which the diagram was based. I_S in this case is smaller in magnitude than I_P because a part of the primary current I_P is required to supply the exciting current I_E.

The turn ratio of a current transformer $\left(\frac{T_S}{T_P}\right)$ is built in at the time of construction and the marked ratio is indicated on the nameplate by the manufacturer. These ratios are fixed and permanent values for a given transformer.

The true ratio of a current transformer is not a single fixed value, since it depends upon the specified conditions of use. These conditions are secondary burden, primary current, frequency, and waveform. Under ordinary conditions, frequency and waveform are practically constant so that the true ratio is primarily dependent upon the secondary burden, the primary current, and the characteristics of the particular current transformer.

The true ratio of a current transformer cannot be marked on the nameplate since it is not a constant value, but a variable which is affected by external conditions. The true ratio is determined by test for the specified conditions under which the current transformer is to be used. (For most practical applications, where no corrections are to be applied, the true ratio is considered to be equal to the marked ratio under specified ANSI standard accuracy tolerances and burdens.)

Thus, it might be found that the true ratio of a current transformer having the marked ratio of 120:1 was 119.796:1 under the specified conditions. However, the true ratio is not usually written in this way because this form is difficult to evaluate and inconvenient to use. The figure 119.796 may be broken into two factors and written 120 × 0.9983. Note that 120 is the marked or nominal ratio of the current transformer which is multiplied by the factor 0.9983.

This factor, by which the marked ratio must be multiplied to obtain the true ratio is called the ratio correction factor, or RCF. It has exactly the same meaning when applied to the current transformer as previously given for the voltage transformer. True Ratio = Marked Ratio × Ratio Correction Factor.

$$\text{RCF} = \frac{\text{True Ratio}}{\text{Marked Ratio}}$$

Phase Angle

It is apparent from Fig. 11-9 that the reversed secondary current I_S is not in phase with the impressed primary current I_P. The angle beta (β) between these phasors is known as the phase angle of the current transformer and is usually expressed in minutes of arc (60 minutes of arc is equal to one degree).

In the ideal current transformer, the secondary current I_S would be exactly 180 degrees out of phase with the impressed primary current I_P. The polarity markings automatically correct for this 180-degree reversal. The reversed secondary current $-I_S$ would therefore be in phase with the impressed primary current I_P and the phase angle β would be zero.

In the actual current transformer, the phase angle β represents a phase shift between the primary and secondary currents in addition to the normal 180-degree phase shift. The 180-degree shift is corrected by the reversal that occurs when the polarity markings are followed, but the phase angle β remains. This uncorrected phase shift can cause errors in measurements where exact phase relations must be maintained.

ANSI C57.13 defines the phase angle of an instrument transformer as the phase displacement, in minutes, between the primary and secondary values. The phase angle of a current transformer is designated by the Greek letter beta (β) and

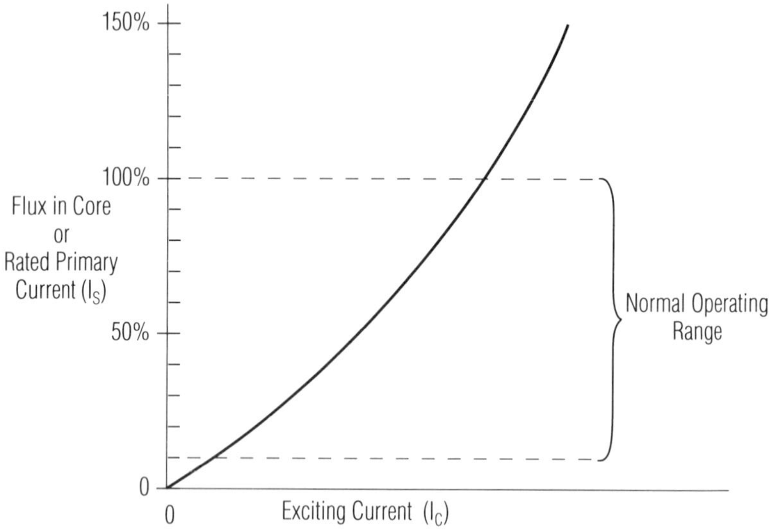

Figure 11-10. Typical Exciting Current Curve for a Current Transformer.

is positive when the current leaving the identified secondary terminal leads the current entering the identified primary terminal.

The phase angle of a current transformer is not a single fixed value, but varies with burden, primary current, frequency, and waveform. It results from the component of the primary current required to supply the exciting current I_E as shown in Fig. 11-9. Under ordinary conditions where frequency and waveform are practically constant, the phase angle is primarily dependent upon the secondary burden, the primary current, and the characteristics of the particular current transformer.

Effects of Secondary Burden on Ratio and Phase Angle

An increase of secondary burden, which for a current transformer means an increase in the burden impedance Z_B, requires an increase in the secondary voltage V_S if the secondary current I_S is to remain the same. (See Fig. 11-9.) (Note that in a *voltage* transformer an increase of secondary burden requires an increase in the secondary *current* if the secondary voltage is to remain the same.) Increasing the secondary current requires an increase in the induced secondary voltage E_S which can only be produced by an increase in the flux ϕ.

To provide an increased flux, the magnetizing current I_M must increase and the core loss current I_W also increases. This results in an increase in the exciting current I_E. Thus the result of increasing the burden is to cause an increase in the exciting current. Since the exciting current is the primary cause of the ratio and phase angle errors in the current transformer, these errors are affected by any change in the secondary burden.

Effect of Primary Current on Ratio and Phase Angle

Unlike the voltage transformer which operates at a practically constant primary voltage, the current transformer must operate over a wide range of primary currents from zero to rated current, and above rated current in special cases, such as the operation of protective relays. This means that with a constant secondary burden the flux in the core must vary over a wide range as the primary current is changed. To produce this varying flux, the exciting current must also vary over a wide range. If the flux ϕ varied in exact proportion with the exciting current I_E, then the changes in primary current would not affect the ratio and phase angle. However, current transformers are designed to operate at low flux densities in the core and under these conditions the flux is not directly proportional to the exciting current. Figure 11-10 shows a typical exciting current curve for the magnetic core of a current transformer.

Note that the change of exciting current over the normal operating range from 10 to 100 percent rated primary current is not a linear function of the primary current. The shape of the saturation curve for the current transformer is actually similar to the curve for the voltage transformer, Fig. 11-4, but only an expanded portion of the lower end of the curve is shown in Fig. 11-10. With normal secondary burdens, saturation does not occur until the primary current reaches 5 to 20 times the rated value. Thus the saturation point is not shown in Fig. 11-10.

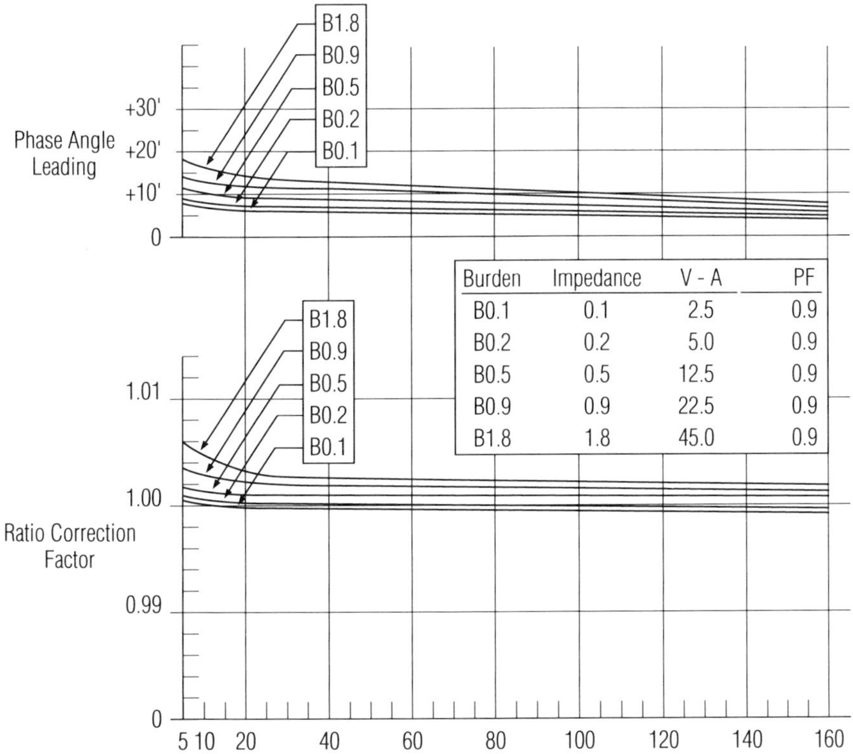

Figure 11-11. Characteristic Ratio and Phase Angle Curves for a Typical Current Transformer at 60 Hertz.

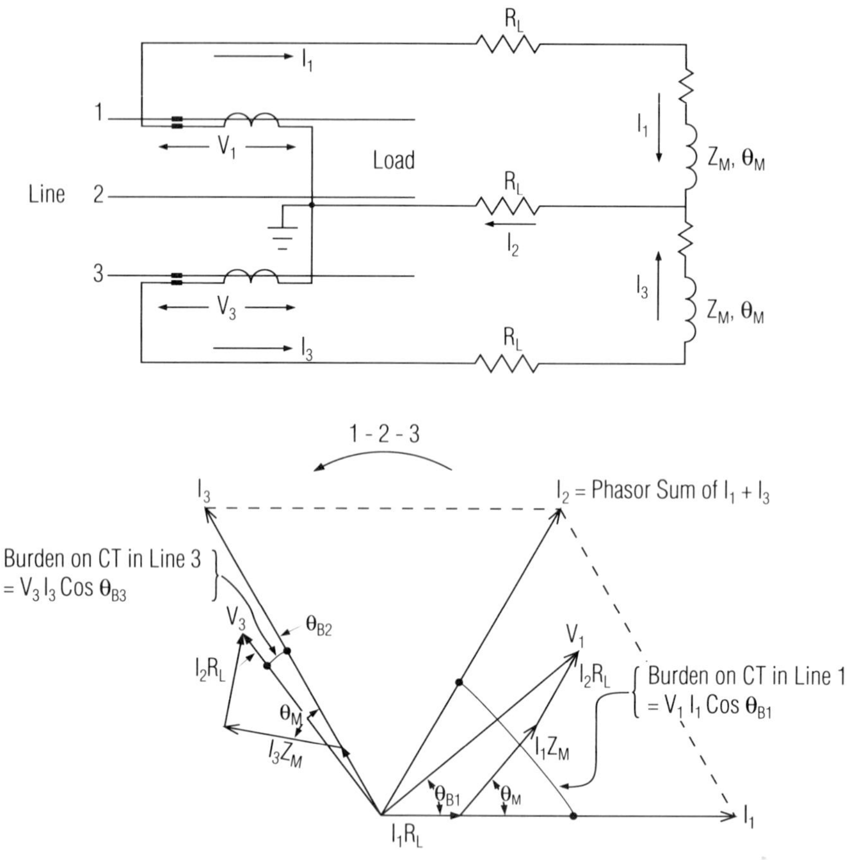

Z_M = Impedance of Meter Coil
θ_M = Phase Angle of Meter Impedance
R_L = Resistance of Each Secondary Lead
I_1 = Secondary Current of Current Transformer in Line 1
I_3 = Secondary Current of Current Transformer in Line 3
I_2 = Current In Common Secondary Lead
V_1 = Terminal Voltage of Current Transformer in Line 1
V_3 = Terminal Voltage of Current Transformer in Line 3
θ_{B1} = Phase Angle of Burden on Current Transformer in Line 1
θ_{B3} = Phase Angle of Burden on Current Transformer in Line 3

Figure 11-12. Effect of Common Secondary Lead on Burdens of Current Transformers.

Since the exciting current does not change in an exact proportion with the primary current, the true ratio and phase angle vary to some extent with the primary current. The ratio and phase angle errors are usually greater at 10 percent primary current than at 100 percent primary current, though this depends upon the burden and the compensation of the particular current transformer. Figure 11-11 shows the metering accuracy curves of a current transformer at standard burdens.

Effects of Frequency on Ratio and Phase Angle

The effect of frequency variation on the ratio and phase angle of a current transformer is less than that on a voltage transformer, primarily because of the low flux density. Current transformers may be designed to have reasonable accuracy over a range from 25 to 133 hertz. There will, however, be some slight variation with frequency in this range.

Effects of Waveform on Ratio and Phase Angle

Waveform distortion in the primary current may have slight effects on the ratio and phase angle, but in general such effects are negligible. Even a large amount of third harmonic in the primary current wave is reasonably well reproduced in the secondary, thus causing little error. Higher harmonics could cause errors, but these are not normally present in sufficient magnitude to be significant.

Effects of Secondary Leads on Ratio and Phase Angle

In the current transformer, the secondary current I_S must be the same in all parts of the secondary circuit, including the burden, since it is a series circuit. Thus, the secondary current and, therefore, the true ratio and phase angle, will be the same whether measured at the transformer or at the meter at the end of the secondary leads. The only effect of the secondary leads is their effect on the burden. With long secondary leads, the leads may constitute the major portion of the secondary burden. In all cases the secondary leads must be included in all tests and calculations as part of the secondary burden.

Effects of Common Secondary Lead

In a polyphase circuit where two or three current transformers are used, it is normal practice to use one wire as the common secondary lead for all of the current transformers.

This fact must be taken into account when measuring or calculating the effect of the leads as part of the secondary burden. If the current transformers are connected in wye as shown in Fig. 11-24, the neutral secondary lead carries no current if the primary load current is balanced. In this case, the resistance of the common lead is not part of the burden on any of the current transformers. If the two current transformers are connected open delta as shown in Fig. 11-12, the common secondary lead carries a current whose magnitude is the same as the other leads under conditions of balanced line currents and an open-delta burden as shown. However, the current in the common lead is not in phase with the current in either of the other two leads. Thus, the lead resistance of the common lead does not affect the burden on the two current transformers equally.

Figure 11-12 is a schematic and a phasor diagram of a two-stator polyphase meter whose current elements are connected to two current transformers. Note that if the lead resistance R_L is an appreciable part of the burden, the burden on the two current transformers is not the same because of the effect of the common lead resistance. The effective burdens differ in both magnitude and phase. The burden on one current transformer is $V_1I_1 \cos \theta_{B1}$ and the other is $V_3I_3 \cos \theta_{B3}$. The secondary currents I_1 and I_3 were assumed to be equal, but the terminal voltages V_1 and V_3 are not equal. In addition, the phase angles of the two burdens θ_{B1} and θ_{B3} are not equal. Thus, the effect of the common secondary lead resistance, in this case, results in unequal burdens on the two current transformers even though the two elements of the meter are identical.

If a burden of 2.1 VA at 0.60 power factor lagging and a lead resistance of 0.1 ohm (100 feet of No. 10 wire) are assumed, the burdens on the two transformers would be 6.34 VA at 0.79 power factor lagging on the current transformer in line 1 and 5.05 VA at 0.996 power factor leading on the current transformer in line 3.

Actually these small differences in burden would have little effect on the ratio and phase angle of a modern current transformer. However, if a much longer common secondary lead with a resistance of 0.3 ohm or more were used, the effect might cause significant error unless appropriate corrections were applied.

In most installations, the common lead resistance is kept low so that the resulting error is insignificant. In the most accurate work, if long secondary leads must be used and exact corrections must be applied, the current transformers can be tested under actual three-phase conditions. If the common lead is eliminated by using separate return leads for each transformer, the calculations of burden are simplified.

Difficulties with Low-Ampere-Turn Designs

With a given current transformer core, the number of ampere-turns needed to excite this core to a certain flux density is essentially a constant value. The exciting-current ampere-turns must be taken from the primary ampere-turns and the remainder supplies the secondary ampere-turns.

As the total ampere-turns of the primary becomes lower, the exciting ampere-turns become a greater percentage of the total, thus increasing the errors. When the primary ampere-turns are less than about 600, it becomes difficult to design current transformers with small errors. Only by using special core materials and compensation methods can the errors be reduced to reasonable values.

Dangers Due to Open Secondary

The secondary circuit of a current transformer must never be opened when current is flowing in the primary. With an open secondary, the secondary impedance becomes infinite, the flux rises to saturation, and the voltage drop in the primary is increased due to the reflected secondary impedance. The primary voltage is stepped up by the ratio of the transformer and the secondary voltage rises to dangerously high values. Voltages of several thousand volts are possible under open-circuit conditions. Such voltages are dangerous to personnel and can damage the transformer.

Permanence of Accuracy

The accuracy of a current transformer does not change appreciably with age. It may be permanently changed by mechanical or electrical failure and it may be temporarily changed by magnetization.

INSTRUMENT TRANSFORMER CORRECTION FACTOR

Ratio Correction Factor and Related Terms

The marked ratio, the true ratio, and the ratio correction factor have been defined and discussed. In addition to the ratio correction factor, the terms percent ratio (or percent marked ratio), ratio error, and percent ratio error are often used when stating the errors in ratio of instrument transformers. It is unfortunate that there are four numerically different terms used to describe the same phenomenon, as they are easily confused. Table 11-1 defines these and related terms by means of simple algebraic formulas which provide means of converting from one term to another. Of the four, the ratio correction factor is most easily understood and has the least chance of misapplication, since neither percentage nor plus or minus signs are involved. The ratio correction factor is the only one of the four terms defined in the ANSI Standard C57.13 and it is therefore recommended as the preferred term.

Table 11-1. Definitions of Instrument Transformer Ratio, RCF, and Related Terms.

Term	Formula	Application
Marked (Nominal) Ratio	$\frac{\text{Marked Primary Voltage}}{\text{Marked Secondary Voltage}}$	VT
Marked (Nominal) Ratio	$\frac{\text{Marked Primary Current}}{\text{Marked Secondary Current}}$	CT
True Ratio	$\frac{\text{True Primary Voltage}}{\text{True Secondary Voltage}}$	VT
True Ratio	$\frac{\text{True Primary Current}}{\text{True Secondary Current}}$	CT
True Primary Voltage	True Secondary Voltage × Marked Ratio × RCF	VT
True Primary Current	True Secondary Current × Marked Ratio × RCF	CT
Ratio Correction Factor (RCF)	$\frac{\text{True Ratio}}{\text{Marked Ratio}}$	VT, CT
Ratio Correction Factor (RCF)	$\frac{\text{True Primary Voltage}}{\text{True Secondary Voltage} \times \text{Marked Ratio}}$	VT
Ratio Correction Factor (RCF)	$\frac{\text{True Primary Current}}{\text{True Secondary Current} \times \text{Marked Ratio}}$	CT
Percent Ratio	100 × RCF	VT, CT
Ratio Error	$\frac{\text{True Ratio} - \text{Marked Ratio}}{\text{Marked Ratio}} = (\text{RCF} - 1)$	VT, CT
Percent Ratio Error	$100 \times (\text{RCF} - 1)$ $= 100 \times \frac{\text{True Ratio} - \text{Marked Ratio}}{\text{Marked Ratio}}$	VT, CT

Examples:

If RCF is 1.0027, the percent ratio is 100.27 percent, the ratio error is +0.0027, and the percent ratio error is +0.27 percent.

If the RCF is 0.9973, the percent ratio is 99.73 percent, the ratio error is –0.0027, and the percent ratio error is –0.27 percent.

Note that the proper sign (+ or –) must be used for the ratio error or the percent ratio error and the word percent or a percent sign (%) must be used with the percent ratio and the percent ratio error.

Combined Ratio Correction Factor (RCF_K)

Where both a voltage and a current transformer are used for the measurement of watts or watt-hours, the combined ratio correction factor is

$$RCF_K = RCF_E \times RCF_I$$

where RCF_K is the combined ratio correction factor

RCF_E is the ratio correction factor of the voltage transformer

RCF_I is the ratio correction factor of the current transformer

The combined ratio correction factor RCF_K corrects for the ratio error of both the voltage and current transformers, but does not correct for the effects of phase angles.

Approximate Method of Multiplying Two Numbers Close to One by Addition

The problem of multiplication of ratio and phase angle correction factors occurs often in calculating instrument transformer corrections. Since the numerical value of these correction factors is close to one (1), the work may be simplified by using an approximate method involving addition and subtraction rather than multiplication. The correction factors can be represented by $(1 \pm a)$ and $(1 \pm b)$, where a and b are small decimal fractions. The multiplication can then be written:

$$(1 \pm a)(1 \pm b) = 1 \pm a \pm b \pm ab$$

This is approximately equal to $1 \pm a \pm b$ since ab is the product of two small numbers and is therefore extremely small.

Example:

$$(0.9987)(1.0025) = 1.00119675 \text{ exactly}$$

Using the approximate method:

$$0.9987 = 1 - 0.0013, \quad \text{hence a} = -0.0013$$
$$1.0025 = 1 + 0.0025, \quad \text{hence b} = +0.0025$$
$$1 - 0.0013 + 0.0025 = 1.0012$$

Note: $ab = -0.0013 \times 0.0025 = -0.00000325$
and $1.0012 - 0.00000325 = 1.00119675$.

However, the figures beyond the fourth decimal place are not significant so the answer given by the approximate method is as accurate as is justified by the original figures.

If the correction factors are within 1 percent of 1.0000 (0.9900 to 1.0100), the error in this approximation will not exceed 0.01 percent (0.0001).

Another way to use this approximation is to add the two values and subtract 1.

$$0.9987 + 1.0025 = 2.0012$$

$$2.0012 - 1.0000 = 1.0012$$

The multiplication of correction factors can be done by inspection using this approximation.

Phase Angle Correction Factor (PACF)

Figure 11-13 shows the schematic and phasor diagrams of a meter connected to a high-voltage line using a voltage and a current transformer.

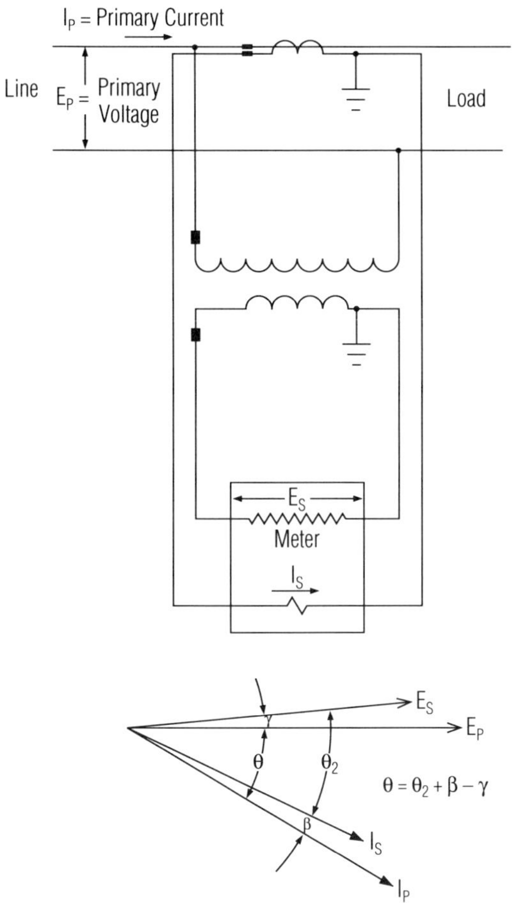

Current Transformer:

N_I = Marked (Nominal) Ratio
RCF_I = Ratio Correction Factor
β = Phase Angle

Load:

E_P = Primary Voltage
I_P = Primary Current
$\text{Cos}\ \theta$ = Primary (True) Power Factor

Meter:

E_S = Secondary Voltage
I_S = Secondary Current
$\text{Cos}\ \theta_2$ = Secondary (Apparent) Power Factor

Voltage Transformer:

N_E = Marked (Nominal) Ratio
RCF_E = Ratio Correction Factor
γ = Phase Angle

W_P = Primary Power (True Power Supplied to Load) = $E_P I_P \text{Cos}\ \theta$

W_S = Secondary Power (Power Measured by the Meter) = $E_S I_S \text{Cos}\ \theta_2$

$$E_P I_P \text{Cos}\ \theta = (E_S I_S \text{Cos}\ \theta_2)(N_E)(N_I)(RCF_E)(RCF_I)\left[\frac{\text{Cos}\ (\theta_2 + \beta - \gamma)}{\text{Cos}\ \theta_2}\right]$$

Figure 11-13. Relations Between Primary and Secondary Power.

The primary power W_P is equal to the product of the primary voltage E_P, the primary current I_P, and the true power factor of the primary circuit (cos θ):

$$W_P = E_P I_P \cos\theta$$

The secondary power W_S measured by the meter is equal to the product of the secondary voltage E_S, the secondary current I_S, and the power factor of the secondary circuit (cos θ_2):

$$W_S = E_S I_S \cos\theta_2$$

The power factor of the secondary circuit (cos θ_2) is called the apparent power factor and differs from the primary power factor (cos θ) because of the effect of the phase angles beta (β) and gamma (γ) of the current and voltage transformers respectively.

If the instrument transformers had one-to-one ratios and no errors due to ratio, then the subscripts could be omitted. For this condition:

$$W_P = EI\cos\theta$$

$$W_S = EI\cos\theta_2$$

In this special case, the primary power W_P would be equal to the secondary power W_S were it not for the difference between cos θ and cos θ_2 which is due to the phase angles of the instrument transformers.

The phase angle correction factor (*PACF*) is defined by ANSI as the ratio of the true power factor to the measured power factor. It is a function of both the phase angles of the instrument transformer and the power factor of the primary circuit being measured.

Note that the phase angle correction factor is the factor that corrects for the phase displacement of the secondary current or voltage, or both, due to the instrument transformer phase angles.

The measured watts or watt-hours in the secondary circuits of instrument transformers must be multiplied by the phase angle correction factor and the true ratio to obtain the true primary watts or watt-hours.

The combined phase angle correction factor ($PACF_K$) is used when both current and voltage transformers are involved. $PACF_I$ is used when current transformers but no voltage transformers are involved.

Therefore, for this special case of one-to-one ratio and no ratio errors:

$$W_P = W_S(PACF_K)$$

and

$$PACF_K = \frac{W_P}{W_S} = \frac{EI\cos\theta}{EI\cos\theta_2} = \frac{\cos\theta}{\cos\theta_2}$$

The combined phase angle correction factor is therefore equal to the ratio of the true power factor (cos θ) to the apparent power factor (cos θ_2). However, this equation for the phase angle correction factor is not directly usable, since, in general, cos θ_2, the apparent power factor, is known, but the exact value of the true power factor, cos θ, is not.

From the phasor diagram in Fig. 11-13, it is apparent that $\theta = \theta_2 + \beta - \gamma$. Note that in this phasor diagram all the angles shown have a plus (+) sign. The secondary current and voltage phasors have been drawn so that they lead their respective primary phasors. Therefore, β and γ are both positive by definition. The angles θ

and θ_2 between the voltage and current phasors are considered positive (+) when the current phasors are lagging the voltage phasors (lagging power factor). Hence, θ and θ_2 are both positive as drawn. Substituting $\theta = \theta_2 + \beta - \gamma$ into the previous equation:

$$PACF_K = \frac{\cos(\theta_2 + \beta - \gamma)}{\cos \theta_2}$$

When a current transformer is used alone, $PACF_I$ may be determined by using the formula for $PACF_K$ with the $-\gamma$ term deleted.

If $\cos \theta_2$, β, and γ are known, the combined phase angle correction factor ($PACF_K$) can now be calculated using trigonometric tables. Care must be taken to use the proper signs for θ, β, and γ, as previously noted.

Example:

Given: $\cos \theta_2 = 0.80 \text{ lag}, \beta = -13', \gamma = +10'$

Then:

$$\theta_2 = \cos^{-1} 0.80 = 36°52'$$
$$\theta = \theta_2 + \beta - \gamma$$
$$\theta = (+36°52') + (-13') - (+10') = 36°29'$$
$$\cos \theta = \cos 36°29' = 0.804030$$
$$PACF_K = \frac{0.804030}{0.800000} = 1.0050$$

This method of evaluating the phase angle correction factor is straight forward and fundamental, but too time consuming for practical work. Therefore, Tables 11-2 and 11-3 have been calculated by this method to give the phase angle correction factor directly in terms of the apparent power factor ($\cos \theta_2$) and the combined value of the phase angles ($\beta - \gamma$).

Use of Tables 11-2 and 11-3 to Find the Phase Angle Correction Factor

In the example just given, $\cos \theta_2 = 0.80$ lagging and $\beta - \gamma = (-13 \text{ minutes}) - (+10 \text{ minutes}) = -23$ minutes. Hence, Table 11-3 must be used as indicated by the heading "For Lagging Current When $\beta - \gamma$ is Negative." At the intersection of the 0.80 power factor column and the 23-minute row the phase angle correction factor is found to be 1.0050.

Two precautions are necessary when using these tables:

1. The algebraic signs of the phase angles and the minus sign in the formula must be carefully observed when calculating $\beta - \gamma$.
2. Care must be used in selecting either Table 11-2 or 11-3 according to the notes heading these tables regarding leading or lagging power factors and the resultant sign of $\beta - \gamma$.

For an installation where a current transformer is used, but no voltage transformer is used:

$$PACF_I = \frac{\cos(\theta_2 + \beta)}{\cos \theta_2}$$

Tables 11-2 and 11-3 can still be used to find the phase angle correction factor using the value of β itself for $\beta - \gamma$, since γ is not involved.

The combined phase angle correction factor ($PACF_K$) depends upon the phase angles of the instrument transformers (β and γ) and on the apparent power factor

Table 11-2. Phase Angle Correction Factors (PACFs).

For Lagging Current When ($\beta - \gamma$) is Positive
For Leading Current When ($\beta - \gamma$) is Negative

$\beta - \gamma$	Apparent Power Factor (Cos θ_2)																				$\beta - \gamma$
	.05	.10	.15	.20	.25	.30	.35	.40	.45	.50	.55	.60	.65	.70	.75	.80	.85	.90	.95	.99	
1'	.9942	.9971	.9981	.9986	.9989	.9991	.9992	.9993	.9994	.9995	.9996	.9996	.9997	.9997	.9997	.9998	.9998	.9999	.9999	1.0000	1'
2'	.9884	.9942	.9962	.9971	.9977	.9981	.9984	.9987	.9988	.9990	.9991	.9992	.9993	.9994	.9995	.9996	.9996	.9997	.9998	.9999	2'
3'	.9826	.9913	.9942	.9957	.9966	.9972	.9977	.9980	.9983	.9985	.9987	.9988	.9990	.9991	.9992	.9993	.9995	.9996	.9997	.9999	3'
4'	.9768	.9884	.9923	.9943	.9955	.9963	.9969	.9973	.9977	.9980	.9982	.9984	.9986	.9988	.9990	.9991	.9993	.9994	.9996	.9998	4'
5'	.9709	.9855	.9904	.9929	.9944	.9954	.9961	.9967	.9971	.9975	.9978	.9981	.9983	.9985	.9987	.9989	.9991	.9993	.9995	.9998	5'
6'	.9651	.9826	.9885	.9914	.9932	.9944	.9953	.9960	.9965	.9970	.9973	.9977	.9980	.9982	.9985	.9987	.9989	.9992	.9994	.9997	6'
7'	.9593	.9797	.9866	.9900	.9921	.9935	.9945	.9953	.9960	.9965	.9969	.9973	.9976	.9979	.9982	.9985	.9987	.9990	.9993	.9997	7'
8'	.9535	.9768	.9847	.9886	.9910	.9926	.9938	.9947	.9954	.9960	.9965	.9969	.9973	.9976	.9979	.9983	.9986	.9989	.9992	.9997	8'
9'	.9477	.9739	.9827	.9872	.9899	.9917	.9930	.9940	.9948	.9955	.9960	.9965	.9969	.9973	.9977	.9980	.9984	.9987	.9991	.9996	9'
10'	.9419	.9711	.9808	.9857	.9887	.9907	.9922	.9933	.9942	.9950	.9956	.9961	.9966	.9970	.9974	.9978	.9982	.9986	.9990	.9996	10'
11'	.9361	.9682	.9789	.9843	.9876	.9898	.9914	.9927	.9936	.9945	.9951	.9957	.9963	.9967	.9972	.9976	.9980	.9984	.9989	.9995	11'
12'	.9303	.9653	.9770	.9829	.9865	.9889	.9907	.9920	.9931	.9939	.9947	.9953	.9959	.9964	.9969	.9974	.9978	.9983	.9988	.9995	12'
13'	.9245	.9624	.9751	.9815	.9853	.9880	.9899	.9913	.9925	.9934	.9943	.9950	.9956	.9961	.9967	.9972	.9976	.9982	.9987	.9995	13'
14'	.9186	.9595	.9731	.9800	.9842	.9870	.9891	.9907	.9919	.9929	.9938	.9946	.9952	.9958	.9964	.9969	.9975	.9980	.9987	.9994	14'
15'	.9128	.9566	.9712	.9786	.9831	.9861	.9883	.9900	.9913	.9924	.9934	.9942	.9949	.9955	.9961	.9967	.9973	.9979	.9986	.9994	15'
16'	.9070	.9537	.9693	.9772	.9820	.9852	.9875	.9893	.9908	.9919	.9929	.9938	.9945	.9952	.9959	.9965	.9971	.9977	.9985	.9993	16'
17'	.9012	.9508	.9674	.9758	.9808	.9843	.9868	.9887	.9902	.9914	.9925	.9934	.9942	.9949	.9956	.9963	.9969	.9976	.9984	.9993	17'
18'	.8954	.9479	.9655	.9743	.9797	.9833	.9860	.9880	.9896	.9909	.9920	.9930	.9939	.9946	.9954	.9961	.9967	.9975	.9983	.9992	18'
19'	.8896	.9450	.9636	.9729	.9786	.9824	.9852	.9873	.9890	.9904	.9916	.9926	.9935	.9943	.9951	.9958	.9966	.9973	.9982	.9992	19'
20'	.8838	.9421	.9616	.9715	.9775	.9815	.9844	.9867	.9884	.9899	.9911	.9922	.9932	.9940	.9949	.9956	.9964	.9972	.9981	.9992	20'
21'	.8780	.9392	.9597	.9701	.9763	.9806	.9836	.9860	.9879	.9894	.9907	.9918	.9928	.9937	.9946	.9954	.9962	.9970	.9980	.9991	21'
22'	.8721	.9363	.9578	.9686	.9752	.9796	.9829	.9853	.9873	.9889	.9903	.9914	.9925	.9935	.9943	.9952	.9960	.9969	.9979	.9991	22'
23'	.8663	.9334	.9559	.9672	.9741	.9787	.9821	.9846	.9867	.9884	.9898	.9911	.9922	.9932	.9941	.9950	.9958	.9967	.9978	.9990	23'
24'	.8605	.9305	.9540	.9658	.9729	.9778	.9813	.9840	.9861	.9879	.9894	.9907	.9918	.9929	.9938	.9947	.9956	.9966	.9977	.9990	24'
25'	.8547	.9276	.9520	.9643	.9718	.9768	.9805	.9833	.9855	.9874	.9889	.9903	.9915	.9926	.9936	.9945	.9955	.9965	.9976	.9989	25'
26'	.8489	.9247	.9501	.9629	.9707	.9759	.9797	.9826	.9850	.9869	.9885	.9899	.9911	.9923	.9933	.9943	.9953	.9963	.9975	.9989	26'
27'	.8431	.9218	.9482	.9615	.9696	.9750	.9789	.9820	.9844	.9864	.9880	.9895	.9908	.9920	.9930	.9941	.9951	.9962	.9974	.9989	27'
28'	.8373	.9189	.9463	.9601	.9684	.9741	.9782	.9813	.9838	.9859	.9876	.9891	.9904	.9917	.9928	.9939	.9949	.9960	.9973	.9988	28'
29'	.8315	.9160	.9444	.9586	.9673	.9731	.9774	.9806	.9832	.9854	.9872	.9887	.9901	.9914	.9925	.9936	.9947	.9959	.9972	.9988	29'
30'	.8256	.9131	.9424	.9572	.9662	.9722	.9766	.9800	.9826	.9848	.9867	.9883	.9898	.9911	.9923	.9934	.9946	.9957	.9971	.9987	30'

Table 11-2 (Concluded). Phase Angle Correction Factors (PACFs).

For Lagging Current When $(\beta - \gamma)$ is Positive
For Leading Current When $(\beta - \gamma)$ is Negative

$\beta - \gamma$	Apparent Power Factor ($\cos \theta_2$)																			$\beta - \gamma$
	.05	.10	.15	.20	.25	.30	.35	.40	.45	.50	.55	.60	.65	.70	.75	.80	.85	.90	.95	.99	
31'	.8198	.9102	.9405	.9558	.9650	.9713	.9758	.9793	.9821	.9843	.9863	.9879	.9894	.9908	.9920	.9932	.9944	.9956	.9970	.9987	31'
32'	.8140	.9073	.9386	.9544	.9639	.9704	.9750	.9786	.9815	.9838	.9858	.9875	.9891	.9905	.9917	.9930	.9942	.9954	.9969	.9986	32'
33'	.8082	.9044	.9367	.9529	.9628	.9694	.9743	.9780	.9809	.9833	.9854	.9872	.9887	.9902	.9915	.9928	.9940	.9953	.9968	.9986	33'
34'	.8024	.9015	.9348	.9515	.9616	.9685	.9735	.9773	.9803	.9828	.9849	.9868	.9884	.9899	.9912	.9925	.9938	.9952	.9967	.9985	34'
35'	.7966	.8986	.9328	.9501	.9605	.9676	.9727	.9766	.9797	.9823	.9845	.9864	.9880	.9896	.9910	.9923	.9936	.9950	.9966	.9985	35'
36'	.7908	.8958	.9309	.9486	.9594	.9666	.9719	.9760	.9792	.9818	.9840	.9860	.9877	.9893	.9907	.9921	.9935	.9949	.9965	.9985	36'
37'	.7850	.8929	.9290	.9472	.9583	.9657	.9711	.9753	.9786	.9813	.9836	.9856	.9874	.9890	.9905	.9919	.9933	.9947	.9964	.9984	37'
38'	.7791	.8900	.9271	.9458	.9571	.9648	.9704	.9746	.9780	.9808	.9832	.9852	.9870	.9887	.9902	.9916	.9931	.9946	.9963	.9984	38'
39'	.7733	.8871	.9252	.9444	.9560	.9639	.9696	.9739	.9774	.9803	.9827	.9848	.9867	.9884	.9899	.9914	.9929	.9944	.9962	.9983	39'
40'	.7675	.8842	.9232	.9429	.9549	.9629	.9688	.9733	.9768	.9798	.9823	.9844	.9863	.9881	.9897	.9912	.9927	.9943	.9961	.9983	40'
41'	.7617	.8813	.9213	.9415	.9537	.9620	.9680	.9726	.9763	.9793	.9818	.9840	.9860	.9878	.9894	.9910	.9925	.9942	.9960	.9982	41'
42'	.7559	.8784	.9194	.9401	.9526	.9611	.9672	.9719	.9757	.9788	.9814	.9836	.9856	.9875	.9892	.9908	.9924	.9940	.9959	.9982	42'
43'	.7501	.8755	.9175	.9386	.9515	.9601	.9664	.9713	.9751	.9783	.9809	.9832	.9853	.9872	.9889	.9905	.9922	.9939	.9958	.9981	43'
44'	.7443	.8726	.9156	.9372	.9503	.9592	.9657	.9706	.9745	.9778	.9805	.9829	.9850	.9869	.9886	.9903	.9920	.9937	.9957	.9981	44'
45'	.7384	.8697	.9136	.9358	.9492	.9583	.9649	.9699	.9739	.9772	.9800	.9825	.9846	.9866	.9884	.9901	.9918	.9936	.9956	.9980	45'
46'	.7326	.8668	.9117	.9344	.9481	.9574	.9641	.9693	.9734	.9767	.9796	.9821	.9843	.9863	.9881	.9899	.9916	.9934	.9955	.9980	46'
47'	.7268	.8639	.9098	.9329	.9470	.9564	.9633	.9686	.9728	.9762	.9791	.9817	.9839	.9860	.9878	.9897	.9914	.9933	.9954	.9980	47'
48'	.7210	.8610	.9079	.9315	.9458	.9555	.9625	.9679	.9722	.9757	.9787	.9813	.9836	.9857	.9876	.9894	.9912	.9931	.9953	.9979	48'
49'	.7152	.8581	.9060	.9301	.9447	.9546	.9618	.9672	.9716	.9752	.9783	.9809	.9832	.9854	.9873	.9892	.9911	.9930	.9952	.9979	49'
50'	.7094	.8552	.9040	.9286	.9436	.9536	.9610	.9666	.9710	.9747	.9778	.9805	.9829	.9851	.9871	.9890	.9909	.9929	.9951	.9978	50'
51'	.7036	.8523	.9021	.9272	.9424	.9527	.9602	.9659	.9705	.9742	.9774	.9801	.9825	.9848	.9868	.9888	.9907	.9927	.9950	.9978	51'
52'	.6978	.8494	.9002	.9258	.9413	.9518	.9594	.9652	.9699	.9737	.9769	.9797	.9822	.9845	.9865	.9885	.9905	.9926	.9949	.9977	52'
53'	.6919	.8465	.8983	.9244	.9402	.9509	.9586	.9646	.9693	.9732	.9765	.9793	.9819	.9842	.9863	.9883	.9903	.9924	.9948	.9977	53'
54'	.6861	.8436	.8963	.9229	.9390	.9499	.9578	.9639	.9687	.9727	.9760	.9789	.9815	.9839	.9860	.9881	.9901	.9923	.9947	.9976	54'
55'	.6803	.8407	.8944	.9215	.9379	.9490	.9571	.9632	.9681	.9722	.9756	.9785	.9812	.9836	.9858	.9879	.9900	.9921	.9946	.9976	55'
56'	.6745	.8378	.8925	.9201	.9368	.9481	.9563	.9625	.9675	.9717	.9751	.9781	.9808	.9832	.9855	.9877	.9898	.9920	.9945	.9975	56'
57'	.6687	.8349	.8906	.9186	.9356	.9471	.9555	.9619	.9670	.9711	.9747	.9778	.9805	.9829	.9852	.9874	.9896	.9918	.9944	.9975	57'
58'	.6629	.8320	.8887	.9172	.9345	.9462	.9547	.9612	.9664	.9706	.9742	.9774	.9801	.9826	.9850	.9872	.9894	.9917	.9943	.9975	58'
59'	.6571	.8291	.8867	.9158	.9334	.9453	.9539	.9605	.9658	.9701	.9738	.9770	.9798	.9823	.9847	.9870	.9892	.9915	.9942	.9974	59'
60'	.6512	.8262	.8848	.9143	.9323	.9444	.9531	.9599	.9652	.9696	.9733	.9766	.9794	.9820	.9845	.9868	.9890	.9914	.9941	.9974	60'

Table 11-3. Phase Angle Correction Factors (PACFs).

For Lagging Current When ($\beta - \gamma$) is Negative
For Leading Current When ($\beta - \gamma$) is Positive

$\beta - \gamma$	Apparent Power Factor (Cos θ_2)																			$\beta - \gamma$
	.05	.10	.15	.20	.25	.30	.35	.40	.45	.50	.55	.60	.65	.70	.75	.80	.85	.90	.95	.99	
1'	1.0058	1.0029	1.0019	1.0014	1.0011	1.0009	1.0008	1.0007	1.0006	1.0005	1.0004	1.0004	1.0003	1.0003	1.0003	1.0002	1.0002	1.0001	1.0001	1.0000	1'
2'	1.0116	1.0058	1.0038	1.0028	1.0023	1.0018	1.0016	1.0013	1.0012	1.0010	1.0009	1.0008	1.0007	1.0006	1.0005	1.0004	1.0004	1.0003	1.0002	1.0001	2'
3'	1.0174	1.0087	1.0058	1.0043	1.0034	1.0028	1.0023	1.0020	1.0017	1.0015	1.0013	1.0012	1.0010	1.0009	1.0008	1.0007	1.0005	1.0004	1.0003	1.0001	3'
4'	1.0232	1.0116	1.0077	1.0057	1.0045	1.0037	1.0031	1.0027	1.0023	1.0020	1.0018	1.0016	1.0014	1.0012	1.0010	1.0009	1.0007	1.0006	1.0004	1.0002	4'
5'	1.0291	1.0145	1.0096	1.0071	1.0056	1.0046	1.0039	1.0033	1.0029	1.0025	1.0022	1.0019	1.0017	1.0015	1.0013	1.0011	1.0009	1.0007	1.0005	1.0002	5'
6'	1.0349	1.0174	1.0115	1.0085	1.0068	1.0055	1.0047	1.0040	1.0035	1.0030	1.0026	1.0023	1.0020	1.0018	1.0015	1.0013	1.0011	1.0008	1.0006	1.0002	6'
7'	1.0407	1.0203	1.0134	1.0100	1.0079	1.0065	1.0054	1.0047	1.0040	1.0035	1.0031	1.0027	1.0024	1.0021	1.0018	1.0015	1.0013	1.0010	1.0007	1.0003	7'
8'	1.0465	1.0232	1.0153	1.0114	1.0090	1.0074	1.0062	1.0053	1.0046	1.0040	1.0035	1.0031	1.0027	1.0024	1.0020	1.0017	1.0014	1.0011	1.0008	1.0003	8'
9'	1.0523	1.0260	1.0173	1.0128	1.0101	1.0083	1.0070	1.0060	1.0052	1.0045	1.0040	1.0035	1.0031	1.0027	1.0023	1.0020	1.0016	1.0013	1.0009	1.0004	9'
10'	1.0581	1.0289	1.0192	1.0142	1.0113	1.0092	1.0078	1.0067	1.0058	1.0050	1.0044	1.0039	1.0034	1.0030	1.0026	1.0022	1.0018	1.0014	1.0010	1.0004	10'
11'	1.0639	1.0318	1.0211	1.0157	1.0124	1.0102	1.0086	1.0073	1.0063	1.0055	1.0049	1.0043	1.0037	1.0033	1.0028	1.0024	1.0020	1.0015	1.0010	1.0005	11'
12'	1.0697	1.0347	1.0230	1.0171	1.0135	1.0111	1.0093	1.0080	1.0069	1.0060	1.0053	1.0046	1.0041	1.0036	1.0031	1.0026	1.0022	1.0017	1.0011	1.0005	12'
13'	1.0755	1.0376	1.0249	1.0185	1.0146	1.0120	1.0101	1.0087	1.0075	1.0065	1.0057	1.0050	1.0044	1.0039	1.0033	1.0028	1.0023	1.0018	1.0012	1.0005	13'
14'	1.0813	1.0405	1.0268	1.0199	1.0158	1.0129	1.0109	1.0093	1.0081	1.0070	1.0062	1.0054	1.0048	1.0041	1.0036	1.0030	1.0025	1.0020	1.0013	1.0006	14'
15'	1.0871	1.0434	1.0288	1.0214	1.0169	1.0139	1.0117	1.0100	1.0086	1.0075	1.0066	1.0058	1.0051	1.0044	1.0038	1.0033	1.0027	1.0021	1.0014	1.0006	15'
16'	1.0930	1.0463	1.0307	1.0228	1.0180	1.0148	1.0124	1.0107	1.0092	1.0081	1.0071	1.0062	1.0054	1.0047	1.0041	1.0035	1.0029	1.0022	1.0015	1.0007	16'
17'	1.0988	1.0492	1.0326	1.0242	1.0191	1.0157	1.0132	1.0113	1.0098	1.0086	1.0075	1.0066	1.0058	1.0050	1.0043	1.0037	1.0031	1.0024	1.0016	1.0007	17'
18'	1.1046	1.0521	1.0345	1.0256	1.0203	1.0166	1.0140	1.0120	1.0104	1.0091	1.0079	1.0070	1.0061	1.0053	1.0046	1.0039	1.0032	1.0025	1.0017	1.0007	18'
19'	1.1104	1.0550	1.0364	1.0271	1.0214	1.0176	1.0148	1.0126	1.0110	1.0096	1.0084	1.0074	1.0064	1.0056	1.0049	1.0041	1.0034	1.0027	1.0018	1.0008	19'
20'	1.1162	1.0579	1.0383	1.0285	1.0225	1.0185	1.0156	1.0133	1.0115	1.0101	1.0088	1.0077	1.0068	1.0059	1.0051	1.0043	1.0036	1.0028	1.0019	1.0008	20'
21'	1.1220	1.0608	1.0402	1.0299	1.0236	1.0194	1.0163	1.0140	1.0121	1.0106	1.0093	1.0081	1.0071	1.0062	1.0054	1.0046	1.0038	1.0029	1.0020	1.0009	21'
22'	1.1278	1.0637	1.0422	1.0313	1.0248	1.0203	1.0171	1.0146	1.0127	1.0111	1.0097	1.0085	1.0075	1.0065	1.0056	1.0048	1.0039	1.0031	1.0021	1.0009	22'
23'	1.1336	1.0665	1.0441	1.0328	1.0259	1.0213	1.0179	1.0153	1.0133	1.0116	1.0101	1.0089	1.0078	1.0068	1.0059	1.0050	1.0041	1.0032	1.0022	1.0009	23'
24'	1.1394	1.0694	1.0460	1.0342	1.0270	1.0222	1.0187	1.0160	1.0138	1.0121	1.0106	1.0093	1.0081	1.0071	1.0061	1.0052	1.0043	1.0034	1.0023	1.0010	24'
25'	1.1452	1.0723	1.0479	1.0356	1.0281	1.0231	1.0194	1.0166	1.0144	1.0126	1.0110	1.0097	1.0085	1.0074	1.0064	1.0054	1.0045	1.0035	1.0024	1.0010	25'
26'	1.1510	1.0752	1.0498	1.0370	1.0293	1.0240	1.0202	1.0173	1.0150	1.0131	1.0115	1.0101	1.0088	1.0077	1.0066	1.0056	1.0047	1.0036	1.0025	1.0010	26'
27'	1.1569	1.0781	1.0517	1.0384	1.0304	1.0249	1.0210	1.0180	1.0156	1.0136	1.0119	1.0104	1.0092	1.0080	1.0069	1.0059	1.0048	1.0038	1.0026	1.0011	27'
28'	1.1627	1.0810	1.0537	1.0399	1.0315	1.0259	1.0218	1.0186	1.0161	1.0141	1.0123	1.0108	1.0095	1.0083	1.0071	1.0061	1.0050	1.0039	1.0026	1.0011	28'
29'	1.1685	1.0839	1.0556	1.0413	1.0326	1.0268	1.0225	1.0193	1.0167	1.0146	1.0128	1.0112	1.0098	1.0086	1.0074	1.0063	1.0052	1.0040	1.0027	1.0012	29'
30'	1.1743	1.0868	1.0575	1.0427	1.0338	1.0277	1.0233	1.0200	1.0173	1.0151	1.0132	1.0116	1.0102	1.0089	1.0077	1.0065	1.0054	1.0042	1.0028	1.0012	30'

Table 11-3 (Concluded). Phase Angle Correction Factors (PACFs).

For Lagging Current When $(\beta - \gamma)$ is Negative
For Leading Current When $(\beta - \gamma)$ is Positive

$\beta - \gamma$	Apparent Power Factor (Cos θ_2)																			$\beta - \gamma$
	.05	.10	.15	.20	.25	.30	.35	.40	.45	.50	.55	.60	.65	.70	.75	.80	.85	.90	.95	.99	
31'	1.1801	1.0897	1.0594	1.0441	1.0349	1.0286	1.0241	1.0206	1.0179	1.0156	1.0137	1.0120	1.0105	1.0092	1.0079	1.0067	1.0055	1.0043	1.0029	1.0012	31'
32'	1.1859	1.0926	1.0613	1.0456	1.0360	1.0296	1.0249	1.0213	1.0184	1.0161	1.0141	1.0124	1.0108	1.0095	1.0082	1.0069	1.0057	1.0045	1.0030	1.0013	32'
33'	1.1917	1.0955	1.0632	1.0470	1.0371	1.0305	1.0256	1.0219	1.0190	1.0166	1.0145	1.0128	1.0112	1.0097	1.0084	1.0072	1.0059	1.0046	1.0031	1.0013	33'
34'	1.1975	1.0984	1.0651	1.0484	1.0383	1.0314	1.0264	1.0226	1.0196	1.0171	1.0150	1.0131	1.0115	1.0100	1.0087	1.0074	1.0061	1.0047	1.0032	1.0014	34'
35'	1.2033	1.1012	1.0671	1.0498	1.0394	1.0323	1.0272	1.0233	1.0202	1.0176	1.0154	1.0135	1.0119	1.0103	1.0089	1.0076	1.0063	1.0049	1.0033	1.0014	35'
36'	1.2091	1.1041	1.0690	1.0512	1.0405	1.0332	1.0280	1.0239	1.0207	1.0181	1.0158	1.0139	1.0122	1.0106	1.0092	1.0078	1.0064	1.0050	1.0034	1.0014	36'
37'	1.2149	1.1070	1.0709	1.0527	1.0416	1.0342	1.0287	1.0246	1.0213	1.0186	1.0163	1.0143	1.0125	1.0109	1.0094	1.0080	1.0066	1.0052	1.0035	1.0015	37'
38'	1.2207	1.1099	1.0728	1.0541	1.0427	1.0351	1.0295	1.0253	1.0219	1.0191	1.0167	1.0147	1.0129	1.0112	1.0097	1.0082	1.0068	1.0053	1.0036	1.0015	38'
39'	1.2265	1.1128	1.0747	1.0555	1.0439	1.0360	1.0303	1.0259	1.0224	1.0196	1.0172	1.0151	1.0132	1.0115	1.0099	1.0084	1.0070	1.0054	1.0037	1.0016	39'
40'	1.2323	1.1157	1.0766	1.0569	1.0450	1.0369	1.0311	1.0266	1.0230	1.0201	1.0176	1.0154	1.0135	1.0118	1.0102	1.0087	1.0071	1.0056	1.0038	1.0016	40'
41'	1.2382	1.1186	1.0785	1.0584	1.0461	1.0379	1.0318	1.0273	1.0236	1.0206	1.0180	1.0158	1.0139	1.0121	1.0104	1.0089	1.0073	1.0057	1.0038	1.0016	41'
42'	1.2440	1.1215	1.0805	1.0598	1.0472	1.0388	1.0326	1.0279	1.0242	1.0211	1.0185	1.0162	1.0142	1.0124	1.0107	1.0091	1.0075	1.0058	1.0039	1.0017	42'
43'	1.2498	1.1244	1.0824	1.0612	1.0484	1.0397	1.0334	1.0286	1.0247	1.0216	1.0189	1.0166	1.0145	1.0127	1.0110	1.0093	1.0077	1.0060	1.0040	1.0017	43'
44'	1.2556	1.1273	1.0843	1.0626	1.0495	1.0406	1.0342	1.0292	1.0253	1.0221	1.0194	1.0170	1.0149	1.0130	1.0112	1.0095	1.0079	1.0061	1.0041	1.0017	44'
45'	1.2614	1.1302	1.0862	1.0640	1.0506	1.0415	1.0349	1.0299	1.0259	1.0226	1.0198	1.0174	1.0152	1.0133	1.0115	1.0097	1.0080	1.0063	1.0042	1.0018	45'
46'	1.2672	1.1330	1.0881	1.0655	1.0517	1.0425	1.0357	1.0306	1.0265	1.0231	1.0202	1.0178	1.0156	1.0136	1.0117	1.0099	1.0082	1.0064	1.0043	1.0018	46'
47'	1.2730	1.1359	1.0900	1.0669	1.0529	1.0434	1.0365	1.0312	1.0270	1.0236	1.0207	1.0181	1.0159	1.0139	1.0120	1.0102	1.0084	1.0065	1.0044	1.0019	47'
48'	1.2788	1.1388	1.0919	1.0683	1.0540	1.0443	1.0373	1.0319	1.0276	1.0241	1.0211	1.0185	1.0162	1.0141	1.0122	1.0104	1.0086	1.0067	1.0045	1.0019	48'
49'	1.2846	1.1417	1.0938	1.0697	1.0551	1.0452	1.0380	1.0326	1.0282	1.0246	1.0215	1.0189	1.0166	1.0144	1.0125	1.0106	1.0087	1.0068	1.0046	1.0019	49'
50'	1.2904	1.1446	1.0958	1.0711	1.0562	1.0461	1.0388	1.0332	1.0288	1.0251	1.0220	1.0193	1.0169	1.0147	1.0127	1.0108	1.0089	1.0069	1.0047	1.0020	50'
51'	1.2962	1.1475	1.0977	1.0726	1.0573	1.0471	1.0396	1.0339	1.0293	1.0256	1.0224	1.0197	1.0172	1.0150	1.0130	1.0110	1.0091	1.0071	1.0048	1.0020	51'
52'	1.3020	1.1504	1.0996	1.0740	1.0585	1.0480	1.0404	1.0345	1.0299	1.0261	1.0229	1.0201	1.0176	1.0153	1.0132	1.0112	1.0093	1.0072	1.0049	1.0020	52'
53'	1.3078	1.1533	1.1015	1.0754	1.0596	1.0489	1.0411	1.0352	1.0305	1.0266	1.0233	1.0204	1.0179	1.0156	1.0135	1.0114	1.0094	1.0073	1.0049	1.0021	53'
54'	1.3136	1.1562	1.1034	1.0768	1.0607	1.0498	1.0419	1.0359	1.0310	1.0271	1.0237	1.0208	1.0182	1.0159	1.0137	1.0117	1.0096	1.0075	1.0050	1.0021	54'
55'	1.3194	1.1591	1.1053	1.0782	1.0618	1.0507	1.0427	1.0365	1.0316	1.0276	1.0242	1.0212	1.0186	1.0162	1.0140	1.0119	1.0098	1.0076	1.0051	1.0022	55'
56'	1.3252	1.1619	1.1072	1.0797	1.0630	1.0517	1.0435	1.0372	1.0322	1.0281	1.0246	1.0216	1.0189	1.0165	1.0142	1.0121	1.0100	1.0078	1.0052	1.0022	56'
57'	1.3310	1.1648	1.1091	1.0811	1.0641	1.0526	1.0442	1.0379	1.0328	1.0286	1.0250	1.0220	1.0192	1.0168	1.0145	1.0123	1.0101	1.0079	1.0053	1.0022	57'
58'	1.3368	1.1677	1.1111	1.0825	1.0652	1.0535	1.0450	1.0385	1.0333	1.0291	1.0255	1.0224	1.0196	1.0171	1.0147	1.0125	1.0103	1.0080	1.0054	1.0023	58'
59'	1.3427	1.1706	1.1130	1.0839	1.0663	1.0544	1.0458	1.0392	1.0339	1.0296	1.0259	1.0227	1.0199	1.0174	1.0150	1.0127	1.0105	1.0082	1.0055	1.0023	59'
60'	1.3485	1.1735	1.1149	1.0853	1.0674	1.0553	1.0466	1.0398	1.0345	1.0301	1.0263	1.0231	1.0203	1.0177	1.0152	1.0129	1.0107	1.0083	1.0056	1.0023	60'

of the load (cos θ_2). Thus the phase angle correction factor varies with the apparent power factor of the load. In actual practice the difference between the apparent power factor (cos θ_2) and the true power factor (cos θ) is so small that for ordinary values of phase angle either power factor can be with Tables 11-2 and 11-3, to find the phase angle correction factor. The value of $\beta - \gamma$ must however, be accurately known. Note in Tables 11-2 and 11-3 that the $PACF_K$ increases rapidly at low power factors.

Tables 11-2 and 11-3 cover values of $\beta - \gamma$ from zero to one degree by minutes and from 0.05 to 1.00 power factor in steps of 0.05 power factor. Interpolation between values may be done, but will rarely be required with these tables. Values of $\beta - \gamma$ greater than 60 minutes are rarely encountered with modern instrument transformers.

Transformer Correction Factor (TCF)

The correction factor for the combined effect of ratio error and phase angle of an instrument transformer is called the transformer correction factor (*TCF*). It is the factor by which the reading of a wattmeter or the registration of a watthour meter must be multiplied to correct for the effect of ratio error and phase angle.

$$TCF = RCF \times PACF$$

then

$$TCF_I = RCF_I \times \frac{\cos(\theta_2 + \beta)}{\cos\theta_2} = RCF_I \times PACF_I$$

$$TCF_I = RCF_I \times PACF_I$$

and

$$TCF_E = RCF_E \times \frac{\cos(\theta_2 - \gamma)}{\cos\theta_2}$$

$$TCF_E = RCF_E \times PACF_E$$

Where both current and voltage transformers are used, the combined phase angle correction factor should be determined for the combination in one step as previously shown and not calculated separately and combined. The product of the two separate phase angle correction factors is not exactly equal to the true value of the overall phase angle correction factor.

Final Correction Factor (FCF)

The correction factor for the combined effects of ratio error and phase angle, where both current and voltage transformers are used, is called the final correction factor (*FCF*). It is also referred to as the instrument transformer correction factor. It is the factor by which the reading of a wattmeter, or the registration of a watthour meter, operated from the secondaries of both a current and voltage transformer must be multiplied to correct for the effect of ratio errors and phase displacement of the current and voltage caused by the instrument transformers.

$$FCF = RCF_K \times PACF_K$$

The Nominal Instrument Transformer Ratio

If the marked (or nominal) ratio of the voltage transformer is N_E and the marked (or nominal) ratio of the current transformer is N_I, the product of these two marked ratios is the nominal instrument transformer ratio, N_K.

$$N_K = N_E \times N_I$$

Table 11-4. Summary of Fundamental Relations for Single-Phase Metering Installations Involving Instrument Transformers.

Primary Power = *Primary Volts* × *Primary Amperes* × *Primary Power Factors*

$$W_P = E_P \times I_P \times \cos\theta$$

$$W_P = E_S N_E (RCF_E) \times I_S N_I (RCF_I) \times \cos(\theta_2 + \beta - \gamma)$$

These terms can be rearranged to give:

$$W_P = E_S I_S \times N_E N_I \times (RCF_E)(RCF_I) \times \cos(\theta_2 + \beta - \gamma)$$

Multiplying the first term by $\cos\theta_2$ and dividing the last term by $\cos\theta_2$ is the same as multiplying by $\frac{\cos\theta_2}{\cos\theta_2} = 1$, which does not change the product, which gives:

$$W_P = E_S I_S \cos\theta_2 \times N_E N_I \times (RCF_E)(RCF_I) \times \frac{\cos\theta_2 + \beta - \gamma}{\cos\theta_2}$$

$$W_P = W_S \times N_K \times RCF_K \times PACF_K$$

$$W_P = W_S \times N_K \times FCF$$

Primary Power = Secondary Power × Nominal Instrument Transformer Ratio × Final Correction Factor

W_P = Primary power (watts)
E_P = Primary voltage
I_P = Primary current
$\cos\theta$ = Primary power factor
θ = Angle between E_P and I_P
N_E = Marked (Nominal) ratio of voltage transformer
N_I = Marked (Nominal) ratio of current transformer
β = Phase angle of current transformer
γ = Phase angle of voltage transformer
W_S = Secondary power (watts)
E_S = Secondary voltage
I_S = Secondary current
$\cos\theta_2$ = Secondary (apparent) power factor
θ_2 = Angle between E_S and I_S
RCF_E = Ratio Correction Factor of voltage transformer
RCF_I = Ratio Correction Factor of current transformer
RCF_K = Combined Ratio Correction Factor, voltage transformer and current transformer
$PACF_K$ = Phase Angle Correction Factor, voltage transformer and current transformer
FCF = Final Correction Factor

See Fig. 11-13 for the corresponding schematic and phasor diagrams.

Summary of Basic Instrument Transformer Relationships

Table 11-4 is a summary, in concise form, of the relation of primary and secondary values in a single-phase metering installation using instrument transformers. This table can be used as a reference that ties together most of the factors which have been covered in detail in the preceding pages. Table 11-4 is in terms of the primary and secondary power in watts. If both sides of all of these equations are multiplied by time in hours they would then apply equally well in terms of energy in watthours. All of the equations in this table apply to the metering installation whose schematic and phasor diagrams are shown in Fig. 11-13.

Compensating Errors

It is apparent from the equation for the transformer correction factor ($TCF = RCF \times PACF$) that for some values of *RCF* and *PACF*, their product could be closer to one than either separately. For example, (1.0032)(0.9970) = 1.0002. Thus, under some conditions the overall effect of the error in ratio may be offset by an opposite effect due to the phase angle.

This fact is used as a basis for the tolerance limits of the ANSI accuracy classifications where the specified tolerances of ratio and phase angle are interdependent. These classifications are set up on the basis of a maximum over-all tolerance in terms of transformer correction factor (TCF) for power factors from unity to 0.6 lagging. This is covered later under the subheading "ANSI Accuracy Classes for Metering Service."

Where both a current and a voltage transformer are used, the combined ratio correction factor can be improved by matching transformers with opposite ratio errors since $RCF_K = RCF_E \times RCF_I$.

To reduce the effect of phase angle errors, which are dependent upon $\beta - \gamma$, current and voltage transformers can be selected having phase angles of the same sign (that is, both positive or both negative) thus reducing the over-all phase angle error.

Current and voltage transformers are not usually matched to balance errors in this manner, but occasionally these methods may be useful.

APPLICATION OF CORRECTION FACTORS

When Correction Factors Should Be Applied

In most commercial metering installations using instrument transformers, no corrections need be applied if modern instrument transformers, meeting ANSI accuracy specifications, are used within the burden and power-factor limits of these specifications and the secondary leads are short enough so that they cause no appreciable error. Under such conditions, the error contributed by any single instrument transformer should not exceed the ANSI accuracy class. Where both a current and a voltage transformer are used, their combined error could theoretically reach the sum of the maximum errors represented by the accuracy classes of the two transformers, but will in most cases be much less. In polyphase metering, the total error is the weighted average of the combined errors of the current and voltage transformer on each phase and can never be greater than the maximum errors on the worst phase. For 0.3 percent ANSI Accuracy Class transformers the maximum errors, under the ANSI specified conditions, may be summarized as follows:

Maximum Percent Errors For Combinations of 0.3 Percent ANSI Accuracy Class Instrument Transformers Under ANSI Specified Conditions of Burden, and Load Power Factors Between 1.00 and 0.6 Lag

	Percent Error at 100% Load	Percent Error at 10% Load
Current Transformers	0.3	0.6
Voltage Transformers	0.3	0.3
Maximum Percent Error	0.6	0.9

These maximum errors would occur only rarely in an actual combination of instrument transformers. There is a good probability that the errors would be less than 0.3 to 0.5 percent, which would be acceptable for most commercial metering.

Special cases may arise, however, that make the application of instrument transformer corrections necessary or desirable. Such cases could be due to the use of older types of instrument transformers that do not meet the ANSI accuracy specifications, the necessity of using heavier burdens than specified by ANSI, the use of long secondary leads, power factor of the load below 0.6 lagging, power factor of the load leading, and requirements for higher than normal accuracy for special installations, such as large wholesale installations, interchange metering between power companies, or measurement of total generator output during efficiency tests of power station generators and turbines.

The decision as to when instrument transformer corrections should be applied is a matter of policy that must be decided by each utility company on the basis of both technical and economic considerations. In general, most utilities do not apply instrument transformer corrections for routine work and may or may not apply corrections in special cases.

If the meter is to be adjusted to compensate for the errors of the instrument transformers, great care must be taken to make this adjustment in the proper direction. An error in the sign of the correction applied results in doubling the over-all error instead of eliminating it. The best precaution against this type of mistake is the use of prepared forms which are set up to show each step in the process.

With a good prepared form, correction factors can be applied very easily. The actual field work may involve nothing more than adding the percent error caused by the transformers to the percent error of the meter. Several methods of applying corrections will be shown.

Determining the Meter Adjustment in Percent Registration to Correct for Instrument Transformer Errors — Calculations Based on Tables 11-2 and 11-3

It has already been shown in Table 11-4 that:

Primary Power = Secondary Power × Nominal Instrument Transformer Ratio × Final Correction Factor

Multiplying both sides by hours gives:

True Primary Watthours = True Secondary Watthours $\times N_K \times$ FCF

But the indicated primary watthours are:

Indicated Primary Watthours = Indicated Secondary Watthours $\times N_K$

The over-all percent registration of the installation, or primary percent registration, is:

$$\text{Primary Percent Registration} = \frac{(\text{Indicated Primary Watthours})(100)}{\text{True Primary Watthours}}$$

Substituting equivalent secondary values gives:

$$\text{Primary Percent Registration} = \frac{(\text{Indicated Secondary Watthours})(N_K)(100)}{(\text{True Secondary Watthours})(N_K)(FCF)}$$

$$\text{Primary Percent Registration} = \frac{(\text{Indicated Secondary Watthours})(100)}{(\text{True Secondary Watthours})(FCF)}$$

$$\text{Primary Percent Registration} = \frac{\text{Secondary Percent Registration}}{FCF}$$

$$\text{Primary Percent Registration} = \frac{\text{Percent Registration Meter Only}}{FCF}$$

Thus, the over-all percent registration may be obtained by dividing the percent registration of the meter by the final correction factor.

Example:

Given: Percent Registration of Meter Alone = 99.75% and *FCF* = 1.0037

Then: Primary (or over-all) Percent Registration = $\frac{99.75}{1.0037}$ = 99.38%

To divide using the approximate method for numbers close to 1, add one to the numerator and subtract the denominator. 0.9975 divided by 1.0037 is approximately equivalent to 1.9975 minus 1.0037 = 0.9938, and 0.9938 × 100 = 99.38.

It is apparent that the primary (overall) percent registration can be made 100.00 percent if the percent registration of the meter is adjusted to 100 times the final correction factor.

Thus, if the meter in the preceding example were adjusted to 100.37 percent registration, then

$$\text{Primary (overall) Percent Registration} = \frac{100.37}{1.0037} = 100.00\%$$

Table 11-5 shows a standard form that can be used to determine the required meter adjustment by this method. This method is particularly useful where meter tests are made with a fixed routine, such as light-load, full-load, and inductive-load, made respectively with 10 and 100 percent rated current at 1.0 power factor and with 100 percent rated current at 0.5 power factor lagging.

This method is applicable to installations with current and voltage transformers, or either, and the calculations are simplified by using addition and subtraction for the multiplication of quantities near unity, as previously explained. Ratio correction factors and phase angles are used directly and the result is the accuracy performance to which the meter should be adjusted to compensate for instrument transformer errors.

Table 11-5. Calculation of Meter Accuracy Settings.

Item	Symbol	Load: Light	Load: Heavy	Load: Inductive
		Power Factor 1.0		P.F. 0.5 Lag
		10% Cur.	100% Current	
Phase Angle, Minutes				
Current Transformer	β	+10	−2	−2
Voltage Transformer	γ	+8	+8	+8
Combined	$\beta-\gamma$	+2	-10	-10
Ratios				
Current Transformer	RCF_I	1.0043	0.9992	0.9992
Voltage Transformer	RCF_E	0.9976	0.9976	0.9976
Effect of Combined Phase Angle	$PACF_K$	1.0000	1.0000	1.0000
By Addition		3.0019	2.9968	3.0018
Subtract (No. of Terms Added Minus 1)		2.	2.	2.
Final Correction Factor	FCF	1.0019	0.9968	1.0018
Meter Accuracy Setting Required to Compensate = (100)(FCF)	Percent Registration	100.19%	99.68%	100.18%
% Error Caused by Inst. Trans. ≈ (1-FCF)(100)	% E	0.19%−	0.32%+	0.18%−
% Meter Adjustment Required to Compensate	% E	0.19%+	0.32%−	0.18%+

The ratio correction factors and phase angles are taken from test data on the instrument transformers or from the manufacturers' certificates. These values must be the values that apply at the terminals of the meter and be based on the actual burdens. If long secondary leads are used from the voltage transformer to the meter, the effect of the lead drop on the ratio and phase angle as seen at the meter must be included. This can be determined by test or calculation as previously explained. If the available instrument transformer data are not based on the actual burden, the desired value may be determined by interpolation or calculation by methods to be explained later.

The appropriate ratio correction factors and phase angles are then entered in Table 11-5 as shown. The phase angle correction factor at unity power factor is

1.0000, within 0.02 percent or less, for all values of $(\beta - \gamma)$ up to 60 minutes. At 0.50 power factor lagging, the phase angle correction factor is read from Table 11-3 as 1.0050 for a value of $(\beta - \gamma)$ of -10 minutes. The operations indicated in Table 11-5 are performed and the meter accuracy settings in percent registration are determined as shown. The bottom two lines show the percent errors caused by the instrument transformers and the percent errors to which the meter should be set to compensate. This method is discussed in the next section. The meter is then adjusted within the desired tolerance of these settings and the compensation has been accomplished.

The calculations in this table have been carried to 0.01 percent, as it is normal practice to carry one more place in calculations of this kind than is used in the final result. If the final overall accuracy of the installation were to be reported, it would normally be rounded off to the nearest 0.1 percent.

The same setup may be used where only a current transformer or a voltage transformer is used. It is only necessary to enter "0" under phase angle and "1.0000" under ratio correction factor in the places where no transformer is used and make the additions and subtractions indicated. For polyphase installations, where correction factors and phase angles are not widely divergent, the ratio correction factors and phase angles for the current transformers for all phases may be respectively averaged and the average values of ratio correction factor and phase angle of the current and voltage transformers respectively used for the calculations.

Alternatively, calculations may be made on each stator using the ratio correction factors and phase angles for the transformers connected to that stator. For precise work, where either voltage or current transformer phase angles materially differ, this method is preferred.

Overall Percent Error Caused by the Instrument Transformers Alone

The percent error due to the instrument transformers may be derived as follows:

$$\text{True Primary Watthours} = \text{True Secondary Watthours} \times N_K \times FCF$$

$$\text{Indicated Primary Watthours} = \text{Indicated Secondary Watthours} \times N_K$$

$$\text{Over-all Percent Error} = \frac{\text{Indicated} - \text{True}}{\text{True}} \times 100 =$$

$$\frac{(\text{Indicated Secondary Watthours})(N_K) - (\text{True Secondary Watthours})(N_K)(FCF)}{(\text{True Secondary Watthours})(N_K)(FCF)} \times 100$$

Over-all Percent Error =

$$\frac{(\text{Indicated Secondary Watthours}) - (\text{True Secondary Watthours})(FCF)}{(\text{True Secondary Watthours})(FCF)} \times 100$$

This is the overall percent error of the installation including both meter and transformer errors. To find the errors due to the transformers alone, assume that the meter is correct. Then, Indicated Secondary Watthours = True Secondary Watthours, and substituting in the preceding equation:

Percent Error Caused by Instrument Transformer =

$$\frac{(\text{True Secondary Watthours}) - (\text{True Secondary Watthours})(FCF)}{(\text{True Secondary Watthours})(FCF)} \times 100$$

Percent Error Caused by Instrument Transformer =

$$\frac{1 - FCF}{FCF}(100) \approx (1 - FCF) \times 100$$

Where ≈ means "is approximately equal to."

The second or approximate form is the most convenient to use and will not be in error by more than 0.01 percent for values of *FCF* between 0.9900 and 1.0100 or more than 0.02 percent for values of *FCF* between 0.9800 and 1.0200.

Example:

Given *FCF* = 0.9853

Percent Error Caused by Instrument Transformer =

$$\frac{1 - 0.9853}{0.9853} \times 100 = \frac{(0.0147)(100)}{0.9853} = 1.49\% \text{ using the exact method.}$$

Percent Error Caused by Instrument Transformer ≈ (1 − 0.9853)100 = 0.0147 × 100 = 1.47% using the approximate method.

Note that the sign of the error will be minus (−) for values of *FCF* greater than 1.

A form such as Table 11-5 can be used to determine the final correction factor (*FCF*) from which the percent error caused by the instrument transformers is determined. This is shown on the next to bottom line of Table 11-5.

Determining the Overall Percent Error by Adding the Percent Errors Caused by Instrument Transformers and the Meter

It can be shown that: Overall Percent Error ≈ Percent Error Caused by the Instrument Transformer + Percent Error of the Meter.

This expression is an approximation that is good only when the percent errors are small. When adding percent errors up to ± 1.0 percent, the error in this approximation will not exceed 0.01 percent. When adding percent errors up to ±2.0 percent, the error in this approximation will not exceed 0.04 percent. This expression is convenient to use and may be used for errors up to two or three percent without significant error.

Example:

Meter Error	+	*Instrument Transformer Error*	=	*Over-All Error*
+ 0.32%		− 0.15%		+ 0.17%

The required compensation can be made by adjusting the percent error of the meter to the same magnitude as the percent error caused by the instrument transformers, but with the opposite sign. This is shown in the last line of Table 11-5.

A Graphical Method of Determining the Percent Error Caused by the Instrument Transformers and the Required Compensation

The percent error caused by the instrument transformers and the required meter adjustment to compensate may be determined by using the chart shown in Fig. 11-14.

A straight edge is placed on the chart so that one end intercepts the ratio correction factor scale on the left at the desired value of RCF and the other end intercepts the phase angle scale on the right at the desired value of ($\beta - \gamma$). The percent error, or percent meter adjustment, is read from the center scale that represents the desired power factor. The proper half of the phase angle scale to be used depends upon the load power factor and the sign of $(\beta - \gamma)$ and this is indicated in the headings for this scale. The sign of the error caused by the instrument transformers and the sign of the percent error of the required compensating meter adjustment is indicated in the blocks between the 100 and 95 percent power factor scales. The chart is designed to give percent errors for a current and voltage transformer combined, by using the combined ratio correction factor RCF_K and the combined phase angle $(\beta - \gamma)$.

To use the chart for an installation involving a current transformer only, use RCF_I on the RCF scale and β in place of $(\beta - \gamma)$. For polyphase values, the percent errors may be determined separately for each phase, or average values of RCF_K and

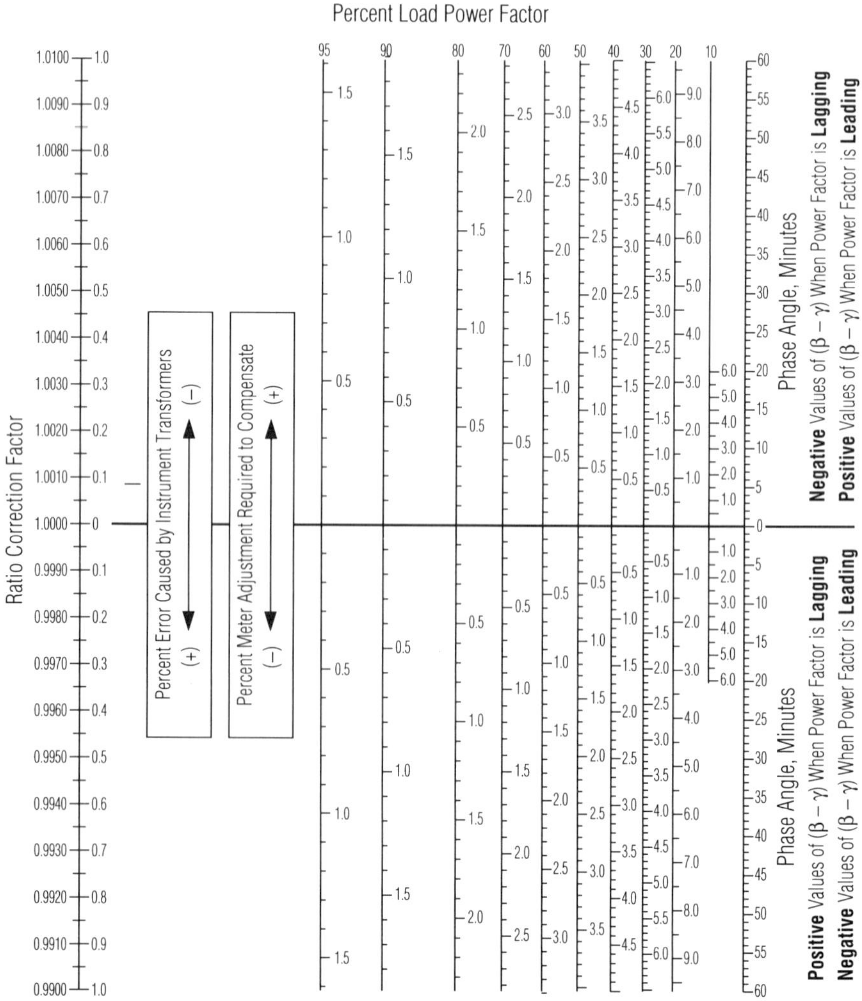

Figure 11-14. Percent Error Calculation Chart for Effects of Instrument Transformer Ratio and Phase Angle.

($\beta - \gamma$) may be used to obtain the total percent error in one step. The chart is based on the approximate formula for the percent error caused by the instrument transformer previously discussed. Thus, the results read from the chart may differ by a few hundredths of a percent from the values computed from Tables 11-2 and 11-3.

Example:

For a load power factor 70 percent, lagging:

Current Transformer:	RCF_I	=	1.0043	β	=	+12
Voltage Transformer:	RCF_E	=	1.0012	γ	=	+7
Combined Values:	RCF_K	=	1.0055	$\beta - \gamma$	=	+5

One end of the straight edge is placed on the ratio correction factor scale at 1.0055 and the other end on the lower half of the phase angle scale at 5. The straight edge then intercepts the 70 percent power factor scale at 0.40 percent in the upper half of the chart. Therefore, the error caused by the instrument transformers is -0.40% and the meter must be adjusted to +0.40% (fast) to compensate.

Application of Instrument Transformer and Rotating Standard Corrections in One Step to a Three-Phase, Three-Stator, Four-Wire, Wye Metering Installation

Most special installations justifying the application of corrections for the instrument transformers will be three-phase. If the load is reasonably balanced, the work may be greatly simplified by averaging the corrections. In addition, the corrections for the calibration errors of the rotating standards may also be included. The required total percent error caused by all of the instrument transformers and all of the rotating standards can be calculated for any load and power factor. These calculations may be made and checked before going into the field to test the meter. The actual work in the field then simply requires the addition of these percent errors to the apparent percent error of the meter as determined by test.

To use this method, forms such as Tables 11-6, 11-7, and 11-8 are prepared and filled in as needed. For a three-phase, three-stator, four-wire, wye installation, the procedure is as follows:

The procedure is shown for a test method using three rotating standards and a special three-phase phantom load, such that the meter is tested under actual three-phase conditions. It is also suitable for a three-phase customer's load test using three rotating standards, one in series with each meter stator respectively. This second method is limited to the system load and power factor of the installations at the time of test, but is occasionally useful for installations having a relatively constant load. This procedure can also be used to apply corrections when using the usual standard test methods requiring only one rotating standard to make single-phase series tests on the three-phase meter. The method is therefore adaptable to any test procedure desired.

The ratio correction factors and phase angles of the three current transformers at various values of secondary current are entered in the proper spaces in Table 11-6 as shown. These values would be available from certificates or test data. The averages of all these values are computed and entered.

The ratio correction factors and phase angles of the voltage transformers are entered and averaged. The average values of these are recopied into the additional spaces as shown, so that they may be combined with the current transformer values.

Table 11-6. Average Ratio and Phase Angle Calculation Sheet for Polyphase Installations.

Location: APPLEGATE SUB. **Circuit:** SEA VIEW LINE - NO. 1
C. T. Ratio: 400:5 **V. T. Ratio:** 16,500:110 **V. T. Sec. Voltage:** 120
Date: 12-17-91 **Calculated By:** C. E. P. **Checked By:** G. T.

Ratio		Secondary Amperes					
C.T. Serial Nos.	**Symbol**	**0.5**	**1**	**2**	**3**	**4**	**5**
675 932	RCF_I	1.0037	1.0013	1.0007	1.0002	0.9997	0.9990
675 933	RCF_I	1.0035	1.0010	1.0006	1.0001	0.9996	0.9998
675 934	RCF_I	1.0030	1.0007	1.0003	0.9998	0.9992	0.9981
Average	RCF_I	1.0034	1.0010	1.0005	1.0000	0.9995	0.9987
V.T. Serial Nos.	**Symbol**						
183 276	RCF_E	1.0021					
183 283	RCF_E	1.0017					
183 341	RCF_E	1.0025	**Enter Average RCF_E In All Spaces Below**				
Average	RCF_E	1.0021	1.0021	1.0021	1.0021	1.0021	1.0021
Combined Average (RCF_I) (RCF_E) = RCF_K		1.0055	1.0031	1.0026	1.0021	1.0016	1.0008

Phase Angle, Minutes		Secondary Amperes					
C.T. Serial Nos.	**Symbol**	**0.5**	**1**	**2**	**3**	**4**	**5**
675 932	β	−10	−4	+2	+4	+6	+7
675 933	β	−13	−5	+1	+3	+5	+6
675 934	β	−8	−2	+3	+6	+8	+9
Average	β	−10	−4	+2	+4	+6	+7
V.T. Serial Nos.	**Symbol**						
183 276	γ	+11					
183 283	γ	+12					
183 341	γ	+7	**Enter Average γ In All Spaces Below**				
Average	γ	+10	+10	+10	+10	+10	+10
Combined Average $\beta - \gamma$		−20	−14	−8	−6	−4	−3

The combined average ratio correction factor (RCF_K) and average phase angle (β and γ) are computed and entered as shown.

These combined values will apply to this installation indefinitely unless the instrument transformers or burdens are changed. A form similar to Table 11-7 is now filled out. First the desired three-phase power factors and test voltages are entered in the spaces to the left. In the example shown, power factors of 1.00, 0.87 lag, and 0.50 lag at 120 V are shown. Other values can be used as required.

Table 11-7. Watthour Meter Test, Combined Error Calculation Sheet
For Three-Stator, Three-Phase Meters Tested Three-Phase Using Three Rotating Standards or Single-Phase Series Using One Rotating Standard.

Location: APPLEGATE SUB. **Circuit:** SEA VIEW LINE - NO. 1
Date: 12-17-91 **Calculated By:** C. E. P. **Checked By:** G. T.

	Secondary Amperes	Calibrated Percent Errors of Rotating Standards No. 367	No. 368	No. 369	Ⓐ Combined Average Percent Error of Rotating Standards	Combined Average Ratio Correction Factor (RCF_K) ×	Combined Average Phase Angle Cor. Fact. ($PACF_K$) =	Combined Average Final Correction Factor (FCF)	Ⓑ % Error Caused by Instrument Trans. (1-FCF)(100)	Combined % Error Inst. Trans. Plus Rot. Stds. Ⓐ + Ⓑ
P.F. 1.00	0.5*	.08–	.06–	.08–	.07–	1.0055	1.0000	1.0055	.55–	.62–
Volts 120	1*	.12–	.02–	.10–	.08–	1.0031	1.0000	1.0031	.31–	.39–
	2	.07–	.05–	.10–	.07–	1.0026	1.0000	1.0026	.26–	.33–
	3	.03–	.02–	.03–	.03–	1.0021	1.0000	1.0021	.21–	.24–
	4	.05–	.08+	.02–	0	1.0016	1.0000	1.0016	.16–	.16–
	5	.02–	.03+	.02–	0	1.0008	1.0000	1.0008	.08–	.08–
P.F. .87 Lag.	0.5*	.12–	.10–	.15–	12–	1.0055	1.0033	1.0088	.88–	1.00–
Volts 120	1*	.17–	.08–	.17–	.14–	1.0031	1.0023	1.0054	.54–	.68–
	2	.09+	.13+	.13–	.03+	1.0026	1.0013	1.0039	.39–	.36–
	3	.05+	.03+	.07–	.0	1.0021	1.0009	1.0030	.30–	.30–
	4	.02–	.02+	0	0	1.0016	1.0007	1.0023	.23–	.23–
	5	.17–	.02+	.01+	.05–	1.0008	1.0005	1.0013	.13–	.18–
P.F. .50 Lag.	0.5*	.15–	.20–	.20–	.18–	1.0055	1.0101	1.0156	1.56–	1.74–
Volts 120	1*	.25–	.14–	.20–	.20–	1.0031	1.0070	1.0101	1.01–	1.21–
	2	.14+	.21+	.20+	.18+	1.0026	1.0040	1.0066	.66–	.48–
	3	.10+	.07+	.10+	.09+	1.0021	1.0030	1.0051	.51–	.42–
	4	.02–	.02+	.02+	.01+	1.0016	1.0020	1.0036	.36–	.35–
	5	.23–	.10+	.03+	.03–	1.0008	1.0015	1.0023	.23–	.26–

* On 1 ampere coil of rotating standard watthour meter. All other points on on 5 ampere coil.

The percent errors of the three rotating standards to be used for the test are then entered in the spaces provided. These values are determined by tests of the rotating standards at the current, voltage, and power factors to be used. Since on a balanced load all three stators operate at a single-phase power factor equal to the three-phase load power factor, the three rotating standards will be running at the same speed and power factor. The errors can therefore be averaged and entered in Column A. Where only one rotating standard is used for a single-phase series test of a polyphase meter, the errors of the rotating standard should be entered directly in Column A, as no average is involved. In this case the preceding three columns are not needed.

The combined average ratio correction factors from Table 11-6 are now entered in the proper column of Table 11-7. These are the same at all power factors.

The combined phase angle correction factor ($PACF_K$) is determined from Tables 11-2 or 11-3 for the desired values of load power factor as shown in Table 11-7 and the average values of ($\beta - \gamma$) previously determined in Table 11-6. The product of the average RCF_K and the average $PACF_K$ gives the average final correction factor *FCF.* The percent error caused by the instrument transformers is equal to $(1 - FCF) \times 100$. This is entered in Column B. The values in Columns A and B are added algebraically and entered in the final column to give the combined percent error caused by the instrument transformers and the rotating standards.

If correction for the rotating standards is not desired, this can be omitted, in which case the values in Column A would be zero.

The values in Column B could also be obtained directly from Table 11-6 and the chart shown in Fig. 11-14. This is a simpler, but slightly less accurate method.

Table 11-8 is a watthour meter test form suitable for this method. The revolutions of the three rotating standards for each test run are entered and added as shown. (Where only one rotating standard is used for a single-phase series test of a polyphase meter, its revolutions should be entered directly in the column for the total revolutions. In this case, the preceding three columns are not needed.) The indicated percent error is computed from the total revolutions and entered as shown. The percent error caused by the instrument transformers and rotating standards from the last column of Table 11-7 is entered as shown in Table 11-8. This value, plus the percent error indicated, is equal to the over-all percent error. Only the values at 0.87 power factor have been shown on Table 11-8. Values at other power factors would be obtained in the same manner. Meter adjustments are made as required to reduce the over-all percent error to the desired tolerances. Table 11-8 has been filled in to show an "as left" curve at 0.87 power factor lagging, taken after all adjustments had been made. The "as found" tests and adjustments would be on previous sheets and are not shown in Table 11-8.

This method is simple and rapid in actual use as the corrections are precalculated before starting the meter tests. The forms reduce the whole operation to simple bookkeeping and allow the calculations to be checked at any time. If only standard single-phase series tests are made on polyphase meters, the forms shown in Tables 11-7 and 11-8 may be simplified to one column for the rotating standard data.

Application of Instrument Transformer and Rotating Standard Corrections in One Step to a Three-Phase, Two-Stator, Three-Wire, Delta Metering Installation

It can be shown mathematically that the following statement is true: In a three-phase, three-wire metering circuit having balanced voltages, currents, and burdens, using two voltage transformers having equal ratio and phase angle errors and two current transformers having equal ratio and phase angle errors, the true primary power may be determined by applying the instrument transformer corrections separately to the single-phase power in each meter stator at the single-

Table 11-8. Watthour Meter Test.

Location: APPLEGATE SUB. **Circuit:** SEA VIEW LINE - NO. 1

Meter Nameplate Data

Meter Code No. P014-7581 Mfg. G.E. Type DS-20 Volts 120 Amps 2.5 ϕ 3 EL. 3 W. 4

Meter Ser. No. 15321762 Kh .09 Dial K 1 Rr 111 1/9 Mult. By 12,000

C. T. Ratio 400:5 V. T. Ratio 16,500:110 Demand : Time — Rev. — Chart Div. —

As Found: Pri. K.W. Load — Sec. A — P. F. — Reg. Read 4321.2 Time 9:10 AM

As Left: Pri. K.W. Load — Sec. A — P. F. — Reg. Read 4321.2 Time 2:30 PM

Test Volts 120 Creep No Remarks No lead on line at time of test. Test made with 3 phase phantom load.

As Found Seal T - 32 - 90

As Left Seal T - 32 - 91

Test Data

Sec. Amps.	Cur. Coil of Rot. Std.	P. F.	Rev. Whm / Rev. Std.	Adj.	Rot. Std. No. 3.67 Rev.	Rot. Std. No. 3.68 Rev.	Rot. Std. No. 3.69 Rev.	Total Rev. of Rotating Standards	% E Indicated	% Error Caused by Inst. Trans. and Rot. Std.	% Error Over-All
.05	1	.87 Lag	2/15		4.95 4.95	4.96 4.95	4.99 4.99	14.90 14.89	.67+ .74+	1.00–	.30–
1	1	.87 Lag	2/15	"As left" Curve	4.97 4.96	4.99 4.98	4.97 4.97	14.93 14.91	.47+ .60+	.68–	.15–
2	5	.87 Lag	20/30		9.97 9.97	9.95 9.95	9.96 9.96	29.88 29.88	.40+ .40+	.36–	.04+
3	5	.87 Lag	20/30		9.98 9.98	9.95 9.96	9.96 9.96	29.89 29.90	.37+ .33+	.30–	.05+
4	5	.87 Lag	20/30		9.99 9.99	9.96 9.96	9.97 9.97	29.92 29.92	.27+ .27+	.23–	.04+
5	5	.87 Lag	20/30		10.02 10.02	9.94 9.94	9.98 9.98	29.94 29.94	.20+ .20+	.18–	.02+

Tested By: G. T. **Witness:** John Doe **Page** 3 **of** 3 **Pages**

phase power factor of each stator, or the instrument transformer corrections may be applied in one step to the total three-phase secondary power at the three-phase power factor of the circuit.

For the three-phase, two-stator, three-wire delta installations, if the errors of the instrument transformers on both phases are reasonably similar, the instrument transformer ratio correction factors and phase angles may be averaged and the

Table 11-9. Watthour Meter Test, Combined Error Calculation Sheet For Two-Stator, Three-Phase Meters Tested Three-Phase Using Two Rotating Standards.

Location: APPLEGATE SUB. **Circuit:** SEA VIEW LINE - NO. 2

Date: 12-17-91 **Calculated By:** C. E. P. **Checked By:** G. T.

	Secondary Amperes	% Error Tot. Std. No. 367 in A Phase, Operating at .87 P.F. Lag	Ⓐ $(\%E)\left(\frac{PF_1}{PF_1+PF_2}\right)$	% Error Tot. Std. No. 367 in C Phase, Operating at .87 P.F. Lead	Ⓑ $(\%E)\left(\frac{PF_2}{PF_1+PF_2}\right)$	Ⓒ Combined Weighted Average Percent Error of Rotating Standards Ⓐ + Ⓑ	Combined Average Ratio Correction Factor (RCF_K)	Combined Average Phase Angle Cor. Fact. ($PACF_K$)	Combined Average Final Correction Factor (FCF)	Ⓓ % Error Caused by Instrument Trans. (1-FCF)(100)	Combined % Error Inst. Trans. Plus Rot. Stds.
							(RCF_K)	× ($PACF_K$)	= (FCF)	(1-FCF)(100)	Ⓒ + Ⓓ
3 PH. P.F. 1.00	0.5*	.12–	.06–	.13+	.06+	0					
Volts 120	1*	.17–	.08–	.20+	.10+	.02+					
	2	.09+	.04+	.15–	.07–	.03–					
	3	.05+	.02+	.30–	.15–	.13–					
	4	.02–	.01–	.25–	.12–	.13–					
	5	.17–	.08–	.10+	.05+	.03–					
Stator P.F.s →		.50 Lag	→	1.00		**					
3 PH. P.F. .87 Lag	0.5*	.15–	.05–**	.08–	.05–**	.10–					
Volts 120	1*	.25–	.08–	.10–	.07–	.15–					
	2	.14+	.05+	.10–	.07–	.02–					
	3	.10+	.03+	.03–	.02–	.01+					
	4	.02–	.01–	.02–	.02–	.03–					
	5	.23–	.08–	.02–	.02–	.10–					
Stator P.F.s →		0	→	.87 Lag							
3 PH. P.F. .50 Lag	0.5*	—	0	.15–	.15–	.15–					
Volts 120	1*	—	0	.17–	.17–	.17–					
	2	—	0	.13–	.13–	.13–					
	3	—	0	.07–	.07–	.07–					
	4	—	0	0	0	0					
	5	—	0	.01+	.01+	.01+					

*** On 1 ampere coil of rotating standard watt hour meter. All other points on 5 ampere coil.**

**** Examples of Weighting Calculation**

$$(\%E)\left(\frac{PF_1}{PF_1+PF_2}\right) = (.15-)\left(\frac{.50}{.50+1.00}\right) = (.15-)(1/3) = .05-$$

$$(\%E)\left(\frac{PF_2}{PF_1+PF_2}\right) = (.08-)\left(\frac{1.00}{.50+1.00}\right) = (.08-)(2/3) = .05-$$

total error of the instrument transformers at the three-phase power factor determined in exactly the same manner as for the three-stator meter using Tables 11-6 and 11-7. This method does not involve appreciable error if the errors of the instrument transformers on both phases are reasonably similar. The only difference is that only two current and two voltage transformers will now be shown on Table 11-6.

However, the rotating standard corrections must be weighted before averaging, as the two rotating standards are running at different speeds. Also, the corrections entered for the rotating standards must be at the single-phase power factor of each stator.

This is easily done using a prepared form such as Table 11-9. If the three-phase power factor is cos θ, then the two-stator power factors, for balanced loads, are $\cos(\theta + 30°)$ and $\cos(\theta - 30°)$.

Since the speed of each rotating standard is proportional to the single-phase power factor at which it is running, their percent errors must be weighted before averaging by the factors

$$\frac{PF_1}{PF_1 + PF_2} \quad \text{and} \quad \frac{PF_2}{PF_1 + PF_2}$$

where PF_1 and PF_2 are the two single-phase power factors of the stators involved. This is illustrated clearly in the column headings and in the example shown in Table 11-9. The remaining columns of Table 11-9 would be filled in similarly to Table 11-7 using the three-phase power factor to determine the phase angle correction factor ($PACF_K$).

The same form (Table 11-8) may be used for the watthour meter test as was used for the three-stator meter. In this case, only two columns for the revolutions of the two rotating standards will be used. Otherwise, the procedure is identical to the procedure for the three-stator meter.

This method is quite satisfactory for the three-phase phantom load test using two rotating standards, since balanced loads are applied. It can be used for a customer's load test using two rotating standards if the load on the circuit is reasonably balanced. For customers' load tests with badly unbalanced loads, this method cannot be used. In such cases, the corrections must be applied to each stator and rotating standard separately.

Where only one rotating standard is used for a single-phase series test of a two-stator polyphase meter, both stators operate during test at the same single-phase power factor and it is not necessary to use Table 11-9 at all. Table 11-7 is used and the rotating standard error entered directly in Column A.

Table 11-10. Summary of Basic Formulas for Applying Instrument Transformer Corrections.

Overall (Primary) Percent Registration	$\frac{\text{Percent Registration of Meter Only}}{FCF}$
Required Percent Registration of Meter Only to Compensate for Instrument Transformer Errors	$FCF \times 100$
Percent Error Caused by Instrument Transformers Only	$\approx (1 - FCF) \times 100$
Required Percent Error Adjustment of Meter Only to Compensate for Instrument Transformer Errors	$\approx (FCF - 1) \times 100$
Overall (Primary) Percent Error	≈ Percent Error Caused by Instrument Transformers Only + Percent Error of Meter Only

Summary of Basic Formulas for Applying Instrument Transformer Corrections

Table 11-10 summarizes the basic formulas for applying instrument transformer corrections in a form for convenient reference.

Individual Stator Calculations

In the preceding discussion, we have assumed that the voltage transformers as well as the current transformers are reasonably matched, that is, have nearly similar ratio and phase angle errors. If each of the current transformers is of the same make, model, and type, it is usually found that they will have similar accuracy characteristics. This is also true of voltage transformers. In these cases, the procedures for applying corrections previously described will lead to no significant errors.

However, at times it is necessary to use instrument transformers with widely dissimilar correction factors. Where a high degree of accuracy is required, calculation of the effect of instrument transformer errors on each individual meter stator should be made. Correction factors may be calculated by referring to the basic meter formula and comparing meter registration to true power.

For a three-phase, three-wire delta circuit,

$$\text{True Power} = \sqrt{3}\ EI\cos\theta$$

Meter Registration, for a balanced symmetrical load, =

$$EI \times \text{Ratio Correction Factor}_A \times \cos(\theta_2 - 30° + \beta_A - \gamma_A)$$
$$+ EI \times RCF_B \times \cos(\theta_2 + 30° + \beta_B - \gamma_B).$$

It has previously been shown that the phase angle error depends on the apparent power factor of the load. Because of the phase voltage and the line current displacement as seen by each stator, the power factor under which stator A operates differs from that of stator B. Hence, when either the current or voltage transformer phase angle errors differ widely, calculation of correction factors for individual stators may be advisable. Differences in signs particularly may lead to unsuspected errors.

In the following example, although values of β and γ would average to zero, signs have been applied to give maximum error.

Given the following conditions:

Three-phase power factor	=	0.866 (30 degrees lagging), balanced load
Combined RCF, stator A	=	0.997
Combined RCF, stator B	=	1.001
β, stator A	=	–12 minutes
β, stator B	=	+ 12 minutes
γ, stator A	=	+ 12 minutes
γ, stator B	=	–12 minutes

Secondary Meter Registration =

$$EI \times RCF_A \times \cos(\theta_2 - 30° + \beta_A - \gamma_A) + EI \times RCF_B \times \cos(\theta_2 + 30° + \beta_B - \gamma_B)$$

With the transformer errors listed above, and at 5 amperes, 120V secondary:

$$\begin{aligned}\text{Secondary Meter Registration} &= 120 \times 5 \times 0.997 \times \cos[30° - 30° + (-12') - (+12')] \\ &\quad + 120 \times 5 \times 1.001 \times \cos[30° + 30° + 12' - (-12')] \\ &= 598.2 \cos 24' + 600.6 \cos 60°\,24' \\ &= 598.2 + 296.6 \\ &= 894.8\end{aligned}$$

$$\text{True Power} = \sqrt{3} \times 120 \times 5 \times \cos 30^\circ = 900$$

$$FCF = \frac{900}{894.8} = 1.0058$$

Cosines have been used in this calculation to make clear the phase angle errors possible. Similar results may be obtained by use of PACF, Tables 11-2 and 11-3.

BURDEN CALCULATIONS

Voltage Transformer Burdens

The secondary burdens of voltage transformers are connected in parallel across the secondary of the transformer. The volt-ampere burden is equal to $\frac{E^2}{Z}$ where Z is the impedance.

Usually voltage transformer burdens are expressed as volt-amperes at a given power factor. To calculate the total burden on the secondary of a voltage transformer, the burden of each device should be divided into in-phase and quadrature-phase components and added. (Volt-amperes cannot be added directly unless they are all at the same power factor.)

The in-phase component is:

$$\text{Watts} = (\text{Volt-Amperes})(\text{Cos } \theta) = VA \times PF$$

The quadrature-phase component is:

$$\text{Vars} = (\text{Volt-Amperes})(\text{Sin } \theta) = VA\sqrt{(1 - PF)^2}$$

$$\text{Total Volt-Amperes} = \sqrt{(\text{Total Watts})^2 + (\text{Total Vars})^2}$$

$$\text{Power factor of combined burden} = \text{Cos } \theta = \frac{\text{Total Watts}}{\text{Total Volt-Amperes}}$$

Current Transformer Burdens

When more than two instruments or meters together with the required wiring are connected in series with the secondary of a current transformer, the total burden impedance is:

$$\text{Total Burden Impedance } (Z) = \sqrt{(\text{Sum of resistances})^2 + (\text{Sum of reactances})^2}$$

The volt-ampere burden on a current transformer is equal to I^2Z. Burdens are usually computed at 5-amperes rated secondary current. It may be necessary to convert burdens stated in meter manuals at 2.5 amperes to 5.0 amperes.

Where the burdens are expressed in volt-amperes at a given power factor, the burden of each device and the secondary conductors should be divided into in-phase and quadrature-phase components, and added.

The in-phase component is:

$$\text{Watts} = (\text{Volt-Amperes})(\text{Cos } \theta) = VA \times PF = I^2R$$

The quadrature-phase component is:

$$\text{Vars} = (\text{Volt-Amperes})(\text{Sin } \theta) = VA\sqrt{(1 - PF)^2} = I^2X$$

Where X is the inductive reactance, $X = 2\pi fL$, L is the inductance in henries, and f is the frequency in hertz.

$$\text{Total Volt-Amperes} = \sqrt{(\text{Total Watts})^2 + (\text{Total Vars})^2}$$

$$\text{Power Factor of Combined Burden} = \text{Cos } \theta = \frac{\text{Total Watts}}{\text{Total Volt-Amperes}} = \frac{R}{Z}$$

It should be particularly noted that the secondary lead resistance must be included in the burden calculations for current transformers.

The basic formulas for burden calculations are summarized in convenient form in Table 11-11 for both current and voltage transformers.

Polyphase Burdens

Where the secondary burdens of instrument transformers are interconnected, as is often the case in polyphase metering, no simple method of computing the burdens on each transformer is applicable in all cases. Such combinations of burden must be computed phasorially on the basis of the actual circuit.

For wye-connected burdens on wye-connected instrument transformers, each transformer is affected by the burden directly across its terminals from the polarity to the neutral secondary leads. Thus, each transformer "sees" only the burden on its own phase and burdens are easily calculated. The same situation is true for an open-delta burden on transformers connected open delta. These are the normal arrangements for metering burdens.

Unusual cases, such as wye-connected burdens on open-delta-connected instrument transformers, delta-connected burdens on wye-connected instrument transformers, and complex combinations of single-phase and three-phase burdens must be analyzed individually. Since such analysis is complex, this type of burden should be avoided in metering applications where possible.

The Circle, or Farber Method, for Determination of Voltage Transformer Accuracy

The accuracy of a voltage transformer is primarily affected by the burden connected to the secondary of the transformer. This burden is usually expressed in terms of volt-amperes and percent power factor.

Table 11-11. Methods of Expressing Burdens of Instrument Transformers

	Conversion to Watts and Vars	
Voltage Transformers		
Burden Expression	*Watts at 120 V*	*Vars at 120 V, 60 Hz*
VA = Volt-Amperes at 120 V, 60 Hz PF = Burden power factor	$VA \times$ PF	$VA \times \sqrt{(1 - PF)^2}$
Current Transformers		
Burden Expression	*Watts at 5 A*	*Vars at 5 A, 60 Hz*
R = Resistance in ohms L = Inductance in millihenries	$25 \times R$	$9.43 \times L$
Z = Inductance in ohms, 60 Hz PF = Burden power factor	$25 \times Z \times PF$	$25 \times Z \times \sqrt{(1 - PF)^2}$
VA = Volt-Amperes at 5 A, 60 Hz PF = Burden power factor	$VA \times PF$	$VA \times \sqrt{(1 - PF)^2}$

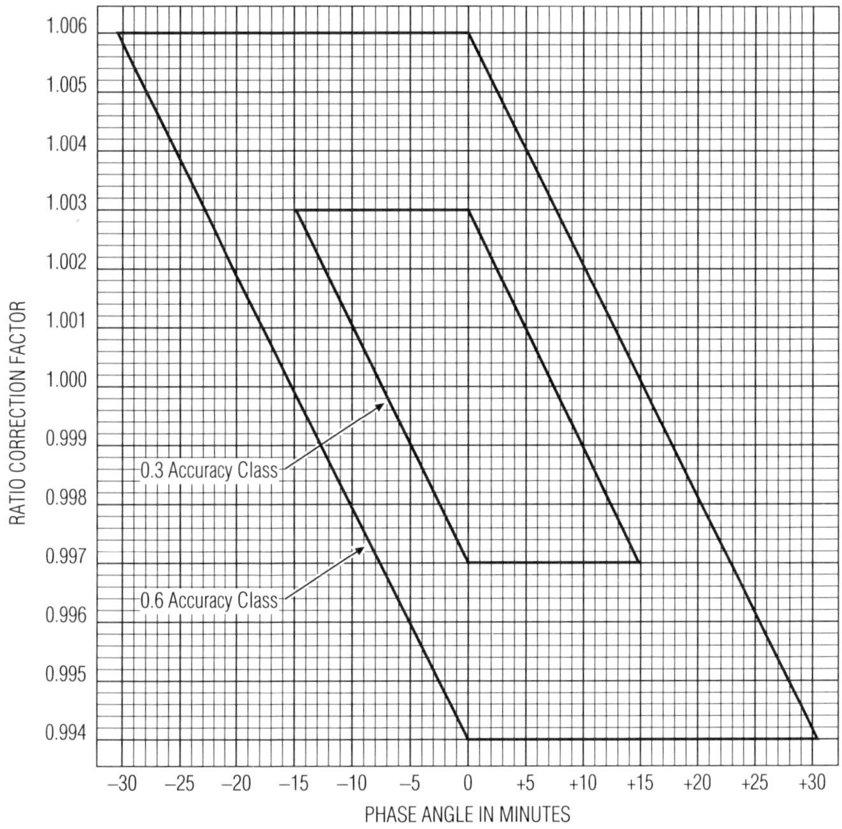

Figure 11-15(a). Sample of Graph Paper Specifically Scaled for Voltage Transformer Circle Diagram.

The circle, or Farber Method, copyrighted 1960 by Westinghouse Electric Corp., provides an easy method for determining the accuracy of a voltage transformer at any desired burden by using only the phase angle and ratio correction factor of the transformer at zero burden and one other known burden. Normally the manufacturer furnishes this information with the transformer.

Graph paper is scaled so that a given distance represents 0.0010 units on the RCF axis and 3.438 minutes on the phase angle. A sample is shown in Fig. 11-15(a).

An example of the use of the circle method is shown in Fig. 11-15(b). The following data are test results at 120 secondary volts for a 2400:120 volt voltage transformer: At 0 volt-ampere burden, RCF = 0.9979 and phase angle = +2.0 minutes; At 50 volt-amperes, 85 percent power-factor burden, RCF = 1.0040 and phase angle = +1.0 minute. Performance at other volt-ampere and power-factor burdens can be plotted by making radii proportional to volt-amperes and angles equal to burden power-factor angles.

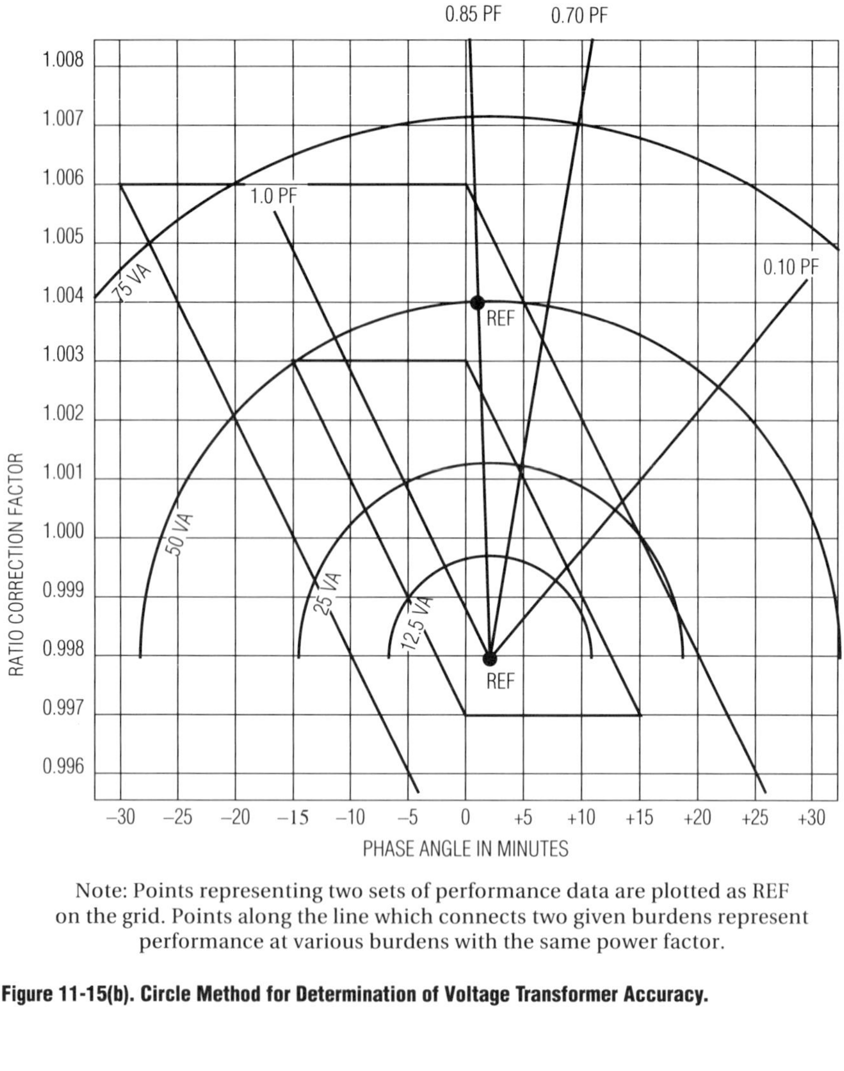

Note: Points representing two sets of performance data are plotted as REF on the grid. Points along the line which connects two given burdens represent performance at various burdens with the same power factor.

Figure 11-15(b). Circle Method for Determination of Voltage Transformer Accuracy.

The circle, or Farber Method, is a graphical method in which volt-amperes are represented by arcs, and percent power factors by straight lines which are plotted on a special graph paper that has *RCF* as the vertical axis and phase angle as the horizontal axis. The special graph paper and protractor required are available from Westinghouse Electric Corp. together with instructions for their use.

ANSI ACCURACY CLASSES FOR METERING SERVICE

The accuracy classifications of instrument transformers for metering service are based on the requirement that the transformer correction factor, *TCF*, shall be within the stated limits over a specified range of power factor of the metered load and with specified secondary burdens. Note that the requirement is in terms of

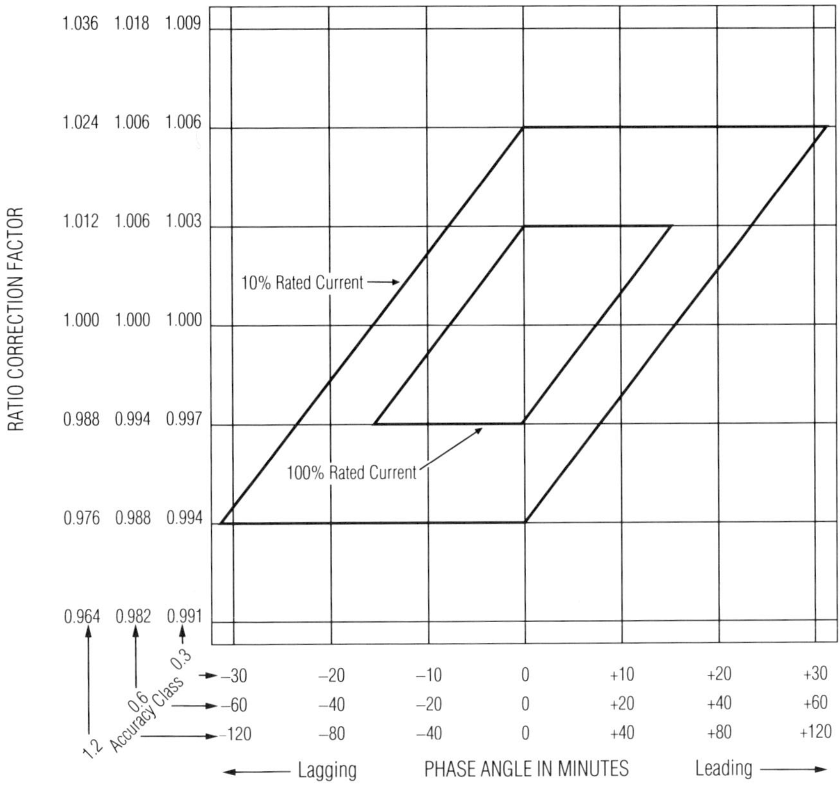

Note: The accuracy requirements for 100 percent rated current also apply at the continuous-thermal-current rating of the transformer.

Figure 11-16(a). Parallelograms Showing Graphical Equivalent of ANSI Accuracy Classes 0.3, 0.6, and 1.2 for Current Transformers for Metering Service.

the transformer correction factor, rather than in terms of either of its components, ratio correction factor or phase angle correction factor. Since at 1.0 power factor the *PACF* is insignificant, the *TCF* is equal to *RCF*. The *PACF* is limited to values and direction (±) such that its effect on the *TCF* does not cause the latter to exceed the limits of its stated class at power factors other than unity. Hence, transformer accuracy classes can best be shown by parallelograms as is done in Fig. 11-16(a) for current transformers and Fig. 11-16(b) for voltage transformers. Note that the inclination of the accuracy class parallelogram for voltage transformers is opposite to that for current transformers. It may also be noted that the current transformer allowable TCF at 10 percent current is double that at 100 percent current.

It has been shown that an instrument transformer correction factor is not a constant but depends on the secondary burden. Hence, the accuracy class is designated by the limiting percent error caused by the transformer followed by the standard burden designation at which the transformer accuracy is determined. For instance, for a current transformer the accuracy class may be written:

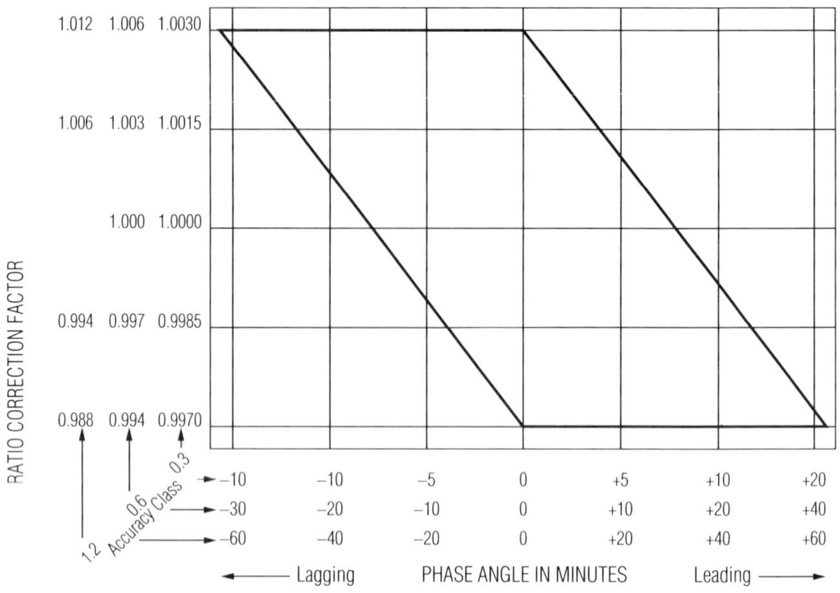

Note: The transformer characteristics shall lie within the limits of the parallelogram for all voltages between 90 percent and 110 percent of rated voltage.

Figure 11-16(b). Parallelograms Showing Graphical Equivalent of ANSI Accuracy Classes 0.3, 0.6, and 1.2 for Voltage Transformers for Metering Service.

0.3 *B*-0.5, 0.6 *B*1.8. This means that at burden *B*-0.5 the transformer would not affect the meter accuracy more than ± 0.3 percent at 100 percent rated current or ±0.6 percent at 10 percent rated current, and at burden *B*1.8 the transformer would not affect the meter accuracy more than ± 0.6 percent at 100 percent rated current or ± 1.2 percent at 10 percent rated current, when the power factor of the metered load is between 0.6 and 1.0 lagging.

Likewise the accuracy of a voltage transformer could be given as 0.3 *X*, 0.3 *Y*, 1.2 *Z*, with similar meanings. Accuracy classes of voltage and current transformers are shown in Tables 11-12 and 11-14.

Table 11-12. ANSI Accuracy Classes for Voltage Transformers.

Accuracy Class	Limits of Ratio Correction Factor and Transformer Correction Factor	Limits of Power Factor (Lagging) of Metered Power Load
1.2	1.012-0.998	0.6-1.0
0.6	1.006-0.994	0.6-1.0
0.3	1.003-0.997	0.6-1.0

Table 11-13. ANSI Standard Burdens for Voltage Transformers.

Burden	Volt-Amperes	Burden Power Factor
W	12.5	0.10
X	25.	0.70
M	35.	0.20
Y	75.	0.85
Z	200.	0.85
ZZ	400.	0.85

Table 11-14. ANSI Accuracy Classes for Metering Current Transformers.

Accuracy Class	Limits of Ratio Correction Factor and Transformer Correction Factor: 100% Rated Current Minimum	100% Rated Current Maximum	10% Rated Current Minimum	10% Rated Current Maximum	Limits of Power Factor (Lagging) of Metered Power Load
1.2	0.988	1.012	0.976	1.024	0.6-1.0
0.6	0.994	1.006	0.988	1.012	0.6-1.0
0.3	0.997	1.003	0.994	1.006	0.6-1.0

Table 11-15. ANSI Standard Burdens for Current Transformers with 5 Ampere* Secondaries.

Burden Designation**	Resistance (Ohms)	Inductance (mH)	Impedance (Ohms)	Volt-Amperes (at 5 A)	Power Factor
		Metering Burdens			
B-0.1	0.09	0.116	0.1	2.5	0.9
B-0.2	0.18	0.232	0.2	5.0	0.9
B-0.5	0.45	0.580	0.5	12.5	0.9
B-0.9	0.81	1.04	0.9	22.5	0.9
B-1.8	1.62	2.08	1.8	45.0	0.9
		Relaying Burdens			
B-1	0.5	2.3	1.0	25	0.5
B-2	1.0	4.6	2.0	50	0.5
B-4	2.0	9.2	4.0	100	0.5
B-8	4.0	18.4	8.0	200	0.5

* If a current transformer is rated at other than 5 amperes, ohmic burdens for specification and rating may be derived by multiplying the resistance and inductance in the table by $[5/(\text{ampere rating})]^2$, the VA at rated current and the power factor remaining the same.
** These standard burden designations have no significance at frequencies other than 60 Hz.

The standard burdens for both voltage and current transformers are precisely defined by ANSI. Standard burdens and characteristics are given in Tables 11-13 and 11-15.

The use of the ANSI accuracy classifications permits the installation of instrument transformers with reasonable assurance that errors will be held within known limits provided that burden limitations are strictly adhered to and that secondary connections introduce no additional error.

TYPES OF INSTRUMENT TRANSFORMERS

General Types

Indoor

An indoor transformer is constructed for installations where it is not exposed to the weather. This construction is generally limited to circuits of 15,000 volts (110 kV BIL) or less.

Outdoor

An outdoor instrument transformer is constructed for installations exposed to the weather.

Indoor-Outdoor

For circuits rated at 24 kV or above, common designs of instrument transformers are suitable for either indoor or outdoor use. At lower voltages, units for outdoor use are provided with additional protection, particularly against moisture. Spacings between high-voltage terminals, and between these terminals and ground, are generally increased for outdoor types.

Types of Insulation

Both liquid-filled and dry-type transformer construction are used for transformers rated 69 kV (350 kV BIL) and lower. Above 69 kV, liquid-filled and gas-filled (SF-6) construction transformers are used.

Dry-Type

The core and coils are embedded in a body of material which serves as the insulation, case, and the bushings. This construction is usually employed for individual current or voltage transformers.

The materials used to insulate and furnish mechanical construction are generally butyl rubber, epoxies, and thermoplastic elastomers (TPE or EPDM). Other plastics may be used, but their application is not widespread. These materials are adaptable to molding of the transformer.

Dry-type construction, in which the core and the coils are wrapped in layers of insulating paper or material and then impregnated with an asphaltic or similar compound, has not been popular since the introduction of molded transformers.

Compound-Filled

In the compound-filled construction the core and coils are wrapped and impregnated in the same manner as for the dry-type construction. The element is then mounted in the case and the case filled with a compound which has a high dielectric strength. These units are also designed for voltages not exceeding 15 kV.

Liquid-Filled

In the liquid-filled construction, the core and coils are insulated and then mounted in the tank which is filled with the insulating liquid. This construction is common for three-phase metering units below 69 kV.

The higher voltage transformers often have two windings, a secondary and a tertiary, on a common core. Sometimes one or both windings have a tap providing other phase angle voltage sources. The maximum (VA) burden capability may depend mostly on the core and therefore is the maximum of both windings. Sometimes the burden capability and accuracy are different or they may provide an even proportion of the total VA capability.

Gas-Filled

Sulfur hexafluoride gas (SF-6) is used to insulate the core and coils of instrument transformers in the higher voltage ranges (above 69 kV). Insulating gas is used instead of liquid insulation and requires special coil insulation and other precautions. The gas eliminates paper and oil which have a tendency to explode violently in the event of an internal fault. Their use is generally limited to installations where space is at a premium.

Butyl and Epoxy

The butyl-rubber or epoxy-resin transformers have the core and coils embedded in a body of butyl rubber or epoxy resin which serves as insulation, case, and bushings.

Voltage Transformers

Voltage transformers are made in a number of methods of winding. They are usually wound for single ratio but may also be arranged for two ratios.

For special purposes, taps may be taken off at various points on the secondary winding. While these taps are usually marked by a tag or otherwise, great care should be used in connecting such a transformer to be sure that the proper tap is used.

Autotransformers

An autotransformer is one having only one coil, this having taps brought out at the proper points in the coil to give the voltages desired. Any portion of the coil may be used as the line-voltage connection and any other portion as the load connection. The ratio of such a transformer is approximately:

$$\frac{\text{Line voltage}}{\text{Load voltage}} = \frac{\text{number of turns used for line winding}}{\text{number of turns used for load winding}}$$

Autotransformers are used for special purposes, as in the phasing transformers used with varhour meters.

Current Transformers

Wound (Wound Primary) Type

This type has the primary and secondary windings completely insulated and permanently assembled on the core. The primary is usually a multi-turn winding.

Three-Wire Transformers

The primary winding is in two equal sections, each of which is insulated from the other and from ground so that the transformer can be used for measuring total power in the conventional three-wire, single-phase power service. Three-wire transformers are used on low voltage only since it is difficult to provide the necessary insulation between the two primary windings. Two two-wire current transformers are commonly used for three-wire metering.

Window-Type

This type is similar in construction to the wound type except that the primary winding is omitted and an opening is provided through the core through which a bus or primary conductor may be passed to serve as the primary winding. Complete insulation for such primary is not always provided by the transformer.

By looping the primary conductor through the core, a number of different ratios may be obtained. For instance, if a transformer had a ratio of 1,200:5 or 240:1 with a single turn, it would have a ratio of 120:1 with two turns, 80:1 with three turns, etc. In other words, the ratio with any number of turns would be:

$$\text{Ratio} = \frac{\text{Original Ratio}}{\text{Turns}}$$

The number of turns in the primary is the number of times the conductor passes through the hole in the core and not the number of times the cable passes some point on the outside.

Bar-Type

The bar-type is similar to the window type but with an insulated primary provided. The bar in bar-type may be removable or fixed in place.

Window-Type as a Three-Wire

This is done by passing one wire of a three-wire, single-phase service through the window in one direction and the other line through in the other direction. The ratio of the current transformer would be one-half the marked ratio.

Double Ratio

A current transformer may have the primary divided into two sections; such a transformer may be used as a double-ratio transformer. The principal advantage of such a transformer is obtained in high-voltage metering where the capacity of the meter may be increased or decreased without changing the current transformer. A 200/400-ampere transformer, for instance, has a ratio of 200:5 when the primary coils are connected in series and 400:5 when connected in multiple.

Split-Core-Type

This type has a secondary winding completely insulated and permanently assembled on the core but has no primary winding. It may or may not have insulation for a primary winding. Part of the core is separable or hinged to permit its encircling a primary conductor or an uninsulated conductor operating at a voltage within the voltage rating of the transformer.

The exciting current of this type of current transformer may be relatively large and the losses and, consequently, the ratio error and phase angle may also be relatively large.

Double or Triple Secondary

When it is necessary or desirable to operate two or three separate burdens from a single current transformer, it is necessary to supply a complete secondary winding and magnetic circuit for each burden, the individual magnetic circuits being linked by a common primary winding. A double-secondary current transformer has two separate and complete magnetic circuits with a complete secondary winding for each and one common primary winding.

Each secondary function is entirely independent of the other one.

Small Transformers

These transformers are exceptionally small, not larger than a 4-inch cube, for use in metering low-voltage circuits. The typical continuous current rating factors are 1, 2, 3, and 4 times the nominal rating. Each type and ratio of CT may have a different rating factor.

With rated current in the primary, the open secondary voltage is low and may be considered non-hazardous. Current transformers should never be open-circuited, even through small CTs may not develop enough voltage before they saturate and damage themselves. An open circuit can cause the CT to be magnetized, and in larger units can cause a violent explosion as well as voltages in excess of 2500 volts.

Bushing-Type

This type has a secondary winding completely insulated and permanently assembled on a ring-type core but has no primary winding or insulation for a primary winding. The circuit breaker or power transformer bushing with its conductor or stud becomes the completely insulated single-turn primary winding of the bushing-type current transformer.

For metering application, considerable improvement in accuracy over the range of primary current is obtained by using special core materials and compensated secondary windings. Bushing-type current transformers, at ratings above 100 amperes, may have accuracies within acceptable revenue metering limits. The burden capability of these meters is directly related to core cross-section, and inversely related to ratio.

SELECTION AND APPLICATION OF INSTRUMENT TRANSFORMERS

Before specifying instrument transformers for any installation, the characteristics of the transformers must be taken into account to make sure that the units proposed meet all requirements. Certain types of installations present no unusual features and standard units may be specified; others require careful study before final decision is made. For detailed specifications, see ANSI C57.13.

Voltage Transformers

Basic Impulse Insulation Level (BIL)

The BIL rating of a voltage transformer indicates the factory dielectric test that the transformer insulation is capable of withstanding. ANSI C57.13 gives the dielectric test values and minimum creepage distances associated with each BIL, as well as the appropriate BIL level for each primary voltage rating and conditions for transformer application. In a wye system, with voltage transformers connected line to grounded neutral, the transformer may be subjected to 1.73 times normal voltage during a ground fault. Hence the distinction between the various groups must be maintained to avoid over-stressing transformer insulation under such conditions.

Insulation must be de-rated when transformers are installed at altitudes greater than 3300 feet above sea level. See ANSI C57.13.

The BIL level of voltage transformers should be coordinated with associated equipment. For instance, in a substation with a BIL level of 110 kV, it is considered poor practice to use voltage transformers of 95 kV BIL. When deciding on the insulation level to be used, questions such as whether the power get-aways are overhead or underground and adequacy of lightning arrester protection should be considered. For long 2400 volt lines in rural areas where lightning storms are frequent, some companies require voltage transformers rated at 75 kV BIL instead of the standard 60 kV.

Burden Capability

It must be remembered that whether the transformer remains within its accuracy class depends upon the burden connected. The accuracy burden capability is not the same as the thermal burden rating. The thermal burden rating of a voltage transformer is the volt-amperes which the transformer will carry continuously at rated voltage and frequency without causing the specified temperature limits to be exceeded. It has practically nothing to do with the burden ratings at which accuracies are established.

Current Transformers

Continuous Thermal Rating Factor

Current transformers may carry a rating factor of 1.0, 1.33, 1.5, 2.0, 3.0, or 4.0. This means that their nameplate current rating may be multiplied by the rating factor applicable to give the maximum current the transformers can carry continuously in an ambient temperature not exceeding 30° C. High-voltage current transformers generally have a rating factor of 1.0 or 1.5. For altitudes above 3,300 feet (1,000 meters) or 24-hour temperatures appreciably different from 30° C, refer to ANSI/IEEE C57.13 Section 4.4.

Most modern small current transformers have a continuous thermal rating factor of 1.5 or higher.

ANSI/IEEE accuracy classifications for current transformers apply throughout the current range defined by the continuous current thermal rating factor.

Basic Impulse Insulation Level (BIL)

The BIL is a useful guide in selecting current transformers for installation in critical locations. Current transformers should not be rated at a lower level than the other station or service equipment.

Short-Time Ratings—Thermal

The thermal short-time current rating of a current transformer is the rms, symmetrical primary current that may be carried for a stated period, 5 seconds or less, with the secondary winding short-circuited, without exceeding a maximum specified temperature in any winding.

The thermal rating, as designated for current transformers, is the factor by which the primary current is multiplied to obtain the maximum allowable primary current without exceeding temperature rise standards and accuracy requirements. For example, if a 400:5 CT has a TRF of 4.0, the CT will continuously accept 400 ×

4 = 1,600 primary amperes with 5 × 4 = 20 amperes from the secondary. The thermal burden rating of a voltage transformer is specified in terms of the maximum burden in volt-amperes that the transformer can carry at rated secondary voltage without exceeding a given temperature rise.

Short-Time Ratings—Mechanical

This rating is a number, which indicates how many times normal primary current, the current transformer can conduct for one second, without failure. The mechanical failure possible is the distortion of the primary winding. Hence, the bar-type or through-type has practically unlimited mechanical rating.

Where indoor current transformers are in locations critical to public safety, it is sometimes necessary to use a higher rated transformer than the circuit calls for, to obtain the necessary mechanical and thermal short-time ratings. Both of these short-time ratings should be matched to possible fault currents in the circuit.

Relay Applications

For relaying applications, current transformers must meet requirements that greatly differ from those of metering. Since relays operate under abnormal conditions, high-current characteristics are important.

Current transformers for relaying service are given accuracy class ratings by letters and numbers, such as T200, which describe their capabilities up to twenty times normal current rating. The standard relay accuracy indicates the RCF will not exceed the 10% accuracy limits at loads from 0 to 20 times nominal, and at rated burden. The letters T and C indicate whether performance is based on tests (T) or calculated (C). For example, a 5 amperes, T200 current transformer, at 20 times normal current rating, has a 2-ohm burden and 50 VA.

$$\begin{aligned} \text{Maximum secondary voltage} &= 20 \times 5 \text{ amperes} \times \text{impedance} \\ \text{T200 or 200 volts} &= 100 \text{ amperes} \times \text{impedance} \\ \text{Impedance} &= 2 \text{ ohms} \end{aligned}$$

Using I^2R, with 5 amperes a 2-ohm burden = 50 VA.

Transformer Installation Procedures

Usually each utility develops its own standard for metering installations, based on local requirements and the type of test facilities desired. These local standards are published and made available for installation guidance. In this *Handbook*, it is not intended to describe the many practices followed for physically mounting and locating instrument transformers.

Current Transformer Secondaries

The secondaries of current transformers should be kept shorted during storage and installation until the secondary leads and burden have been connected. This is to avoid the dangers of high voltage that could occur on an open secondary if the primary were energized. Some utilities make the shorting of secondaries a rule for all current transformers. Some utilities have relaxed this rule for the small current transformers, as these will saturate before the secondary voltage reaches a dangerous value. (See discussion under a subsequent subheading, "Wire Tracing with Instruments.")

Precautions in Routing Secondary Leads

The secondary leads for a set of current or voltage transformers comprising one metering installation should be routed to avoid the pick-up of induced voltages from other conductors. Such induced voltages could cause errors in the metering.

The effects of induced voltages can be reduced by running the secondary leads in a group as a cable or together in a single conduit. In addition, the leads should be kept well away from other conductors carrying heavy current and should not be run in the same conduit with such conductors. Cabling will reduce the effects of stray fields by a partial cancellation of the induced voltages. Iron conduit will provide some magnetic shielding against stray fields. The polarity and neutral secondary wires from a given instrument transformer should never be run in different conduits or by different routes. This could produce a loop that would be sensitive to induced voltages.

INSTRUMENT TRANSFORMER CONNECTIONS

Voltage Transformers

Single-Phase Circuits

Figure 11-17 shows the connection for one voltage transformer supplying single-phase voltage to the potential element of a meter. Note standard polarity designations H_1 and H_2 for the primary and X_1 and X_2 for the secondary. Note that the non-polarity secondary lead X_2 is grounded at the transformer. The numbers at the meter terminals show the secondary voltage corresponding to the original line voltage 1-2.

Three-Phase, Three-Wire, Delta, Three-Wire Secondary

Figure 11-18 shows the connection which is most commonly used for three-wire delta polyphase metering. Note that both the primary and secondary of the transformer across lines 2 and 3 have been reversed. This is the usual practice as it avoids a physical crossover of the high-voltage jumper between the transform-

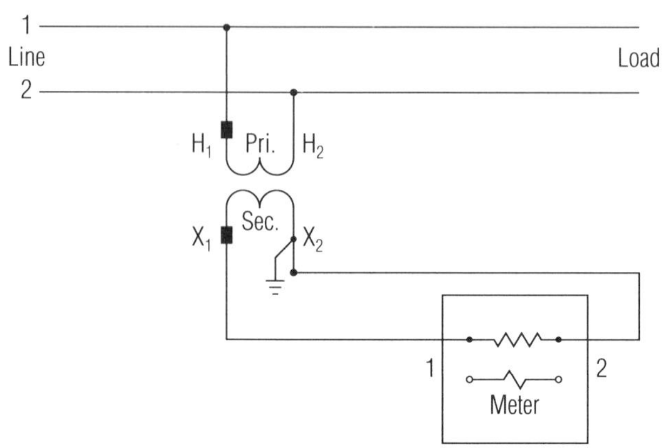

Figure 11-17. Single-Phase.

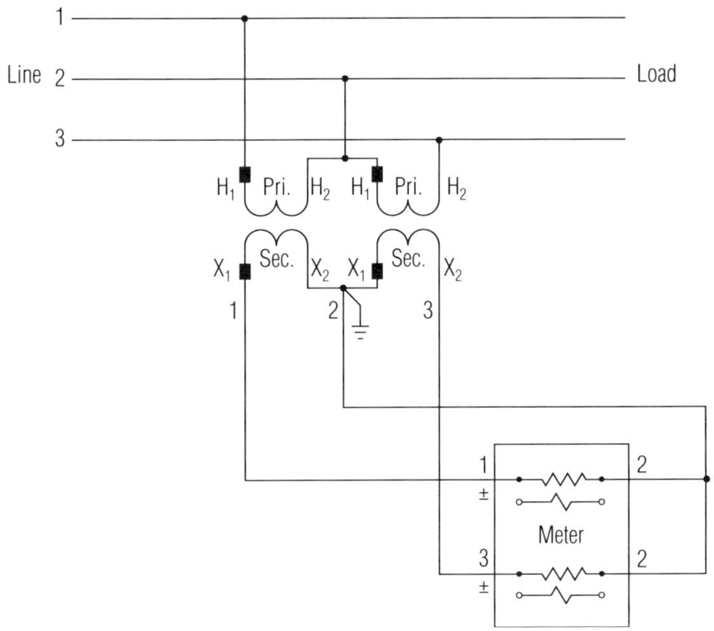

Figure 11-18. Three-Phase, Three-Wire Open-Delta, Three-Wire Secondary.

ers. The two adjacent high-voltage bushings of the two transformers are tied together and to line 2. The secondaries are likewise tied together and grounded at the transformer.

The number 2 secondary lead is common to both transformers and carries the phasor sum of the currents drawn by coils 1-2 and 3-2. This leads to some difficulty in very precise metering, particularly if long secondary leads are used, as it is difficult to calculate the exact effect of this common lead resistance. (See previous subsections of this chapter for effect of lead resistance.) Generally the common lead will not produce any significant error for commercial metering and saves one wire.

Three-Phase, Three-Wire, Delta with Four-Wire Secondary

Where long secondary leads are used and where correction factors are to be applied, the four-wire secondary shown in Fig. 11-19 is preferred.

Three-Phase, Wye-Wye, Four-Wire Secondary

Figure 11-20 shows this connection which is the usual one for a four-wire primary system. Here the neutral secondary wire again carries the phasor sum of the burden currents, but for a balanced voltage and burden this sum is zero. Hence, single-phase tests may be made using the lead resistance of a single lead.

Grounding—Primary

The primary of a voltage transformer is not normally grounded independent of the system.

Grounding—Secondary

It is standard practice to ground the common or neutral secondary wire or wires at the transformer. Figures 11-17 through 11-20 show the normal grounding. Secondary grounding is necessary for safety to prevent a high static potential in secondary leads and as a safeguard in case of insulation failure which could cause high voltage to appear on the secondary leads. There should be only one ground on the circuit, preferably at the transformer. Additional grounds should be avoided due to the indeterminate resistance and voltage gradients in the parallel ground path.

Grounding—Cases

Transformer cases normally should be grounded for safety from static potential or insulation failure. In overhead construction, grounding may be prohibited by local regulations in order to keep overhead fault potentials away from sidewalks or streets. For safety to operators, any standard on grounding voltage transformer cases must be adhered to, as the operators depend on the fact that these cases are either grounded or isolated without exception.

Primary Fuses

The use of primary fuses on a voltage transformer is a highly controversial subject. The primary fuse should protect the transformer from damage due to high-voltage surges and the system from an outage due to failure of the transformer. To accomplish this purpose, the fuse must be of very small current rating, as the

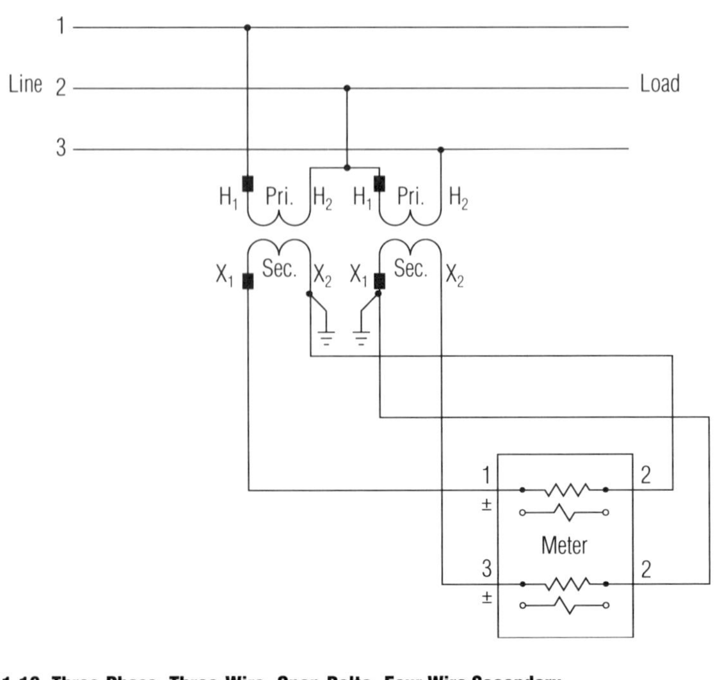

Figure 11-19. Three-Phase, Three-Wire, Open-Delta, Four-Wire Secondary.

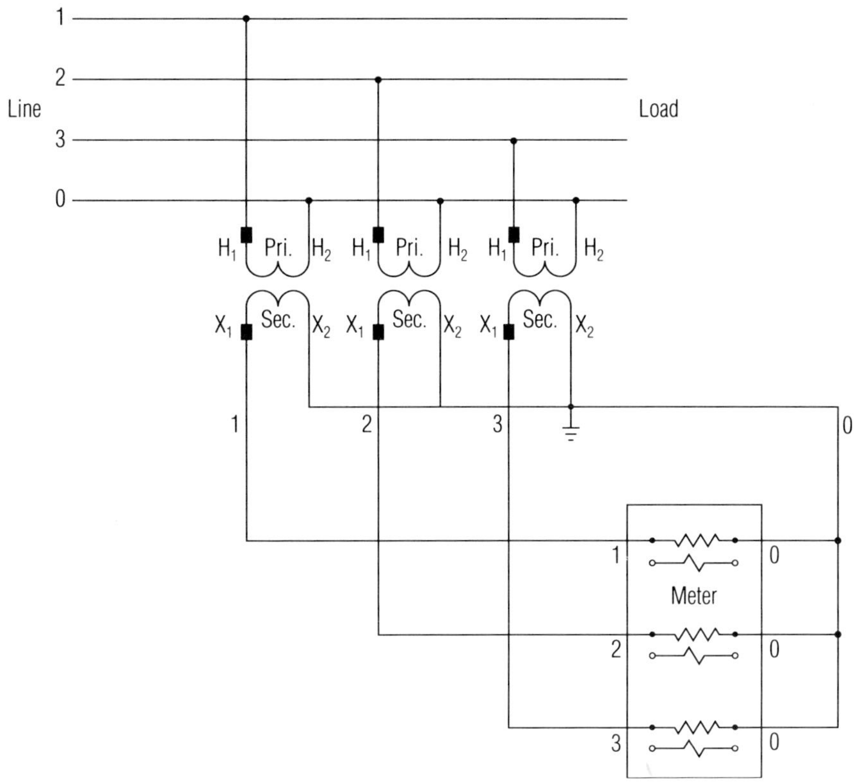

Figure 11-20. Three-Phase, Four-Wire, Wye-Wye, Four-Wire Secondary.

normal primary current of a voltage transformer is exceedingly small. A suitable primary fuse for this application has appreciable resistance which may cause errors in the over-all ratio and phase angle measurements. In addition, such a small fuse may be mechanically weak and may fail due to aging without any transformer failure.

If a primary fuse opens for any reason, the load will be incorrectly metered until the fuse is replaced.

Many companies rely on their circuit protective equipment without the additional fusing of the voltage transformer primaries. If failure should then occur in a voltage transformer, the protective relays function to clear the circuit. In this case the customer would suffer an outage.

Secondary Fuses

The secondary leads of a voltage transformer are often fused. The secondary fuses protect the transformers from short circuits in the secondary wiring. Fuses and fuse clips, however, may introduce sufficient resistance in the circuit to cause metering errors. Where corrosion is present, this effect may become serious. A voltage transformer is not normally subject to overload, as its metering burden is fixed at a value far below the thermal capacity of the transformer. Hence, the only

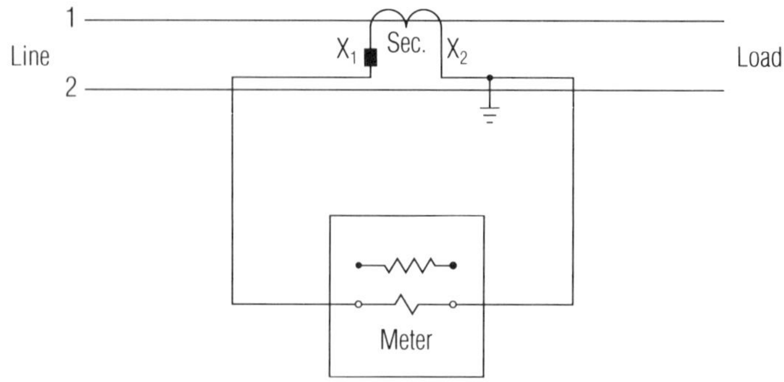

Figure 11-21. Two-Wire, Single-Phase.

value of the fuses is short-circuit protection. Practically, the only chance of short circuit is during test procedures and normally the transformer could stand a momentary short without damage. Where voltage transformers are used for both metering and relay service, an accidental short will operate the relays and cause an interruption. In such cases, the metering circuit can be fused after the division for relaying and for metering have been made.

Effect of Secondary Lead Resistance, Length and Size of Leads

The effect of secondary lead resistance in a voltage transformer circuit is to cause a voltage drop in the leads so that the voltage at the meter is less than the voltage at the terminals of the transformer. (See discussion of "Effects of Secondary Lead Resistance on the Ratio and Phase Angle as Seen by the Meter.")

When the lead resistance exceeds a few tenths of an ohm, this voltage drop can cause errors equal to or greater than the errors due to ratio and phase angle of the transformer.

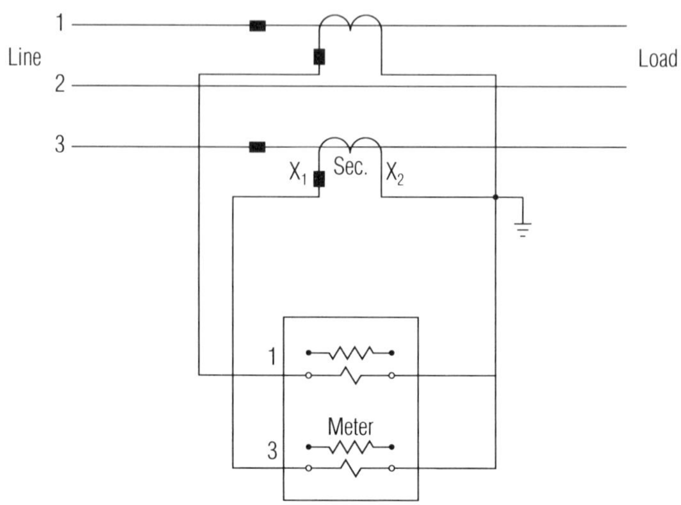

Figure 11-22. Three-Wire, Three-Phase, Three-Wire Secondary.

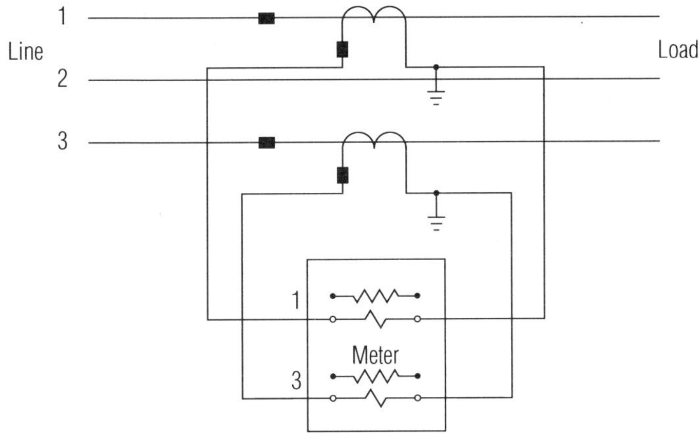

Fig. 11-23. Three-Wire, Three-Phase, Four-Wire Secondary.

Meters can be adjusted to compensate for these errors, but many companies object to upsetting meter calibrations to take care of secondary lead errors. They avoid this problem by limiting secondary leads to a run of not over 100 feet of No. 9 wire. Where greater distances are involved, they use either heavy secondary conductors or meters adjacent to the transformers with contact devices to transmit the intelligence to the station.

To avoid these troubles, the secondary leads to a voltage transformer should be as short as is reasonably possible. Where long leads must be used, the wire size should be increased to keep the lead resistance low. This might require secondary wires as large as No. 2 on a run of 1,000 feet. With normal watthour meter burdens,

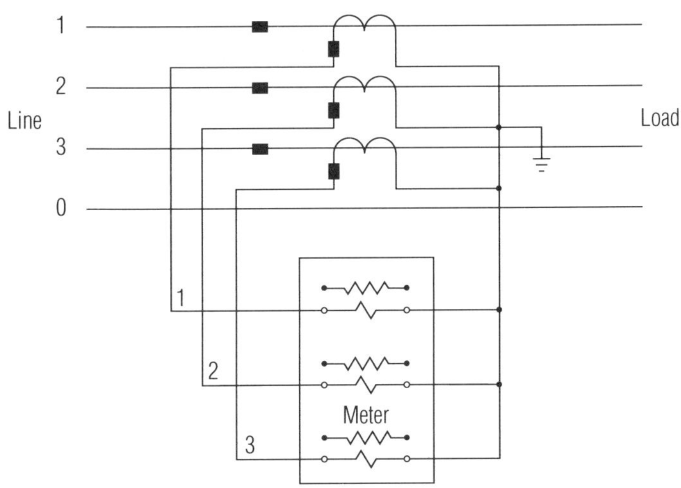

Figure 11-24. Four-Wire, Three-Phase, Four-Wire Secondary.

the error due to the leads will usually be within acceptable limits if the total lead resistance does not exceed 0.3 ohms. If the lead resistance is larger, or if heavy burdens are used, calculations should be made to determine if corrections are necessary.

Current Transformers

Two-Wire, Single-Phase

Figure 11-21 shows the connections for one current transformer supplying single-phase current to the current coil of a meter. Note the grounding of the non-polarity secondary lead at the transformer.

Three-Wire, Three-Phase

This connection is shown in Fig. 11-22. Note the grounding of the common connection at the transformer. The common lead carries the phasor sum of the secondary currents in each transformer. To avoid the problem of applying corrections for the common lead resistance, the connection shown in Fig. 11-23, where four secondary leads are shown is occasionally used.

Four-Wire, Three-Phase with Wye Connected Secondaries

This connection is shown in Fig. 11-24. Note the grounding of the common lead at the transformer. On a balanced load, the common lead carries no current.

Four-Wire, Three-Phase with Delta-Connected Secondaries

Figure 11-25 shows this connection which is sometimes used to provide three-wire metering from a four-wire system. It is often used for indicating and graphic meters and relays and sometimes for watthour metering. The metering is theoretically correct only at balanced voltages, but on modern power systems the

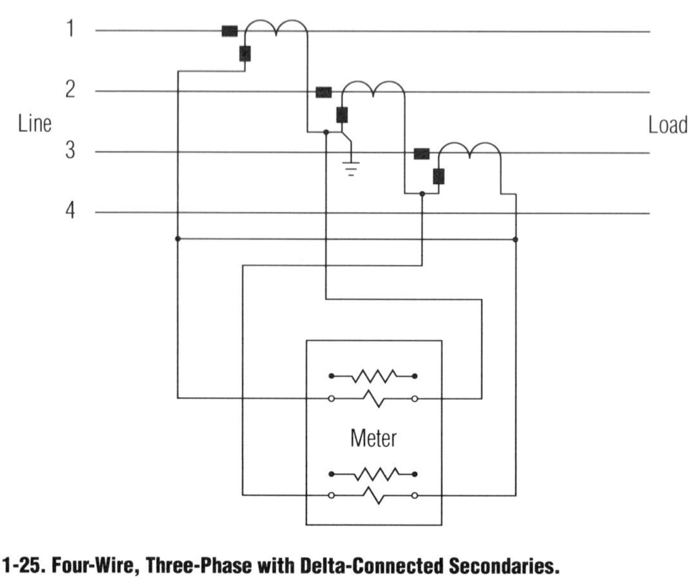

Figure 11-25. Four-Wire, Three-Phase with Delta-Connected Secondaries.

voltage is normally balanced well enough to give acceptable accuracy for billing metering. Note that with delta-connected current transformers, the secondary currents to the meter are displaced 30 degrees from the primary line currents and also increased by the square root of three ($\sqrt{3}$) in magnitude due to the phasor addition. This circuit is equivalent to the 2-1/2-stator meter used by some companies. It permits the use of a standard two-stator meter with none of the test complications that the 2-1/2-stator meter involves.

For connections of meters with instrument transformers, see Chapter 12, "Meter Wiring Diagrams."

Parallel Secondaries for Totalized Metering

The paralleling of current transformer secondaries for totalized metering is covered in Chapter 10, which should be referred to for the details and precautions involved in this method.

With the proper precautions, acceptable commercial metering may be obtained. Without proper consideration of all factors involved, the errors may be excessive, particularly at low current values.

Grounding

It is general practice to ground the non-polarity secondary lead of a current transformer at the transformer. Grounding is a necessary safety precaution for protection against static voltages and insulation failure. Normally, all metal cases of instrument transformers should be grounded. (Local regulations may prohibit such grounding in overhead construction.) There should be only one ground on the secondary circuit, preferably at the transformer. Where current transformer secondaries are connected in parallel, there should be only one ground for the set of current transformers and this should be at the point where the secondary leads are paralleled at the meter. Additional grounds should be avoided due to the indeterminate resistance and voltage gradients in the parallel ground path. On circuits of 250 volts or less, grounding of the current transformer secondary is sometimes not necessary.

Number of Secondary Wires

The use of common secondary wires has been discussed under the various connections. The resistance of a current transformer secondary lead adds to the burden, but unless this added resistance causes the total burden to exceed the burden rating of the transformer, it has a relatively small effect on the transformer accuracy. For most installations, the common lead is used because of the saving of wire. For very precise metering, separate return leads might be justified if the lead resistance is large.

Common Lead for Both Current and Voltage Transformers

The same common lead should never be used for both current and voltage transformers. Here the current in the common lead from the current transformer secondaries could cause a voltage drop that would affect the accuracy of the voltage transformer as seen at the meter. This effect would vary with the magnitude and balance of the secondary currents from the current transformers and no accurate compensations could be made for this effect. For this reason, separate common leads should be used for the current and voltage transformers.

Other connection systems are possible for special problems. Such connections must be analyzed in detail to be sure that they provide correct metering without significant errors.

VERIFICATION OF INSTRUMENT TRANSFORMER CONNECTIONS

When a metering installation using instrument transformers has been installed, it is necessary to verify the connections to insure correct metering. Wrong connections can cause large errors and may go undiscovered during a normal secondary or phantom-load test.

There is no single method of verifying instrument transformer connections that can be used with complete certainty for all possible installations. The best method will depend upon the nature of the particular installation, the facilities and instrumentation available, and the knowledge and ability of the tester. A combination of several methods may often be necessary or desirable. The following methods may be used to verify the instrument transformer connections.

Visual Wire Tracing and Inspection

A reasonably conclusive method of verification of instrument transformer connections is to actually trace each secondary wire from the instrument transformers to the meter. The terminal connections of each lead are checked to see that they conform to an approved meter connection diagram applicable to the installation. The use of color-coded secondary wire greatly facilitates this type of checking.

The primary connections to the instrument transformers must also be checked for conformity with the approved connection diagram. Particular attention must be paid to the relative polarity of the primary and secondary of all instrument transformers. Often some of the instrument transformers in an installation are connected with reversed primary polarity to facilitate a symmetrical primary construction and to avoid unnecessary cross-overs of the primary leads. These must be carefully noted to see that a corresponding reversal has been made at the secondary. If H_2 is used as the primary polarity terminal, then X_2 becomes the secondary polarity terminal.

All modern instrument transformers should have permanent and visible polarity markings. If the polarity is not clearly marked, the visual tracing method will be inconclusive and additional methods will be required.

The nominal ratio of all instrument transformers should be noted from the nameplate and checked against the ratio specified for the installation.

All meter test switches and devices should be checked for proper connection and operation. The installation should also be checked to see that proper secondary grounds have been installed.

If the wiring is sufficiently open to permit a complete visual check, this method is generally reliable, though it is subject to human error. If some of the wiring is concealed, this method can only be used if there is some means of identifying both ends of each concealed wire. The use of color-coded secondary wire makes such identification reasonably certain provided that the colors have not become unrecognizable through fading and that no concealed splices have been made. Where tags or wire markers are used, the reliability of the visual check depends upon the marking being correct.

Wire Tracing with Instruments

When the secondary wiring cannot be traced visually it may be traced electrically. Generally, the secondary windings of the current transformers may be shorted at the transformer terminals so that the secondary leads may safely be removed for test. The utmost precautions must be taken to assure that the secondary winding of a current transformer is never opened while the primary is energized, as dangerously high voltages can be induced in the secondary winding. This voltage is a lethal hazard to personnel and may also damage the current transformer.

The open-circuit voltage of a current transformer has a peaked wave form which can break down insulation in the current transformer or connected equipment. In addition, when the secondary is opened, the magnetic flux in the core rises to an abnormally high value which can cause a permanent change in the magnetic condition of the iron. This change can increase the ratio and phase angle errors of the current transformer. Demagnetization may not completely restore the transformer to its original condition. If the open circuit continues for some length of time, the insulation may be damaged by excess heating resulting from the greatly increased iron losses.

If the shorting of the secondary windings of the current transformers cannot be done with complete safety, then the primary circuits must be de-energized and made safe for work. All standard safety practices and company safety rules covering high-voltage work must be rigorously followed to insure the safety of personnel.

Once the secondary leads have been shorted, secondary leads may then be disconnected, one at a time, from the instrument transformers and the meter and checked out with an ohmmeter or other test device. Each lead is checked for continuity and to verify that it is electrically clear of all other leads and ground. The normal secondary grounds must be lifted for this test. When the leads are reconnected, care must be used to see that all connections are properly made and securely tightened. When the grounds are replaced, they should be tested to see that they properly ground the circuit. Only one lead at a time should be removed to avoid the possibility of a wrong reconnection.

If a good portable resistance bridge is available, the resistance of the secondary leads may be measured. This would check the possibility of poor connections or abnormally high resistance due to any cause.

Particular attention should be paid to all current transformer shorting devices to see that they work properly. If shorting clips in meter sockets are present, they should be tested to be sure that they open when the meter is installed. This type of verification is most conveniently done on a new installation before the service is energized.

When the service is already energized, this wire tracing method requiring the removal of wires from terminals may be impractical.

Interchanging Voltage Leads

This method can be used for a two-stator meter on a three-wire polyphase circuit. First, with normal connections, the meter is observed to see that it has forward rotation. The non-common or polarity voltage leads to each stator are removed and reconnected to the opposite stators. If the rotation ceases or reverses, the original connections may be assumed correct and should be restored. This method gives fairly reliable results if the load on the circuit is reasonably balanced. On unbalanced loads this method is not reliable. Several incorrect connections can cause rotation to cease on this test under special conditions.

Phasor Analysis of Voltages and Currents from Secondary Measurements

With an ammeter, voltmeter, phase rotation and phase angle meter, data may be quickly obtained from which the complete phasor diagram of the secondary currents and voltages may be constructed graphically to scale. First, the phase rotation of the secondary voltages is determined with the phase rotation meter and the magnitude of the voltages measured with the voltmeter. The voltage and current terminals of the meter or test switch are suitably numbered on a connection diagram for identification. One voltage is selected as the zero reference and the magnitude and phase angle of all currents relative to this voltage are measured and plotted to scale on the phasor diagram. Then the phase angles of the other voltages are measured relative to one of the currents and also plotted on the phasor diagram. The phasor voltage and current in each meter stator are now known.

The phasor diagram so constructed is compared with the standard phasor diagram for the type of metering involved. From this comparison, it is usually possible to determine whether the installation has been correctly connected. To make this comparison a positive check on the connections, some knowledge of the load power factor is needed. Usually, an estimate of the load power factor can be made on the basis of the type of load connected. On badly unbalanced loads of completely unknown power factor, this method is not positive. It also may be indeterminate if the secondary currents are too low to give accurate readings on the meters used.

The reliability of this method depends upon the care taken to assure correct identification of each secondary current and voltage measured and upon the tester's ability to correctly analyze the results.

Various other methods have been used to obtain data from secondary measurements from which the phasor diagram may be constructed.

In the classic Woodson check method, three single-phase wattmeters, an ammeter, a voltmeter, a phase rotation indicator, and a special switching arrangement are used to obtain data from which the phasor diagram may be plotted. This method requires two measurements of watts, one measurement of current, and a graphical phasor construction to determine the direction and magnitude of each current phasor. The sum of the wattmeter readings is compared with the watts load on the watthour meter as determined by timing the disk. This gives an additional check.

The Woodson method has been in use by some utilities for over 50 years and is very reliable. On badly balanced loads of completely unknown power factor, it is not positive, having the same limitations in the interpretation of the phasor diagram as the method using the phase angle meter. The method is primarily designed for checking three-phase, three-wire installations, but may, with modifications, be used for other types.

INSTRUMENT TRANSFORMER TEST METHODS

Safety Precautions in Testing Instrument Transformers

All instrument transformer testing involves the hazard of high voltage. Voltage transformers, by their very nature, are high-voltage devices and current transformers can develop dangerously high voltages if the secondary is accidentally opened under load. No one should be allowed to make tests on instrument transformers until he or she has been thoroughly instructed on the hazards involved and the proper safety precautions.

Many safety devices, such as safety tape, warning lights, interlocked foot switches, test enclosures with interlocked gates, and double switches requiring both hands to energize the equipment, may be used to reduce the hazards, but these devices can never be made absolutely foolproof. Ultimately, the responsibility for safety rests with the person making the tests and his or her supervisor.

In testing current transformers, it is particularly important to make all secondary connections mechanically secure so that even a strong pull on the test leads cannot open the circuit. For this reason, spring test clips should never be used on current transformer secondary test leads. Only a solidly screwed or bolted connection can prevent an accidental opening of the secondary circuit with the consequent high-voltage hazard.

The metal cases of voltage transformers and one of the secondary test leads should be solidly grounded to protect the tester from high static voltages and against the danger of a high-voltage breakdown between primary and secondary. All metal-clad test equipment should also be grounded.

Insulation Tests

The insulation of instrument transformers must be adequate to protect the meters and control apparatus, as well as the operators and testers, from high-voltage circuits and to insure continuity of service. The insulation tests should normally precede all other tests for reasons of safety.

Where it is essential to determine the accuracy of instrument transformers removed from service in order to confirm corrections of billing, it may be advisable to make accuracy tests with extreme precautions before any insulation test.

It is recognized that dielectric tests impose a severe stress on the insulations and if applied frequently will hasten breakdown. It is recommended that insulation tests made by the user should not exceed 75 percent of the ANSI standard factory test voltage. When dielectric field tests are made on a periodic basis, it is recommended that the test voltage be limited to 65% of factory test values.

AC Applied Potential (Hi-Pot) Tests, 60-Hertz

The alternating-current test (at 60 hertz) should be made on each instrument transformer by the manufacturer in accordance with ANSI standards. Similar tests are usually made by the user. All insulation tests for liquid-insulated transformers should be made with the transformer cases properly filled.

Hi-pot test sets with fault-current capacities below "Let Go" or "Threshold of Feeling" are a desirable safety precaution. When properly constructed, such equipment does not represent a fatal hazard to the operator. Many small sets of this type are available commercially. These small test sets may not supply the charging current necessary for over-potential tests on high-voltage current and voltage transformers.

Where high-potential testing equipment with larger fault-current capacity is used, it must be handled with all the safety precautions necessary to any other high-voltage power equipment. Such equipment represents a potentially fatal hazard to the operator. Some degree of protection from the hazards of such equipment may be provided by the use of an enclosed test area protected by electrical interlocks that automatically de-energize the equipment when the gate is opened.

The fundamental responsibility for safety lies with the operator who must use the utmost care to de-energize the equipment before approaching the high-voltage terminals. He or she must never fall into the bad habit of depending upon the interlocks, as these could fail.

To protect the transformers being tested, some means should be provided in large-capacity hi-pot equipment to prevent destructive surges and limit the current in case of breakdown. Impedance in the form of choke coils is often used for this purpose.

Where the hi-pot test voltage is very high, a spark gap may be used to prevent the accidental application of voltage above the desired value. Resistors are used in series with the spark gap to limit the current at breakdown and to damp high-frequency oscillation. The gap is set to a breakdown value slightly higher than the desired test value before the transformer to be tested is connected.

The transformer under test is then connected across the gap and its resistors. Should the test voltage then be exceeded, the gap flashes over and prevents the voltage from rising further.

Polarity Tests

The marking of the leads should be carefully checked by a polarity test. Most methods, as well as instrumentation used in checking transformers for ratio and phase angle, automatically check polarity at the same time. When such facilities are not available, the circuits shown in Figs. 11-26 through 11-29 may be used to determine polarity.

Polarity Tests for Voltage Transformers

Figure 11-26, Polarity Test—Voltage Transformer. Voltage H_2 to X_2 is less than voltage H_1 to H_2 if polarity is correct. The reliability of this method is diminished at high ratios.

Figure 11-27, Polarity Test—Voltage Transformer. The standard voltage transformer must have the same nominal ratio as the unknown voltage transformer. Voltmeter reads zero if polarity is correct, and twice the normal secondary voltage if incorrect.

Polarity Tests for Current Transformers

Figure 11-28, Polarity Test—Current Transformer. Polarity is correct if the ammeter reads less when X_2 secondary lead is connected to the line side of the ammeter. The reliability of this method is diminished at high ratios.

Figure 11-29, Polarity Test—Current Transformer. The standard current transformer must have the same nominal ratio as the unknown current transformer. The ammeter reads zero if the polarity is correct, and twice the normal secondary current if incorrect.

Tests to Verify the Marked Ratio

Voltage Transformers

The marked ratio of a voltage transformer may be verified at the time of the polarity check with either of the circuits shown in Figs. 11-26 or 11-27. In Fig. 11-26, the

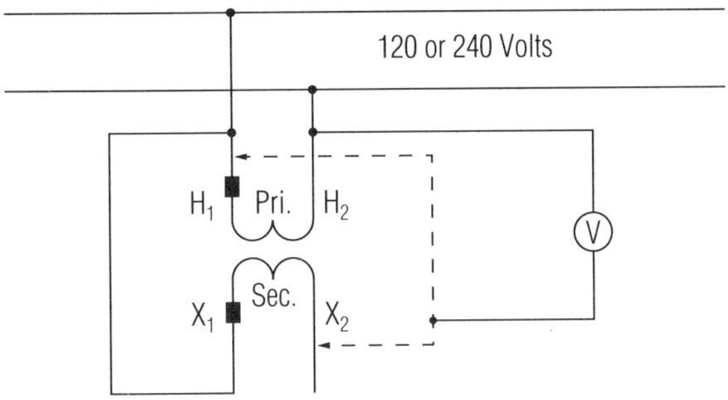

Figure 11-26. Polarity Test, Voltage Transformer.

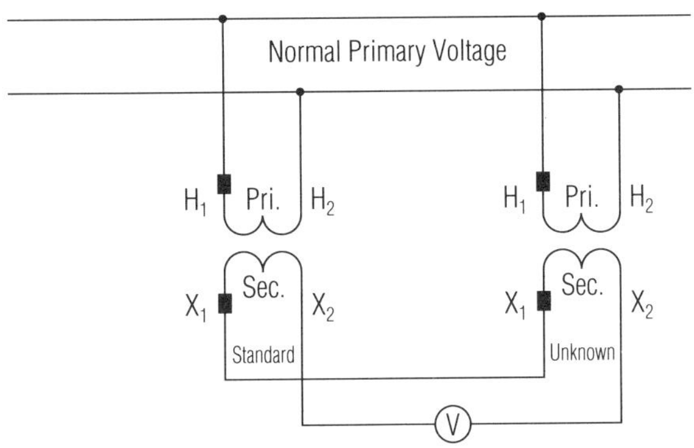

Figure 11-27. Polarity Test, Voltage Transformer.

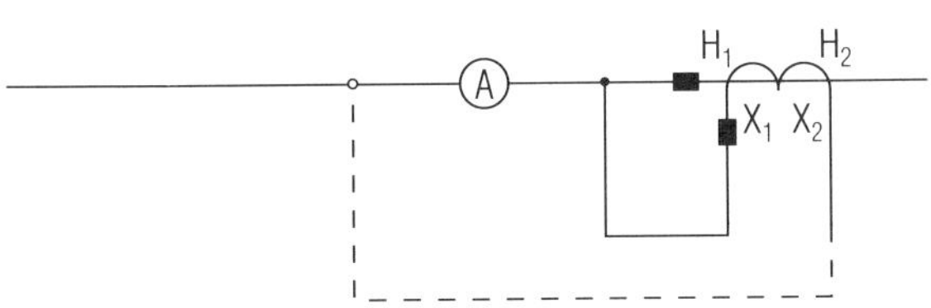

Figure 11-28. Polarity Test, Current Transformer.

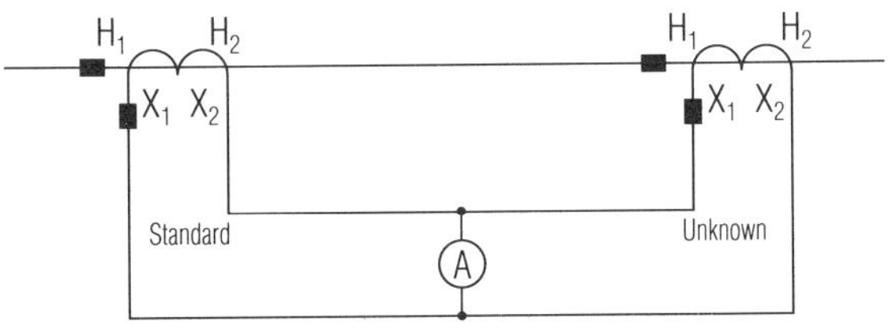

Figure 11-29. Polarity Test, Current Transformer.

voltage measured across H_2 to X_2 should be less than the voltage across H_1 to H_2 by an amount equal to the H_1 to H_2 applied voltage divided by the marked ratio. For example, if 120 V is applied to the primary H_1 to H_2 of a 2,400 to 120 V (20:1) transformer, then the H_2 to X_2 reading should be (120 minus $\frac{120}{20}$), or 114 V. This method may be improved by using two voltmeters so that the two voltages are read simultaneously.

In the circuit shown in Fig. 11-27, the voltage will not be zero unless the unknown transformer has the same ratio as the standard and this automatically verifies its ratio.

A third method that may be used is shown in Fig. 11-30. The secondary voltages will be the same if the ratios of the standard and the unknown are the same. If not, the ratio of the unknown is equal to the ratio of the standard times the secondary voltage of the standard divided by the secondary voltage of the unknown. Care must be used not to apply a primary voltage in excess of the rating of either transformer.

The marked ratio of a voltage transformer may also be checked with a turn-ratio test set, such as the Biddle Model TTR.

Current Transformers

The marked ratio of a current transformer may be checked by measuring the primary and secondary currents directly with ammeters. For large primary values, a standard current transformer must be used and the secondary current of the

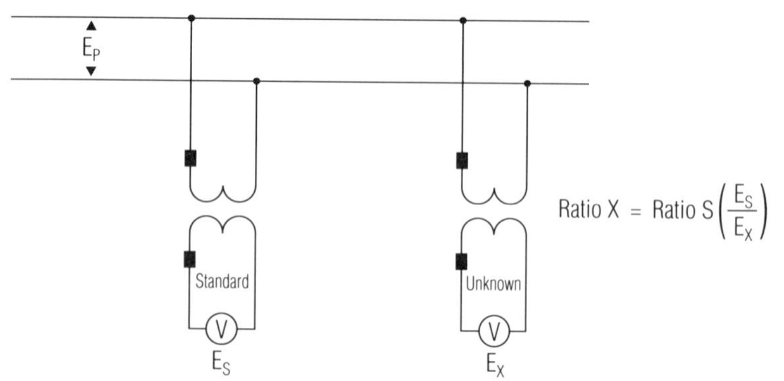

Figure 11-30. Test to Verify Marked Ratio of Voltage Transformer.

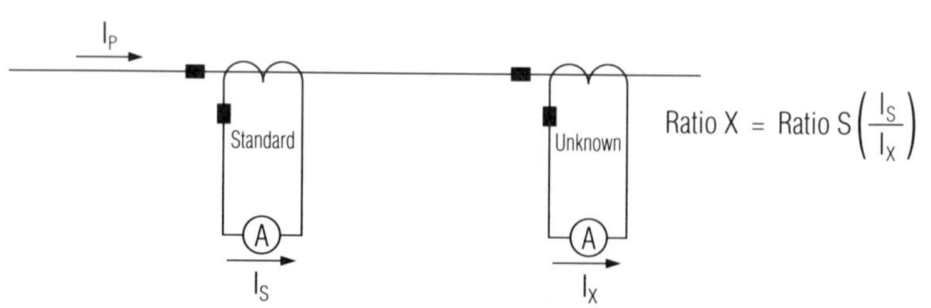

Figure 11-31. Test to Verify Marked Ratio of Current Transformer.

standard is compared with the secondary current of the unknown or "X" current transformer when their primaries are connected in series as shown in Fig. 11-31.

Testing Current Transformers for Shorted Turns with a Heavy Burden

A field method that may be used to detect shorted turns in a current transformer consists of inserting an ammeter and a resistor in series with the secondary circuit. A shorting switch is connected across the resistor. Ammeter readings are taken first with the resistor shorted out and then with the shorting switch open, which adds the burden of the resistor to the circuit. If shorted turns are present, there will be a larger drop in current on the second reading than is normal for a good transformer.

Several precautions are necessary if this method is to provide reliable information on the condition of the transformer. Current transformers vary over a wide range in their ability to maintain ratio under heavy burdens. A burden that has little effect on one type may cause a large drop in secondary current on a different type even though there are no shorted turns. Values ranging from 2 to 60 ohms have been used for this test, but no single value is ideal for all transformers.

To be conclusive, it is necessary to know the effect of the burden used on a good transformer of the same make, model, and current rating. This effect must be known at the same value of secondary current to be used in the test. This can be done by preparing graphs or tables showing the normal effect on all makes, models, and current ratings used.

A simpler method is based on the fact that usually all of the current transformers on a given three-phase installation are of the same make, model, and current rating and the reasonable assumption that they don't all have shorted turns. Thus, if the two or three transformers on the installation are tested by this method, any transformer showing a much larger drop in current than the others, with the addition of the heavy burden, probably has shorted turns.

A commercial test set with two burdens, a multi-range ammeter, and suitable switching is made by the Eastern Specialty Co. (Tesco). With this set, the tests just described may be done quickly and safely.

In addition to shorted turns in the current transformer, the burden test will show shorts in the secondary wiring and grounds in the normally ungrounded wire.

Tests to Determine Ratio and Phase Angle

Instrument transformers may be tested for ratio and phase angle by either direct or comparative methods. Direct methods involve the use of indicating instruments and standard resistors, inductors, and capacitors, while comparative methods require, in addition, a standard instrument transformer of the same nominal ratio whose exact ratio and phase angle have been previously determined.

Direct Methods

Direct methods are necessary for the determination of the ratio and phase angle of instrument transformers in terms of the basic electrical standards. Such methods are used by the National Institute of Science and Technology (NIST) to calibrate their own standard instrument transformers which in turn are used to test instrument transformer standards sent to the NIST for certification.

Direct methods are simple in theory, but they involve so many practical difficulties that they are not suitable for use by utility companies. See Sixth edition of this Handbook for details of direct methods.

Comparative Methods

When calibrated standard instrument transformers are available, the problems of testing instrument transformers are greatly simplified, since only a comparison of nearly equal secondary values is involved.

Deflection Methods

Methods involving the use of indicating instruments connected to the secondary of the standard and unknown transformers suffer from the accuracy limitations of the instruments used. Thus the two-voltmeter or two-ammeter methods, Figs. 11-30 and 11-31, are useful only as a rough check of ratio. If two wattmeters are used, a rough check on phase angle may also be made, though this involves considerable calculation after tests at 1.00 and 0.50 power factor. Accuracy may be somewhat improved by interchanging and averaging the readings of the two instruments, but reading errors still limit the accuracy to about 0.2 percent on ratio.

A modification of the two-wattmeter method makes use of two watthour meters in the form of two rotating standards. This method is capable of fair accuracy but requires excessive time to make the test and compute the results. In addition, it imposes the burden of a rotating standard on the transformer under test, which may not be desirable. Although this method requires unwieldy calculations to determine ratio and phase angle correction factors to the degree of accuracy generally required, it provides a rapid and convenient test method to determine whether transformers meet established accuracy limits. In this case, readings are compared to tables of go and no-go limits without extensive calculations. Some utility companies have adopted this method for testing the commonly used 600 V class of transformers that are not involved in metering large blocks of power. Also, this test confirms polarity and nominal ratio. (See Figs. 11-32 and 11-33.)

Null Methods

Most modern methods of testing instrument transformers are null methods wherein the secondary voltages or currents from the standard and the unknown or "X" transformer are compared and their differences balanced out with suitable circuits to produce a zero or null reading on a detector. After balancing, the ratio and phase angle difference between the "X" transformer and the standard transformer may be read directly from the calibrated dials of the balancing equipment. With suitable equipment of this type, tests for ratio and phase angle may be made rapidly and with high accuracy. Equipment of this type is available commercially.

The Leeds & Northrup Voltage Transformer Test Set

Figure 11-34 is a simplified schematic diagram of the Leeds & Northrup voltage transformer test set. This set has two adjustable dials, one for ratio and one for phase angle. The phase-angle dial moves three sliders which are mechanically coupled. Two of these sliders change the position of a fixed mica capacitor in relation to the resistors to effect a balance for phase angle, while the third slider

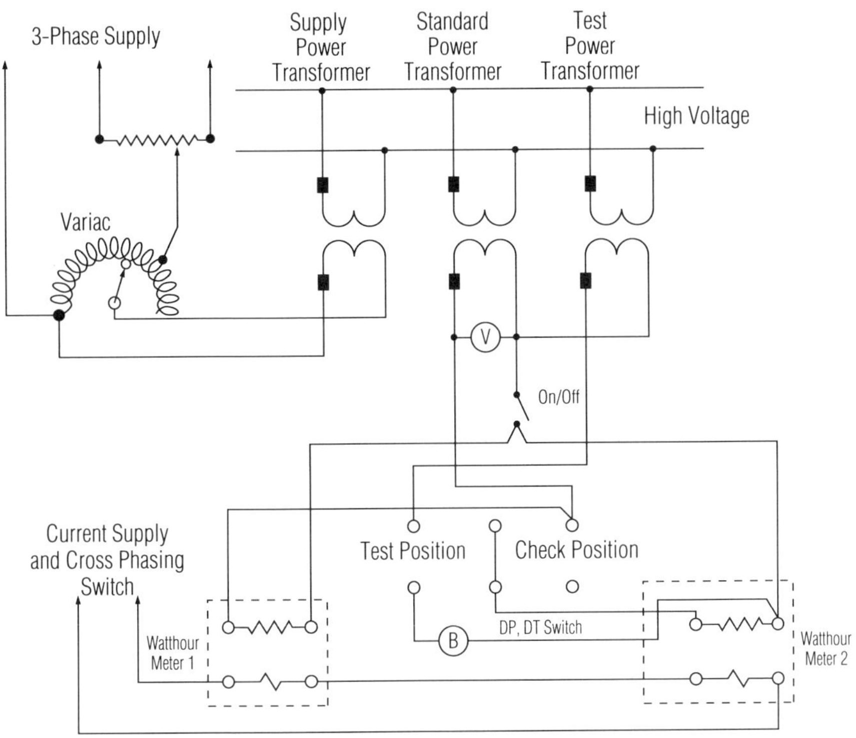

Figure 11-32. Voltage Transformer Test Circuit, Two-Watthour-Meter Method.

rheostat compensates for the change that the first two would also make in the ratio adjustment. The ratio and phase-angle dials are thus made independent so that the adjustment of one does not affect the other.

The balance point is determined by means of a dynamometer-type galvanometer whose field coil is supplied from a phase shifter. This is necessary as the galvanometer would read zero for either zero current (the desired balance point) or zero power factor. The zero current balance is independent of the phase relation of the field, while the zero power factor balance is not. To distinguish the two balance points, the field is supplied first with a voltage of one phase angle and then with a voltage of a different phase angle. If the galvanometer remains balanced for both conditions, then the proper balance has been achieved. For convenience, the galvanometer field flux is set for the in phase and quadrature-phase condition, as this makes the adjustment of the two dials independent. When balance is achieved, the ratio and phase angle of the "X" transformer in terms of the standard may be read directly from the dials.

The ratio dial is calibrated from 95 to 105 percent ratio in divisions of 0.1 percent and the phase-angle dial from -120 minutes to +120 minutes in divisions of 5 minutes. With care, the dials can be read to one-tenth division or 0.01 percent, on ratio and 0.5 minutes on phase angle. Accuracy is stated as ± 0.1

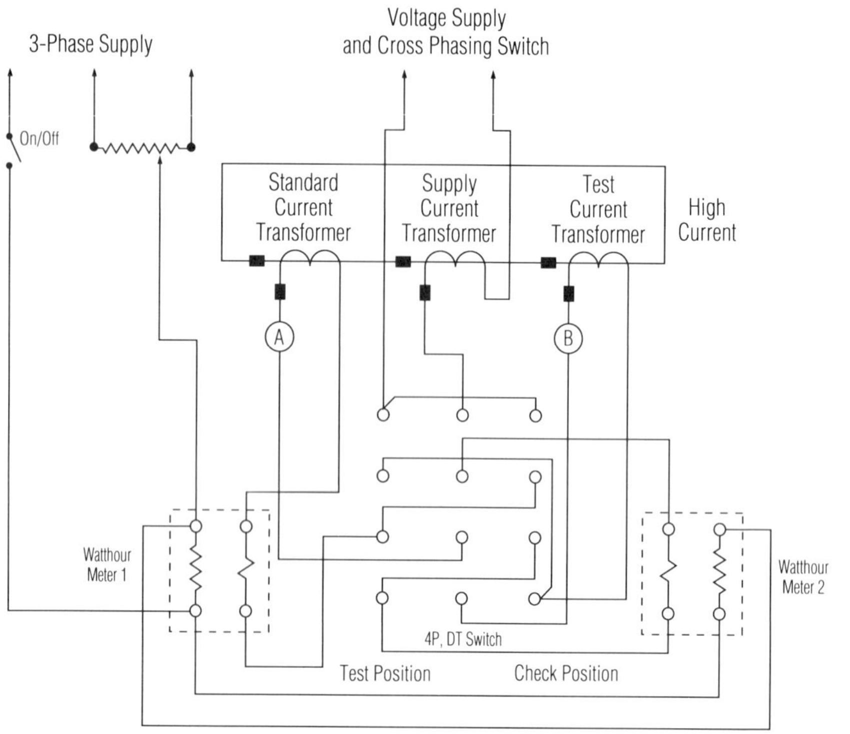

Figure 11-33. Current Transformer Test Circuit, Two-Watthour-Meter Method.

percent on ratio and ± 5 minutes on phase angle. This set can be certified by the NIST who will give corrections to 0.01 percent and 1 minute phase angle and certify them to 0.05 percent and 2 minutes phase angle.

The burden of the standard circuit of the test set on the standard voltage transformer is approximately 3.10 VA at 0.995 power factor leading at 110 volts.

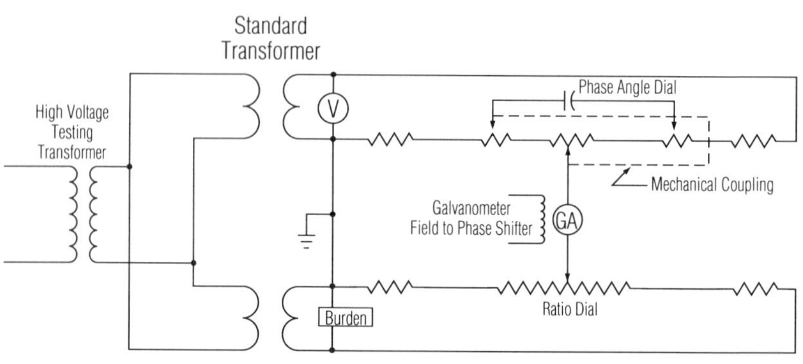

Figure 11-34. Simplified Schematic Diagram of Leeds & Northrup Silsbee Portable Voltage Transformer Test Set.

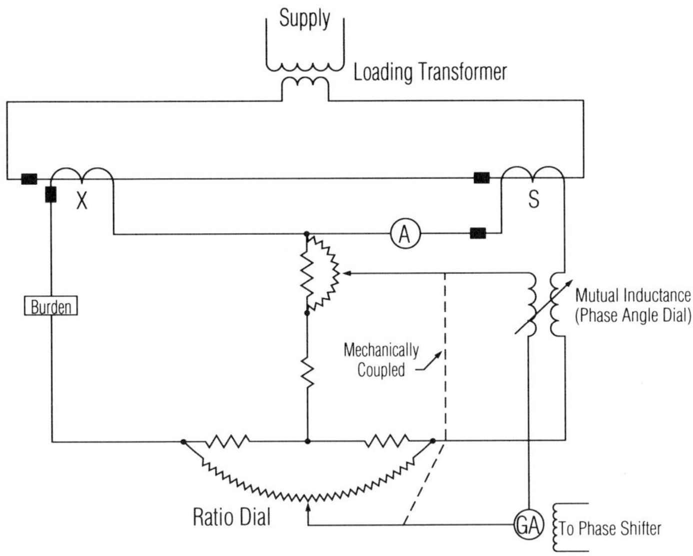

Figure 11-35. Simplified Schematic Diagram of Leeds & Northrup Silsbee Portable Current Transformer Test Set.

The burden of the "X" Circuit on the "X" transformer is approximately 1.18 VA at 1.00 power factor, 110 volts. In most cases, this burden is negligible in regard to the "X" transformer, and the standard circuit burden may be compensated for in the calibration of the standard transformer. The standard transformer burden also includes the burden of the voltmeter used in addition to the burden of the set itself.

The phase shifter used to supply the galvanometer field may have either a single-phase or three-phase primary winding. The three-phase primary winding gives somewhat better voltage regulation.

The high-voltage testing transformer is usually a voltage transformer of the same ratio as the standard and "X" transformers, and is used to step up 120 volts to the primary voltage required.

The Leeds & Northrup Silsbee Current Transformer Test Set

Figure 11-35 is a simplified schematic diagram of the Leeds & Northrup Silsbee current transformer test set. In this set, the ratio adjustment is made by a dial that is coupled mechanically to two variable resistors. The phase angle adjustment is made by a dial that varies the inductance of the air-core mutual inductor. The galvanometer and phase shifter are similar to the ones used for the voltage transformer test set. The same phase shifter may be used for both sets. At balance, the ratio and phase angle of the "X" transformer, relative to the standard transformer, may be read directly from the dials.

The ratio dial is calibrated from 95 to 105 percent ratio in divisions of 0.1 percent and the phase angle dial from -180 to +180 minutes in divisions of 5 minutes. With care, the dials can be read to one-tenth division or 0.01 percent on ratio and 0.5 minutes on phase angle. Accuracy is stated as ± 0.1 percent on ratio and ± 5 minutes on phase angle. This set can be certified by the NIST who will give corrections to 0.01 percent and 1 minute phase angle and certify them to 0.05 percent and 4 minutes phase angle.

The burden of the standard circuit is approximately 0.8 mH and 0.29 ohms. The burden of the "X" circuit varies with the setting and is in the order of 0.01 to 0.02 ohms. In most cases, this burden is negligible in regard to the "X" transformer and the standard burden may be compensated for in the calibration of the standard transformer. The standard transformer burden also includes the burden of the ammeter used in addition to the burden of the set itself.

A test set of this type with reduced ranges is available on special order. This would give better readability in the range of greatest use.

The loading transformer is a stepdown transformer designed to produce the necessary primary test currents at a low voltage.

The air-core mutual inductor used in this set is very sensitive to stray fields and the proximity of magnetic materials. It must be kept several feet from conductors carrying heavy current and well away from any iron or steel. A steel bench top will cause considerable error.

At secondary currents of 0.5 amperes, the galvanometer has only one-tenth the sensitivity that it has with secondary currents of 5 amperes. This makes the exact balance at this point difficult to determine. Another problem is that of inductive action in the galvanometer circuit that may occur when testing miniature current transformers. To overcome these problems, a cathode ray oscilloscope may be used as a detector. This requires shielding of the internal leads in the detector circuit of the Silsbee set and the use of a shielded matching transformer of about 1-3/4-ohm input to 157,000-ohm output to couple the detector circuit to the oscilloscope. Care must be used in the grounding of the shield and secondary circuits to prevent false indications. Grounding at the No. 2 standard terminal of the set has proved satisfactory provided that this is the only ground on the secondary of either the standard or the "X" transformer, and all shield grounds are tied to this point. When the oscilloscope is used, the phase shifter is not needed and adjustments may be made rapidly on both the ratio and phase-angle dials simultaneously to reduce the scope pattern to a minimum peak-to-peak value. This method is rapid and very sensitive. Some transformers will show more third-harmonic content than others at the balance point, but this may be ignored as the balance desired is for the 60 hertz fundamental only. To avoid erroneous balances, several cycles should be displayed along the "X" axis of the oscilloscope and all peaks adjusted for the same height and minimum value.

Testing Current Transformers for Abnormal Admittance

The condition of a current transformer can be tested by monitoring its admittance. (Admittance is the reciprocal of impedance.)

The admittance of a current transformer secondary loop can be measured with or without service current flowing in the secondary. The tester shown in Fig. 11-36 measures admittance by injecting an audio signal into the secondary of an in-service transformer, then measuring the admittance seen by that signal. While any frequency between 1 Khz and 2 Khz could be used, this tester uses 1580 Hz to reduce false readings caused by harmonics of 60 Hz in the secondary.

Metering current transformers have very small errors (typically less than 0.3%) when operated within their specified current and burden ratings. Therefore, the circuit admittance of a current transformer is nearly constant throughout its normal operating range, unless a fault develops. If an admittance measurement shows deviation from normal while in service, it is likely the current transformer has (a) an internal short such as short-circuited turns, (b) an abnormal internal or external resistance such as a high resistance joint, or (c) the current transformer is being operated under abnormal conditions, for example with a DC component in the primary. Most faults are immediately obvious because they produce a high admittance reading, typically greater than 1.5 times a normal reading. Transformers with a wrong ratio, such as those hooked to the wrong tap, also will have readings substantially different from normal readings.

The best way to establish a "normal" admittance reading is to develop a history of measurements. Admittance readings can be taken, for example, before installation, during initial field tests, and during subsequent checks. Admittance values depend on fixed features such as core design, burden rating, and the turns ratio. Changes to admittance which are caused by non-fault conditions are small when compared with changes caused by fault conditions.

Service current transformers can be tested in groups and a high admittance reading by one transformer in the group suggests a fault condition on that transformer. If all readings in the group are high, the cause could be a capacitive load on both sides of the current transformer, or high system noise including harmonics close to the test frequency, or the presence of DC in the primary.

Testing Current Transformers for Abnormal Burden

The condition of a current transformer can be evaluated by measuring the burden of the transformer.

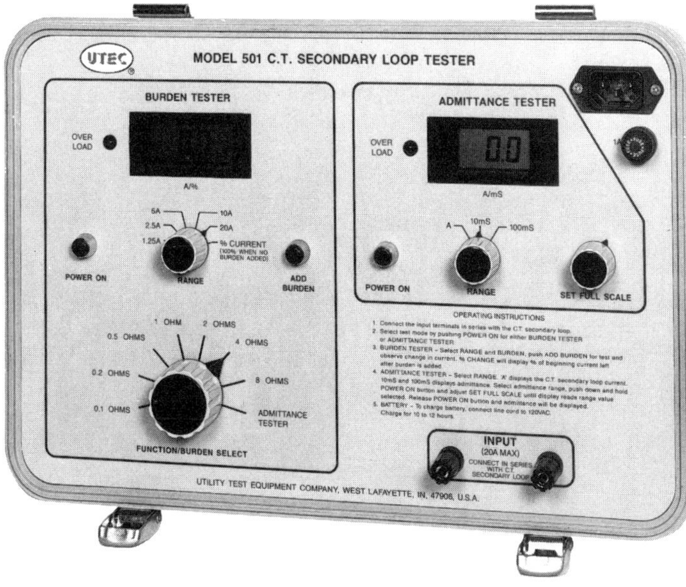

Figure 11-36. Instrument for Measuring Current Transformer Secondary Admittance and Burden.

Current transformers are designed to supply a known current as dictated by the turns ratio, into a known burden, and to maintain a stated accuracy. The principle of a burden tester is to challenge the capability of the current transformer to deliver a current into a known burden.

The total burden of the current transformer secondary loop includes the burden of the watthour meter current coils, the mounting device, the test switch, connection resistances, and the loop wiring. When burden is added which exceeds the design capacity of the current transformer, the transformer cannot supply the same level of current to the increased burden, and results in a drop in current transformer loop current. The tester shown in Fig. 11-36 measures the burden of a current transformer by adding a known ohmic resistance in series with the current transformer secondary loop, and comparing the total burden, including the known resistance, with the burden when the resistance is not in the loop.

The magnitude of the current change depends on several factors and is not absolutely definable. The operating level of current in the current transformer secondary loop can be a significant factor. Current transformers operating at low currents are able to support several times the burden rating, because at low currents the flux density of the core is low, leaving ample head room for additional flux before saturation. Therefore, to obtain accurate readings, these burden tests are performed at full rated secondary current. At the high end of the current range, additional burden quickly pushes the current transformer out of its operating range, resulting in significant drops in operating current.

Form factor affects the burden capability of a current transformer. Transformers with high form factors can support a burden greater than their nameplate specification. For high form factor transformers, it is important to take measurements at full rated secondary current.

Other procedures for testing the burden of current transformers are included in the ANSI C57-13 specification *Requirements for Instrument Transformers.*

THE KNOPP INSTRUMENT TRANSFORMER COMPARATORS

Description

Knopp transformer comparators provide a direct means of measuring phase angle and ratio errors of instrument transformers. These comparators use a refined null method. The procedure is a four-step process: interconnection, precheck, null, and test results.

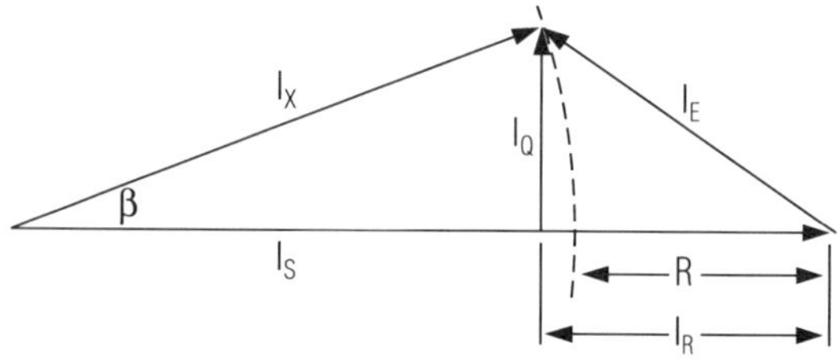

Figure 11-37. Quantities Measured by Current Comparator.

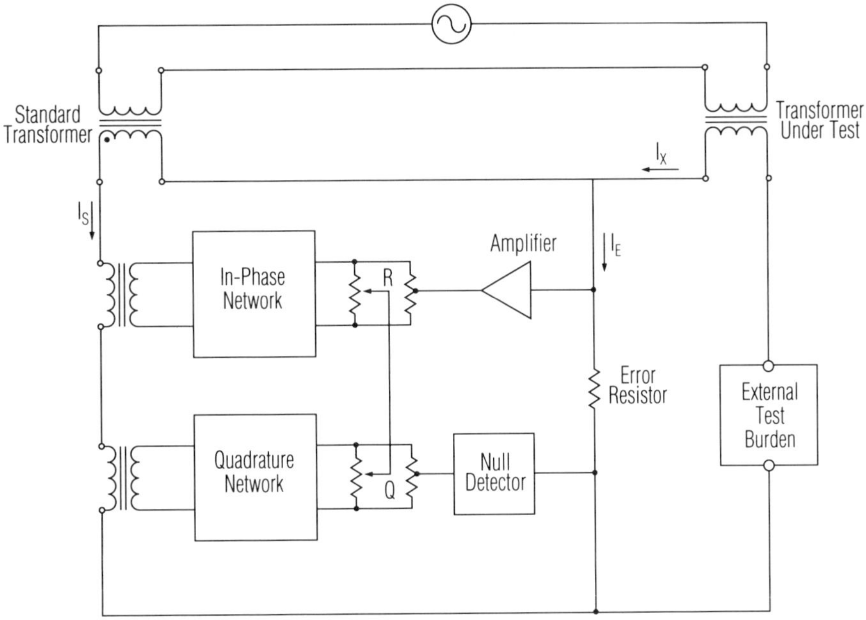

Figure 11-38. Simplified Block Diagram of Knopp Current Transformer Comparator.

During precheck, the test connections, transformer integrity, dial zeroing, secondary output, and power source level are simultaneously verified. Then an appropriate multiplier is selected and the null established with two calibrated dials. The results, phase angle and ratio error, are read directly from these dials.

Null Method

Both the current and voltage comparators measure a transformer's ratio and phase errors with respect to a "standard" transformer. The following discussion of the current comparator illustrates the principles applying to the voltage comparator as well. The quantities being measured by the current comparator are best described by use of the following vector diagram in Fig. 11-37. I_S represents the secondary current in the "standard" transformer. I_S represents the secondary current in the transformer under test, and I_X is the resulting error current. If the transformer under test were identical to the standard, I_E would, of course, be zero. If the vector I_E can be resolved into its in-phase and quadrature (90°) components, the desired quantities (ratio error and phase angle) are produced. For errors such as are encountered in most instrument transformers, I_Q is essentially identical to the arc represented by β (phase angle) and I_R is equal to R (ratio error). The purpose of the comparator then is to resolve this vector error into its in-phase and quadrature components.

The simplified diagram of the current transformer comparator, Fig. 11-38, shows the vector quantities just discussed. The transformer secondaries are connected in series-aiding to provide current flow as illustrated.

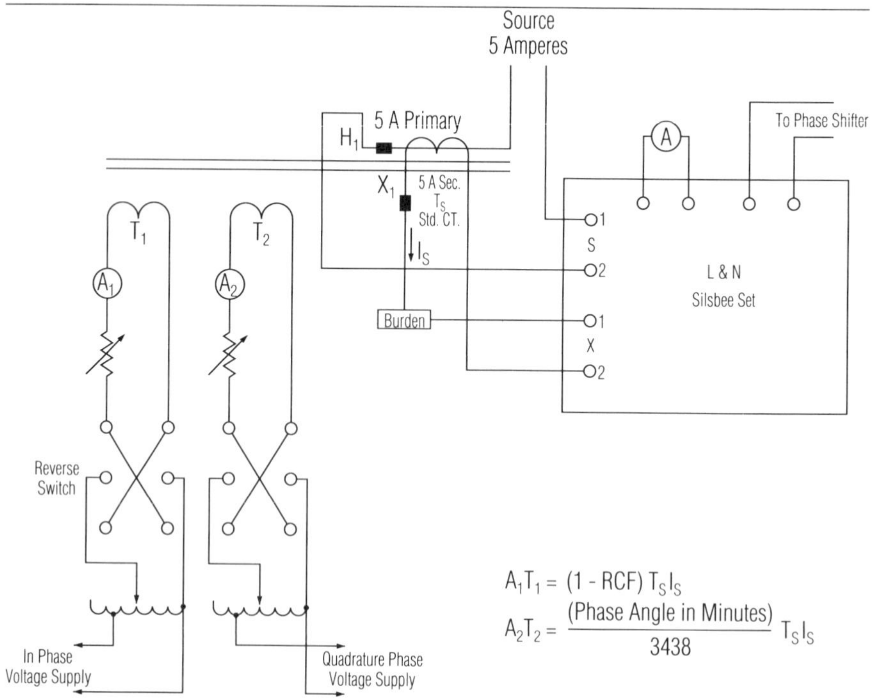

Figure 11-39. Use of Compensated 1-to-1 Standard to Calibrate Silsbee Set.

A portion of the error current I_E is allowed to flow through the "In-Phase" and "Quadrature" networks and the null detector. The two networks allow two other currents to flow in this same path. One is in phase with I_S and the other is in quadrature with I_S. The magnitudes of these currents are varied by potentiometers R and Q until the portion of I_E originally injected is canceled. This cancellation is indicated by a null on the null detector meter. The resulting positions of the potentiometers are translated into ratio error and phase angle by simply reading the calibrated dials attached to the potentiometer shafts.

Although the circuit details for the voltage comparator differ, the approach is fundamentally the same. That is, the vector error is resolved into its in-phase and quadrature components.

Knopp comparators can be supplied with an accessory unit that presents a digital display of phase angle and ratio error (in percent or ratio correction factor) selectable with a front panel switch.

Compensated Standard Method to Calibrate a Current Transformer Test Set

This method may be used to check the calibration of the Silsbee current transformer test set or the Knopp current transformer comparator or other direct-reading test sets. To do this, a window-type standard current transformer having a 5:5 or 1:1 ratio is set up with the line to the standard circuits of the test set and the secondary to the secondary circuits of the test set (Fig. 11-39). With no tertiary current, the set should read the error of the standard current transformer on the 5:5 range with the burden used. By applying in-phase and quadrature-phase tertiary ampere-turns the standard current transformer may be compensated to have any apparent ratio and phase angle desired. These values may be calculated from the original one-to-one test and the formulas in Fig. 11-39. The readings of the test set at balance are then compared with the calculated values.

This method provides a useful check on the accuracy of the test sets. The formulas are approximations based on the assumption of small angles and can introduce slight errors when the phase angle exceeds 60 to 120 minutes.

Precautions in Testing Instrument Transformers

Stray Fields

In instrument transformer testing, precautions must be taken to prevent stray fields from inducing unwanted voltages into the test circuits. Secondary leads are usually twisted into pairs to prevent this. Where test equipment is not shielded, it must be kept well away from conductors carrying heavy current.

Effect of Return Conductor

The location of the return conductor in a heavy-current primary loop can affect the ratio and phase angle of a current transformer. Normally the return conductor should be kept two or three feet away from the current transformer under test. This effect is most pronounced where the primary current is large.

In winding down a window-type current transformer for test, tight loops may give a different result than open loops due to the return-conductor effect. Since normal operation is on a straight bus bar, the results obtained with the open loops will be more nearly comparable to the field conditions.

Demagnetizing

Current transformers should be demagnetized before test to insure accurate results. Demagnetization may be accomplished by bringing the secondary current up to the rated value of 5 amperes by applying primary current and then gradually inserting a resistance of about 50 ohms into the secondary circuit. This resistance is then gradually reduced to zero and then the current is reduced to zero. Caution must be used to avoid opening the secondary circuit at any time during this procedure. A reactor in place of the resistor reduces the possibility of re-magnetizing by accidentally open-circuiting during the procedure.

A current transformer can be magnetized by passing direct current through the windings, by surges due to opening the primary under heavy load, or due to accidental opening of the secondary with load on the primary. Test circuits should provide for a gradual increase and decrease of primary current to avoid surges.

Modern high-accuracy current transformers show relatively little change in accuracy due to magnetization.

Ground Loops and Stray Ground Capacities

In all instrument transformer testing, care must be used to avoid ground loops and stray ground capacities that might cause errors. Usually, only one primary and one secondary ground are used. The location of these grounds must be carefully determined to avoid errors.

Effect of Inductive Action of Silsbee Galvanometer in Testing Small Current Transformers

In the Silsbee current transformer test set the galvanometer field coil and the galvanometer moving coil form, in effect, a small transformer. When the field coil is energized, a small voltage is induced in the moving coil. The moving coil is connected in the differential circuit between the secondaries of the standard and unknown transformer and, therefore, the effect of this induced voltage is to cause a small current to flow in these circuits.

The effect of this current in the galvanometer moving coil is to cause a deflection that is not due to the differential current. This deflection is compensated by means of the electrical zero adjustment which consists of a movable piece of iron which distorts the field flux. In effect, the electrical zero adjustment produces a false zero which eliminates the unwanted deflection of the galvanometer.

When this electrical zero adjustment has been made according to the instructions for the Silsbee set, it normally makes no difference in the results if the polarity of the leads from the phase shifter to the galvanometer field of the Silsbee set is reversed. However, in testing miniature current transformers, it has been found that a difference as high as 0.1 percent at 10 percent current may be found when the galvanometer field is reversed. It appears that this is due to a large difference in impedance between the secondaries of the standard and unknown current transformers. The current set up by the induced voltage in the galvanometer does not divide equally, and this effect is not entirely eliminated by the electrical zero adjustment.

It has been found that the average of the readings with direct and reversed galvanometer fields agrees with the results obtained when using a cathode-ray oscilloscope and matching transformer as a detector.

Accuracy

Accurate testing of instrument transformers requires adequate equipment and careful attention to detail. Readability to 0.01 percent and 0.5 minute phase angle is easily possible with modern equipment, but accuracy to this limit is much more difficult. For the greatest accuracy, the test equipment and standard instrument transformers should be certified by the NIST or by an independent laboratory whose standard accuracies are established by the NIST. With the greatest care, absolute accuracies in the order of 0.04 percent and 1 to 3 minutes phase angle may be achieved.

CHAPTER 12

METER WIRING DIAGRAMS

DIAGRAMS OF INTERNAL CONNECTIONS for watthour meters and the associated form numbers are in accordance with ANSI C12.10.

INDEX FOR ILLUSTRATIONS

Symbols used are in accordance with ANSI Y32.2 except where none are listed.

For typical meter wiring diagrams of kilovar and kilovolt-ampere metering, refer to Chapter 9. Refer to previous editions of the *Electrical Metermen's Handbook* for AC, two-phase, and DC meter wiring diagrams.

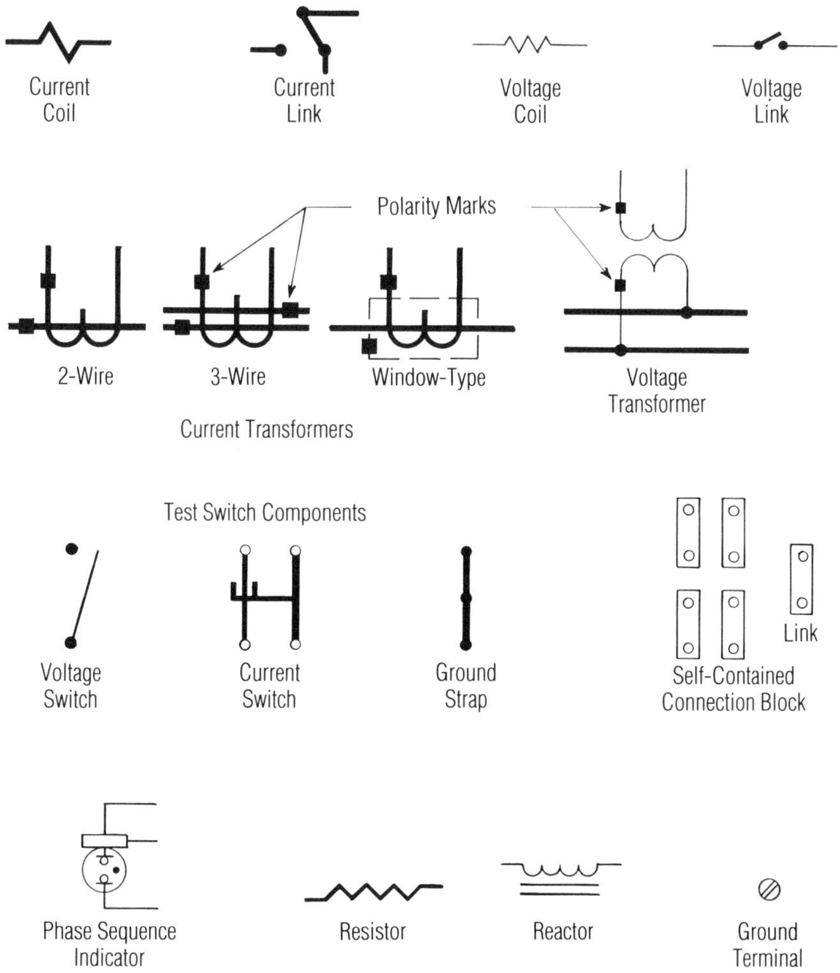

Figure 12-1. Symbols

Internal Connections of Type "S" Meters

Front Views

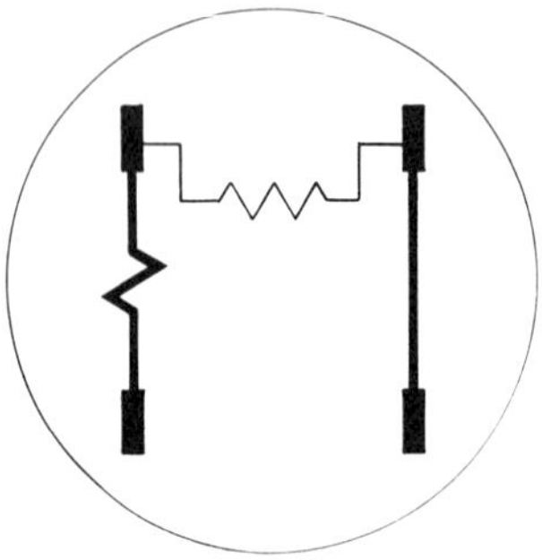

Form 1S
2-Wire Self-Contained

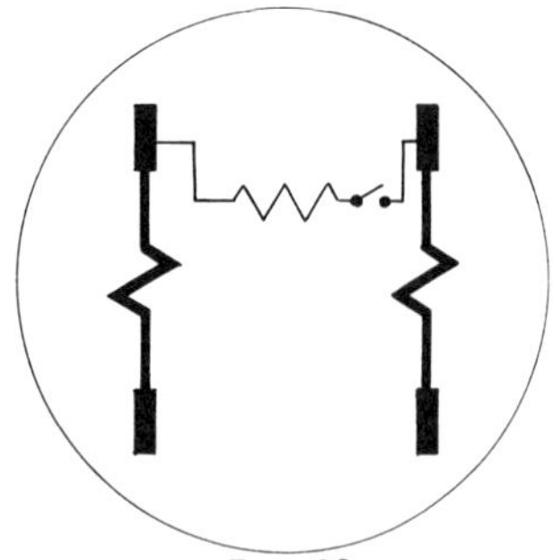

Form 2S
3-Wire Self-Contained

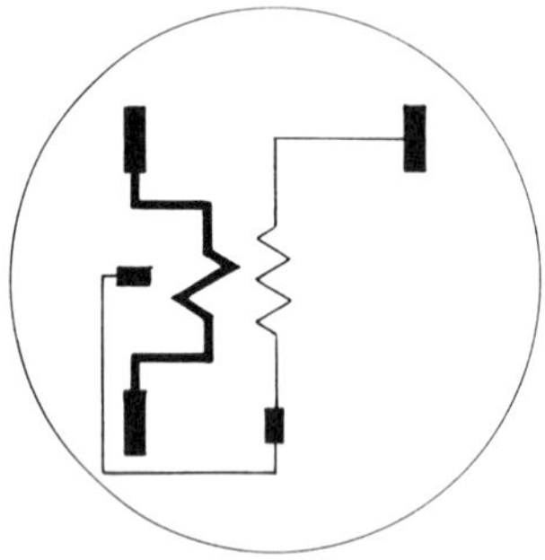

Form 3S
2-Wire Transformer-Rated

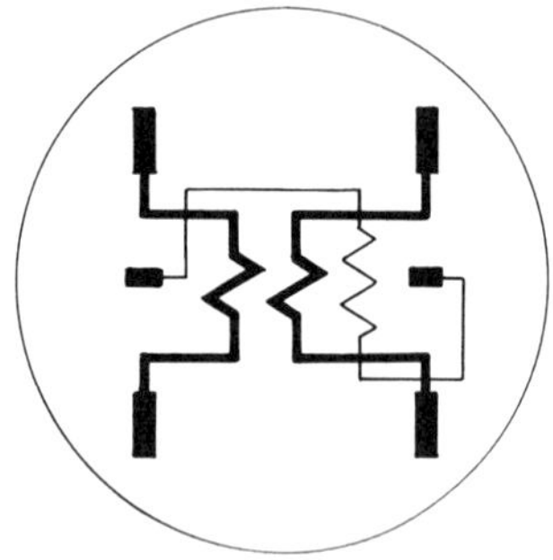

Form 4S
3-Wire Transformer-Rated

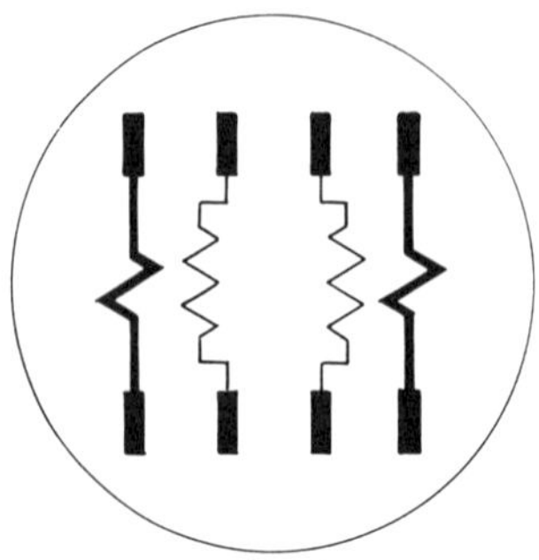

Form 5S
2 Stator Transformer-Rated
2-Phase, 3- or 4-Wire
3-Phase, 3-Wire

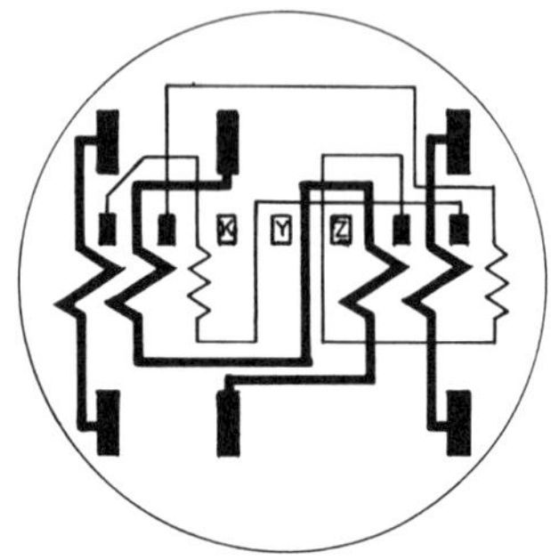

Form 6S
2 Stator Transformer-Rated
3-Phase, 4-Wire Wye

Figure 12-2.

Internal Connections of Type "S" Meters
Front Views

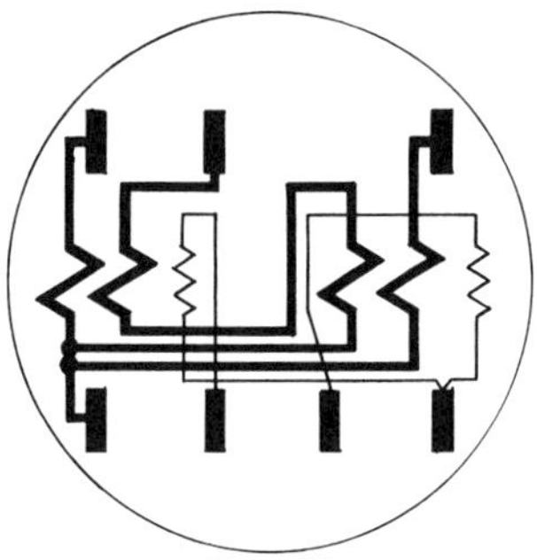

Form 7S
2 Stator Transformer-Rated
3-Phase, 4-Wire Wye
(Alternative)

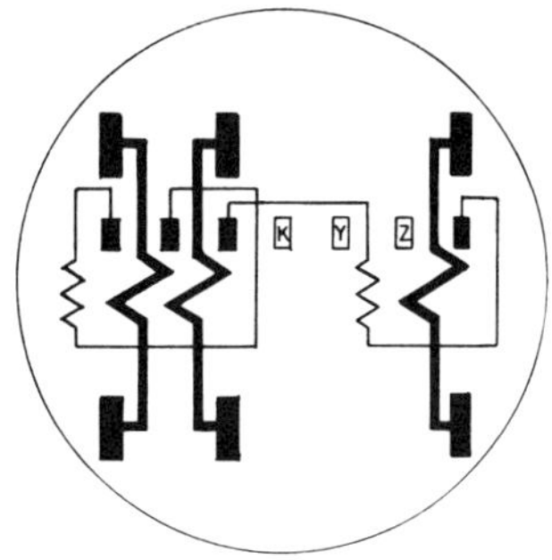

Form 8S
2 Stator Transformer-Rated
3-Phase, 4-Wire Delta

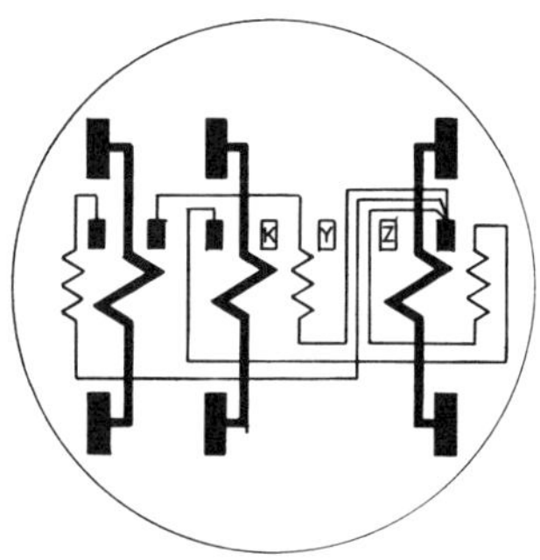

Form 9S
3 Stator Transformer-Rated
3-Phase, 4-Wire Wye

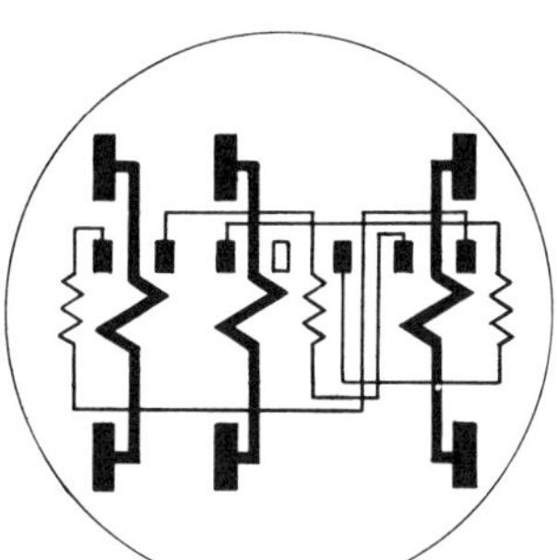

Form 10S
3 Stator Transformer-Rated
3-Phase, 4-Wire Wye
(Alternative)

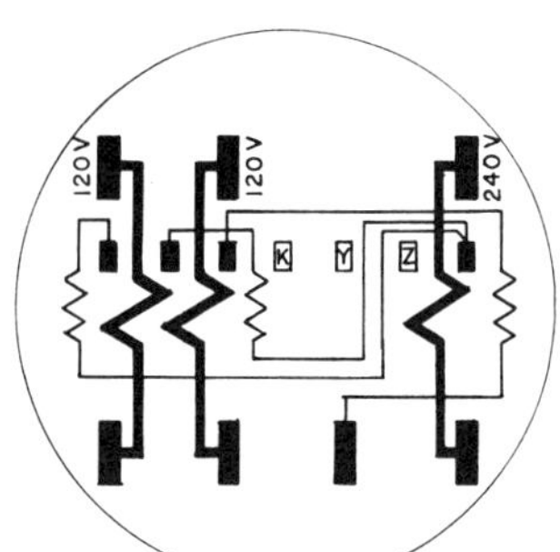

Form 11S
3 Stator Transformer-Rated
3-Phase, 4-Wire Delta

Figure 12-3.

Internal Connections of Type "S" Meters

Front Views

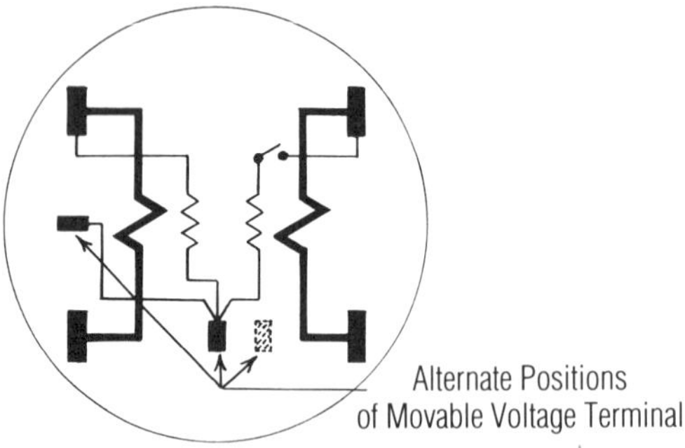

Form 12S
2 Stator Self-Contained
3-Wire Network

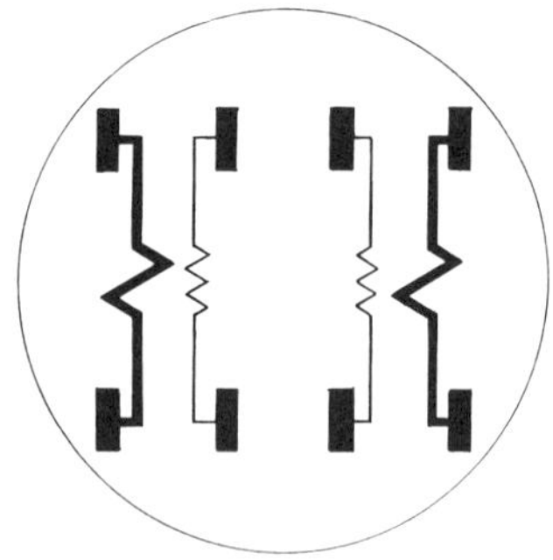

Form 13S
2 Stator Self-Contained
2-Phase, 3- or 4-Wire
3-Phase, 3-Wire

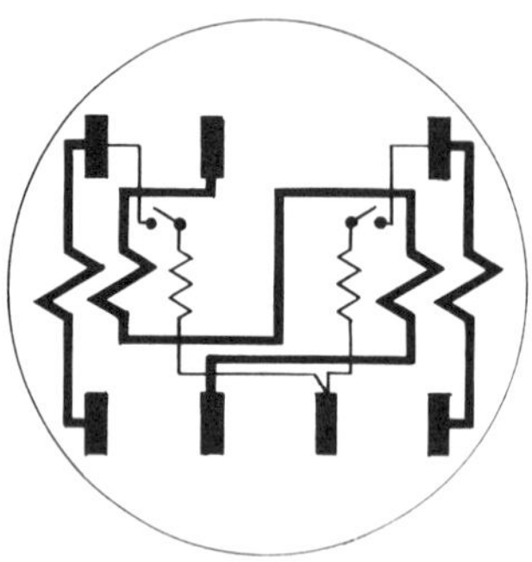

Form 14S
2 Stator Self-Contained
3-Phase, 4-Wire Wye

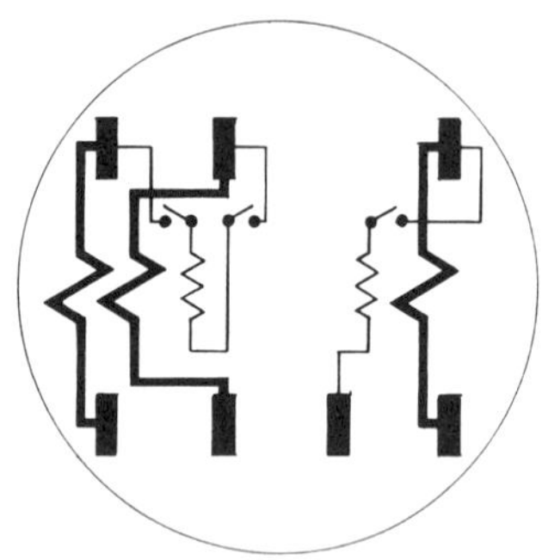

Form 15S
2 Stator Self-Contained
3-Phase, 4-Wire Delta

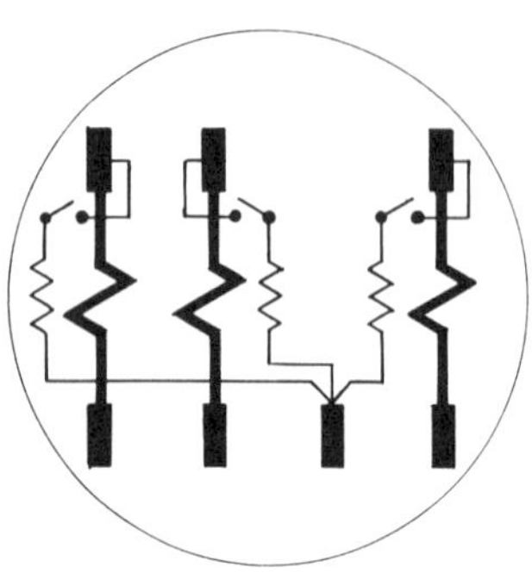

Form 16S
3 Stator Self-Contained
3-Phase, 4-Wire Wye

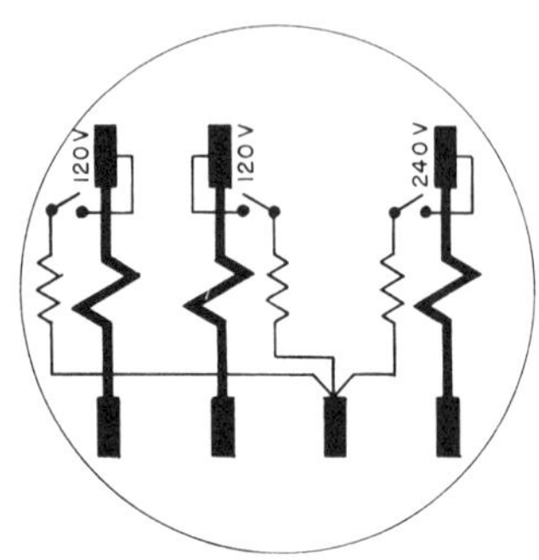

Form 17S
3 Stator Self-Contained
3-Phase, 4-Wire Delta

Figure 12-4.

Internal Connections of Type "S" Meters

Front Views

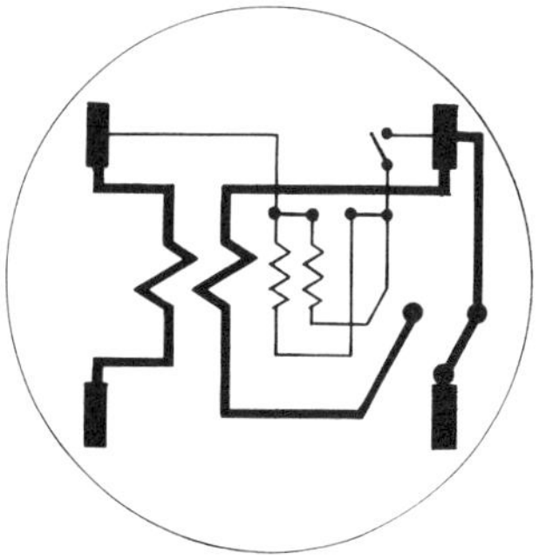

Form 19S-2
2/3-Wire Single Stator
Self-Contained 2-Wire Connection

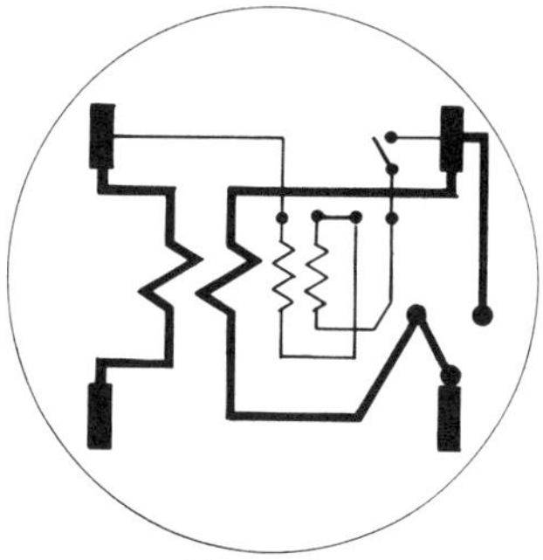

Form 19S-3
2/3-Wire Single Stator
Self-Contained 3-Wire Connection

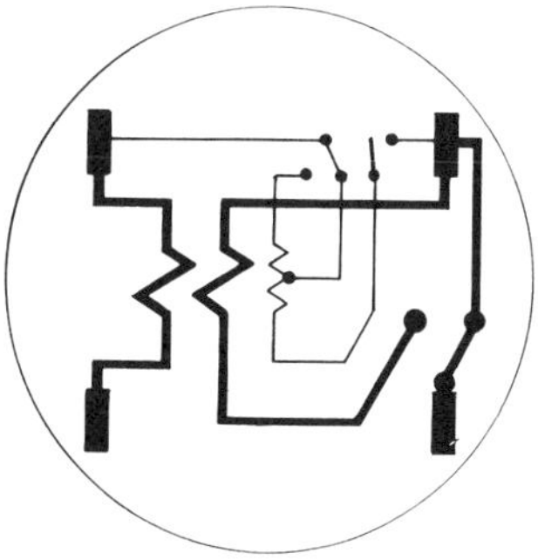

Form 20S-2
2/3-Wire Single Stator
Self-Contained 2-Wire Connection

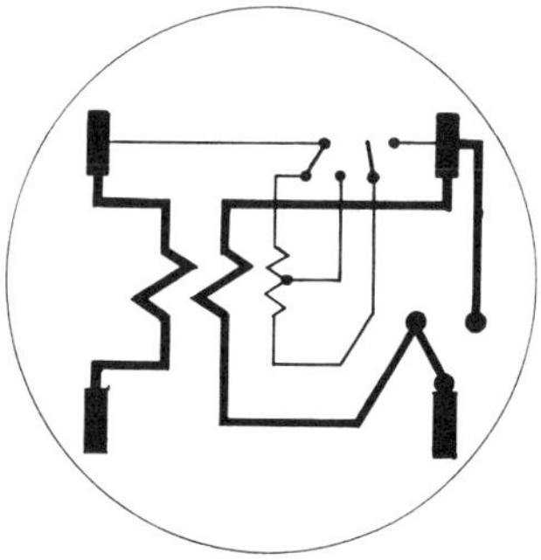

Form 20S-3
2/3-Wire Single Stator
Self-Contained 3-Wire Connection

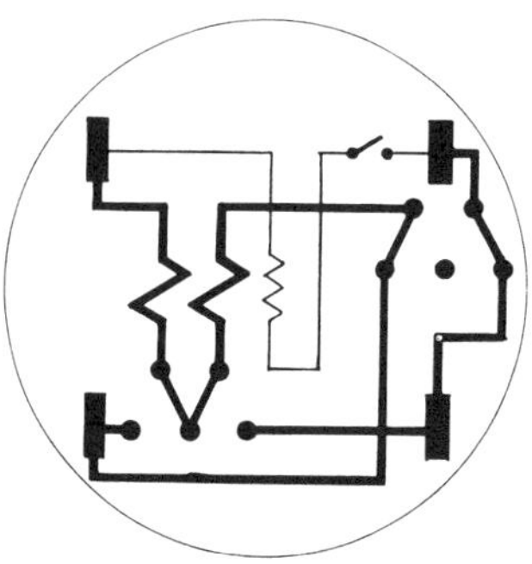

Form 21S-2
2/3-Wire Single Stator
Self-Contained 2-Wire Connection

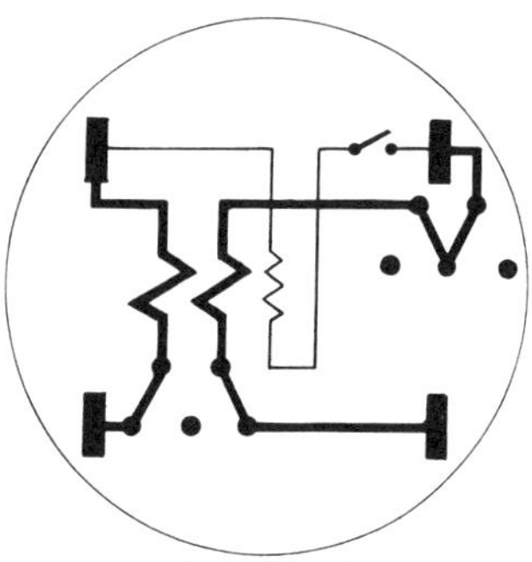

Form 21S-3
2/3-Wire Single Stator
Self-Contained 3-Wire Connection

Figure 12-5.

Internal Connections of Type "S" Meters

Front Views

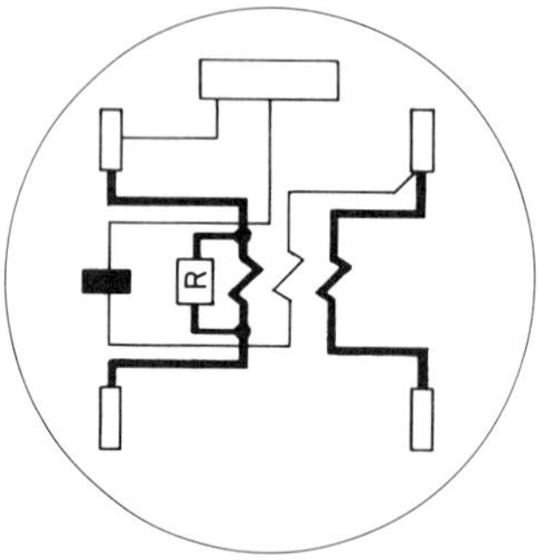

Form 22S
3-Wire Self-Contained
(Network Only)

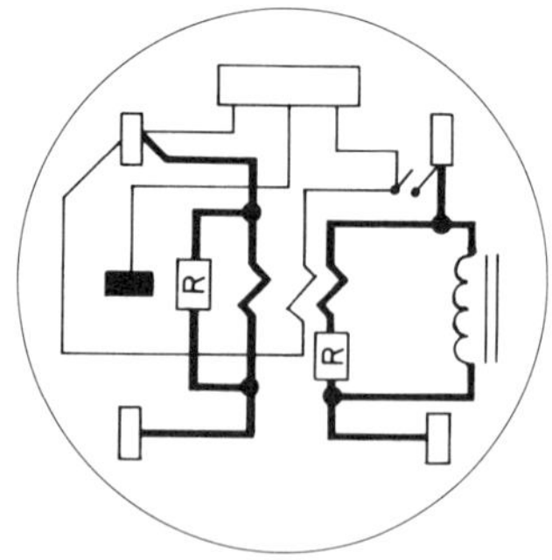

Form 23S
3-Wire Self-Contained
(Network Only)

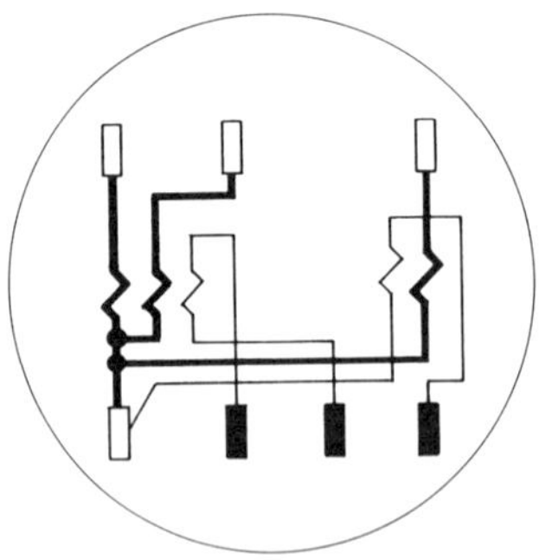

Form 24S
2 Stator Transformer-Rated
3-Phase, 4-Wire Delta
(7 Terminals)

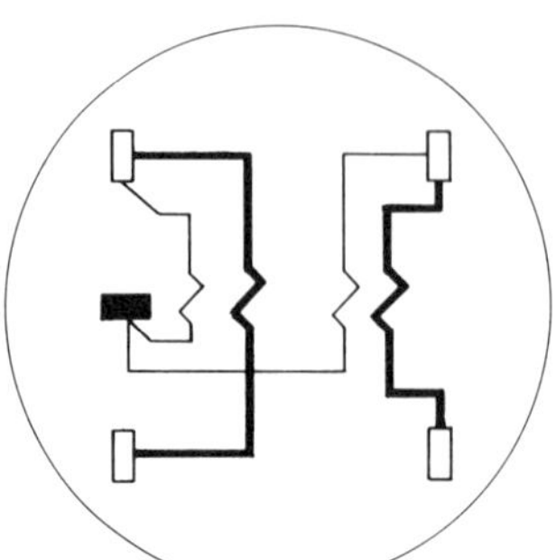

Form 25S
2 Stator Self-Contained
3-Wire Network

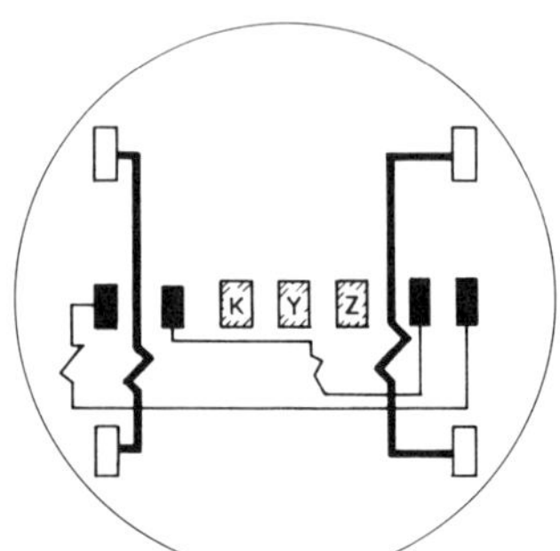

Form 26S
2 Stator Transformer-Rated
3-Phase, 3-Wire

Figure 12-6.

Internal Connections of Type "A" Meters

Front Views (Except Terminals)

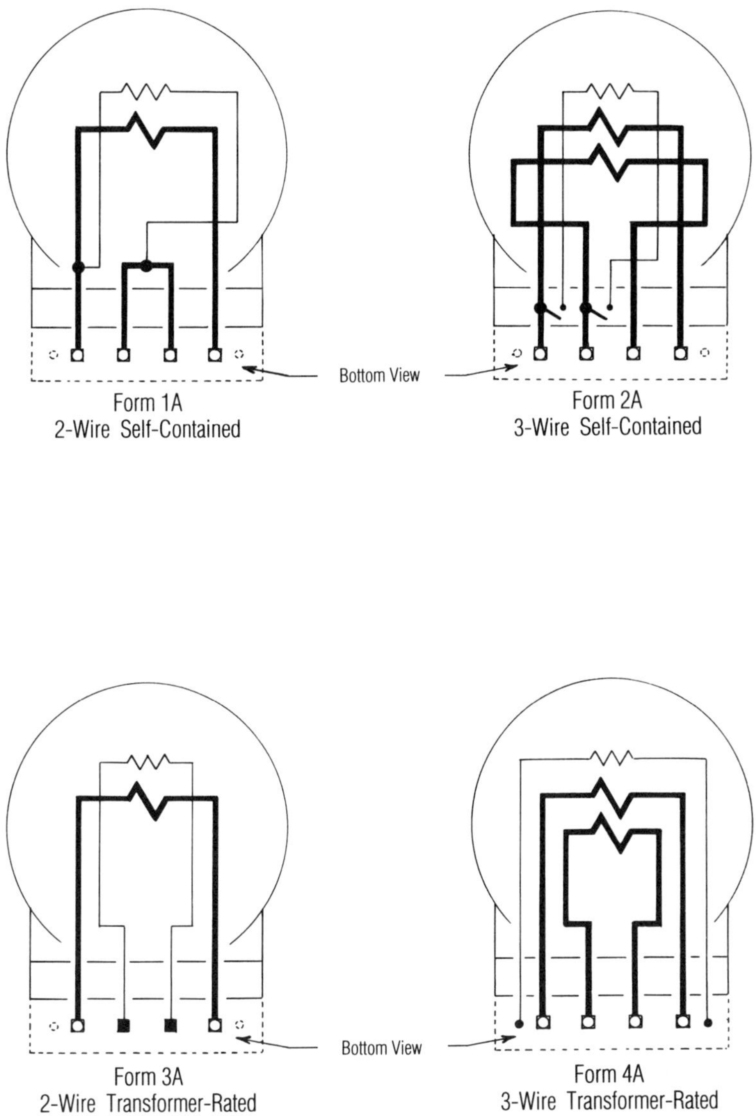

Figure 12-7.

Internal Connections of Type "A" Meters

Front Views (Except Terminals)

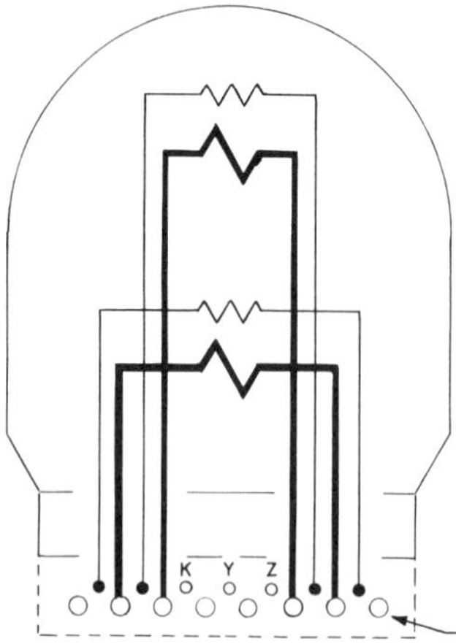

Form 5A
2 Stator Transformer-Rated
2-Phase, 3- or 4-Wire
3-Phase, 3-Wire

Form 6A
2 Stator Transformer-Rated
3-Phase, 4-Wire Wye

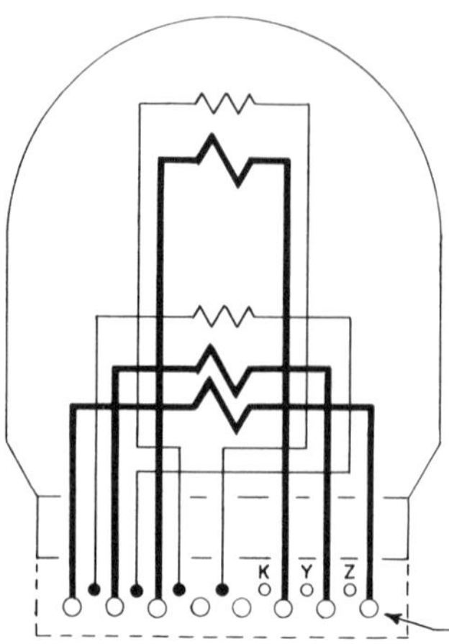

Form 8A
2 Stator Transformer-Rated
3-Phase, 4 Wire Delta

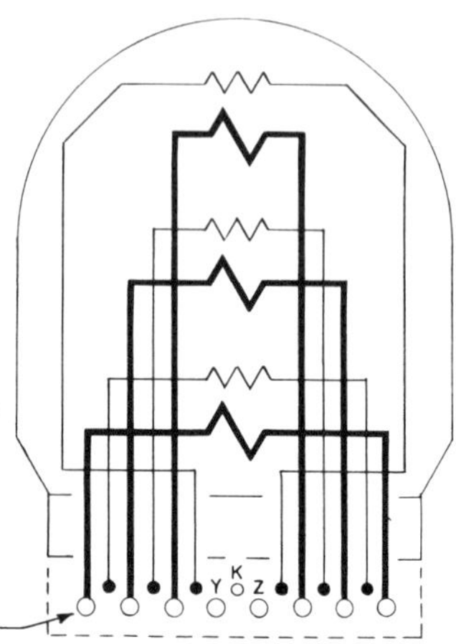

Form 9A
3 Stator Transformer-Rated
3-Phase, 4-Wire Wye

Figure 12-8.

Internal Connections of Type "A" Meters

Front Views (Except Terminals)

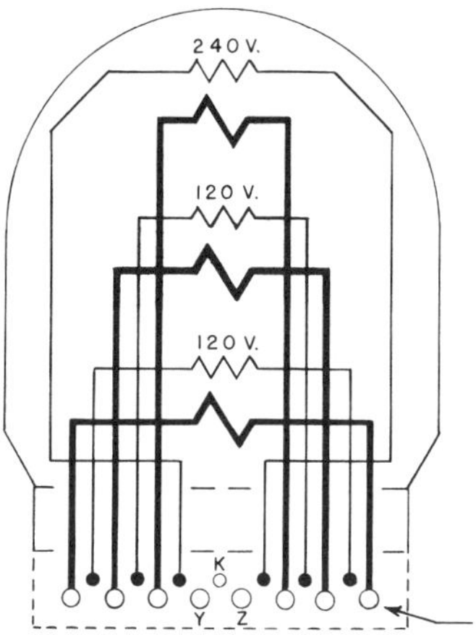

Form 11A
3 Stator Transformer-Rated
3-Phase, 4-Wire Delta

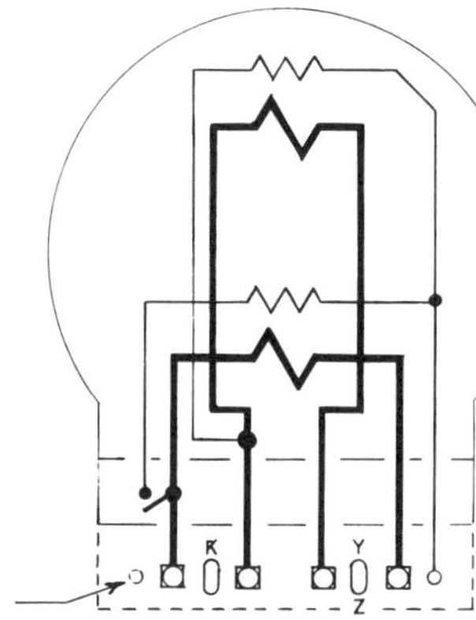

Form 12A
2 Stator Self-Contained
3-Wire Network

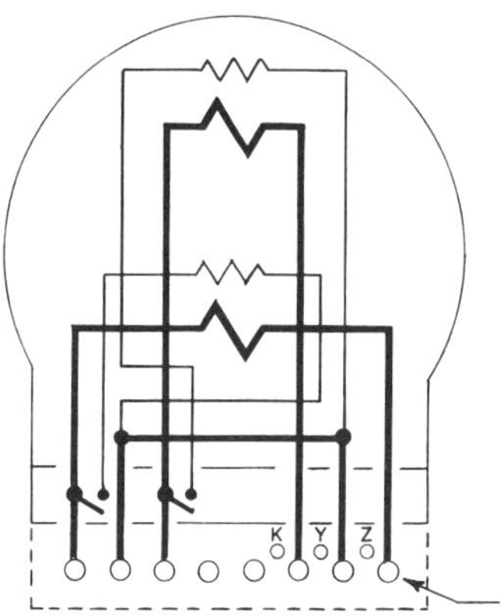

Form 13A
2 Stator Self-Contained
2- or 3-Phase, 3-Wire

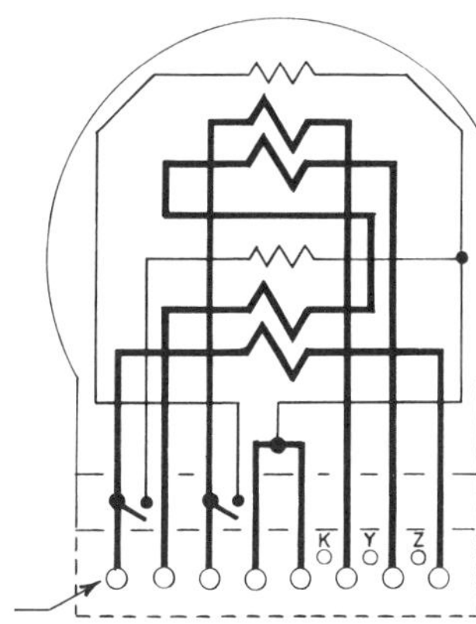

Form 14A
2 Stator Self-Contained
3-Phase, 4-Wire Wye

Figure 12-9.

Internal Connections of Type "A" Meters

Front Views (Except Terminals)

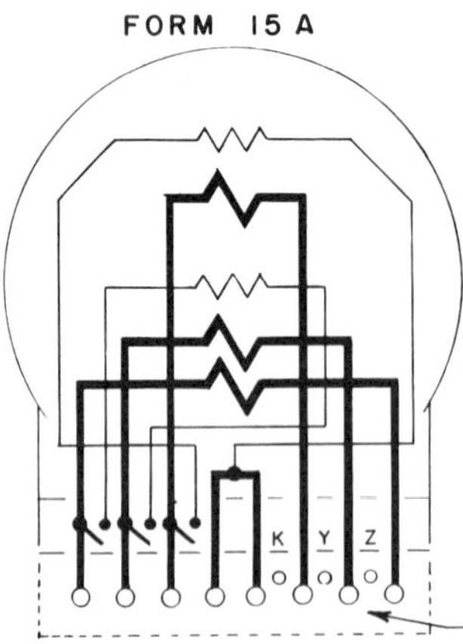

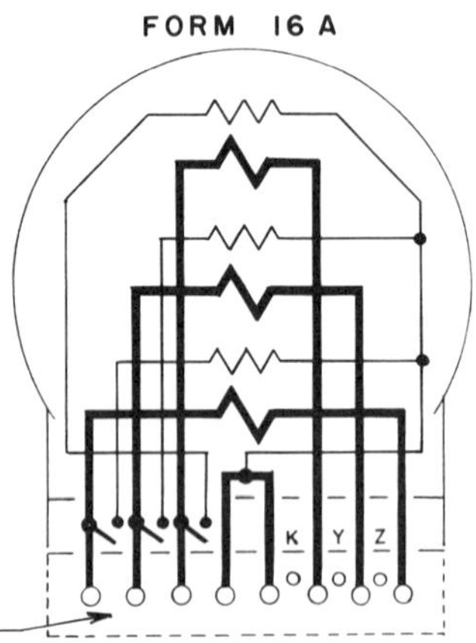

Bottom View

Form 15A
2 Stator Self-Contained
3-Phase, 4-Wire Delta

Form 16A
3 Stator Self Contained
3-Phase, 4-Wire Wye

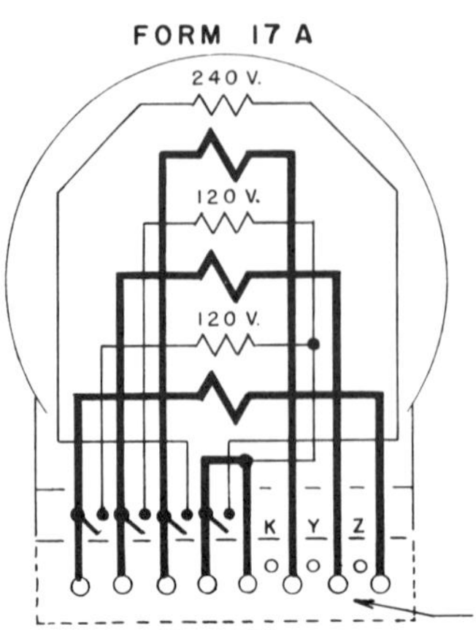

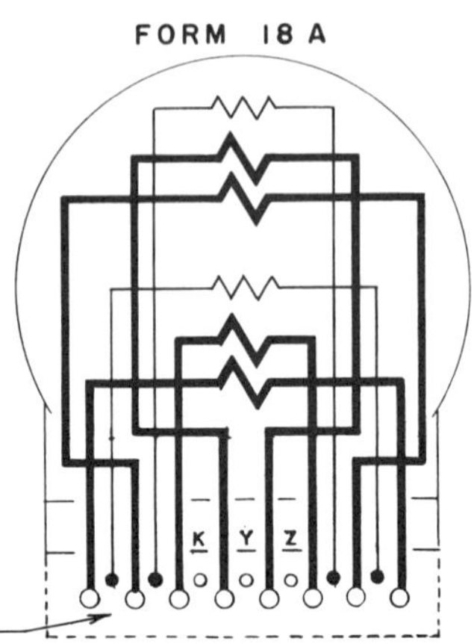

Bottom View

Form 17A
3 Stator Self-Contained
3-Phase, 4-Wire Delta

Form 18A
2 Stator Self-Contained
2-Phase, 5-Wire

Figure 12-10.

Internal Connections of Type "A" Meters

Front Views

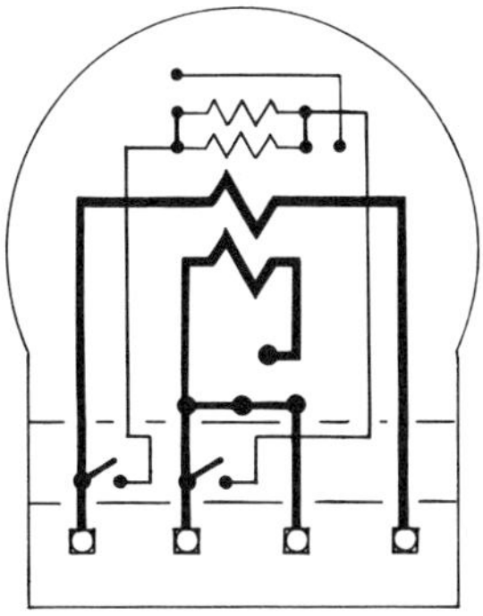

Form 19A-2
2/3-Wire Single Stator Self-Contained
2-Wire Connection

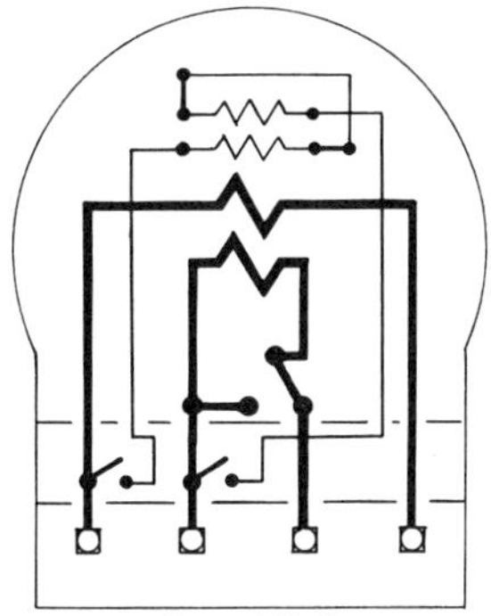

Form 19A-3
2/3-Wire Single Stator
Self-Contained 2-Wire Connection

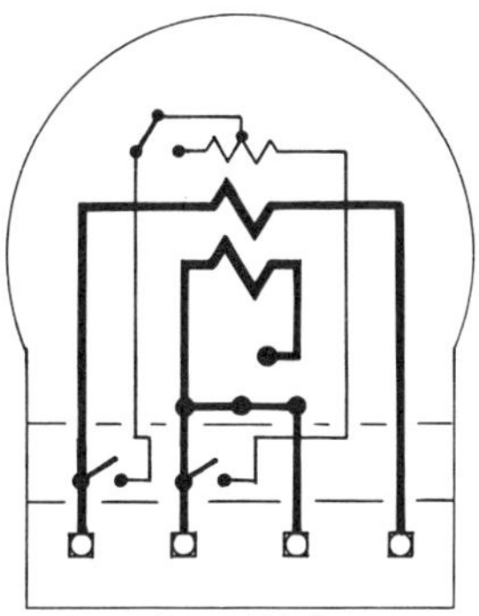

Form 20A-2
2/3-Wire Single Stator
Self-Contained 2-Wire Connection

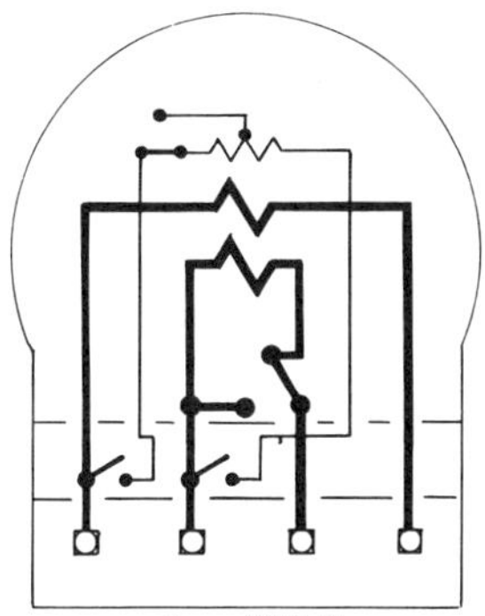

Form 20A-3
2/3-Wire Single Stator
Self-Contained 3-Wire Connection

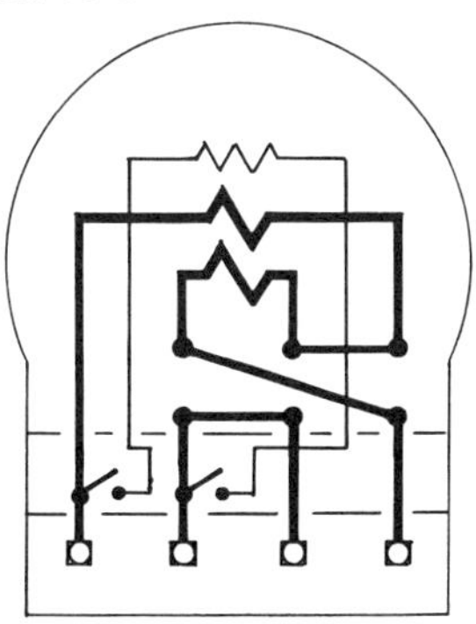

Form 21A-2
2/3-Wire Single Stator
Self-Contained 3-Wire Connection

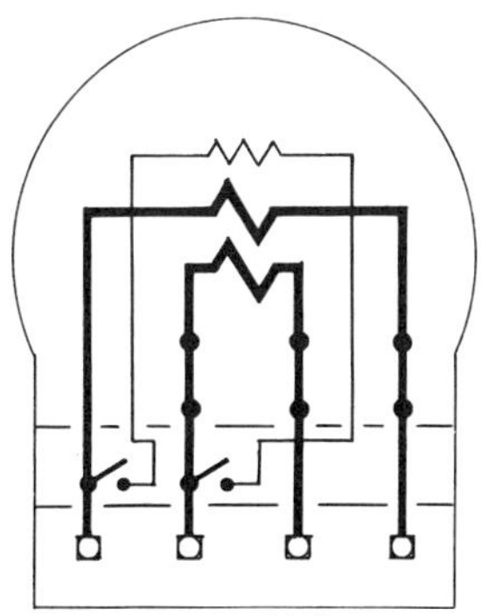

Form 21A-3
2/3-Wire Single Stator
Self-Contained 3-Wire Connection

Figure 12-11.

Single-Phase, 2-Wire Circuits
Front Views

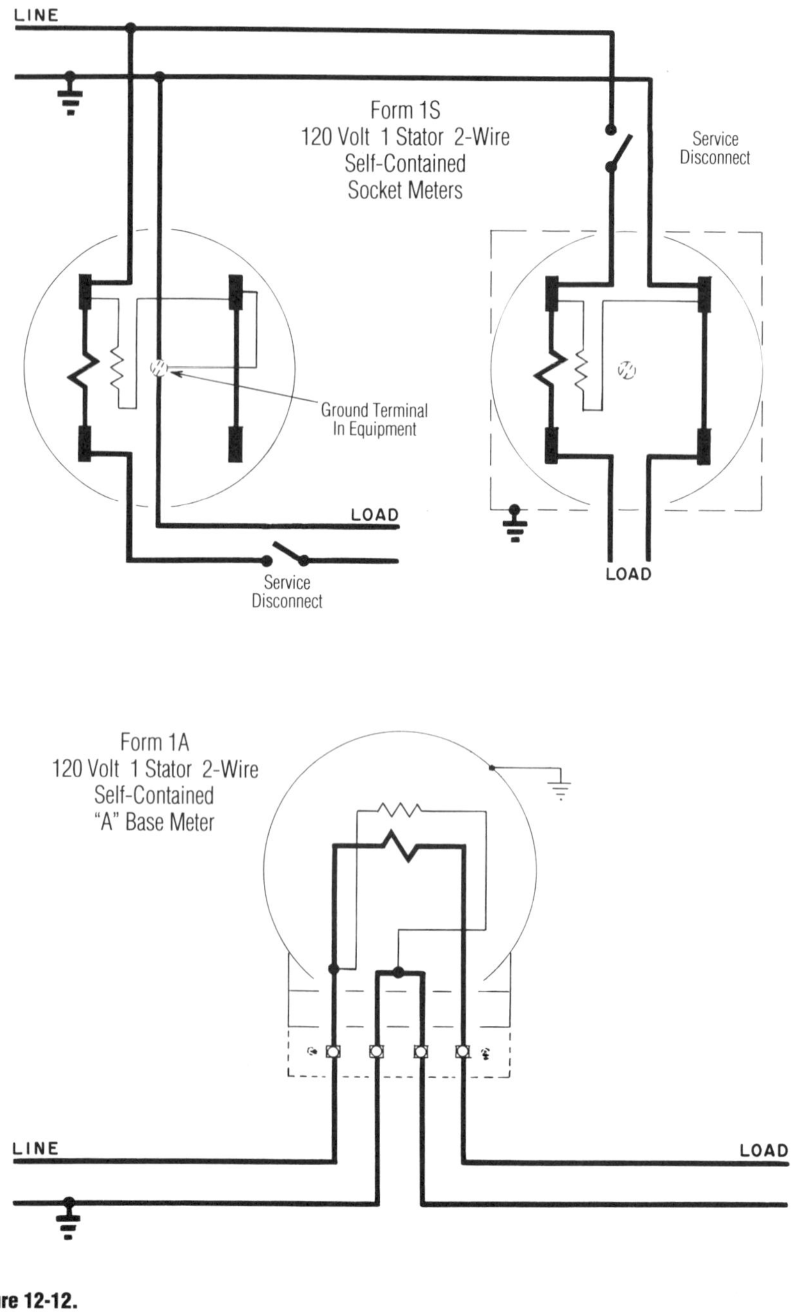

Figure 12-12.

Single-Phase, 2-Wire Circuits
Front Views

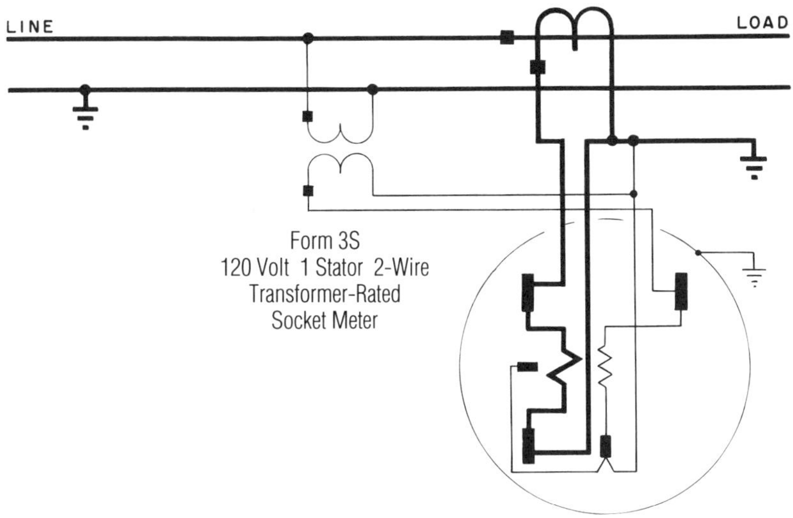

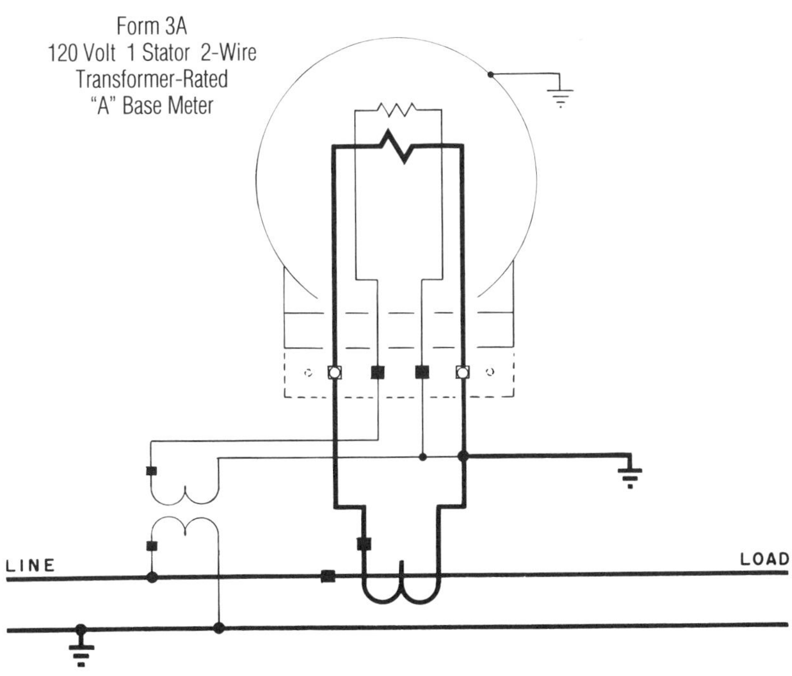

Figure 12-13.

Single-Phase, 2/3-Wire Circuits
Front View

Form 2S
240 Volt 1 Stator 3-Wire
Self-Contained Socket Meter
Connected to Single-Phase 2-Wire 120 Volt Service,
Using an "A" Base Socket Meter Adapter.

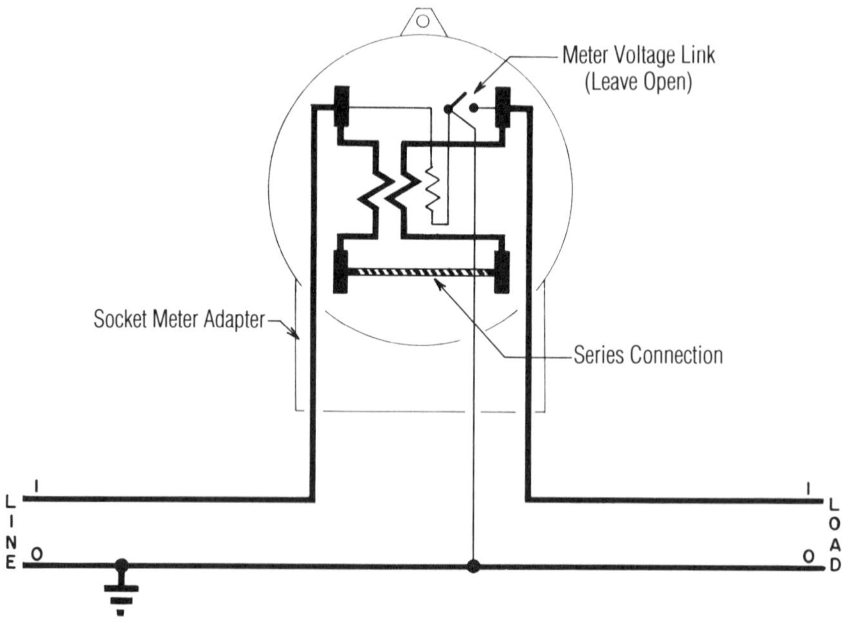

Note: Socket meter adapter to be wired with the line wire (live) connected to the upper-left terminal, the load wire (live) connected to the upper-right terminal. Series-connect lower-left terminal to lower-right terminal. With meter voltage link open, connect neutral (ground) wire to the voltage coil terminal (link) screw on the meter.

Figure 12-14.

Single-Phase, 2/3-Wire Circuits
Front Views

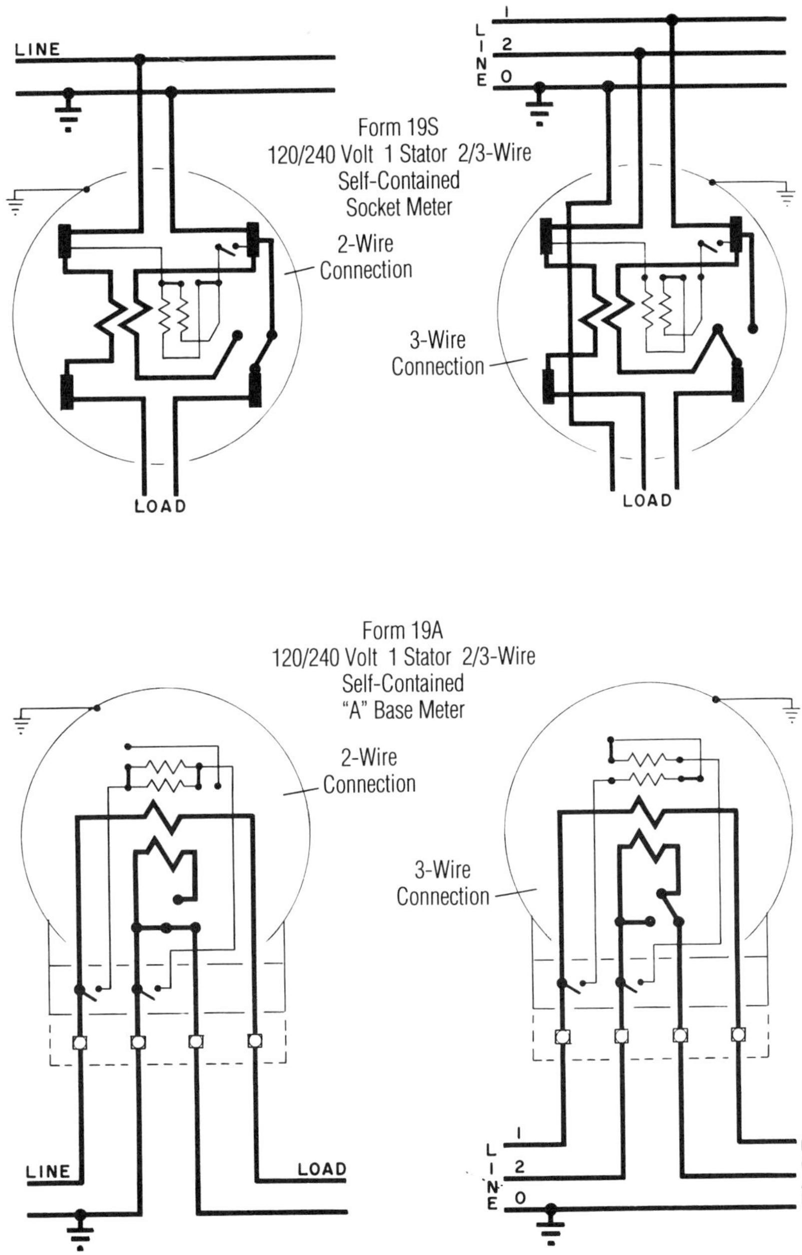

Figure 12-15.

Single-Phase, 2/3-Wire Circuits
Front Views

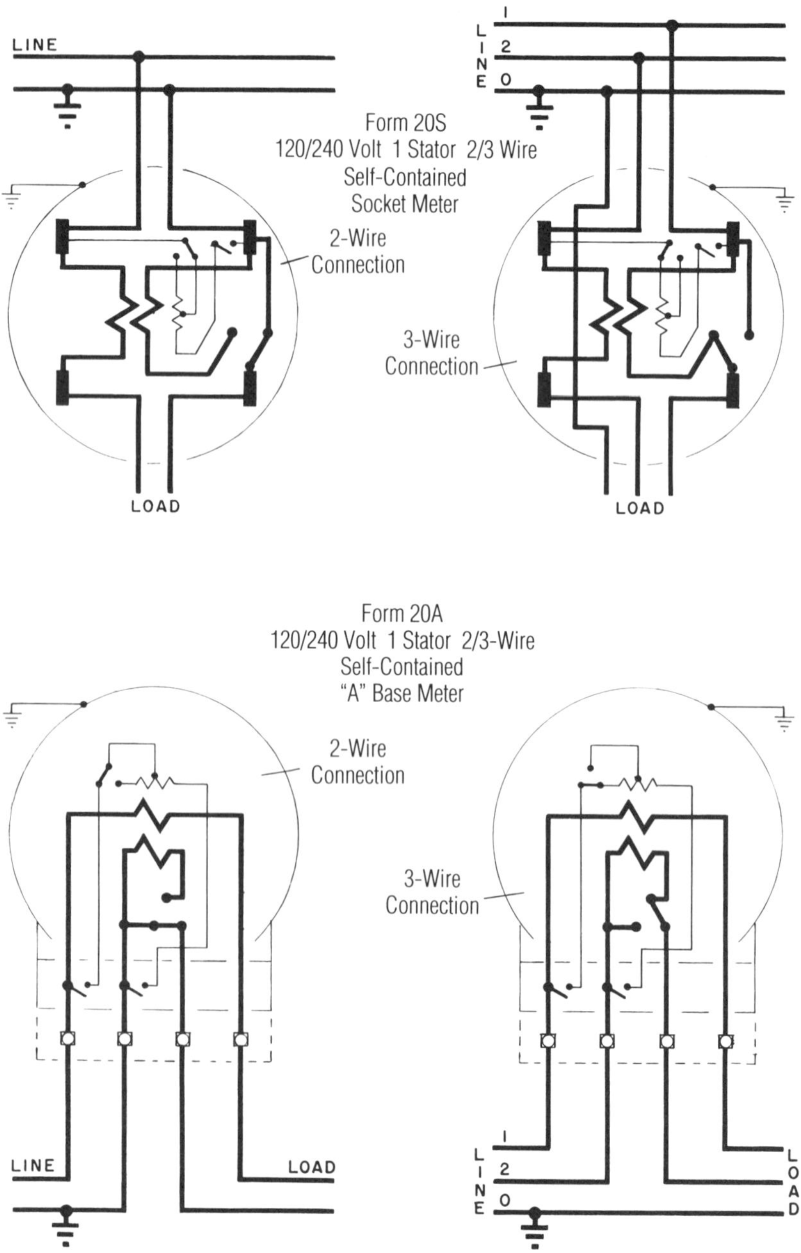

Figure 12-16.

Single-Phase, 2/3-Wire Circuits
Front Views

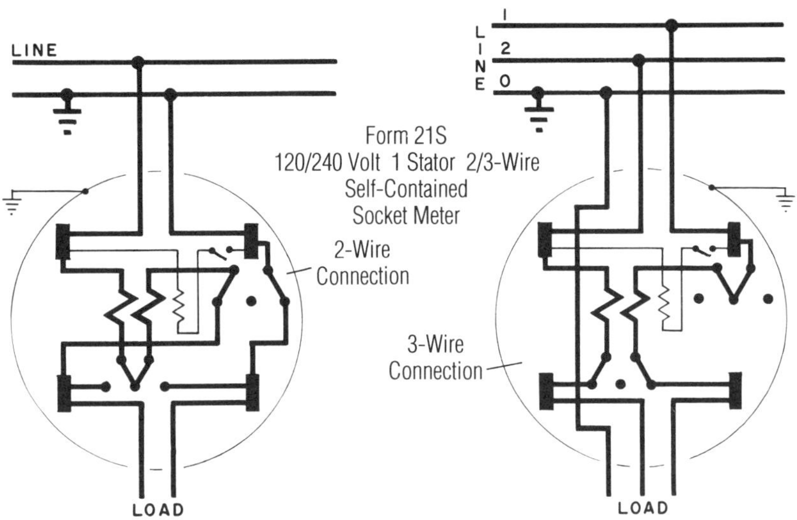

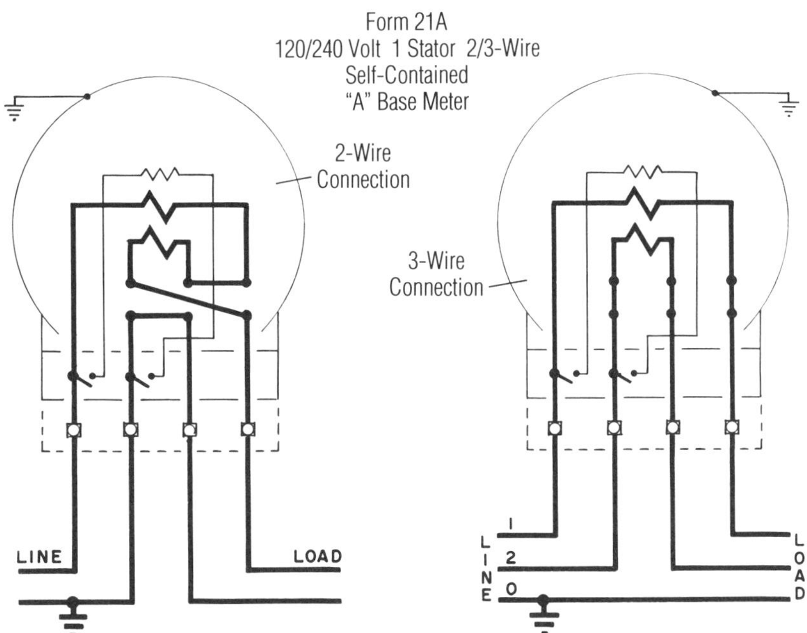

Figure 12-17.

Single-Phase, 3-Wire Circuits
Front Views

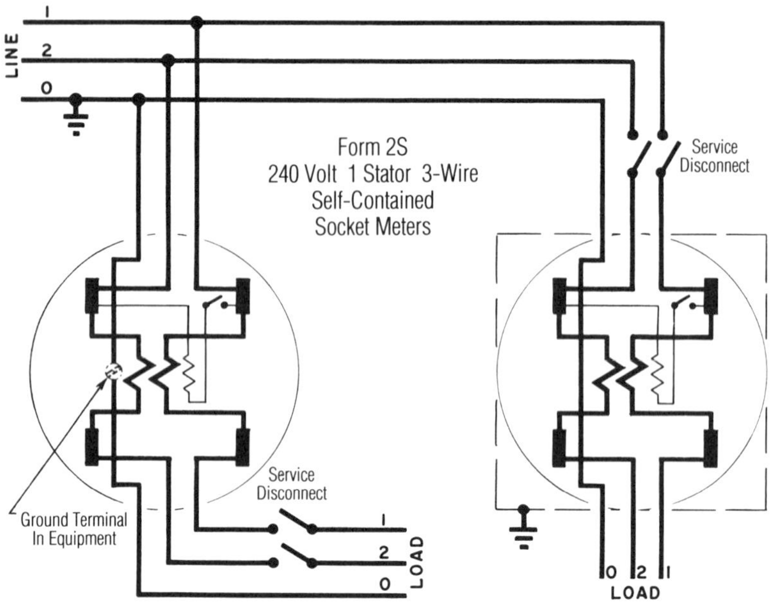

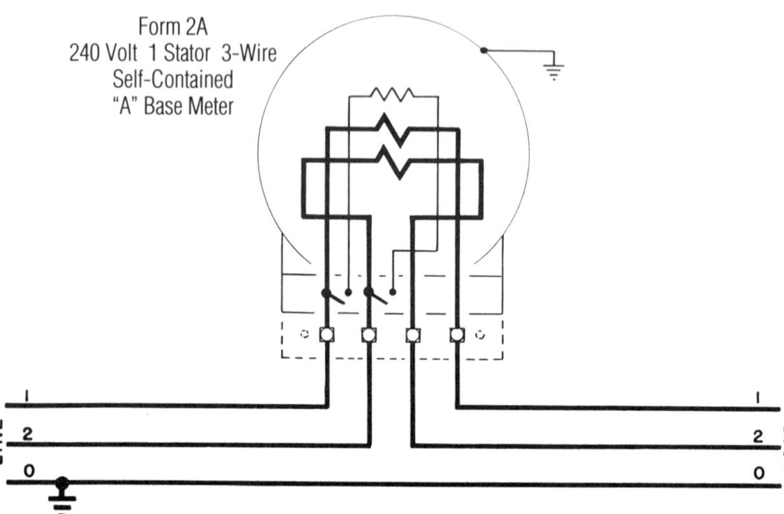

Figure 12-18.

Single-Phase, 3-Wire Circuits
Front Views

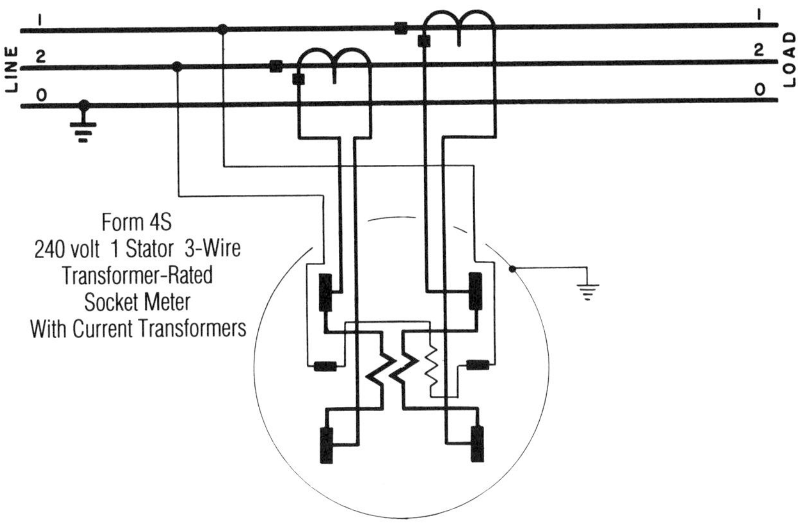

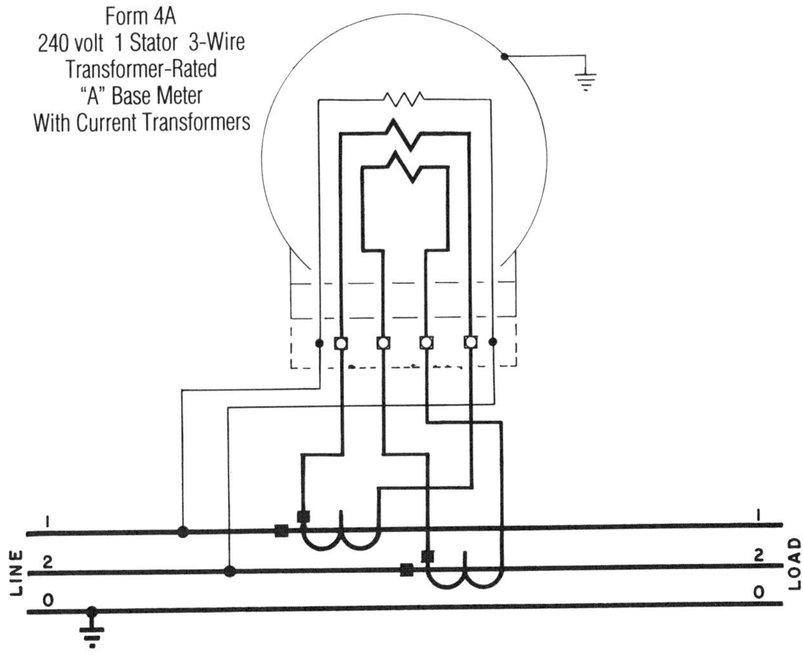

Figure 12-19.

Single-Phase, 3-Wire Circuits
Front Views

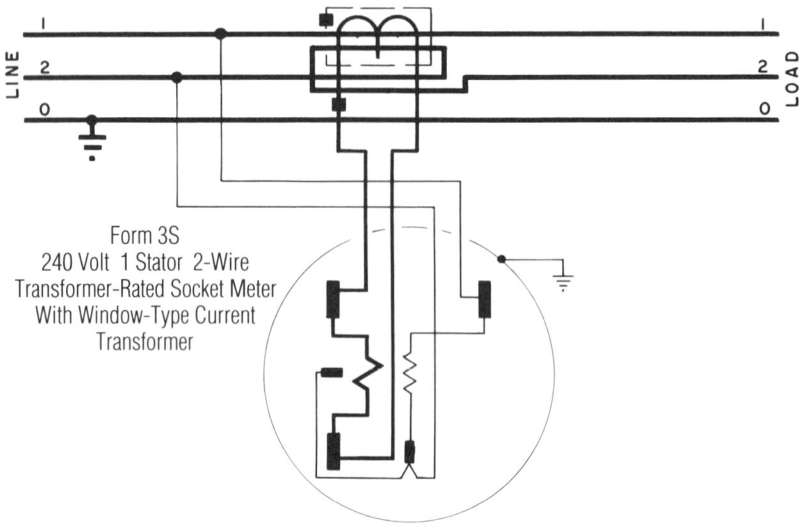

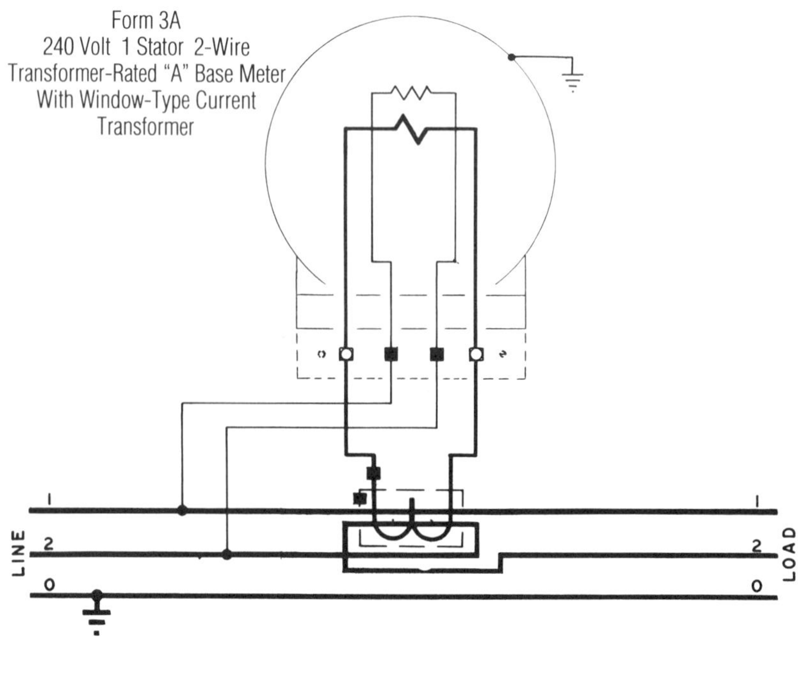

Figure 12-20.

Single-Phase, 3-Wire Circuits
Front Views

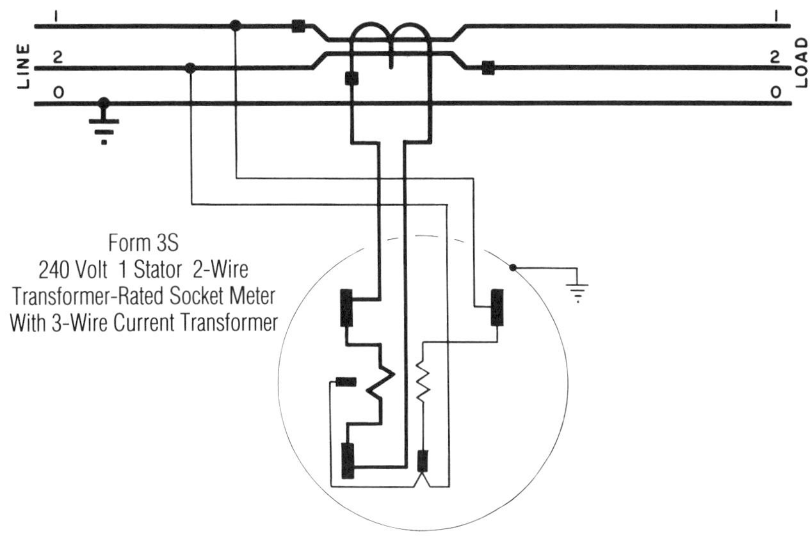

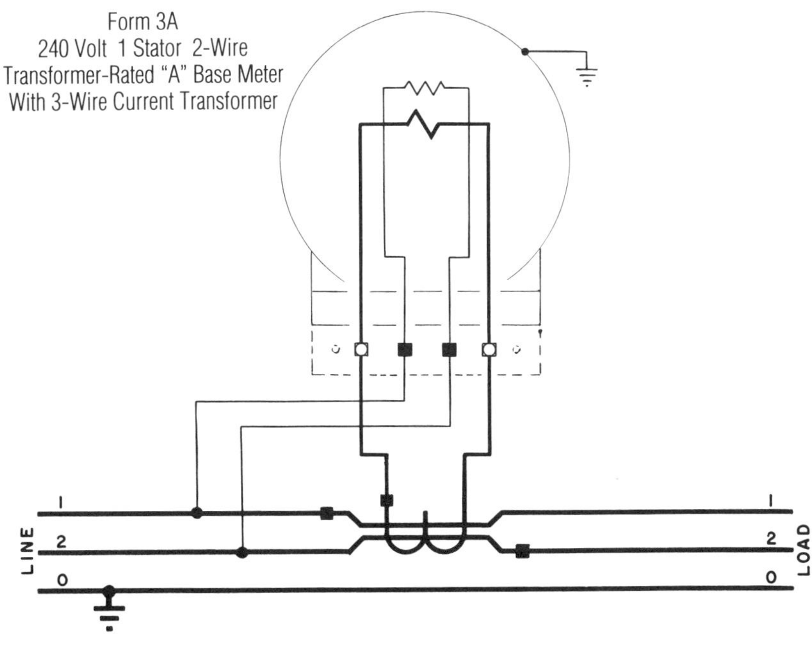

Figure 12-21.

3-Wire Network Circuit
Front Views

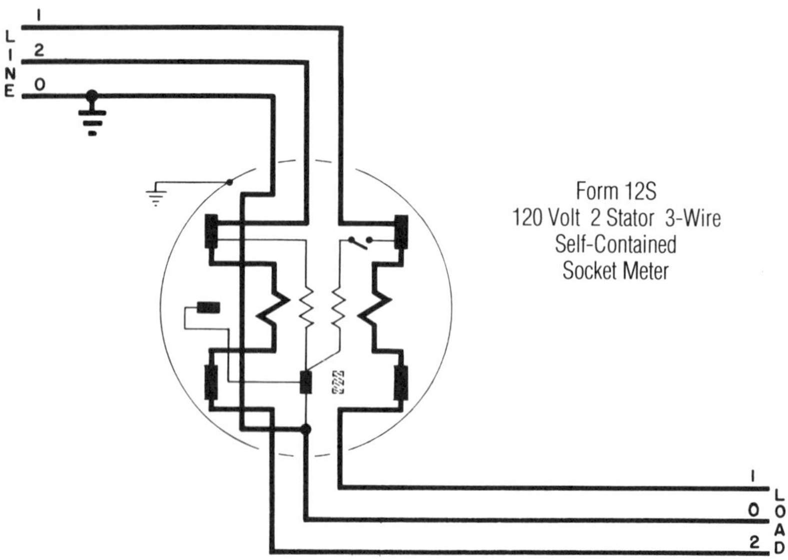

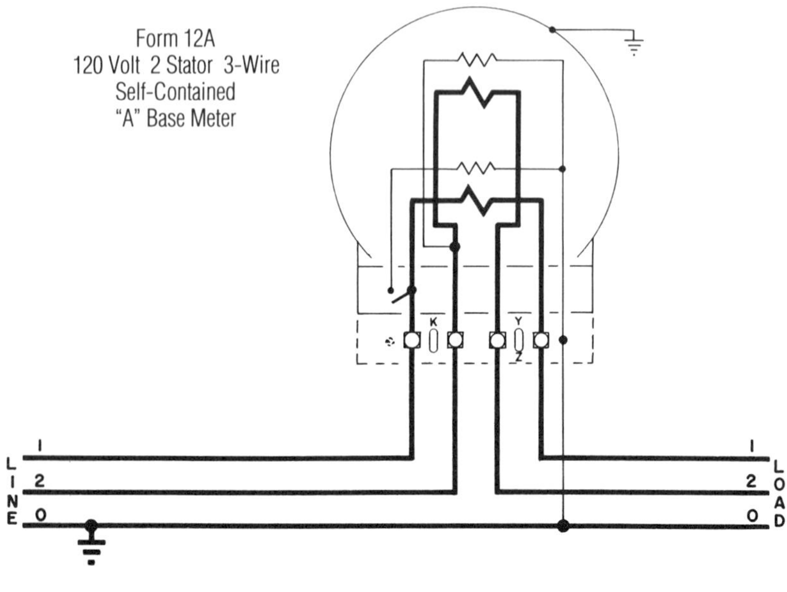

Figure 12-22.

3-Wire Network Circuit
Front Views

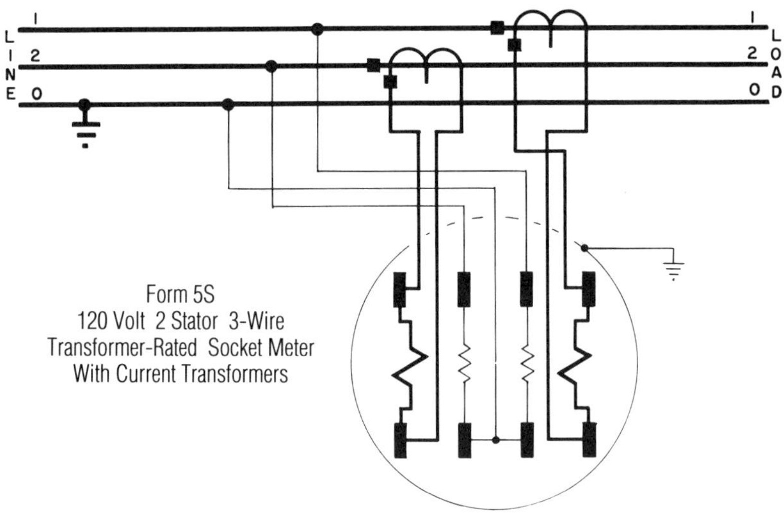

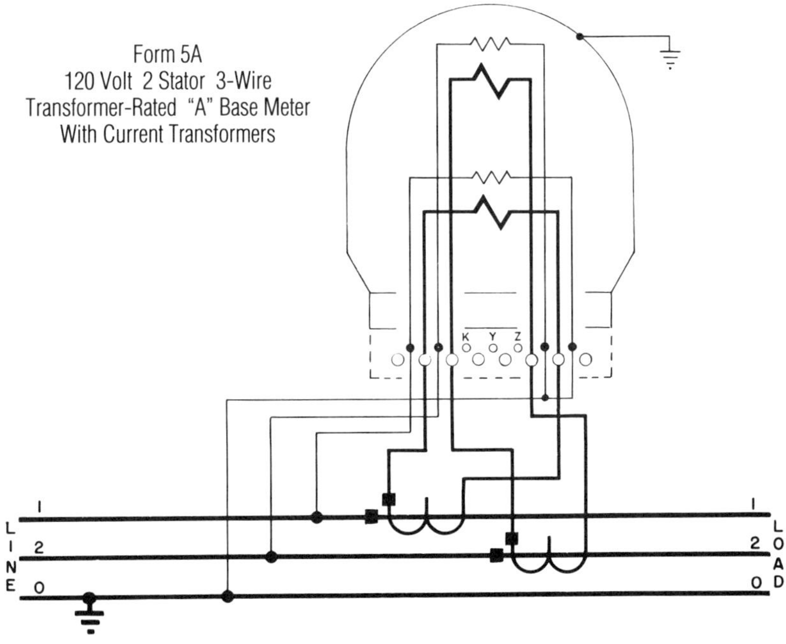

Figure 12-23.

3-Wire Network Circuit
Front Views

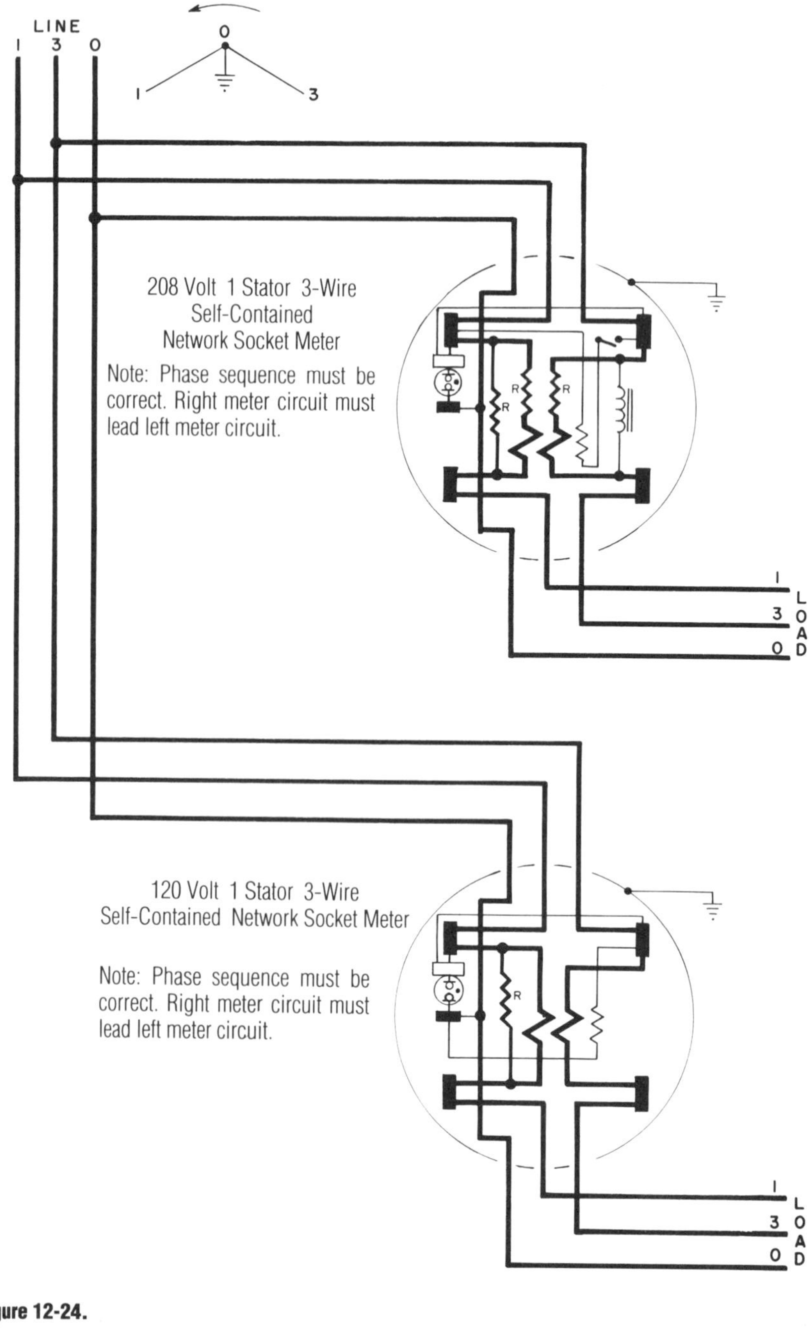

Figure 12-24.

Three-Phase, 3-Wire Circuits
Front Views

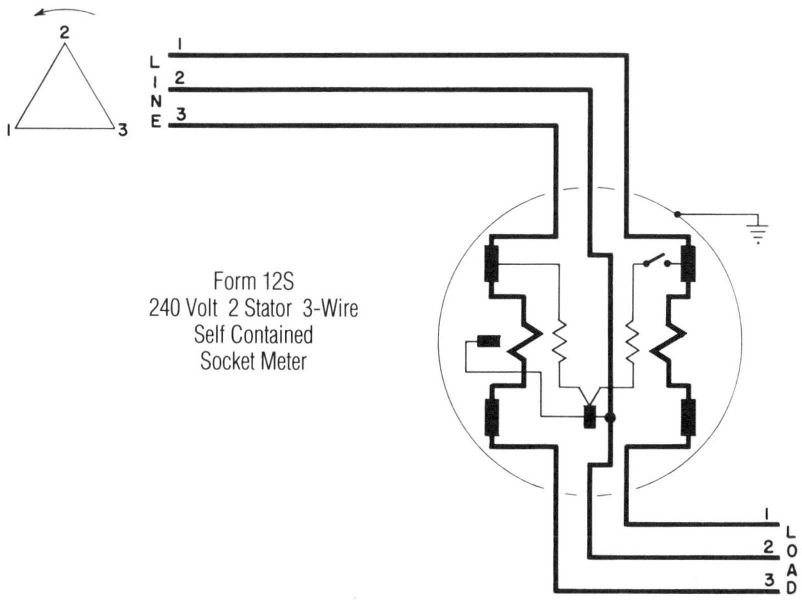

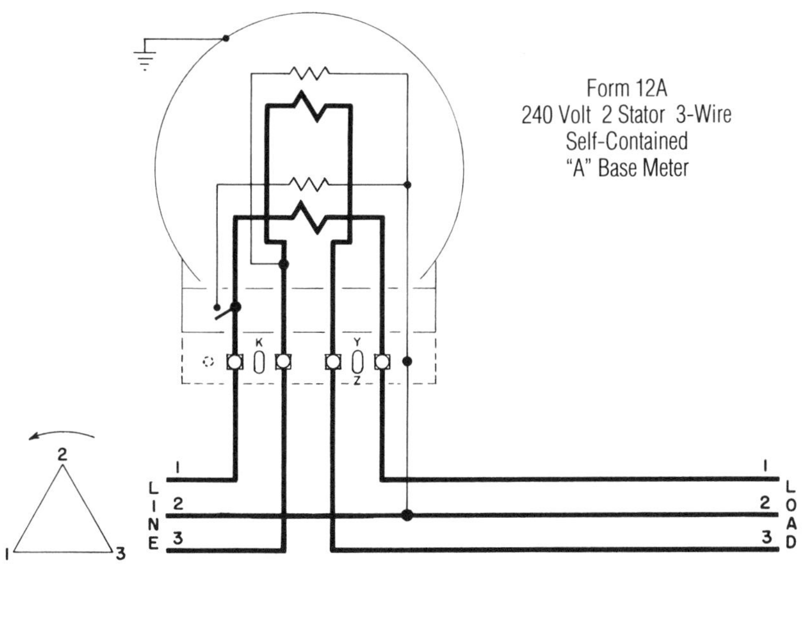

Figure 12-25.

Three-Phase, 3-Wire Circuits

Front Views

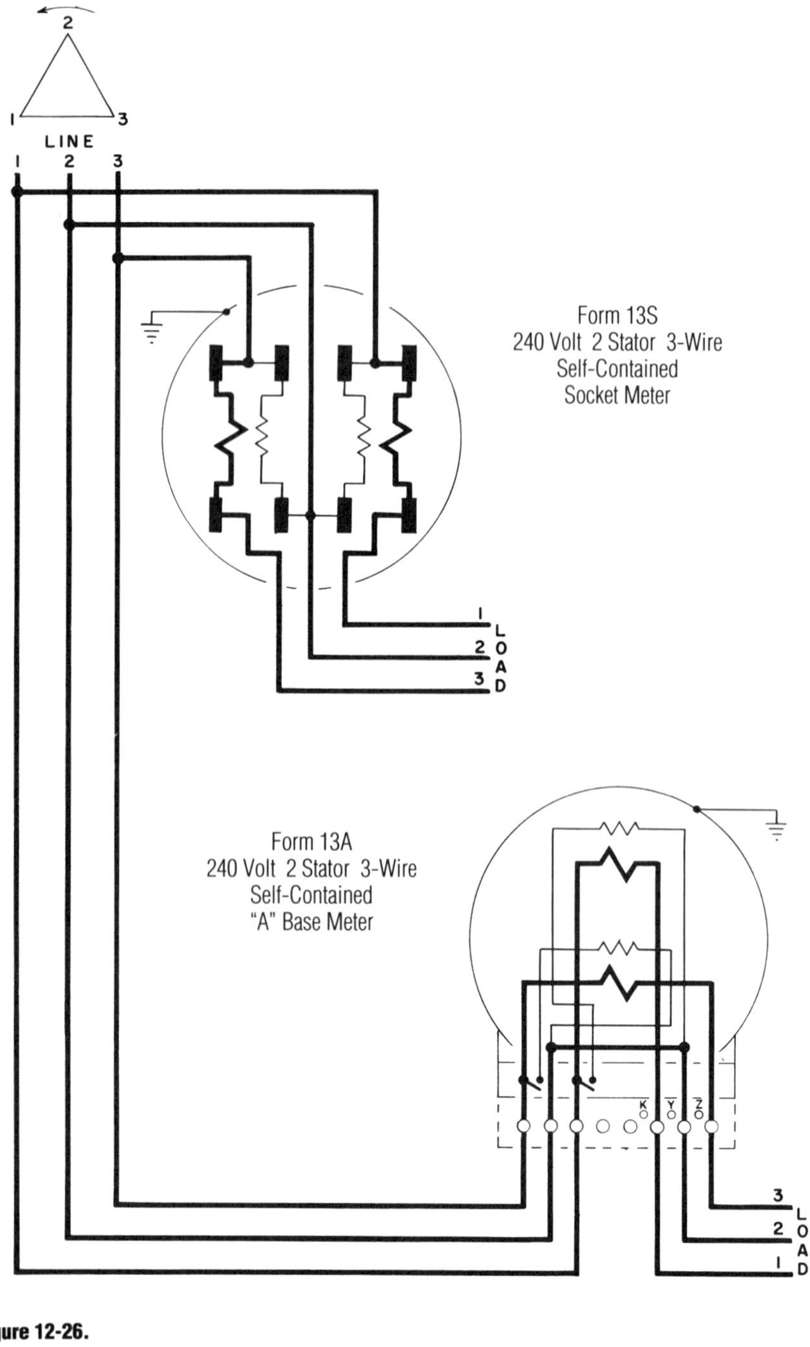

Figure 12-26.

Three-Phase, 3-Wire Circuits
Front View

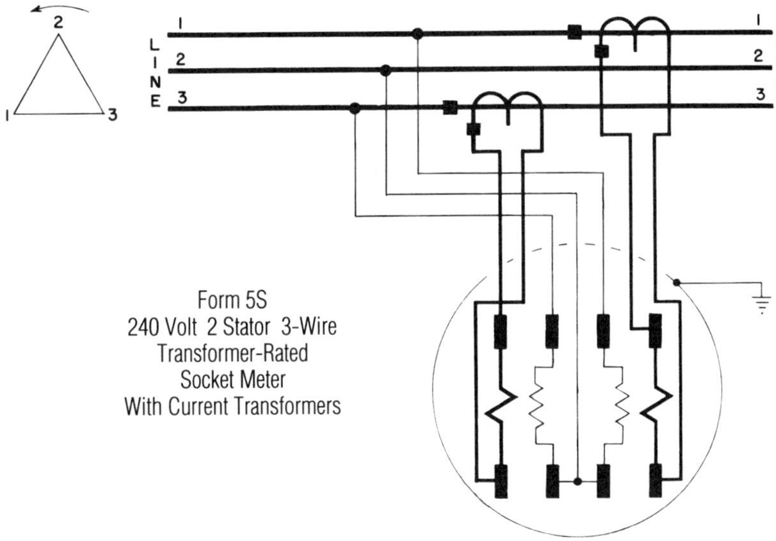

Form 5S
240 Volt 2 Stator 3-Wire
Transformer-Rated
Socket Meter
With Current Transformers

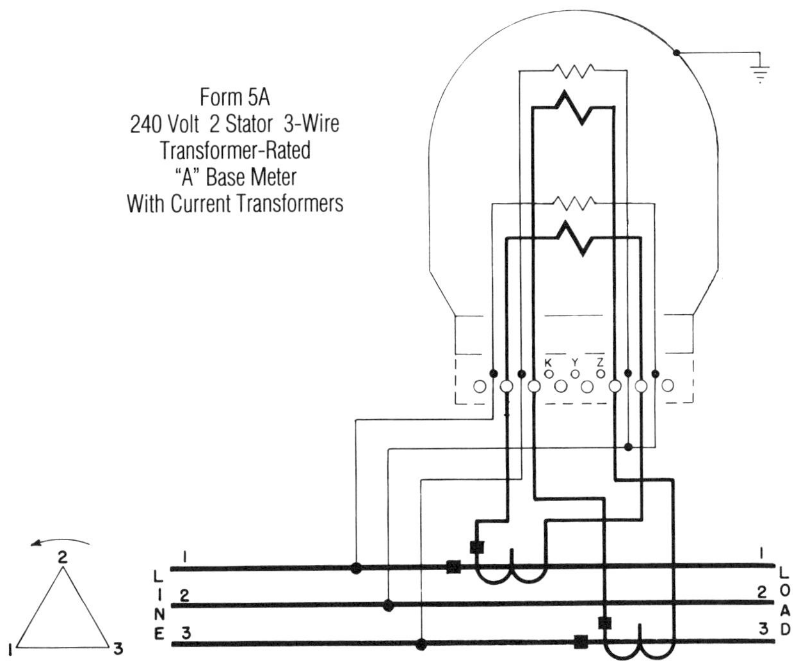

Form 5A
240 Volt 2 Stator 3-Wire
Transformer-Rated
"A" Base Meter
With Current Transformers

Figure 12-27.

Three-Phase, 3-Wire Circuits
Front View

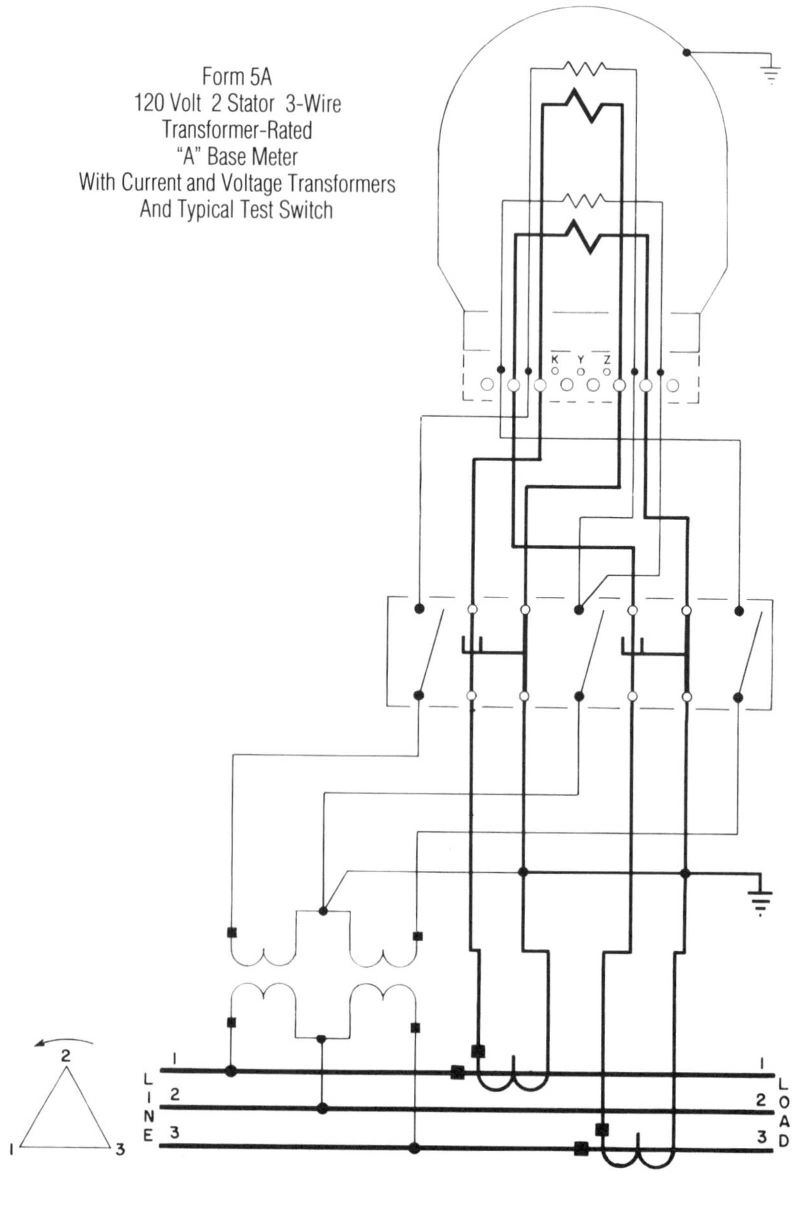

Figure 12-28.

Three-Phase, 3-Wire Circuits
Front Views

Form 5A.
120 Volt 2 Stator 3-Wire Meters
With Current and Voltage Transformers, and Test Switches

Typical circuit of metering "Out" and "In" current flow,
showing current circuits of "In" meter reversed.

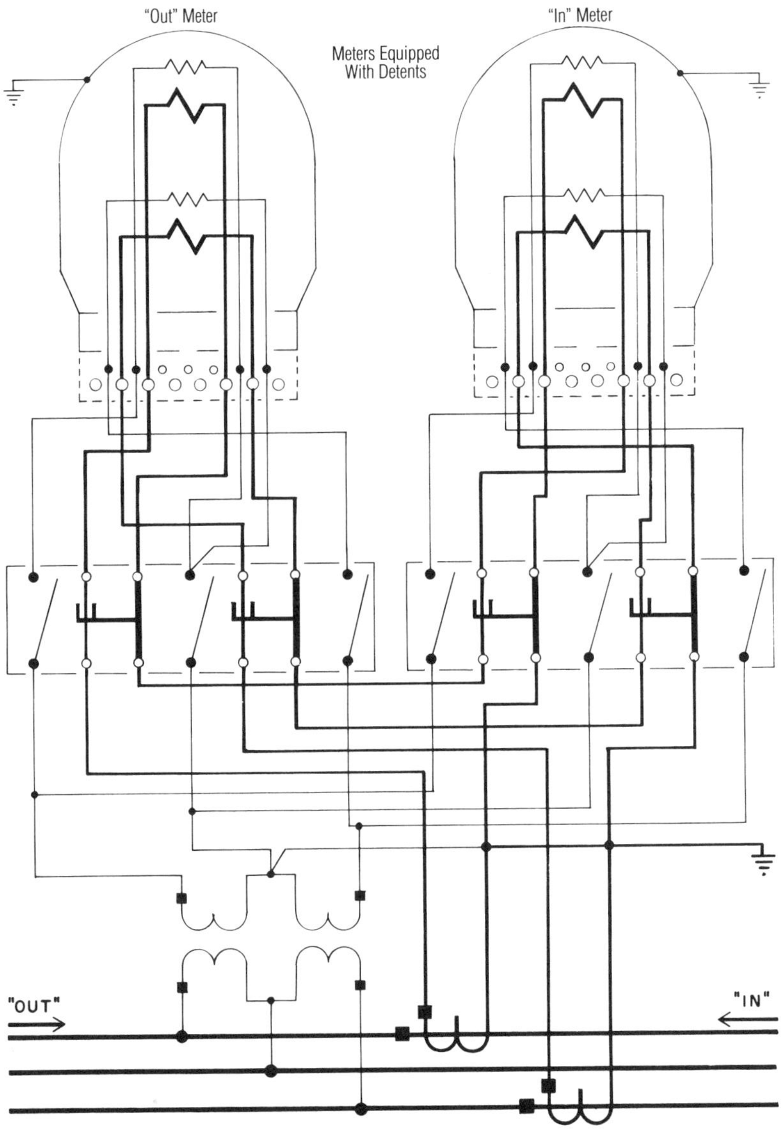

Figure 12-29.

Three-Phase, 4-Wire Wye Circuits

Front Views

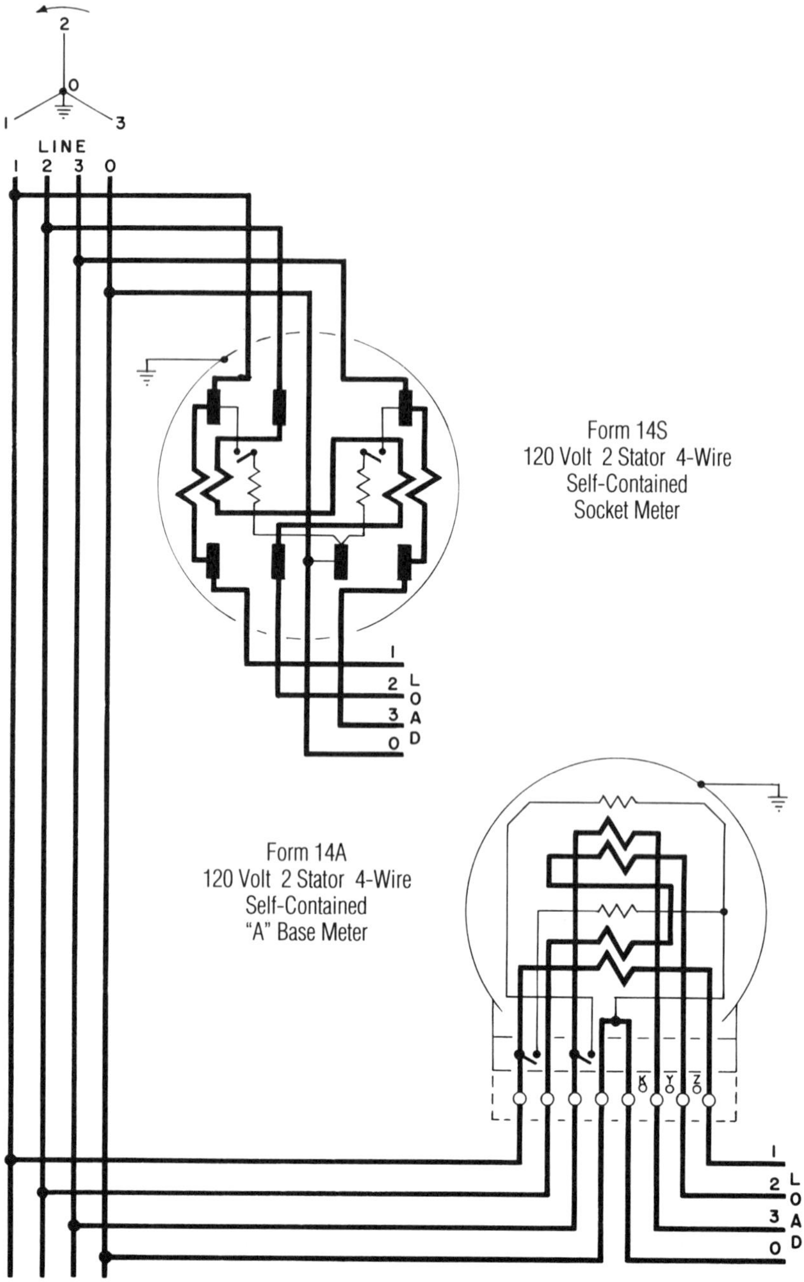

Figure 12-30.

Three-Phase, 4-Wire Wye Circuits
Front View

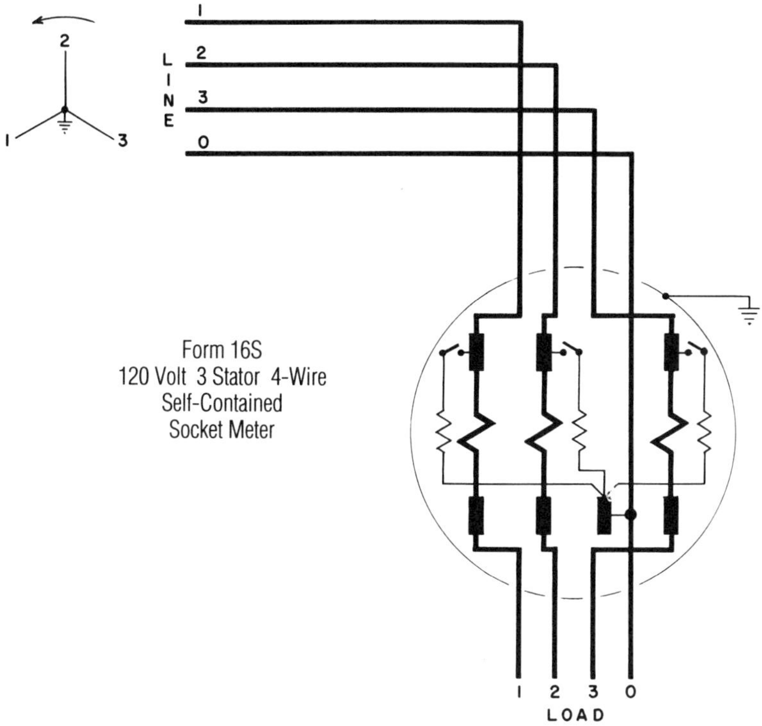

Figure 12-31.

Three-Phase, 4-Wire Wye Circuits
Front Views

Form 16A
120 Volt 3 Stator 4-Wire
Self-Contained
"A" Base Meter
Showing typical self-contained
connection blocks with wiring connected in
Line-Load, Line-Load, Line-Load sequence.

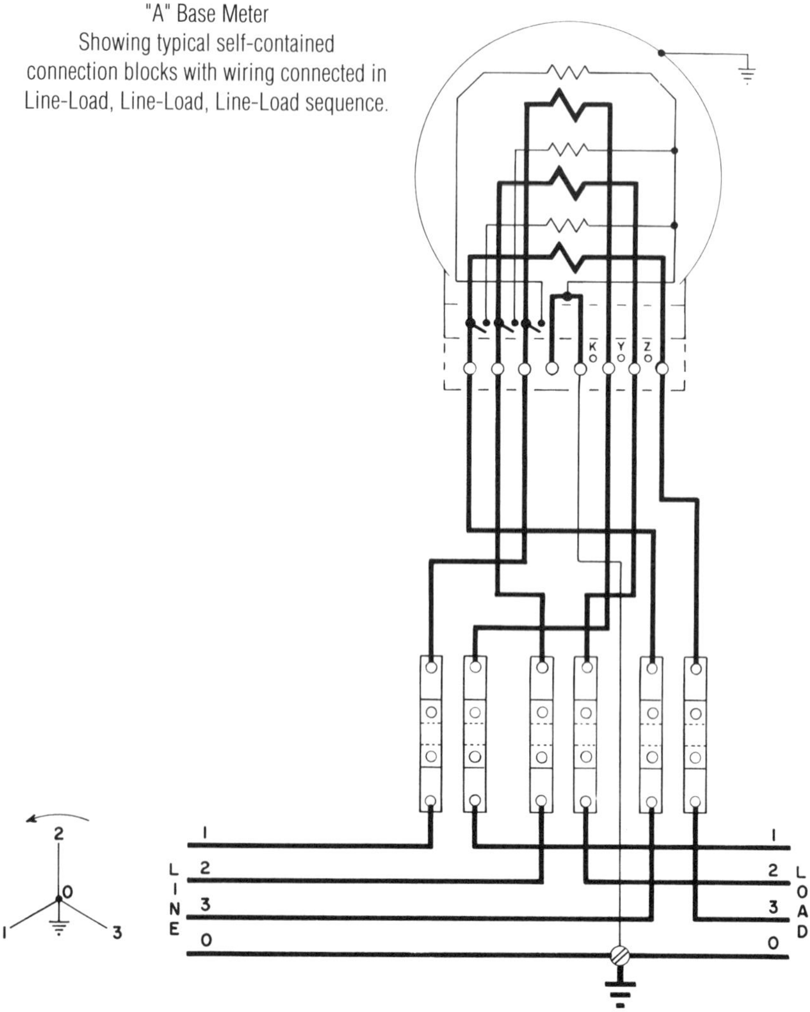

Figure 12-32.

Three-Phase, 4-Wire Wye Circuits
Front View

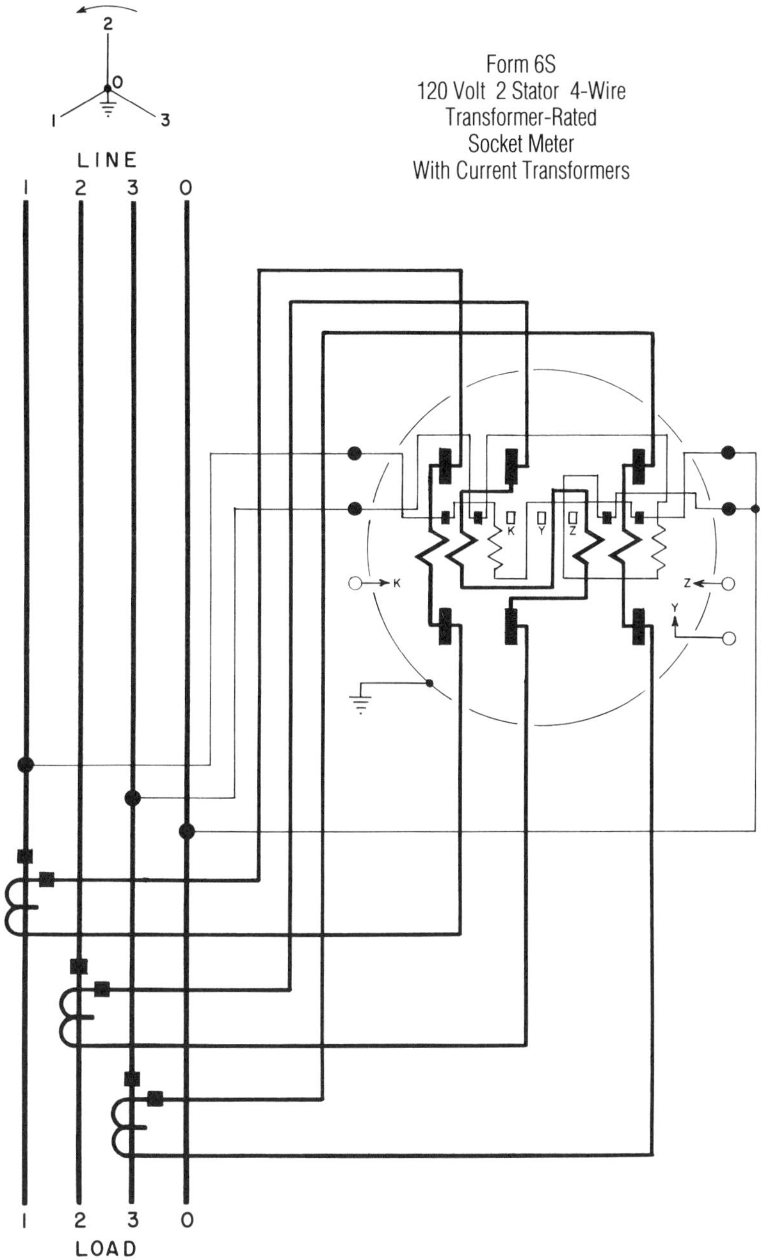

Figure 12-33.

Three-Phase, 4-Wire Wye Circuits
Front View

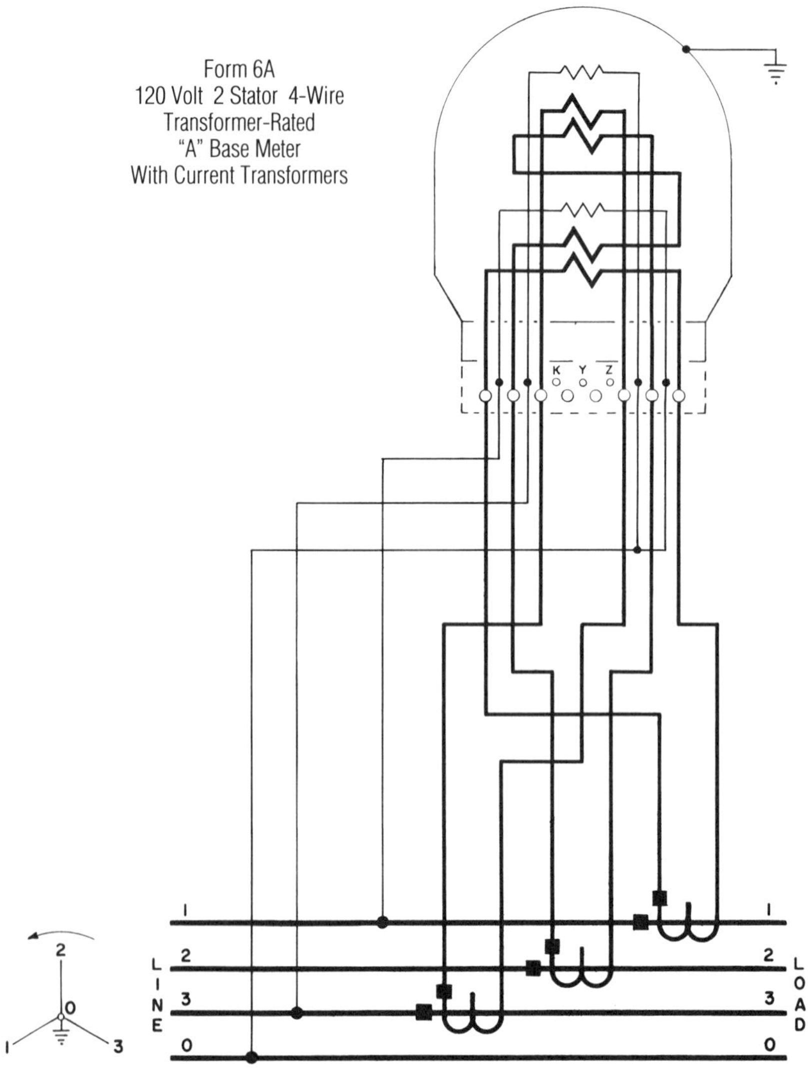

Figure 12-34.

Three-Phase, 4-Wire Wye Circuits
Front View

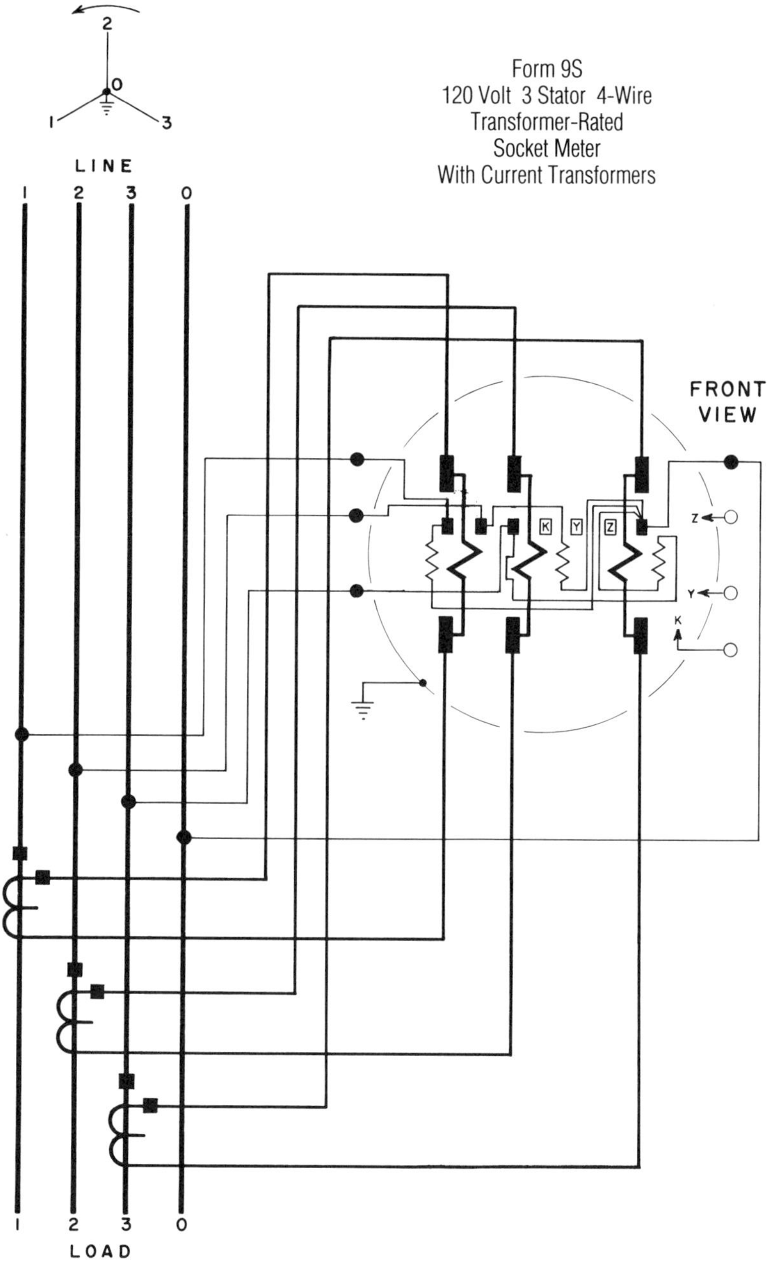

Figure 12-35.

Three-Phase, 4-Wire Wye Circuits
Front View

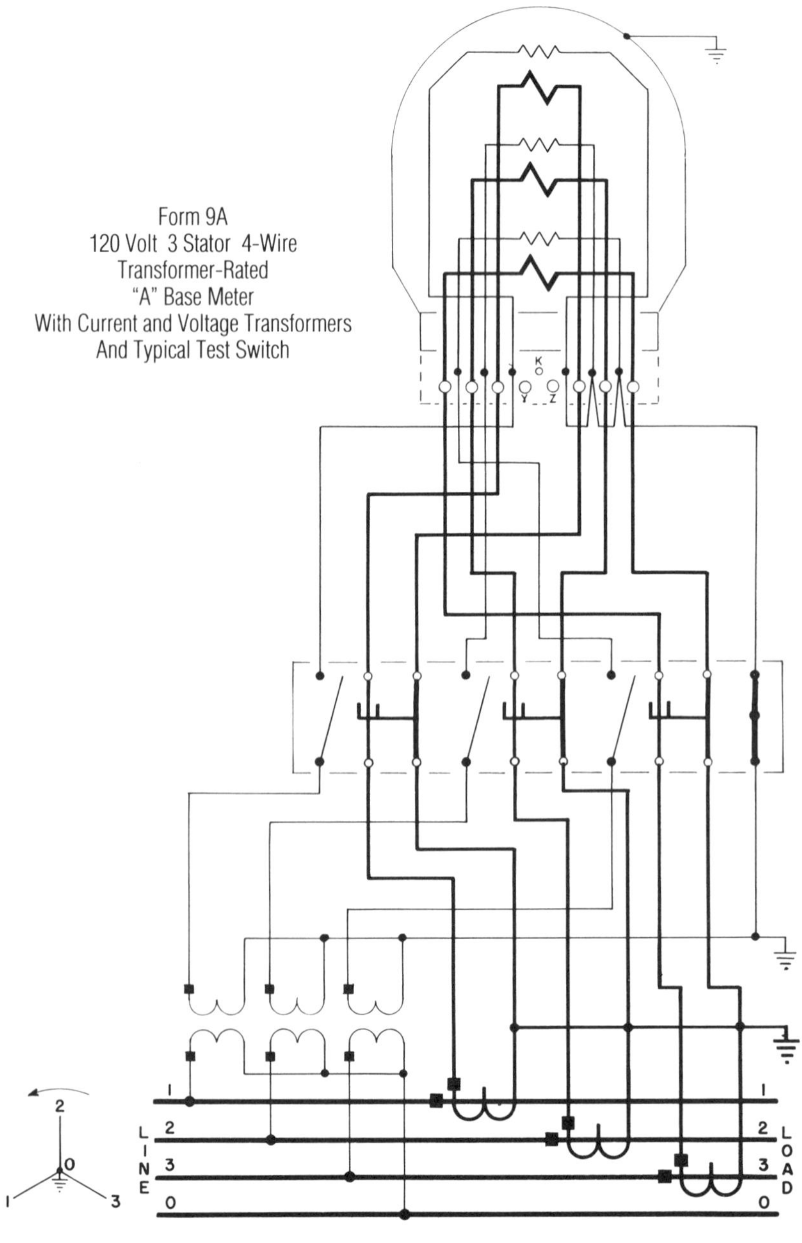

Figure 12-36.

Three-Phase, 4-Wire Delta Circuits
Front Views

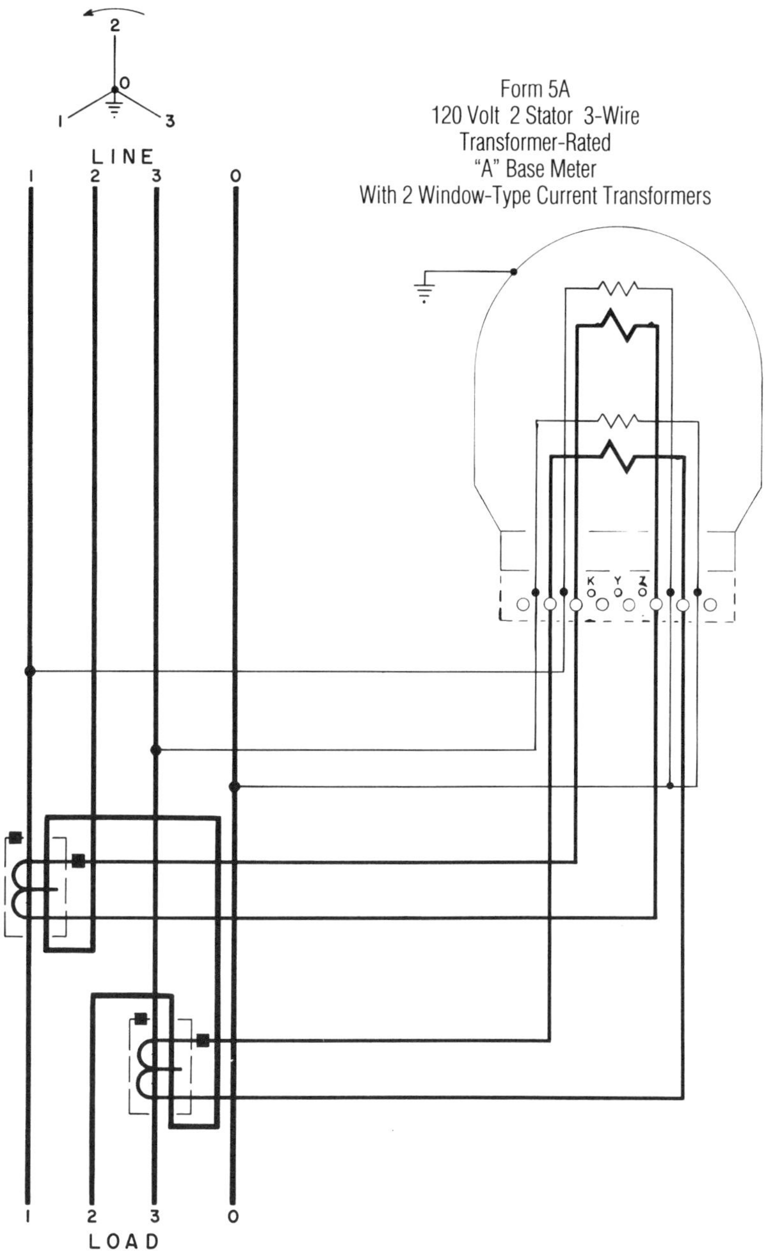

Figure 12-37.

Three-Phase, 4-Wire Wye Circuits
Front View

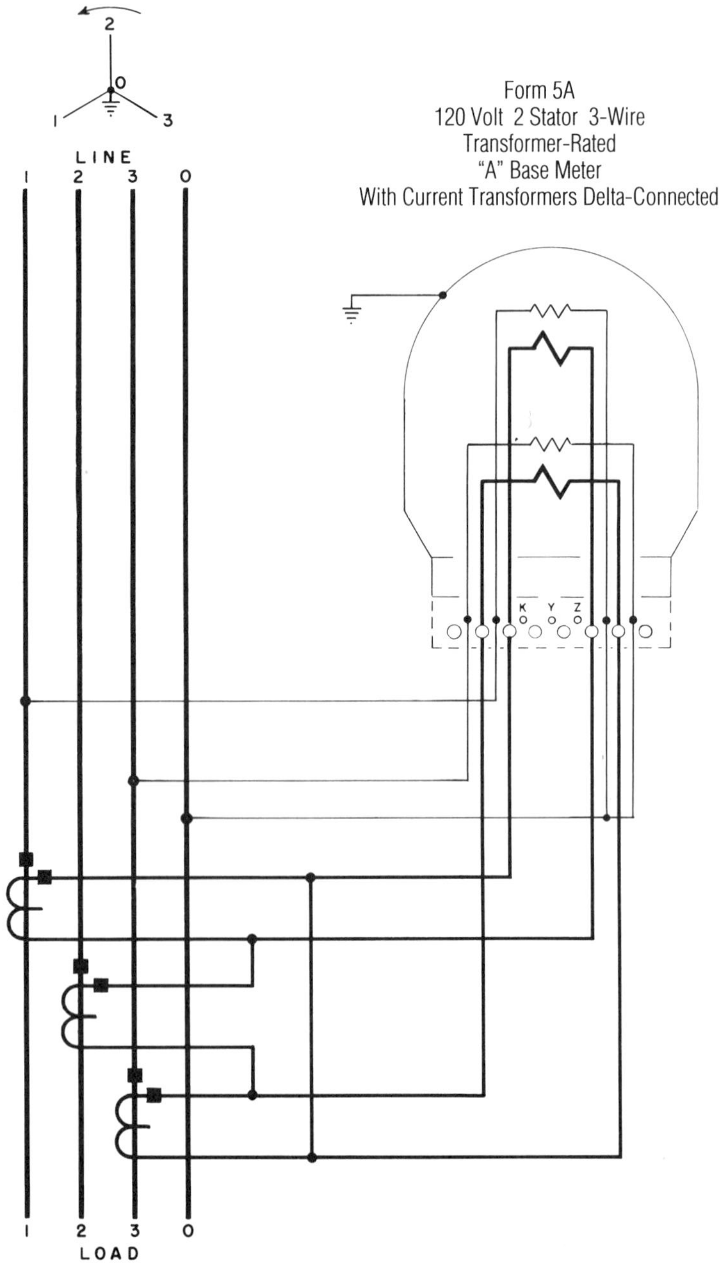

Figure 12-38.

Three-Phase, 4-Wire Delta Circuits

Front Views

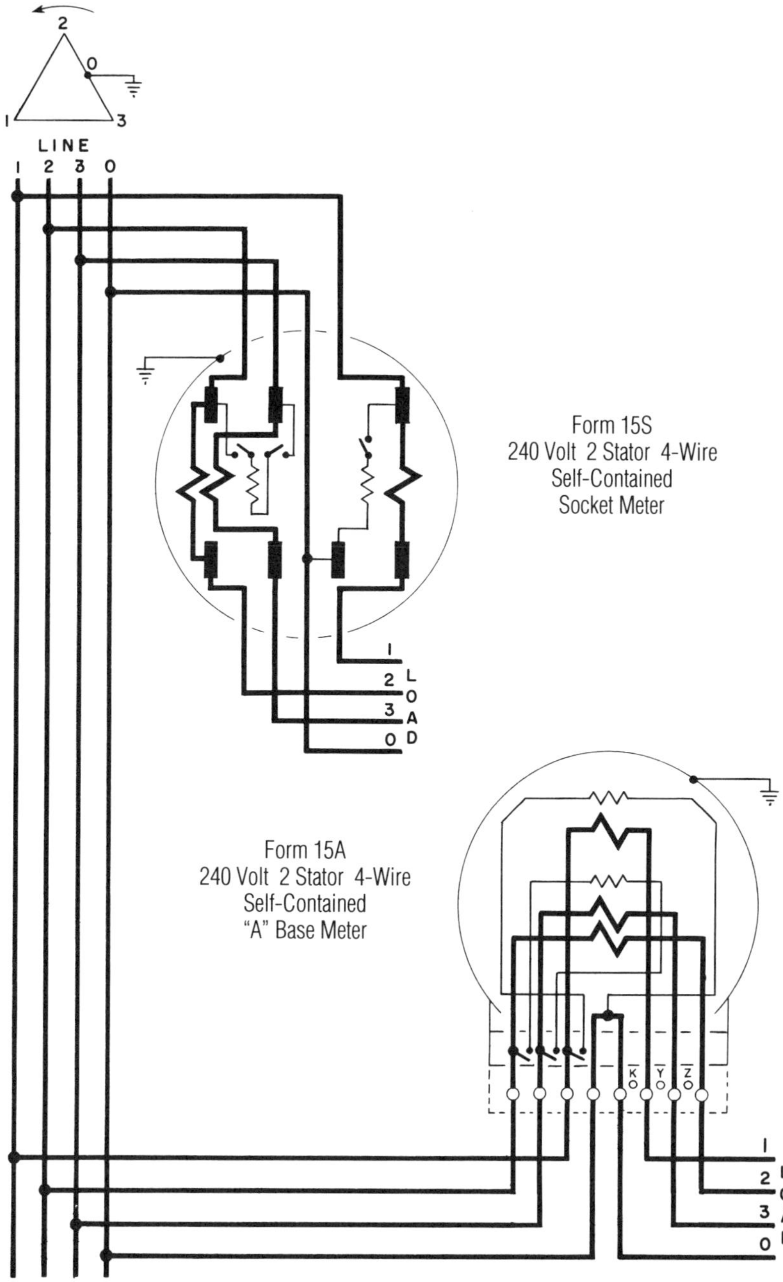

Figure 12-39.

Three-Phase, 4-Wire Delta Circuits
Front Views

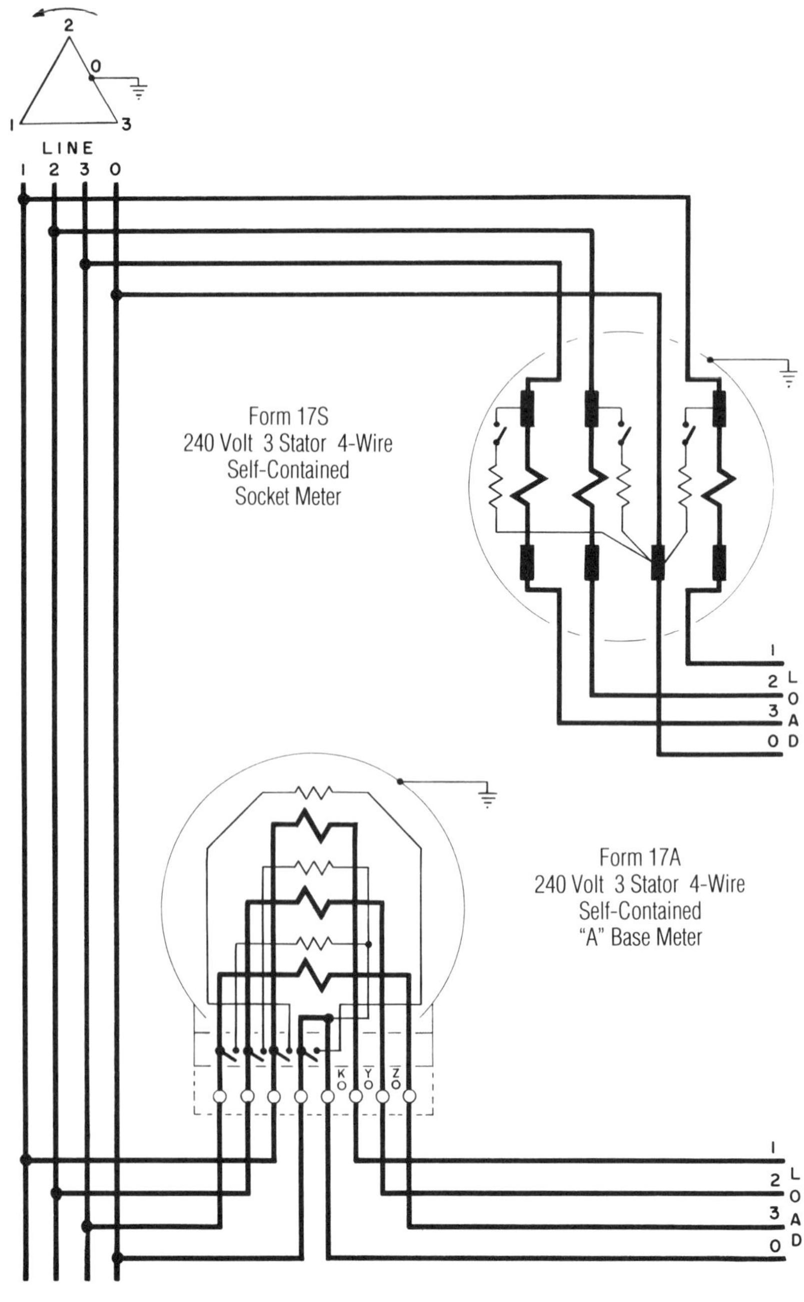

Figure 12-40.

Three-Phase, 4-Wire Delta Circuits

Front Views

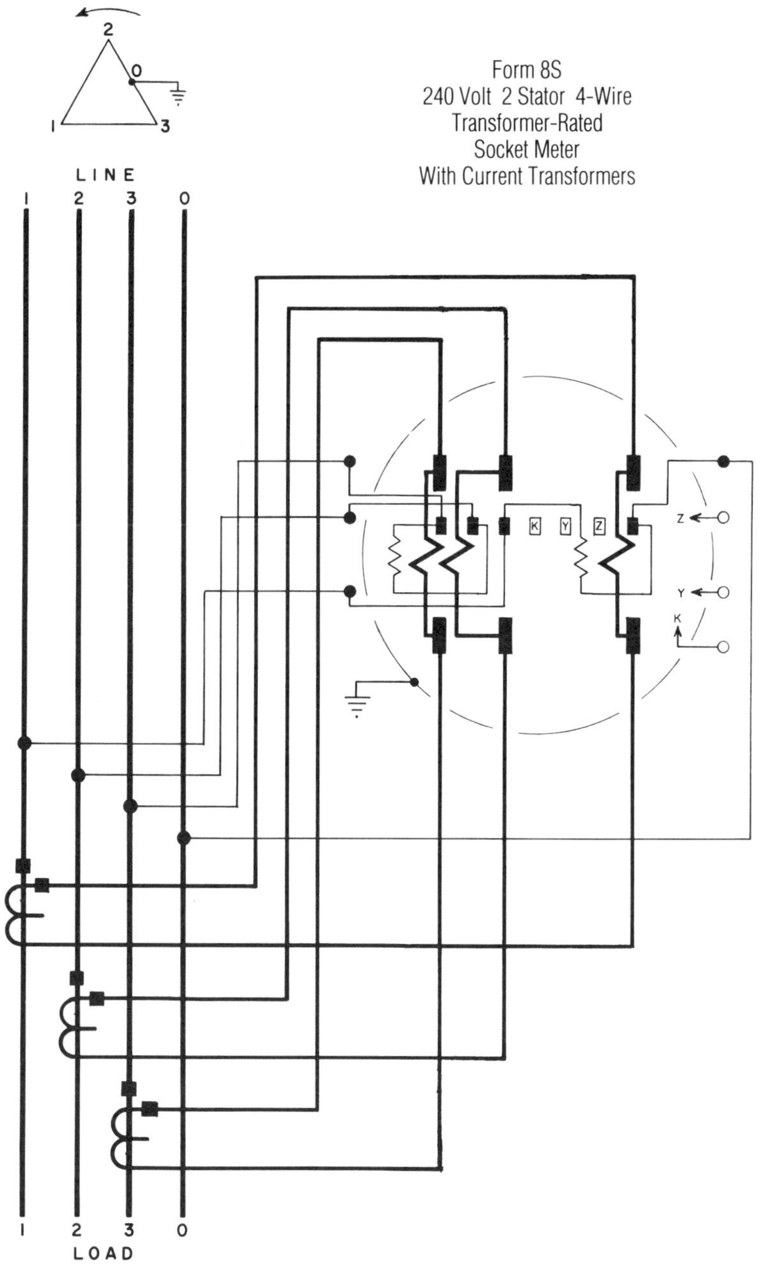

Figure 12-41.

Three-Phase, 4-Wire Delta Circuits
Front View

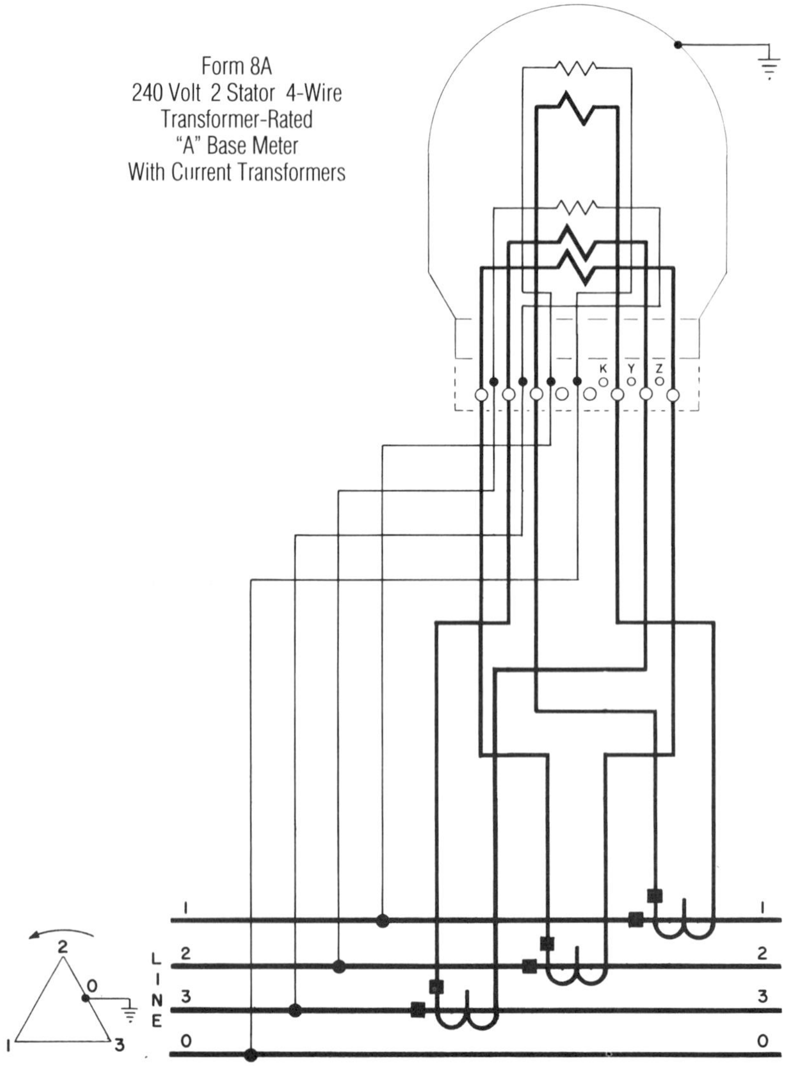

Figure 12-42.

Universal Meter

For Use on 3-Phase 4-Wire Delta, and 3-Phase 4-Wire Wye Circuits
Front Views

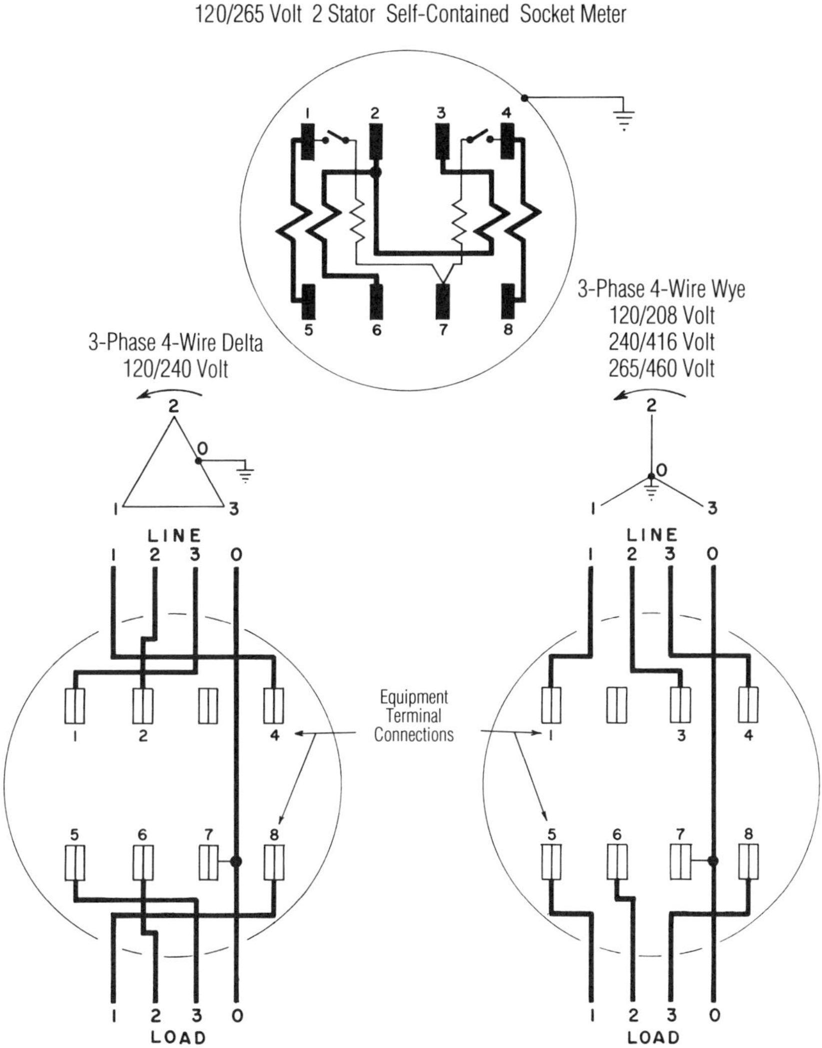

Note: Voltage coils suitable for from 120 volts to 265 volts.

Figure 12-43.

Universal Meter

For Use on 3-Phase 4-Wire Delta, 3-Phase 4-Wire Wye, and 2-Phase 5-Wire Circuits
Front View

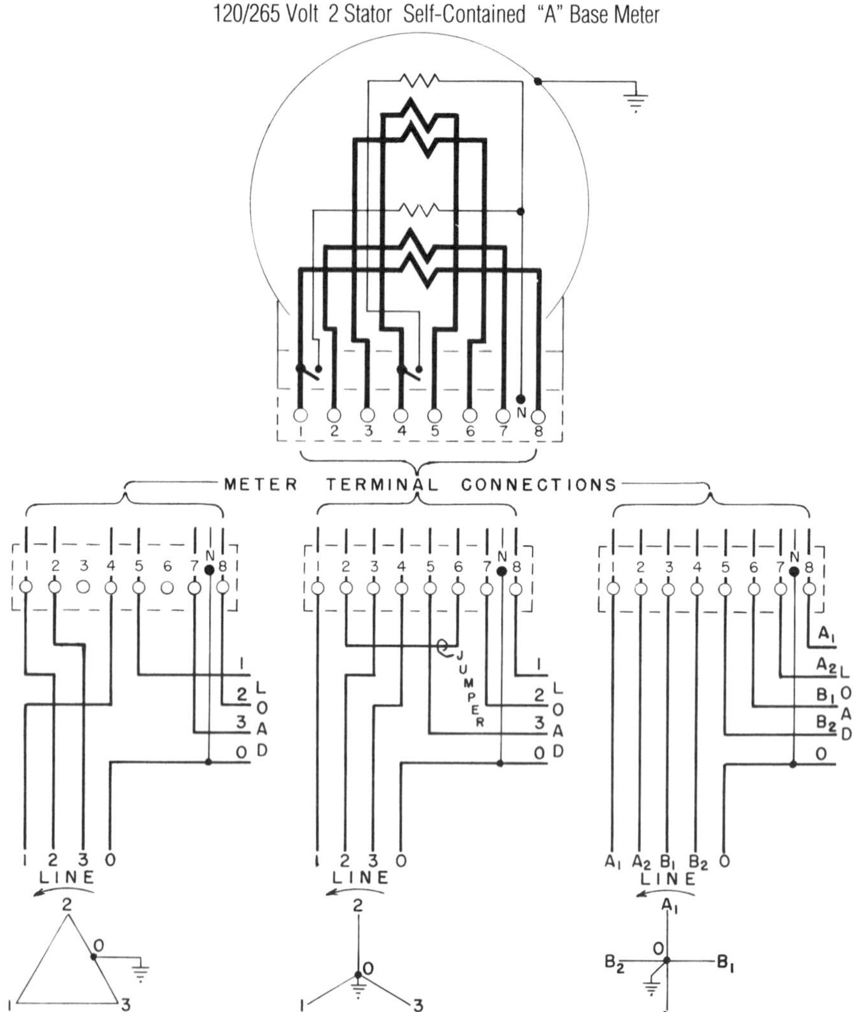

3-Phase 4-Wire Delta
120/240 Volt

3-Phase 4-Wire Wye
120/208 Volt
240/416 Volt
265/460 Volt

2-Phase 5-Wire
120/240 Volt
240/480 Volt

Note: Voltage coils suitable for from 120 volts to 265 volts.

Figure 12-44.

CHAPTER

13

THE CUSTOMERS' PREMISES SERVICE AND INSTALLATIONS

THE PURPOSE OF THIS CHAPTER is to discuss briefly certain fundamentals concerning the utility's service to the customer and to present information which may be of value to the meter installer, tester, or troubleshooter.

Although metering employees are not commonly called on to serve as wiring inspectors and although they should not, except under specific instructions from their supervisors, discuss wiring deficiencies with the customer, a knowledge of the basic principles of service wiring should be of assistance in discovering hazardous conditions and in understanding all aspects of the work. Each utility company has its own service rules which provide specifications of materials and workmanship. It is not intended here to treat such details.

THE CUSTOMER'S SERVICE

The Service Drop

The service drop is the connection between the customer's wiring and the company's distribution line. The drop may be open wire, or cable. Considerations of current-carrying capacity, voltage drop, and mechanical strength control the material and wire size. Mechanical strength is important since the conductors may be subject to ice loading and wind pressure. Some slack is necessary to avoid excessive strain on the service bracket which is the attachment on the customer's house. The point of attachment must be high enough to provide proper clearances. According to the 1990 *National Electrical Code*, services must have the following clearances: 10 feet above sidewalk, finished grade, or platform; 12 feet over residential driveways and commercial areas not subjected to truck traffic; 15 feet over commercial and other areas subjected to truck traffic; 18 feet over public streets, alleys, roads.

The underground service conductor is the connection between the customer's wiring and the electrical company's secondary distribution system. The connection may be owned by the customer or by the electric company. If the customer owns the underground service conductor, it must be installed in accordance with the *National Electrical Code.*

The Service Entrance

The service entrance conductors are those wires, generally on the outside of the building, from the service drop support to the customer's service equipment. The *National Electrical Code* describes minimum specifications for service entrances, and these are often supplemented by rules of city or town inspectors. In some areas, service drop and service entrance conductors are installed as a continuous run of cable without a break at the point of attachment.

It is obviously impossible for the company to assume responsibility for the various hazards that may result from either faulty wiring or improper use of equipment which has been installed and is maintained by the customer. Normally the responsibility of the company ends at the point of attachment of the service drop-wires to the customer's premises or at the terminal point of company-owned equipment. As a protection for the customer, the company has the right to refuse service to any customer where wiring hazards are known to exist.

Classes of Service

Two-wire service may be taken from a two-wire, three-wire, or four-wire distribution system. Almost without exception, one wire of such service is grounded by the company near the transformer and by the electrical contractor or customer at the customer's service equipment.

Three-wire, single-phase service may be supplied from either a single-phase or a polyphase distribution system. One conductor is normally grounded and the nominal voltage for this service is 240 volts between the ungrounded conductors and 120 volts from each of the ungrounded conductors to the grounded conductor.

Three-wire network service is supplied from a four-wire, wye distribution system. One line is always grounded. The nominal voltage for this circuit is 208 volts between the ungrounded conductors and 120 volts between each of the ungrounded conductors to the grounded conductor. Note that 208 volts is the phasor resultant, not the algebraic sum, of the two line-to-ground voltages.

Three-wire, three-phase service may be supplied from either a closed-delta or an open-delta transformer bank. The difference between these two connections is largely one of capacity. The output of an open-delta bank is 58 percent of the output of a closed delta where the individual transformers are similar.

Four-wire, three-phase, delta service is often used for combined power and lighting. It is a delta service with one transformer center tapped to provide 120 volts for lighting. The center tap must be grounded. In this case, the voltage between any two phase wires is 240 volts; the voltage between the grounded wire and either of two phase wires is 120 volts, but between the grounded line and the third phase wire the voltage is 208 volts. The transformer bank may be either open-delta or closed-delta.

Four-wire, three-phase, wye service is also used for combined power and lighting. Under balanced load conditions, the grounded neutral carries no current. If voltage from neutral to phase conductor is 120 volts, the voltage between any two phase wires is 208 volts. Likewise, if the voltage from the neutral to the phase conductor is 277 volts, the voltage between any two phase wires is 480 volts.

Primary service is service above 600 volts where the customer owns the transformers which reduce the service voltage from the primary level to the operating level requested by the customer. The metering is done with instrument transformers, normally installed at the point where the customer facilities are connected to the electric utility.

How to Distinguish between Three-Wire Network and Three-Wire Single-Phase Services

Elsewhere in this *Handbook* it has been shown that the requirements for network metering are not met by the meter normally used on single-phase services.

To distinguish between a three-wire network service and a three-wire, single-phase service, any of three methods may be used:

Voltmeter Check

In a single-phase circuit, the voltage across the ungrounded wires is equal to the sum of the voltages between the grounded conductor and each of the other two lines. In a network service, the voltage across the ungrounded wires is equal to about 87 percent of the sum of the voltages between the grounded conductor and the ungrounded lines.

Example (with unbalanced voltages):

	Three-Wire, Single Phase	Three-Wire Network
Voltage between grounded wire and live leg A	116 volts	116 volts
Voltage between grounded wire and live leg C	122 volts	122 volts
Voltage between A and C	238 volts	207 volts

This test requires a voltmeter rather than a voltage tester, since it is necessary to determine a relatively small difference in voltages.

Phase-Sequence Indicator

In a single-phase circuit, the rotating-disk type of phase-sequence indicator will give no indication. On either three-wire network or three-phase circuits, the disk will rotate.

Test Lamps

One method for identifying a network service is to connect two identical test lamps in series across the two outside wires. If both lamps brighten perceptibly when the center tap between the two lamps is touched to the neutral, the circuit is wye network. This is because each lamp has the voltage applied changed from 104 volts to 120 volts. If the lamps remain at the same rilliancy or if one dims and the other brightens slightly due to unbalanced voltages, the circuit is single-phase. Since test lamp method is not without hazard, this method is no longer commonly used. (See Fig. 13-1.)

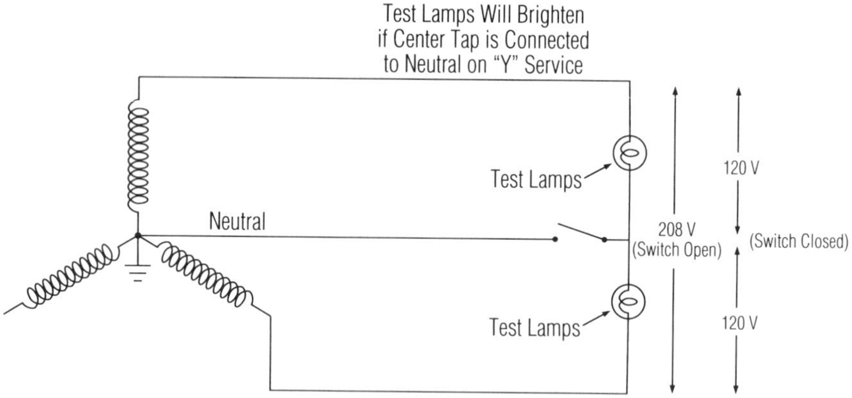

Figure 13-1. Test-Lamp Check.

To Distinguish between Four-Wire Delta and Four-Wire Wye Service

A voltmeter check is sufficient for this purpose. First determine which is the grounded conductor. Then check voltages between the grounded conductor and each of the phase wires and phase to phase.

In a wye service, each of these voltages will be 120 volts phase-to-ground and 208 volts phase-to-phase. In a delta service, only two of these voltages will be 120 volts, the third will be 208 volts to ground and 240 volts phase-to-phase.

On services of voltages higher than 208/120 volts or 240 volts, the voltages measured will be proportionately higher but in the same ratio.

OVERHEAD SERVICE TO LOW HOUSES

The current popularity of ranch-type houses poses a real problem in maintaining proper clearances of the service drop. This is particularly true where the distribution line and the residence are on opposite sides of the street. One method of maintaining clearances is by a service pole set on the house side of all public ways, driveways, and sidewalks. Another is the use of a service mast attached to the building. A pipe mast of rigid steel conduit from 2 to 3 inches in diameter makes a neat extension when properly installed.

CIRCUIT PROTECTION

Lightning Arresters

Lightning arresters are designed to protect circuits and connected equipment by providing a by-pass to ground when the supply circuit voltage rises above safe limits. Basically, the arrester consists of a fixed gap connected between each ungrounded line conductor and ground. Normal line voltages are not capable of arcing across the gap. In electronic meters, metal oxide varistors (MOVs) are often used as protection against lightning.

An example of a simple arrester may be found in the relief gaps of modern socket-type watthour meters. In one type of meter there are two relief-gap pins protruding inside the meter base. When the meter is assembled, the current-coil leads are located near, but not in contact with, these pins. If lightning or a system disturbance of any kind causes a sudden surge of high voltage which might break down the meter insulation, an arc will be established across one or both relief air gaps. The surge will then go through the pins to the stainless-steel straps on the back of the base, into the socket, and then to ground.

Fuses and Circuit Breakers

In general, fuses are designed to open a circuit when the current flowing in it is above a safe limit. The fuse is, then, an over-current protection and must be coordinated with the size of wire it protects. Excessive current may be due to an overload caused by appliances drawing current beyond the current-carrying capacity of the wire or to a short circuit resulting from a failure of insulation. In the latter case, the current may reach very high values. Obviously a fuse must be able to open a circuit under either of these conditions.

Overloaded conductors become a source of fire hazard because of conductor heating. Heat is a form of energy rather than of power and, as such, is a function not only of the current in the wire but of the time during which the current flows.

A fuse consists of a strip of metal which has a low melting temperature, enclosed in a screw plug or a cartridge. This is placed in series with the load. When the current passing through the fuse exceeds the ampere rating of the fuse, the heat produced due to the flow of current will cause the metal to melt, thereby opening the circuit.

Since the fuse link is melted by heat energy and since heat energy is a function of both current and time, the fuse will open either on a very high current existing for a short period of time or a current moderately in excess of rated capacity for a longer period. This time-current characteristic is designated as inverse-time. That is, operation of the fuse occurs in progressively shorter intervals as the current increases. As an example, a 15 ampere fuse may carry 30 amperes for several seconds but will open in about a tenth of a second should 150 amperes occur in the circuit.

Circuit breakers are used in place of fuses as protective devices in many instances. They too will open circuits when overloads or fault currents develop. Breakers have many advantages—one of them being that they can be reset and used again.

Time-Delay Fuses

Many small motors have a starting inrush current several times the normal running current. This inrush current is of such short duration that, although it may cause an annoying voltage dip, it does not result in a serious fire hazard. The time-delay fuse or circuit breaker permits the passage of such starting currents without opening. However, should this high current last for more than the normal starting period of the motor, as might be the case if the motor stalled, the fuse or breaker will open and clear the circuit. Time-delay fuses or circuit breakers will eliminate many "no-light" trouble calls by reducing unnecessary circuit openings.

GROUNDING

A ground is a conducting connection between an electric circuit or equipment and the earth. Such a connection may refer to the grounding of the neutral wire of a circuit or to the connection between such hardware as meter sockets or switch cabinets and the earth.

When and how to ground are questions which do not always have simple answers. Conditions of use, location of equipment, and other considerations have a bearing on methods of grounding. *Grounding at any one location must conform to company policy which has been established for the whole utility system.*

Residential Wiring Grounds

Responsibility for adequate grounding on the customer's premises lies with the customer and not with the utility company.

Grounding is one of the most important subjects covered by the *National Electrical Code* and this code is very clear in regard to the purposes of grounding. It reads:

> "Circuits are grounded to limit excessive voltages from lightning, line surges, or unintentional contact with higher voltage lines and to limit the voltage to ground during normal operation."

The *National Electrical Code* requires the grounding of one conductor of electrical systems in which the voltage to ground does not exceed 150 volts. One of the wires in a single-phase, three-wire system, and also in a four-wire wye system is called the neutral conductor and is grounded within the company distribution system. This same conductor, whenever brought into the customer's premises, must again be connected to a ground at a point as close to the entrance to the structure as possible, but at no other point. Since a two-wire system comprises the neutral conductor and one line conductor, the grounding of the neutral in the two-wire system follows the same rules as those applied in the three-wire system.

Grounding at one point only on the customer's premises guards against the possible hazards due to difference in ground resistance. Also, with only one ground, it is easy to check the presence and adequacy of the grounding circuit.

In residential wiring systems, not only must the neutral conductor be grounded, but exposed metal such as cabinets and conduits which might come in contact with ungrounded current-carrying wires must also be grounded.

Most commonly, the ground connection to the neutral wire is made in the first switch, distribution panel, or meter mounting device installed as a part of the service equipment. The ground connection must be made on the supply side of the customer's switch.

At this point, the neutral wire is connected to the metal enclosure, meter socket, or switch box and grounding of both neutral wire and "non-current-carrying parts" of the system is effected by one grounding conductor. There is, however, an important distinction. On the supply side of the service-disconnecting device, the grounded conductor may be used for grounding the meter socket or trough and service equipment. On the load side, this grounded conductor shall not be used to ground equipment or conductor enclosures. Exceptions to this rule are made only for the frames of 120/240 volt electric ranges and electric clothes dryers which may be grounded to the neutral conductor of a three-wire circuit.

In city areas where there is a public water system, the grounding wire is connected to the water pipe, near the point at which it enters the building.

To avoid corrosion, connectors for use on copper pipe should be made of copper, those for use on galvanized pipe should be made of galvanized iron.

In areas not served with an extensive public water system or where plastic pipe is used, there can be difficulty getting a dependable ground connection. A satisfactory ground connection is one which presents a low resistance to the flow of electric current. Such a ground connection usually consists of a rod, at least 8 feet in length, driven into the ground adjacent to the building served.

Identification of Grounded Conductor

Soil conditions vary widely with location and weather. This means that the resistance of grounds can vary widely and may reach very high ohmic values. Under these conditions the correct identification of the grounded secondary conductor is difficult when the service connection is being completed. If a mistake is made and a "live" line conductor is connected to the grounded service wire, a short circuit will develop when the pole ground resistance drops in value under wet weather conditions. Thus it is essential to correctly and permanently identify the grounded conductor when the service connection is made.

One method of identifying the grounded conductor is by the use of a high-resistance voltmeter, such as a rectifier type with 1,000 ohms per volt.

After the customer's load has been disconnected by opening the main switch or, where the meter is ahead of the switch, removing the meter, voltage checks are made as in Fig. 13-2. Voltmeter readings should be zero for all combinations of service drop to entrance wires except A to N', and B to N'. Readings of approximately 120 volts identify N' as the customer's grounded conductor.

Caution:
If the service is underground, a cable test might indicate ground on all conductors due to capacitance of the cable. Such cases should be checked with a low-resistance meter or a small lamp.

To be effective, grounding should maintain the voltage of the grounded conductor at the same level as that of the ground. Perfect grounding thus implies

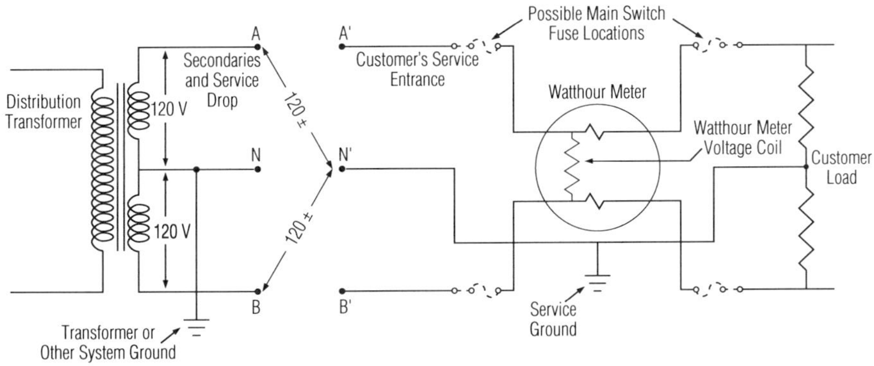

Figure 13-2. Identification of Grounded Conductor.

zero resistance of the grounding conductor, zero voltage, and zero current flowing in this conductor.

Two points in regard to grounding must be remembered:

1. The earth's crust is not a good conductor. Because of this fact, differences in ground voltage, particularly under conditions of high current flow such as lightning, may occur. Such voltage differences may exist within distances of a few feet. For instance, in pole metering installations it is wise to have grounding electrodes completely surrounding the area on which a meterman might stand when working.
2. Grounding is not the answer to all electrical safety problems. In some instances, isolation is safer than grounding. For instance, some companies do not ground pole-top metal enclosures because such grounding may be hazardous to personnel working on live conductors.

CONDUCTOR IDENTIFICATION BY COLOR

The *National Electrical Code* specifies the color marking of certain insulated conductors in interior wiring systems, and these markings may be helpful in identifying conductors. These specifications require that:

1. Grounded insulated conductors of No. 6 or smaller size shall have identification of white or natural gray color.
2. Grounded insulated conductors larger than No. 6 shall have an outer identification of white or natural gray color, or shall be identified by distinctive white marking at terminals during process of installation.
3. Where, on a four-wire, delta-connected secondary, the midpoint of one phase is grounded to supply lighting and similar loads, that phase conductor having the higher voltage to ground (sometimes called "power leg" or "wildcat phase") shall be indicated by painting or other effective means at any point where a connection is to be made if the neutral (grounded) conductor is present.
4. Grounding conductors shall have a continuous identifying marker readily distinguishing it from other conductors. This marker shall show a green color.

METER SEQUENCE

In the early days of electric distribution it was the practice to install meters and service equipment in the sequence: switch-fuse-meter. In this sequence the meter was protected by the customer's fuses. Also, it was possible to open the circuit before working on the meter. With early meters this sequence was highly desirable. Meter coils and terminal chambers were of low dielectric strength requiring fuse protection. Meter terminals were such that meter changes with live lines often were hazardous. Although there are certain disadvantages, this meter sequence is still common.

As meter test blocks were developed, as meter insulation was improved, and particularly when socket meters were introduced, it became feasible to install meters in the sequence: meter-switch-fuse. With this sequence, it became possible to establish a sharp dividing line between company and customer equipment. Connection of new loads to the customer's service equipment could not result in unmetered loads. This freed the company from any inspection requirement of customer's service equipment.

Figure 13-3 gives schematic diagrams of these sequences.

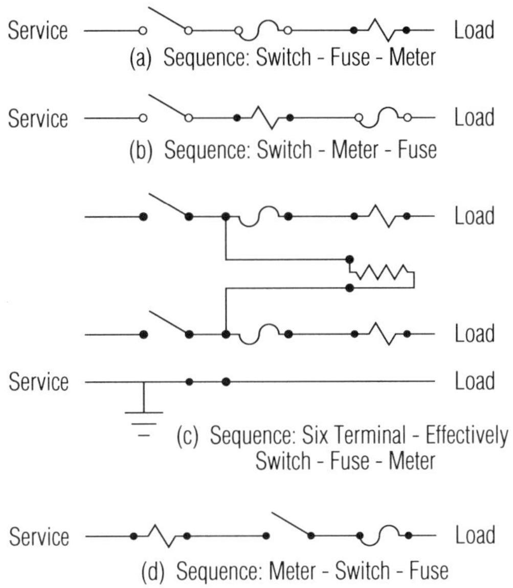

Figure 13-3. Meter and Service Equipment Sequence.

METER CONNECTIONS

Chapter 12, "Meter Wiring Diagrams," of this Handbook gives diagrams of correct meter wiring.

In all meter operations, it must be remembered that the most careful meter test, the most conscientious meter handling in shop and field, can all be made worthless by connection errors.

Employees involved in metering are advised to look out for the following common connection errors.

Connection Errors

With the current coil connected in the grounded service conductor the meter may be partially or completely by-passed.

To correctly measure the power in a two-wire circuit, we must have the current coil in the watthour meter carrying *all* the load current of the circuit. To accomplish this, the current coil must be connected in series with the ungrounded line.

It is worthwhile to elaborate on the importance of having the current coil in the ungrounded leg. Suppose the watthour meter had been installed incorrectly as in Fig. 13-4. Suppose in this faulty connection a ground occurred as shown on the load side of the meter. There are now two paths for current flow as shown by arrows. The meter current coil is effectively shunted by the ground path. The meter is bypassed and the energy used is, to a considerable degree, unmetered and unbilled.

Figure 13-5 shows another condition which might result with the current coil connected to the ground wire and the fuses not blown. Following the arrows, it may be seen that the meter is registering forward without a load being used,

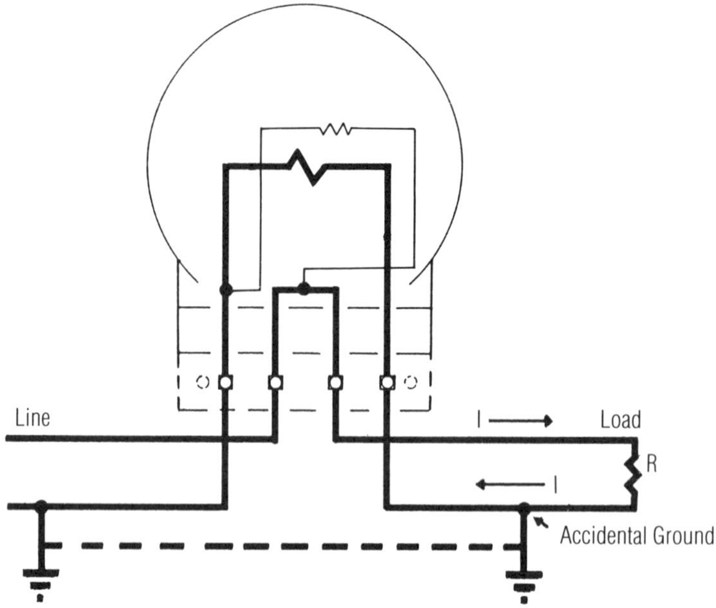

Figure 13-4. Wrong Connection of Two-Wire, Single-Phase Watthour Meter.

because the meter and the load wires present another path to the transformer for loads not properly on this meter. The amount of current thus shunted will depend on the resistance of the neutral conductor. Due to numerous splices, the resistance may be quite high. This will cause all the more current to take the path through the meter.

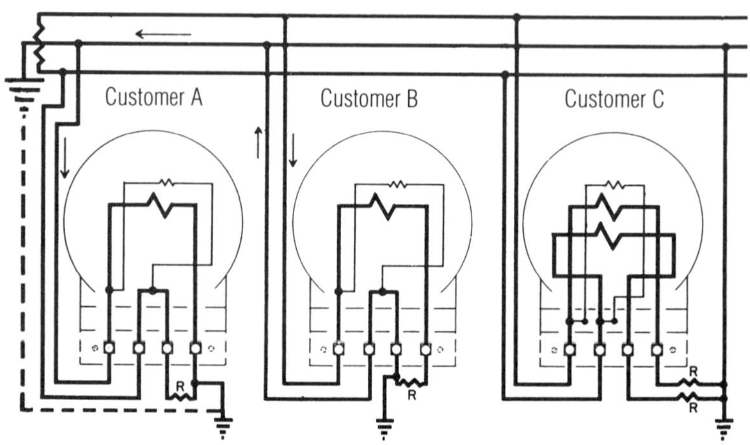

Figure 13-5. Wrong Connection of Two-Wire Meter on Customer "A" with Load on Meter of Customer "B." Arrows Show Possible Direction of Current from Meter "B."

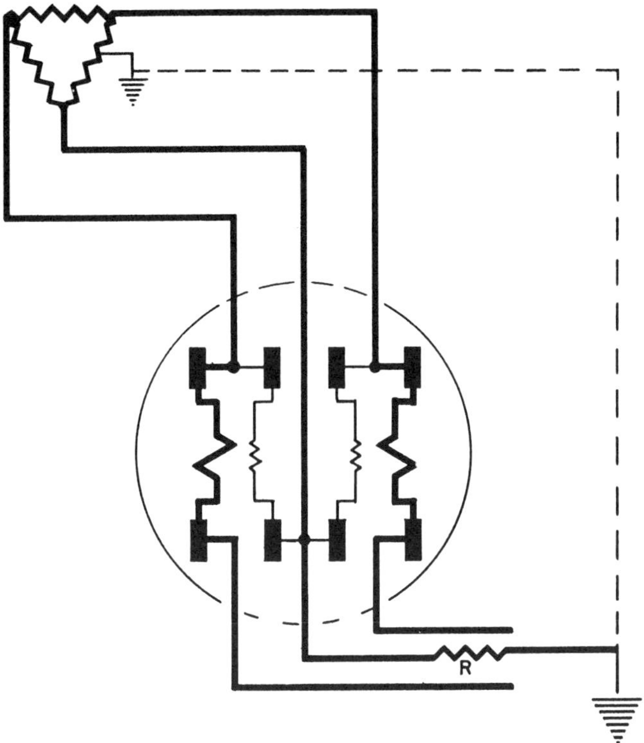

Figure 13-6. Wrong Connection for a Two-Stator, Three-Wire Meter on a Delta Bank with Grounded Transformer Center Tap.

Note that the direction of rotation of this incorrectly wired meter will depend on the direction of current in the neutral. Note further that if from the same service there is taken a three-wire load, as is shown in Fig. 13-5, Customer "C," any unbalance in the three-wire load will tend to produce rotation in the incorrectly wired meter.

Remember that if all meters were properly wired, with the current coil in the ungrounded conductor, such conditions would not be possible.

Similar precautions must be taken when connecting a three-wire meter. If the neutral instead of an ungrounded line is connected to one of the meter coils, there is not only a variable error in the current components but a constant error of 50 percent in the voltage, since 120 volts instead of 240 volts is applied to the voltage coil.

A three-wire, single-phase meter installed on a two-wire circuit, without alteration to fit the conditions, may have one current coil by-passed by customer's ground. Correct metering of a two-wire circuit with a three-wire, single-phase meter is described under the subheading "Metering Two-Wire Service with Three-Wire Meters."

A three-wire, three-phase service is properly metered with a two-stator meter. However if one of the transformers has a center tap grounded, it is possible for energy to be consumed on this service that would not pass through the current coils of the meter. See Fig. 13-6.

Other types of errors are possible. Careless or dishonest persons may connect circuits at some point on the source side of the meter. In multi-family buildings, careless connection of customer's service conductors can result in the registration of one customer's use on both his own and his neighbor's meter. It is also possible that meters measure loads other than those for which they are intended, such as a meter tagged Apartment "A" actually measuring the consumption of Apartment "B". Sometimes an interruption of service is necessary to obtain assurance that these conditions do not exist.

In an apartment house, it must be recognized that both three-wire and two-wire services may exist. The customer's service equipment must be checked before the meter is installed.

The failure to meter the customer's load when one fuse on the supply side of the meter has been opened must be recognized. In this case, branch circuits from the live wire and the neutral are still energized but the meter will not register since its voltage coil is connected across the 240 volts circuit. Tying the outside wires together as in Fig. 13-7 will make all 120 volts branch circuits alive with no meter registration. Only the 240 volts appliances will not be supplied.

Figure 13-8 illustrates a common error. Beware of making this mistake.

The misconnections described are examples of some of the errors to be avoided. Of course, there are many possibilities for wrong connections, especially in polyphase metering with instrument transformers. Methods for checking the correctness of connections are outlined in the discussion of instrument transformer installations.

Metering Two-Wire Services with Three-Wire Meters

In those areas originally supplied by two-wire services but in which load growth and additions of 240 volts appliances necessitate a continuing changeover to three-wire, it is often considered advisable that new meter purchases be adaptable for either two-wire or three-wire use.

There are available watthour meters in which the internal wiring is easily converted from two-wire to three-wire and vice versa. Internal connections for such meters are shown in Fig. 13-9. Note that, in the two-wire connection, the current in the grounded leg is properly not measured. By reconnection the meter constant is not changed but the full-load speed of the meter when connected for two-wire service is one-half that of the meter when connected for three-wire service.

It is often wise to calibrate the meter for the service on which it is to be installed, but the agreement in performance between the two connections is within commercial limits of accuracy.

In areas where there is a continuing changeover from two-wire to three-wire services, this type of meter may prove a practical answer to meter stocking problems and rapid obsolescence of two-wire meters.

With the improvement of voltage coil characteristics in modern meters, it is feasible to use the standard three-wire meter with the current coils connected in

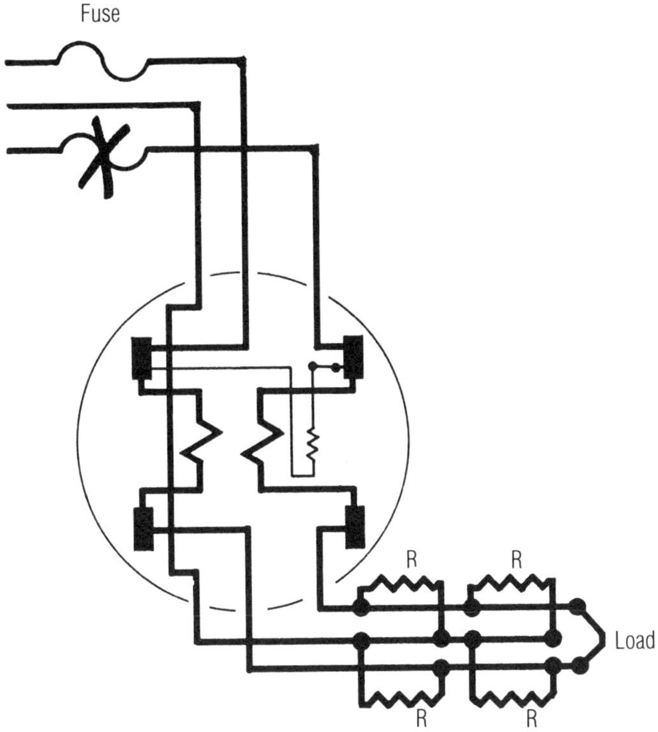

Figure 13-7. Three-Wire, Single Phase Meter with Open Fuse.

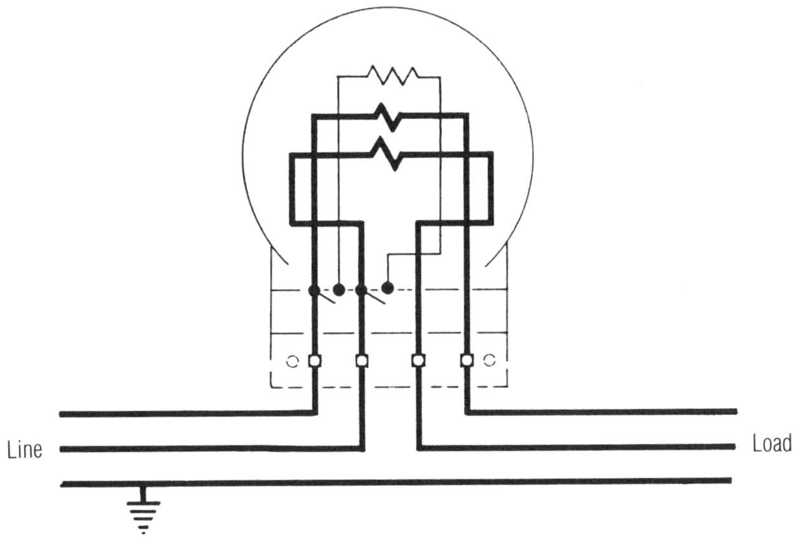

Figure 13-8. Voltage Link Left Open, a Common Error.

Internal Connections for Self-Contained Detachable 2/3-Wire Single-Stator Watthour Meters

Front Views

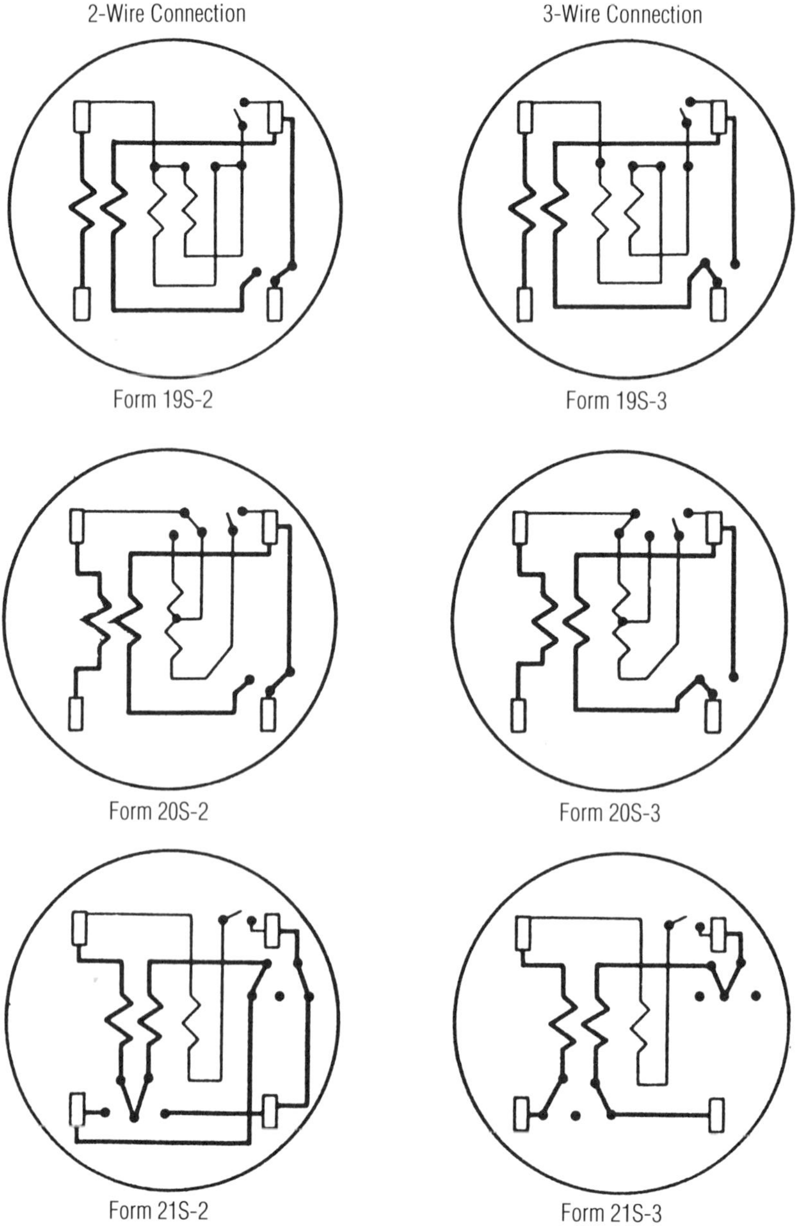

Front Views

Figure 13-9. Convertible Watthour Meters (from ANSI C12.10).

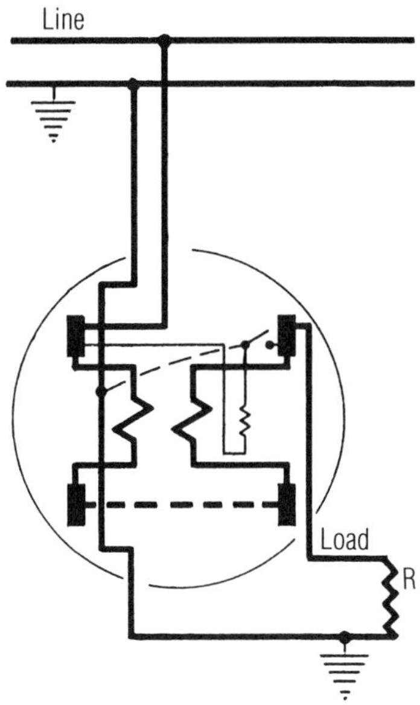

Figure 13-10. Three-Wire Socket Meter Converted for Two-Wire Service.

series for two-wire services. In this connection, one current coil is reversed in order to provide forward torque from both coils. Compare directions of current in a three-wire meter. The voltage coil is not modified and is operated at half voltage.

In a meter socket, as shown in Fig. 13-10, the line wires are run to the top-left jaw and the ground tap. The load wires are taken from the top right jaw and ground tap. The meter voltage coil is connected from the top-left blade to the ground tap with the voltage clip open. The connection to the ground tap may be made by a flexible lead. The meter must be so marked that the tester or installer will know there is a flexible connection between meter and socket. Note that the reconnection of the current coils is necessary to put both coils in the ungrounded leg.

THE NEUTRAL WIRE

In a three-wire, single-phase circuit, the current in the neutral is equal to the difference between the currents in the ungrounded conductors. This can readily be shown by applying Kirchhoff's law of current. With a perfectly balanced load, the neutral in this type of circuit carries zero current.

What Happens if the Neutral is Broken?

Since in a three-wire meter there is no connection to the neutral, the meter will continue to register correctly the energy taken. The effect on the customer's load will depend on the location of the break. Normally, where grounds do not provide a return path for the current, two-wire branch circuits may be dead. If, to a three-wire appliance, such as an electric range, the neutral is opened, the 240-volt units will continue to operate at normal heat, but on any 120-volt units it is possible to get excessive voltage on some units and deficient voltage on others, depending on the load balance among the units. That is, if loads on both sides of the neutral are connected, these loads will be in series across 240 volts and the lower-rated unit will draw too high a current while the higher unit will have too little current to reach expected heat.

Note that, under these conditions, the customer loses the protection of the grounding conductor, the importance of which has already been discussed.

The neutral conductor is not fused because:

1. This conductor provides the connection to ground;
2. An open neutral presents more hazards to equipment than any protection a fuse could provide;
3. Load current in the neutral of a single-phase circuit is never greater than the larger of the currents carried by the fused, ungrounded conductors and therefore the neutral is, in effect, protected by the circuit fuses.

Network Neutral

In the three-wire network circuit (two phase wires and neutral from wye-connected transformers), the condition is somewhat different in that, under balanced load conditions and with only phase-to-neutral loads, the neutral wire carries the same magnitude of current as the phase wires.

If the neutral is broken in a network circuit, the customer loses the protection of his ground. Appliances operating at 208 volts will not be affected. Branch circuits operating at 120 volts will normally be dead, but currents may find a path to the other phase wire, thus putting 120-volt lamps in series across 208 volts.

METER LOCATION

Indoor

Where meters are installed indoors, they should be located near the point of service entrance in order to avoid long runs of unprotected or unmetered conductors. If the meter is to be installed in a partitioned basement, it should be installed in the same area that the service enters. It is good policy to locate meters so that the customer will have as little inconvenience as possible.

Since meters must be read and sometimes tested in place, they should be mounted at a convenient height. Between three and six feet from the floor is generally satisfactory. Space is important. Since meter accessory equipment may have to be changed because of change in customer's load, space sufficient to permit such change must be provided.

The area in which meters are located should be free of corrosive fumes and excessive moisture. Meters should not be installed near furnaces or water heaters, nor should they be mounted under pipes which may drip because of poor joints or condensation. It is extremely important that the wall or panel on which the meter is mounted be free of vibration.

Outdoor

Outdoor meters should not be mounted overhanging driveways or walks. In the one case, they may be damaged by traffic; in the other, they may cause accidents to pedestrians.

Although meters are well temperature-compensated, it is good practice to install them out of the direct sunlight. The north side of the house is generally a good location if the service permits.

As in indoor metering, the height at which the meter is mounted must be convenient. However, in northern areas where snow is a problem, it may be advisable to require a height at least four feet above the ground.

SELECTION OF METER CAPACITY

In determining the capacity of the metering equipment, the current which the customer will demand is the immediate factor. Arriving at the amount of current a customer will use is usually a matter of calculation involving connected load and estimates of diversity and growth. The capacity of the customer's main line switch or service entrance conductors may be used in determining the equipment selected. This keeps the equipment from becoming the weak point in the circuit and from being damaged by overload.

There are self-contained meters available which, with the appropriate sockets and other associated devices, will carry and accurately meter up to 400 amperes. For loads above 400 amperes, current transformers are necessary.

In cases where the load to be metered is only a fraction of the installed service capacity, it is wise to make adequate provision for the future metering of the total installed service capacity. This is particularly important where increases in load may necessitate a change from self-contained to transformer-type metering, or where the increase involves a change in rate schedules requiring a different metering method.

METER SOCKETS

The Type-S meter is designed so that its terminals appear as short, rigid, copper contact blades extending outward from the back of the meter. To connect this meter to line and load wires, an auxiliary mounting device is required. This device is the meter socket.

The socket comprises connectors for line and load conductors, contact jaws to receive the meter terminal blades (thus completing connections between conductors and meter coils), and an enclosure for the whole assembly.

Early sockets were round, cast or drawn, shallow pans with diameters matching those of the meters. In this type socket, wiring space was limited. This limitation led to the development of the rectangular-shaped trough with a round opening the diameter of the meter. A sealing ring, which fitted around meter rim and socket cover rim, secured the meter in place. More recently there has been developed the ringless type of socket in which the socket cover opening fits over the meter after the meter has been installed. The socket cover is then sealed in place to provide protection. In both types the primary functions are to (a) fix the meter firmly on the socket, (b) close the joint between meter and socket rim against weather and tampering, and (c) provide means for sealing the meter against unauthorized removal of meter or cover.

Meter sockets are available in continuous duty current ratings of 20, 80, 100, 120, 150, 160, 200, 320, and 400 amperes, and for one-, two-, or three-stator meters.

The requirements for indoor and outdoor service differ. Meter sockets installed on the outside of the house must not only be weatherproof but must be of a material that is highly resistant to rust and corrosion.

Under some conditions, such as leakage of pipe joints or cable assemblies, sockets will accumulate varying quantities of water. To guard against water accumulations, sockets are provided with means for drainage.

Obviously the dimensions of both meters and sockets must be standardized and closely controlled so that meters of any of the major American manufacturers will fit sockets of any manufacturer. ANSI and NEMA standards have been developed with standards for meter and socket sizes. The Meter and Service Committees of EEI and AEIC have agreed upon certain basic requirements applicable to meters and the associated mounting devices.

The basic requirements are:

1. Interchangeability of all manufacturers' meter mounting devices.
2. Mounting devices to be designed for single-meter or multiple-meter mounting either indoor or outdoor.
3. One seal to serve for both meter and mounting device.
4. Terminals to be inaccessible after the meter is sealed in place.
5. Meter base not to be insulated from the mounting device.
6. Mounting device to have an uninsulated terminal for the service neutral.

The material of the socket jaws is important. A tight contact between the meter connection blade and the contact surfaces of the jaw is necessary. This requires the use of especially high-quality resilient copper alloy, which may be bronze or beryllium. Even with best-quality material, it must be remembered that spreading the jaws by pushing screwdriver blades into them may spring the metal beyond its elastic limit and destroy the tight contact with meter blades.

It is also necessary that the connection between conductors and the line and load terminals be secure and of low resistance. The connectors in the lower-rated sockets may be required to accept conductors as small as No. 6 while the 200- and 400-ampere sockets may be required to accept single or multiple conductors that will carry 200 or 400 amperes. Where aluminum conductors are used, the connectors must be designed for this material; that is, they must not cut the comparatively soft strands and they must not encourage cold flow when the wires are under pressure.

With the growth of domestic loads and the development of self-contained Class 200 meters, the heavy-duty socket also rated at 200 amperes has been introduced. There are two types of such heavy-duty sockets. In one the jaws are made of massive material and sometimes have only one flexible member. This may be spring loaded but will still depend on jaw resiliency for good contacts. In the other type, the jaws are made of nonflexible heavy material and the jaws are wrench tightened or lever tightened after the meter is in place. Either type of jaw can carry 200 amperes continuously without excessive heating.

Circuit-Closing Devices and Bypasses

Certain types of sockets are provided with circuit-closing devices or bypasses. These may be automatic, closing the current circuit as the meter is withdrawn from the socket; or manual, requiring an operation other than meter removal to close the current circuit.

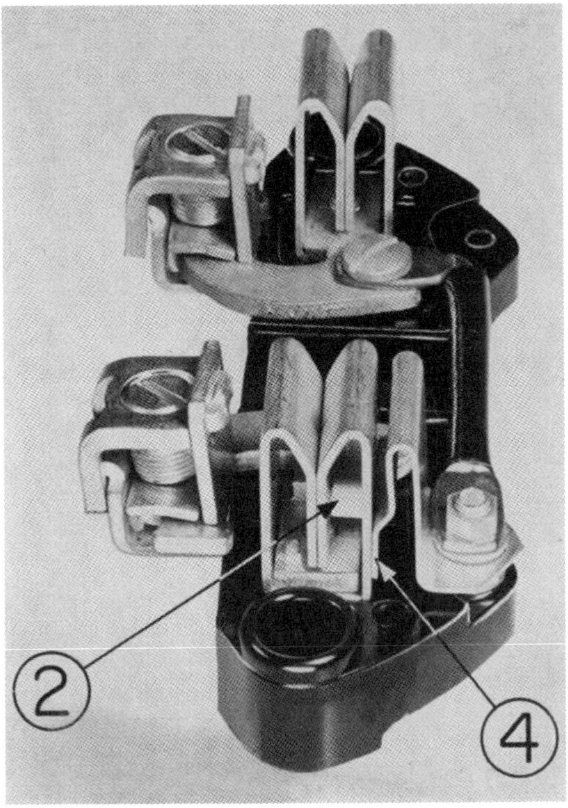

Figure 13-11a. One Type of Automatic Circuit-Closing Device. (No longer manufactured). Circled Numbers are Referred to in Text.

Two types of automatic circuit-closing devices are shown in Fig. 13-11. In Fig. 13-11a, the meter blade is inserted into the socket jaw, causing the bypass contact *4* to be pushed away from the jaw by a small insulating rod *2*. Withdrawal of terminals allows bypass and jaw to spring together to positively short circuit the current transformers.

In Fig. 13-11b, the dual bayonet terminals of the meter current coil spread the dual socket jaws (left) to complete the current transformer circuit. Withdrawal of terminals allows jaws to spring together (right) to positively short circuit the current transformer.

Figure 13-11c is a representative picture of the socket with bypass arrangements as shown in Fig. 13-11b.

Where sockets with automatic closing devices are used, it must be recognized that removal of the meter does not de-energize the customer's service and that a socket with a blank cover instead of a meter may mean no more than that the customer is getting free electricity.

Manual bypasses are often provided in heavy-duty sockets when it is believed unwise to interrupt the customer's load or to pull the meter and break a

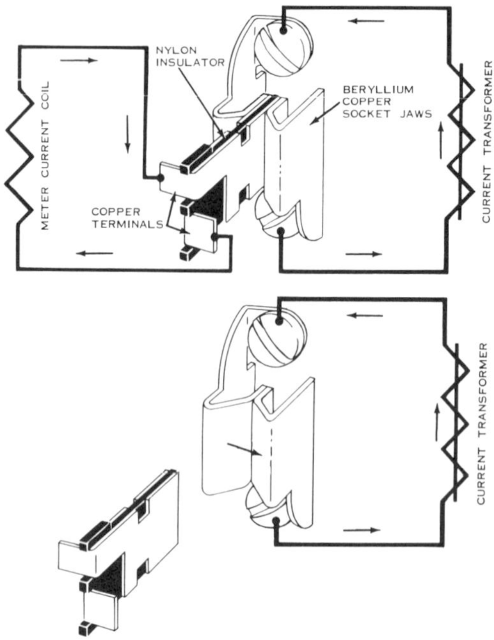

Figure 13-11b. Circuit-Closing Device for 20-Ampere Meter Socket.

Figure 13-11c. Socket with Bypass Arrangement as Shown in Fig. 13-11b.

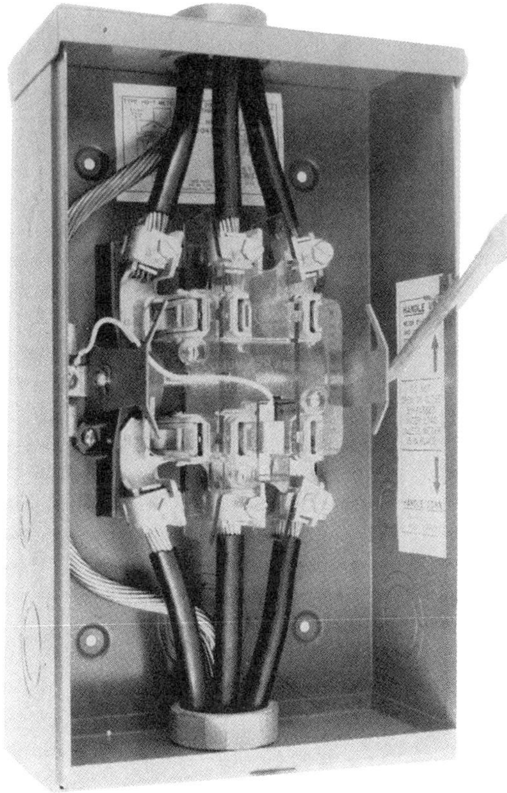

Figure 13-12. Heavy-Duty Socket with Bypass.

circuit which may be carrying 200 or 300 amperes. Such bypasses may be integral with the socket, as in Fig. 13-12, where, as the lever is pulled to release tension on the socket jaws, the bypass is closed, or the bypass may be similar to a meter test block and mounted in an extension of the meter trough.

Another type of socket has facilities for connecting flexible leads to bypass the meter before it is removed as is shown in Fig. 13-13.

METER INSTALLATION AND REMOVAL

Socket Meter

Socket meters are commonly installed in meter-switch-fuse sequence. Where this is the case, it must be remembered that the line terminals (top jaws) are alive. Do not attempt to correct major wiring defects on energized parts in the socket. Report such conditions to the supervisor. If it is necessary to make connections in a live meter socket, protective shields should be installed around the live parts.

Before installing the meter, check to see that voltage clips are closed.

Check correctness of the electrician's wiring. Is the grounded wire connected to the socket case?

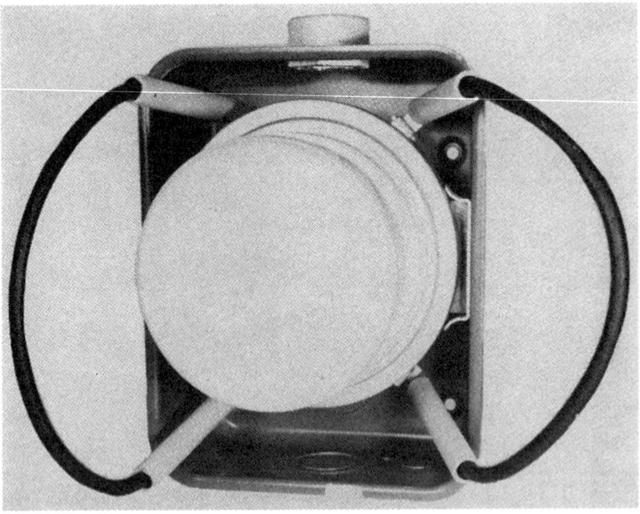

Figure 13-13. Socket and Meter with Flexible Bypasses in Place.

If the socket is not level and plumb, report this condition to the supervisor.

When installing a socket meter, line up the load jaws and meter blades and press these home. Then, using the bottom jaws as a fulcrum, rock the meter into place.

Do not twist the meter in a manner to spring the jaws.

When closing into the line contacts, the action should be positive.

Do not pound the meter in. Cuts from a broken cover can be serious.

Check the meter number against the number on the service order.

Check to see that the meter disk rotates in the proper direction.

When removing socket meters, use the lower jaws as a fulcrum and pull the blades from the lineside jaws with a downward force on the meter before withdrawing the lower blades.

On removal of a socket meter without replacement, the trough or socket opening is closed with a blank cover plate and sealed. This is important because the top terminals are alive and should not be left as a hazard to the curious.

Where the meter covers are broken due to accident or vandalism, great care must be taken to avoid cuts by the sharp glass. Heavy leather gloves should be worn for removing such meters.

When installing a meter, if it is found that the compression of the socket jaws is not sufficient to hold the meter in place, this should be corrected or reported to the supervisor. Poor contacts cause heating, and heating will further destroy the temper of the jaws.

Socket Adapters

Where modernization of customers' services is taking place at the same time that a changeover from bottom-connected to socket-type meters is being made, adapters which permit the replacement by socket meters of bottom-connected

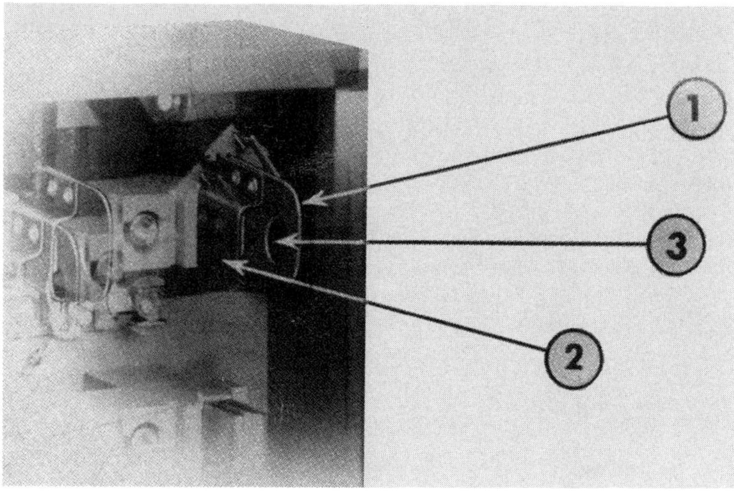

Figure 13-14. Safety Shield for Live Jaws. (I) Tough Phenolic Plate (Fabric Base) Fits Inside the Meter Jaw. Spring Tension of Jaw on Plate Holds Safety Shield in Position. (2) Vulcanized Fiber Shield Attached to Phenolic Plate Effectively Insulates the Meter Jaw from Accidental Contact. (3) Finger Hold for Easy Removal of Shield.

meters may be considered. These adapters essentially give a socket meter a base which permits installation in a location designed for a bottom-connected meter. The adapter may be no more than a low-cost shell to hold the socket meter, with facilities for bringing out flexible leads to the service switch, or it may be equipped with a standard terminal chamber similar to that of the bottom-connected meter. Figures 13-15a and b illustrate this type socket adapter. Note that in Fig. 13-15a, only the top connections are shown. A bridge across the bottom blades of the meter completes the connections by putting the current coils in series.

Bottom-Connected Meters

The handling of "A" type meters for installation or removal plainly involves operations not required with the "S" type meter. All line and load wires must be removed from the several meter terminals before the meter can be lifted from location. Unless the line wires can be "killed" by opening a service switch or removing main fuses, these wires must be handled carefully at all times.

Connection screws must be made up tightly and each wire tested by pulling and shaking it a bit to be sure that there is no looseness of contact. With the smaller sized wires, it should be borne in mind that they can be cut off in the terminal hole by forcing the terminal screws down.

In cases where a service interruption cannot be tolerated for a period long enough to remove and replace a meter, the use of jumpers is necessary. Where test facilities are provided or in certain types of service installations, where line and load terminals are available in the service equipment assembly, this is a simple procedure with test links or with leads equipped with spring-clip terminals.

Figure 13-15a. Skeleton Socket Adapter.

Where the meter is located directly in the line, it becomes necessary to remove insulation from the line and load wires ahead of and beyond the meter terminals. Jumpers are then attached to the spots thus bared.

Placing of such jumpers calls for careful work. If there is any doubt as to where jumpers should be connected, use a voltage tester. A jumper connected across a voltage difference is a short circuit and will result in a flash.

Any bare spots must be properly taped after the job is completed.

Where the meter is mounted above a test block, a change of meter becomes a simple operation since the meter can be by-passed and isolated by test-block links and jumpers. There are two different test-block connections that must not be confused. In one, the test-block terminals are in the same sequence as the meter terminals and, to bypass the meter, flexible jumpers are used to connect each of the line wires to corresponding load wires. The other arrangement is line-load, line-load in which the test block link can be used to short adjacent terminals. Generally, it is easy to distinguish one arrangement from the other, but if there is any doubt, a voltage tester should be used.

Often a bottom-connected meter is mounted directly above the service switch which has test connections involved. Such test connections take many forms and here, particularly, the use of a voltage tester or voltmeter is advisable.

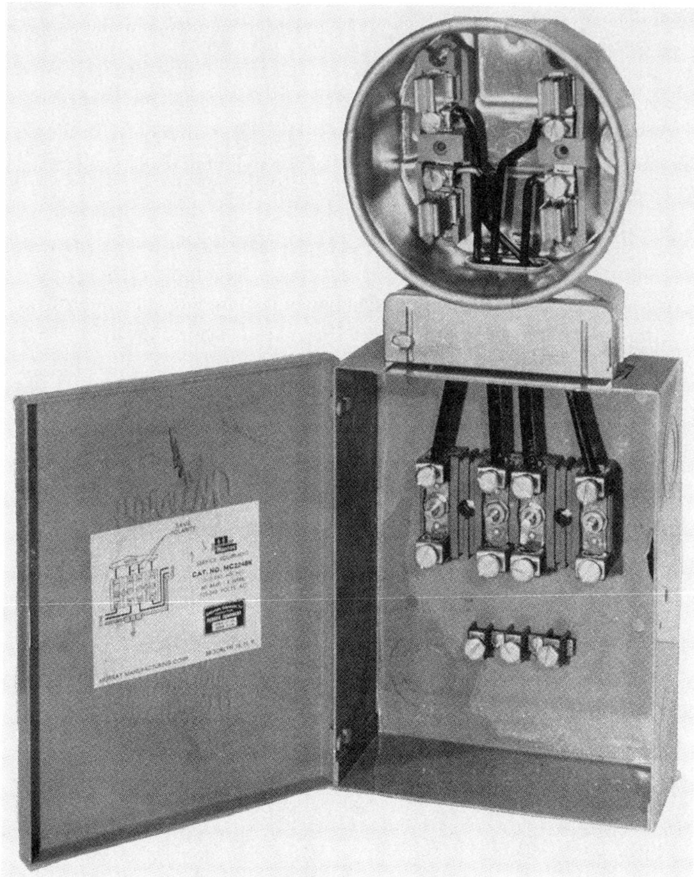

Figure 13-15b. Socket Adapter with Standard Terminal Chamber.

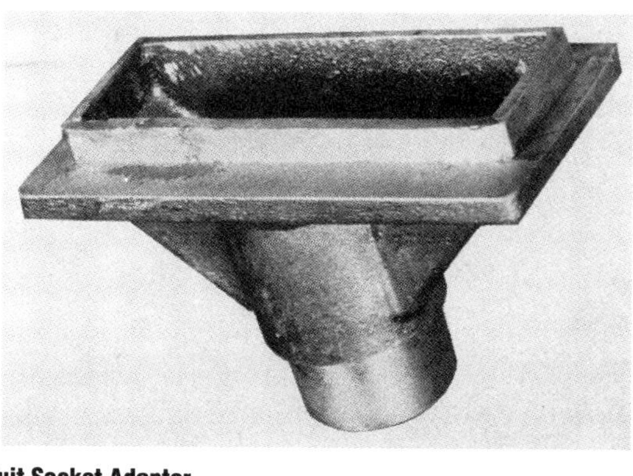

Figure 13-15c. Conduit Socket Adapter.

Outdoor Installations

To adapt the bottom-connected meter to outdoor use, several types of meter enclosures have been developed. These include the complete enclosure, generally with a window through which the meter can be read; the semi-enclosure in which the glass cover of the meter projects through a round opening in the box cover; the "banjo box" which provides a disconnect device and an enclosure for the meter terminals; and conduit pull boxes modified to enclose the meter terminal block.

A conduit pull box with a single cover to enclose both meter terminal chamber and pull box is often designed so that it also encloses a disconnect feature. This permits separation of the meter terminal chamber from a terminal block to which line and load wires are connected. With such construction, the meter can be tested or changed without service interruption.

Installation of Meters with Instrument Transformers

For the use of instrument transformers with meters, see Chapter 11, "Instrument Transformers."

For the economics of primary metering versus secondary metering with compensation, see Chapter 10, "Compensating Metering for Transformer Losses."

INACTIVE AND LOCKED-OUT METERS

It is usual practice to leave meters in place for a reasonable period, at least in certain classes of buildings such as residences, apartments, and offices where occupants are changing so frequently that the prospects of a new tenant at an early date are reasonably sure.

In commercial and industrial classes of business, the practice of leaving meters on vacant premises is not generally followed. The reason is that the new load is usually of a different character and alterations are probable, requiring a relocation or change in the metering equipment.

Inactive services may be left with the meter connected and energy available, with instructions for the incoming customer to notify the utility company of his occupancy of the premises, or the service may be disconnected at the meter switch or socket but with the meter left in place. This latter operation is often called a meter lockout. The disconnection is commonly protected by seals or locks to prevent unauthorized reconnection.

For "A" meter installations, the practice may be to replace the main plug fuses with insulated dummy plugs which require a special tool for removal. In some installations, lockout can be accomplished only by sealing the main switch in the open position.

The service can be disconnected in the meter socket by several means:

1. Some sockets are provided with means for disconnecting the service. In some the meter can be removed and reinserted at an angle, in which case the meter does not make contact with the socket jaws.
2. Disconnection can be effected by use of thin plastic sleeves made for this purpose that can be slipped over the terminal blades of the meter. The meter is then replaced in the socket with the plastic sleeve acting as an insulator between the meter terminals and socket jaws. These sleeves are not used on

a current transformer-rated meter because of the hazard of open secondary of the current transformer.

Where meters are connected in the secondaries of current transformers, it must be remembered that the customer's service is disconnected only by opening the primary circuit. The current transformer secondary must not be opened because of the dangerously high voltage that may occur.

It is essential that all meters left on inactive service be visited periodically to guard against damage and unauthorized use of service. Some companies adopt the practice of continuing to read the meter so a monthly report is received on the conditions existing. Other companies obtain readings less frequently but often enough to assure that the metering equipment has not been disturbed or the service used without authority.

TEST SWITCHES

Test switches are generally used if meters are in the secondaries of instrument transformers. The functions of the test switch are to short-circuit the current transformer secondaries and to isolate the meter so that it may be tested or changed without hazard.

When the test switches are open, the current blades are grounded. The voltage blades, however, may be alive and hence a hazard. Where both voltage and current transformers are used, the low capacity of the voltage transformers limits the hazard to some degree, but where only current transformers are used and the voltage connections are made directly to the line, the whole capacity of the power transformer is back of this circuit and extreme care must be used. A screwdriver or even a connector clip falling across these blades can cause a severe flash. It is necessary to be particularly careful when changing connections on such a live test switch, as when inserting a test instrument. It is because of this danger that many companies install low-current, high interrupting-capacity fuses in the voltage circuits of meters used with current transformers. Another precaution that can be taken is to place insulating blocks over the open voltage blades of the test switch.

Removing or replacing test switch covers must be done carefully. Some nonmetallic covers have metal end walls that can cause a phase-to-ground short.

Where paralleled current transformers are used, it is necessary to make sure that the test switch shorts the secondaries of *all* current transformers connected before disturbing the meter leads.

INSTRUMENT TRANSFORMER METERING IN METALCLAD SWITCHGEAR

In 1958 the Meter and Service Committees of EEI and AEIC issued *Guide for Specifications for Revenue Metering Facilities Installed in Metalclad Switchgear.* In this guide it was stated that the principal objectives to be attained were:

1. That a separate sealable compartment be provided exclusively for revenue metering equipment when mounted within the switchgear.
2. That space be provided within the compartment sufficiently large to accommodate separately the installation of any standard current transformers and any standard voltage transformers required for metering.
3. That space be provided within the compartment for the installation of separate, isolated voltage transformer fuses, where required.

4. That, where required, adequate space and panel facilities be provided within the compartment to permit the installation of all necessary meters, instruments, auxiliary devices, or test facilities, of any type, whether they be front connected, back connected, surface mounted, or flush type.
5. That the arrangements be such that the secondary wiring may be installed in a manner to facilitate checking of connections.

By following these specifications, control of all metering transformers and conductors rests with the utility company.

Where extremely high-capacity current transformers are used, it is essential that spacing of bus bars be adequate to avoid interference between individual transformers.

There are many advantages to be gained by mounting instrument transformers in the customer's switchgear. Protection, appearance, and, in many cases, economy may be the result.

POLE-TOP METERING

Pole-top location of instrument transformers is often necessary. Poles with distribution equipment requiring maintenance mounted on them are generally avoided for metering, since the reduction of climbing space may present hazards to utility personnel.

Instrument transformers may be mounted on crossarms or may be put in place as pre-wired units on cluster mounting brackets. These brackets are generally designed to allow sufficient free climbing space. The transformers used are most often of a type that can be installed in any position. For low-voltage metering, window-type current transformers offer convenience and economy.

Meters may be located on the pole, or the instrument transformer secondary may be extended to a nearby building where a more suitable location may be found. In the latter case, sometimes an underground secondary run is involved. Such a conduit or cable run coming down the pole and extending up the building wall may form a U which will often collect water. When installing cable or conduit, care must be taken to prevent this condition. Comments in Chapter 11, "Instrument Transformers," referring to length of secondary run should be consulted.

Since pole-top metering is generally distant from any extensive water piping system, other forms of grounding must be employed. This is particularly true when meters are installed in an enclosure on the pole, since the enclosure must be well grounded to protect the installer. To guard against high ground resistance, it is good practice to bury the ground wire connecting multiple electrodes in a circle around the pole so that the person working on the meter or enclosure is standing inside the grounding network. All grounding in the area should be bonded together.

Grounding of instrument transformer cases must be in accordance with general company practice. Whether secondary conductors are in metallic or nonmetallic enclosures depends on such grounding practice.

GOOD PRACTICES FOR METERING PERSONNEL

Efficient metering personnel will make good installations. In doing so, they will observe certain practices that will be helpful to both their company and the customers.

These good practices have many benefits. They insure good service by preventing unnecessary outages. They insure good customer relations by preventing damage to the customer's equipment. Also, the meter employee will not have to return to the customer's premises for things forgotten or left undone and thereby undermine the customer's confidence in electric metering. All these benefits, in turn, help the company and the employee.

Following are some of the good metering practices, not necessarily in order of importance.

Competent metering employees will:

Recognize their responsibilities while on the customer's premises

- Take the nearest and safest route to accomplish the work.
- Take care not to damage any of the customer's property.
- Leave the area clean upon completion of the job.
- Report any hazards to the meter department supervisor.

Work in the safest possible manner

- Keep in mind that no job is so important that it cannot be done safely. Inspect all meter wiring connections for correctness
- Check connections to prevent outages, damage to meter installation, damage to customer's property, and personal injury.

Inspect for loose connections

- A loose connection can cause intermittent service or it can cause a complete outage. Loose connections generally arc, causing a fire hazard. Even if there is no fire damage, heating around the connection occurs.

Inspect for good grounding

- Check for equipment ground at the installation. Realize that no ground at the installation is a potential hazard and report it to the supervisor.

Pay attention to details

- Check meter voltage links.

Inspect connections between two dissimilar metals

- Connection between two dissimilar metals often causes corrosion.
- Corrosion can be prevented by using proper connector and by protecting the connector and conductors against oxidation.
- Wires corroded at a joint have the same effect as a loose connection, since corrosion has a high resistance and causes heating, which, in turn, assists the corrosive action.

Check for proper voltages

- Voltage should be checked before installing the meter.
- A reversal of the "power" and "lighting leg" on a four-wire delta system causes excess voltage on customer's equipment. Also, a reversal of "hot leg" and ground has serious consequences.
- Grounded conductors, and the "power leg" of four-wire delta services, should be permanently identified.

Check phase rotation

- Phase rotation on installations which have been disconnected temporarily for service work should be checked. If a reverse phase rotation is connected to the customer's motors, they will reverse, possibly causing extensive damage. This could also mean personal injury.

Check for "single phasing"

- It is possible to prevent damage to the customer's property by disconnecting or warning the customer to disconnect the load on a three-phase service when one phase is out. A running three-phase motor may continue to run on single-phase, but will overheat. A stopped motor may attempt to start but cannot, which causes overheating.

Observe direction of disk rotation

- Whenever possible, try to get a load applied to the meter in order to check for correct disk rotation.

Check for diversion

- Always check for circuits tapped ahead of the meter or current transformers.

Check for correct installation information

- Check for correct phase, amperes, volts, and frequency.
- Check for such details as multipliers, full scales, readings, and similar data.
- Check all written records against actual nameplate data.

Check to see if meter is level

- An out-of-plumb meter may be inaccurate. Besides being inaccurate, it presents an unsightly appearance to the customer and may undermine his confidence in electric metering.

Give the whole job a good once over before leaving it

- Check the job in general for good workmanship and safety before leaving. Be sure area surrounding meter on customer's premises is left clean and neat.

Meter employees are the company in the eyes of many customers. They can make a good impression on the customer by being neat in dress, accurate in work, and courteous at all times. Having equipment and tools in good, clean condition will build the customer's confidence in the company and assure the customer of the employee's skill. Sloppy dress, actions, and equipment leave a poor impression.

There are probably many other practices which are adhered to on local levels throughout the country, but if employees observe those listed, they will turn out a good job. Failure to follow any one of these practices may result in extensive property damage, personal injury, outage of service, or a loss of revenue. And, last but by no means least, it may impair that valuable asset to a public utility—good customer relations.

GUIDE FOR INVESTIGATION OF CUSTOMERS' HIGH-BILL INQUIRIES ON THE CUSTOMERS' PREMISES

As a representative of the company, an employee who is assigned to investigate a customer's inquiry concerning his billing has a dual responsibility. In the first place, he must make absolutely certain that all service used by the customer is being accurately registered on the meter. He should also attempt to explain some of the metering factors to a customer who does not understand them, so that the customer will be assured of the accuracy of the meter.

The following is one suggested plan for conducting metering investigations. Steps 1 and 2 are frequently performed by a representative of the commercial department of the company. In many companies, only Steps 3 and 4 are performed by meter department personnel.

Step 1

A. Upon arriving at the customer's home or place of business, the company representative should introduce himself, show his identification card or badge when requested, and explain the purpose of his visit. He should also advise the customer that, in the process of checking, the electric service might be momentarily interrupted.
B. The meter readings should be compared with those on the last bill. If an error is apparent, it should be explained to the customer and reported to the billing section.
C. If the customer's question concerning his bill cannot be explained from the aforementioned, the company representative should ask the customer if any appliances have been added recently. Inquiry about appliances that affect seasonal loads is particularly important. These would include, among others, air conditioners, dehumidifiers, space heaters, and heating cables. It should also be determined whether the customer has replaced any major appliances recently. The replacement of a refrigerator with a newer, frost-free model may result in an increase in operating cost, and this should be explained to the customer.
D. Ask the customer if there has been an increase in the number of persons living in the household.
E. Ask the customer if situations have occurred that were out of the ordinary routine of the household. These would include such things as more entertaining than usual, guests visiting in the home, an illness in the family, and other factors that would increase the customer's bill.
F. If the customer's inquiry can not be explained after carrying out the parts of Step 1, proceed to Step 2.

Step 2

A. Check for causes of abnormally high consumption, such as:
   1. Dirt- or lint-clogged filters on furnace or air conditioning units.
   2. Leaky hot-water faucets where electric water heaters are used.
   3. Defective water pump.
   4. Use of electric range units for space heating.
   5. Heating water on range.
B. If the preceding investigation is sufficient, discuss these factors with the customer briefly, pointing out any reasons you have found for the increase in the customer's electric bill.
C. If further investigation is necessary, proceed to Step 3.

Step 3

A. If there is a load on the meter, ask the customer to shut off all appliances and lighting at the appliance or light switch. Recheck if meter continues to indicate a load and determine absolutely whether rotation is due to a missed load or loss to ground. If it is determined that there is a loss to ground, this should be checked with a stop watch and the rate of loss established.
B. If possible, it should then be determined which circuit is grounded and, if convenient to the customer, the circuit should be disconnected by removing

the fuse or opening the circuit breaker. The condition should be explained to the customer and he should be advised to have the condition corrected by a wireman before again using the circuit. If the loss to ground is found to be a defective appliance, it should be disconnected and the customer advised to have repairs made before reconnecting the appliance.

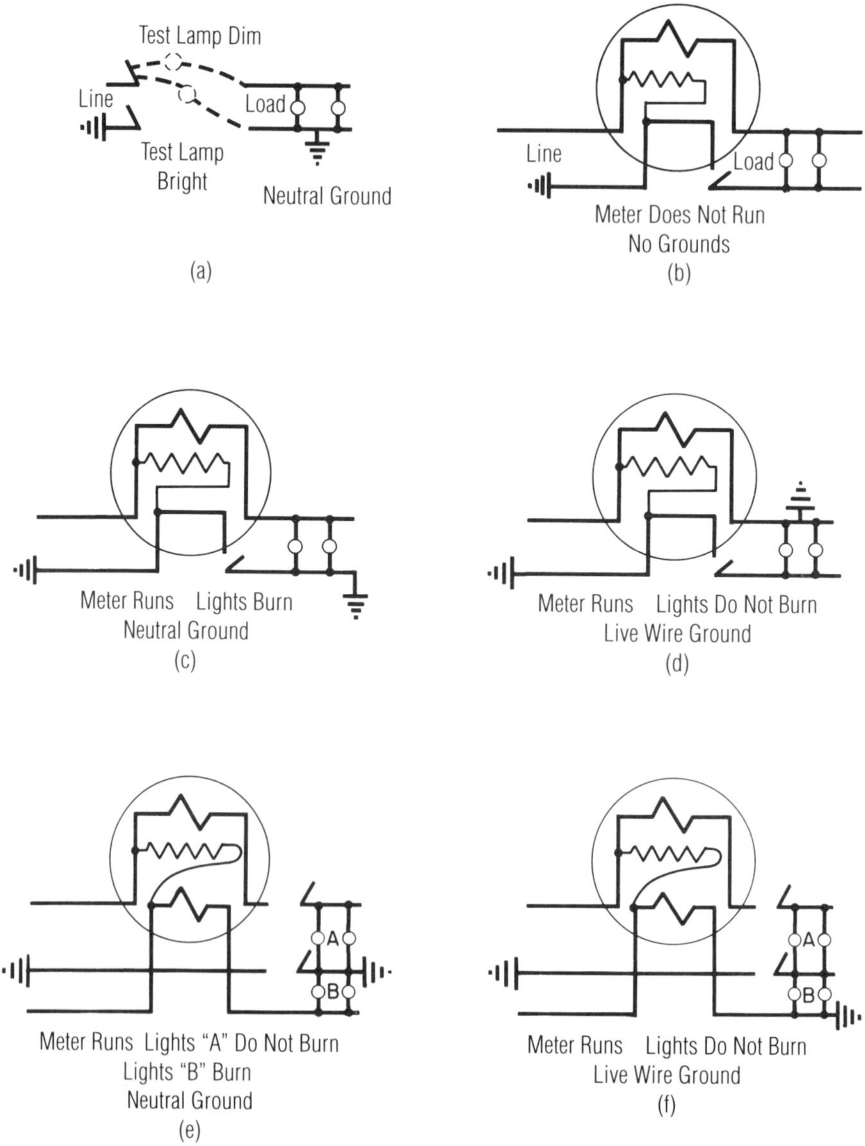

Figure 13-16. Methods Used in Checking Installations for Grounds.

Step 4

If no explanation for the bill has been reached, the meter should then be checked for accuracy, creep, proper constant, and correct register ratio. The results of each check should be noted on the investigation order.

METHODS USED IN CHECKING INSTALLATIONS FOR GROUNDS

In checking the customer's wiring for grounds, all the wiring and circuits should be connected; that is, any wall switches should be turned on, since there may be a ground between the switch and the load it controls. It is not necessary to turn on individual bulbs or floor lamps, because these are always energized up to the socket and any trouble at that point would be a short circuit and not a ground. After finding a ground, the best practice is to turn off all other circuits and leave only a few lamps burning on the grounded circuit. This will make it easier to determine if it is a live-wire ground and, also, easier to check the wattage.

The usual method of checking a small installation (see Fig. 13-16a) is to connect the test lamp from the hot wire of the line side of the opened switch to the load side and determine from which switch blade the brighter light is obtained, that line being the grounded wire. If the test lamp does not light on either wire, there are no grounds. After finding that a ground exists, it is usually necessary to disconnect the various circuits (removing both live and neutral fuses or wires) until the grounded circuit is found. Then turn off all but a few lights on this circuit so that the difference in brilliancy of the test lamp can be noted.

Perhaps the simplest method of testing a two-wire installation is to remove the load neutral wire, but leaving voltage applied to the meter. With a load connected, note if the meter stops. If the meter runs, there is a ground; if the customer's lights burn, it is a neutral ground; and if they do not burn, it may be a live-wire ground. If a neutral, the ground may be very weak so that the lights only burn dimly. In such a case, some lights can be turned off, which will make the remainder burn brighter. See Figs. 13-16b, c, and d.

Using this same principle, a large three-wire installation may be checked by leaving the voltage applied to the meter and disconnecting one live-load wire and the neutral, and then checking each side of the line separately, as in the case of the two-wire. If the meter stops, there are no grounds; if it runs, there is a ground. If the customer's lights burn on one side of the line, it is a neutral ground; if they do not, it may be a live-wire ground. In checking the wattage of a live-wire ground, the following method also presents an easy way. Leave only the grounded hot load wire in the meter and, with the neutral disconnected, check the load on the meter. It is seen that if there is also a neutral ground, the customer's load will also register during the ground check. Therefore, as many circuits and lamps as possible should be turned off, so that the chances of this occurring will be lessened. Figures 13-16e and f illustrate the method of checking a three-wire installation.

CHAPTER 14

WATTHOUR METER TESTING AND MAINTENANCE

INTRODUCTION

Meters are tested in order that their accuracy may be established, that this accuracy may be a matter of record, and to prolong this accuracy by detecting and removing causes which might tend to change it.

A test of a watthour meter consists of determining whether the registration of the meter is correct for a given amount of energy. A watthour meter may be tested by comparing its registration with that of a standard watthour meter or with a meter of known accuracy.

In this chapter, testing of alternating-current watthour meters will be described. For information on demand meter testing, see Chapter 15. For information on direct-current meter testing, see Chapters VI and VII in the 1923 edition of the *Handbook for Electrical Metermen.*

Meters may be tested on the customers' premises, in a mobile test unit, that is, a truck or trailer equipped for testing meters, or brought into a meter shop for test. They may be tested in groups or as individual units, with simple manual controls or with complex automatic controls. Wherever the test location and whatever the method, the principles are the same. In this chapter, first described are basic meter tests and the components necessary for such tests. Procedures for different test locations and complex test equipment will be described after test principles have been established.

To compare the performance of the meter under test with that of some reference standard there are required:

1. A reference standard.
2. A load to consume the energy expended.
3. A power source.
4. Connection devices and controls.

REFERENCE STANDARD

The reference standard may be a wattmeter which reads in terms of power and requires a second reference standard, such as a contact clock or stop watch, to control or measure the element of time. For certain laboratory tests, such equipment is necessary.

The more common reference is a standard watthour meter which reads directly in terms of energy, and does not require an accurately measured or closely controlled time interval. The standard watthour meter is usually referred to as a standard. A standard is a special watthour meter designed for portable use and with a digital readout or a sweephand register to permit reading the meter registration to precision of 0.1 percent or better.

To permit testing meters of various current ratings with one standard, standards are built with several current ranges in the current element. These current ranges may be rated from 1 to 50 amperes. With this span of current ranges to select from, it is practical to use a range which closely matches the actual test current applied, thereby attaining a wattage input at which the standard accuracy is best and at which a registration is achieved which can be read to decimal fractions of 1 percent. For example, a full-load test on a 50-ampere meter can be made with the 50-ampere range of the standard in the circuit; a lightload test (10 percent) on the same meter can then be made with the standard 5-ampere range connected. In all cases, the standard current range best suited to indicate accurately the quantity being measured should be used. The standard current range selected should be such that the test current falls between 25 and 150 percent of the nominal rating of the range. On all except the l-ampere range, this upper limit may be increased to 200 percent without danger, although for most accurate results at this loading the standard should be calibrated at the actual current used.

It is important to remember that the high-accuracy current ranges of the standard meter are not compensated for overload to the degree common in the modern service watthour meter. It is poor practice to overload standard current ranges beyond the limits just suggested. Usually the 1-ampere range of a standard, and sometimes the 5-ampere range, is protected by a fuse. On the other hand, it is rarely necessary to operate a standard at a load below 25 percent of the rating of any of its current ranges. Such practice of loading standard current ranges eliminates one source of lightload testing errors.

Where the standard has no more than four current ranges, the current coils are brought out to separate binding posts, one of these being commoned. With separate binding posts, it is possible to connect the standard to take advantage of full-load, lightload transfer switches provided on many test boards and phantom loads. Some recently-introduced standards have autoranging capabilities. When these standards are used, the requirement to select proper current and voltage ranges is eliminated.

INDUCTION TYPE WATTHOUR STANDARD

Full-load speed of the induction type watthour standard is the same under all connections. That is to say, the rated amperes and the current-coil turns give the same ampere-turns for each rating. The watthour constant varies directly with the current-coil rating. Thus, for a four-current-coil standard we have the following constants:

Coil Rating (Amperes)	*Turns*	*Ampere-Turns*	K_H
1	100	100	0.12
5	20	100	0.6
12.5	8	100	1.5
50	2	100	6.0

Standards are generally equipped with resets to permit starting each test from zero. In one standard, the reset consists of pins mounted on a reset plate which are pressed against heart-shaped cams on the register shafts during the reset action. Resetting should be done smoothly, taking care that the pointers do not bounce from zero as the reset is released.

The normal procedure followed in using a standard to test meters is to start and stop the standard by applying and removing its voltage by means of a special switch while the meter in test continues to run. In calibrating a standard against primary references, it is also started and stopped by means of a voltage switch. To eliminate the possibility of the standard creeping, all standards have disk brakes which are controlled by voltage. When voltage is applied, the brake is lifted and when voltage is removed, the brake drops back against the edge of the disk to stop rotation immediately.

The voltage coil of the standard is usually a double coil, connected in series for use on 240-volt circuits and in parallel for use on 120-volt circuits. For use on

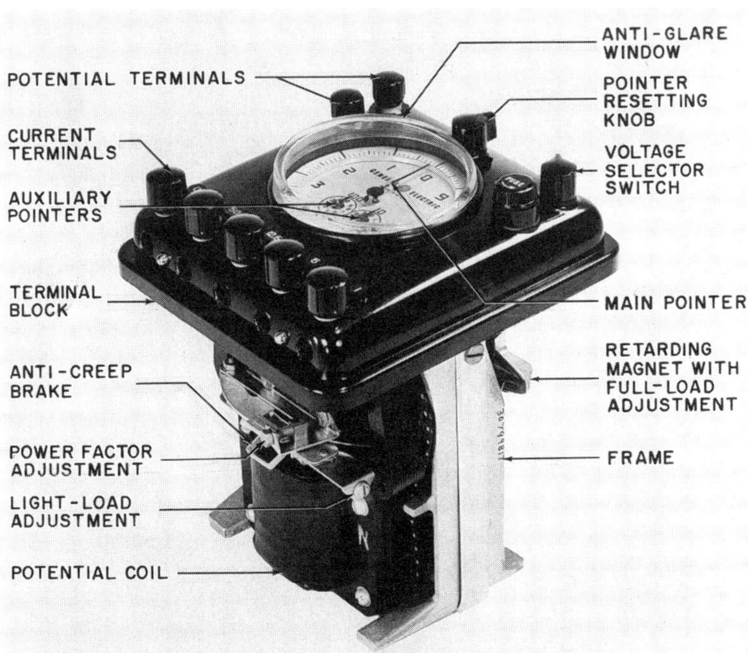

Figure 14-1. Induction Type Watthour Meter Standard.

480- or 600-volt circuits, an external multiplier is supplied. See Fig. 14-1 for the construction of one type of standard.

A standard is a high-accuracy, special watthour meter and should be handled with care both in transportation and in use to avoid mechanical or electrical injury. Although strong magnetic fields have little effect on induction-type standards, care should be taken to keep them away from heavy current-carrying conductors, motors, transformers, or other like apparatus. The effect of masses of iron adjacent to the meter or standard should not be ignored. Such masses of iron may change the reluctance of the magnetic-field path of the standard.

A standard should be level if accuracy is to be maintained.

SOLID-STATE REFERENCE STANDARDS

In the late 1970's, along with the development of the solid-state watthour meter, came the solid-state portable watthour standard. Solid-state standards such as the Scientific Columbus SC-10, Sangamo A7, and Radian RM-10, are packaged to resemble electro-mechanical portable standards for interchangeability within the standard testing circuit, while providing improvements over the rotating standards. For instance, the lack of moving parts makes the need for leveling the standard unnecessary while in use, as well as providing greater accuracy and

Figure 14-2. Scientific Columbus Type SC-10 Solid State Portable Watthour Standard.

ruggedness. The digital readout is visible under poor lighting conditions and can provide greater resolution than electro-mechanical standards. The solid-state standard requires an auxiliary power supply to maintain the digital readout and power the internal electronics of the standard.

Most of the recent solid-state standards provide an auto-reset feature which eliminates the need to manually reset the standard after each test. The auto-reset feature is activated automatically simply by starting the standard. Induction-type mechanical standards and early solid-state standards required opening and closing the voltage circuit to start and stop the standard. However, certain recent solid-state standards use another method for starting and stopping the standard. This new method "freezes" the display, while the standard continues to run. Test results have shown that "freezing" the display provides better accuracy than opening the voltage circuit. Solid-state standards also are typically equipped with a high frequency pulse output which can be used for meter tests and for calibrating the standard.

Type SC-10 Portable Watthour Standard

The Scientific Columbus Type SC-10 Portable Watthour Standard is designed for field testing watthour meters. Features include light weight, built-in 480-volt range, high intensity digital readout, and a pulse output. It may be operated in any position since there are no moving parts. See Fig. 14-2.

The LED six-digit display reads out in "Equivalent Revolutions" corresponding to the calibration of rotating standards, so existing test procedures and calibration tables still apply and present operators require no additional training. The digits are in two colors. Full "revolutions" are displayed in red, and partial revolutions are displayed in yellow, eliminating the need for a decimal point and further enhancing readability.

A pulse output (optically isolated, open collector) is available from a standard BNC connector on the top panel. The pulse rate corresponds to the least-significant digit of the digital readout. This may be used as the reference signal in automatic test tables and for checking its calibration against laboratory standards.

The A7 Portable Standard

The Sangamo A7 solid-state portable watthour standard is designed for meter shop or field testing of watthour meters. It accumulates a pulse-count corresponding to applied watthours, displays that count on a six-digit display, and outputs pulses for external summation.

The A7 and a meter under test are connected to simultaneously measure power dissipated by a single load for a specific period of time. The difference between the A7 reference and the meter under test is used to compute any percent error.

There are two methods of use:

1. A pulse count proportional to the applied watthours is accumulated by the A7 standard, for a specified number of disk revolutions of the meter under test. The pulse count measured is displayed by the A7 readout in "equivalent disk revolutions." Actual and displayed pulse counts can be used to determine the meter error.

2. An external comparator can compare the expected A7 output pulse count for a specific test period, with the actual count obtained, and display any meter error directly. Output pulses can be selected by the user to be either 1,000 pulses, or 1 pulse per kilowatt-hour.

The RM-10 Portable Standard

The RM-10 from Radian Research is a rugged, solid-state portable watthour standard. It utilizes auto-ranging to automatically accept inputs from 80 to 520 volts. The unit has a true watthour display for all current and voltage ranges.

In a typical mode of operation, the RM-10 is connected to the input power by means of a BNC connector. A momentary push of a button on the front panel starts a counter. A second push stops it after the test period, typically after 10 revolutions, and freezes the last reading. A third push zeros the counters for the next test. The output of the RM-10 reads directly in watthours and when compared with the reading of the meter under test, the calculation of meter error can be made.

TEST LOADING METHODS

Any one of three methods of current loading may be used.

Customer's Load

In testing meters in service, the customer's load itself may be used. This has many disadvantages, but is a valuable type of test load under certain conditions which will be described later under the subheading "Testing on Customers' Premises."

Resistance Load

Resistance load has characteristics in common with a customer's lighting load and is adjustable to provide the various test currents desired.

In the resistance-loading method, the current coils of the meter under test, the standard, and the loading resistance are all connected in series. Thus, the current which is permitted to flow by the resistance passes through both the meter and the standard.

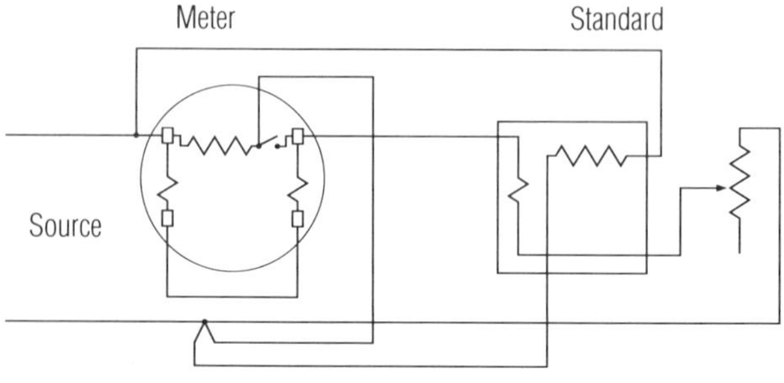

Figure 14-3. Schematic Connections for Resistance Loading.

Schematic circuit connections for resistance loading are shown in Fig. 14-3.

The resistance type of loading device usually consists of a group of fixed resistances of various values which can be connected in any one of several series-parallel networks to give the total resistance which will allow the current flow required for the test. These loading devices are calibrated for specified voltages and the switches are generally marked to indicate the current each allows to flow so that currents can be obtained in steps from zero to the maximum rating of the device.

A real difficulty in this type of test load lies in the problem of dissipating the energy consumed by the I^2R loss at high currents. If the source is 240 volts and a 15-ampere meter is to be tested at the full-load point, a total of 15 × 240 = 3,600 watts must be dissipated in the loading resistance. The source also must be of sufficient capacity to furnish the full 3,600 watts. Furthermore, the weight and bulk of a resistance-loading bank may limit its usefulness as a portable device.

Phantom Loading

Phantom loading reduces the power in the current circuit by reducing the voltage across which the load is connected. A phantom load is basically a small power transformer and an adjustable loading resistance. Test connections apply service voltage—120 volts or 240 volts—to the voltage coils of the meter and the standard. The current circuits of the standard and meter, which are in effect isolated from the voltage circuit by transformer action, are placed in the secondary of the low-voltage, phantom load transformer whose primary is connected across the

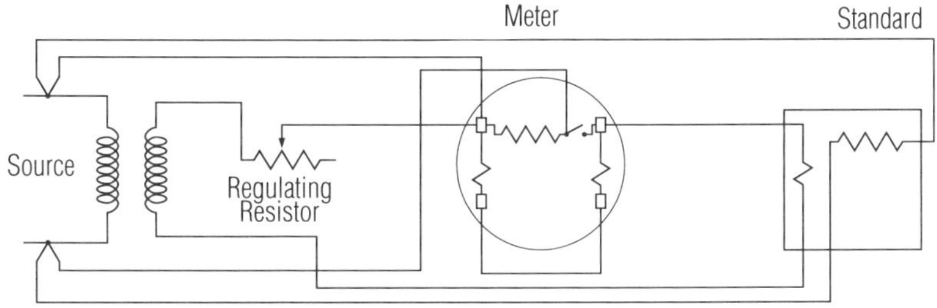

Figure 14-4a. Schematic Connections for Phantom Loading.

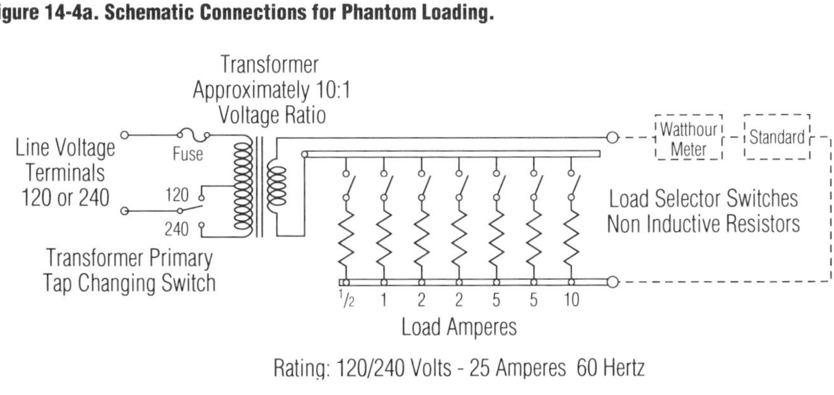

Figure 14-4b. Typical Wiring Diagram, Phantom Load.

line. The regulating resistance of the phantom load is also in series with the meter and standard current coils. See Fig. 14-4. Note that in this type of loading, although the current value is proper for the test being made, the voltage at which this current is supplied is low and, hence, the power has been reduced below that necessary for resistance testing. Assuming the phantom load transformer is rated 240 to 12 volts and the regulating resistor is adjusted to provide 15 amperes to the test circuit, the power of the secondary circuit would be 15 × 12 = 180 watts. The current drawn from the secondary circuit is of course 15 amperes, but from the source is $15 \times \frac{12}{240} = 0.75$ amperes. With the low power requirements, a phantom load can be constructed as a portable device.

The term "phantom load" is usually applied to the portable device which is used particularly for testing on customers' premises. However, the same principle of a loading transformer with low-voltage resistance units is used in practi-

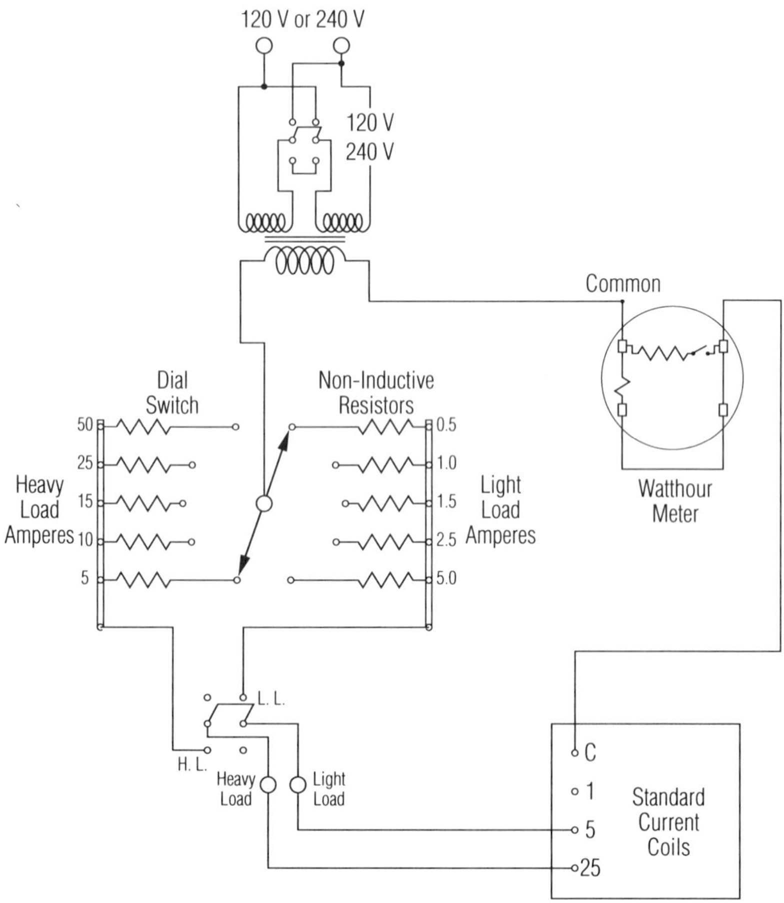

Figure 14-5. Dial Switch, Phantom-Load Circuit.

cally all modern test tables. Note that when the output voltage is lowered, the size of the current regulating resistors in the secondary circuit is also reduced while still maintaining the desired current magnitude .

When using phantom loads, the phase angle of the test circuit should not be ignored. Meter and standard current coil have inductive reactance as well as resistance. Normally, the regulating resistance is large enough to overcome any lagging effect caused by the reactance of the current coils and loading transformer. However, if the regulating resistance is small, the current through the current coils will lag the source voltage by some small angle. This change in phase angle may not be significant at unity power factor tests, but may be of importance at 50 percent power factor tests. This occurs because of the difference in values of cosines near zero and 60 degree angles. As an example, at 60 degrees the cosine is 0.5 and at 61 degrees it is 0.4848, while at zero degrees the cosine is 1 and at 1 degree the cosine is 0.9998. A small change in phase angle at unity power factor causes a much smaller change in cosine than a similar phase angle change at 50 percent power factor. The phantom-load phase angle shift will increase rapidly with increasing secondary burden above rated burden capacity. Furthermore, when a phantom-load transformer is overloaded, the waveform of the output current may be seriously distorted. On the best modern test boards, the loading transformers are designed to avoid these shortcomings.

Figure 14-5 shows a diagram of current connections for testing single-stator meters with one type of phantom load. When using a phantom load for testing a three-wire, single-stator meter, one end of the voltage coil must be disconnected from the incoming current lead by means of the voltage link due to the tying together of the current coils on the load side for test purposes. If this is not done, the current test circuit will be connected across full voltage, with resulting damage to meter coils and phantom load. The disconnected end of the voltage coil is connected to the correct voltage source in order to energize the coil at full voltage.

Control devices consist primarily of the resistances which control the amount of current and the switch which energizes the voltage circuit of the standard. The resistance switches have been briefly described. The voltage switch may be an automatic switch controlled by an optical sensor, or a "click switch," or pendant switch, operated by hand. The optical sensor detects the marked edge of the meter disk or the anti-creep holes on the meter disk. A control circuit uses the signal generated by the optical sensor to start and stop the wattmeter standard. The click switch may be in the form of a single-pole, single-throw switch. Switches made especially for this purpose have a small movement of the switch button to reduce time lag and improve accuracy. The On and Off motions should be similar in magnitude and pressure. Practice in working this switch, especially on fast revolutions of the meter, is necessary to obtain consistent accuracy results. The contact resistance of the snap voltage switch should be checked often to ensure that it is low. If not kept low the high resistance of the switch will cause a significant error.

POWER SOURCE

A steady power source is required, particularly when indicating instruments are used as reference standards. When a standard is the reference, with the current circuits of the standard and of the meter being tested in series and their voltage coils in parallel, minor fluctuations in voltage or current have no significant effect on the test results. In this case both reference standard and meter are affected by minor changes in source or load in the same manner. This does not mean that wide and rapid changes in source voltage should be tolerated. Harmonics do not have the same effect on all meters and waveform distortion should be avoided.

Early meter test boards required a three-wire, three-phase power source. The three-phase feature provided a convenient way to obtain a 50 percent power factor, single-phase test load. Figure 14-6 shows the basic test connections, including cross-phasing to obtain 50 percent power factor. The connections are fundamental and in some form are used for all meter testing.

The preferred source connection is a closed-delta transformer bank. Both open-delta and wye connections may contain unduly large harmonics and will not therefore provide a wave-form as close to a pure sine wave as may be obtained from a closed-delta transformer bank. The closed-delta bank also provides better voltage regulation than the other two types of connections.

The source transformer bank and conductors must be of such size that the test load does not cause any material voltage drop from the source to the test equipment. Heavy loads, particularly motor loads, such as elevators, which cause excessive voltage fluctuations, should not be connected to the meter-test transformer bank.

Late-model electronic test boards usually are powered by 120-volt single phase sources. Using electronic instruments, a variety of test waveforms are generated. These "manufactured" waveforms include harmonics, a variety of power factors, as well as three-phase waveforms.

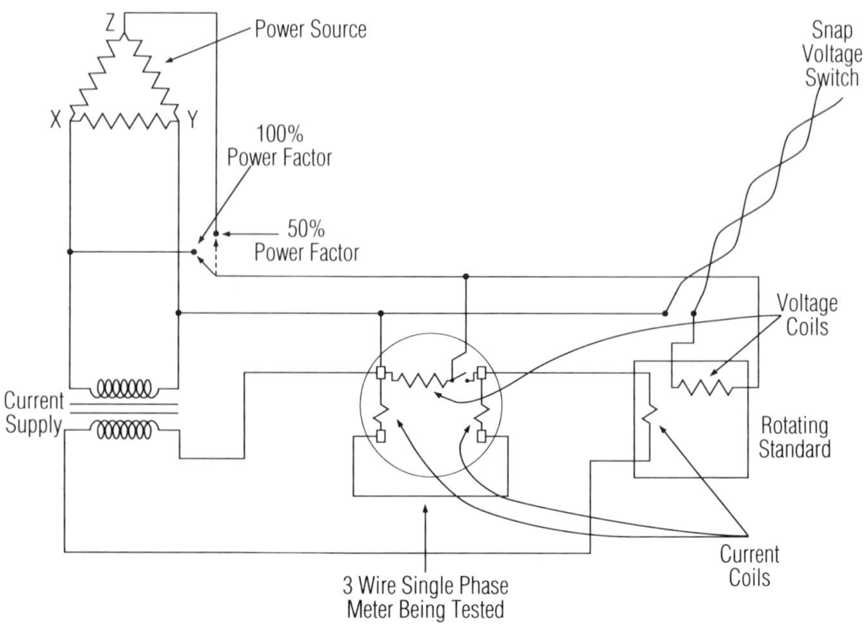

Figure 14-6. Fundamental Meter Test Circuit.

BASIC INDUCTION-TYPE WATTHOUR METER TEST — SINGLE-STATOR

Before suggesting procedures for the various classes of meter test and the several locations at which tests may be made, it is important to describe those practices common to all types of test. Some of these steps will be expanded later.

Step 1

Check meter number and meter rating. Record this data.

Step 2

Check for creep.

Creep may occur either backward or forward. When all load is removed, a meter disk may rotate for a part of a revolution before coming to rest. This is not creep. All measurements of the amount of creep should be based upon at least one complete revolution.

Although only an unusually rapid rate of creeping will result in an appreciable registration, as a matter of principle, no meter in service should be allowed to remain creeping or with a tendency to creep.

In most induction meters, creeping is prevented by two holes or slots cut in the disk on opposite sides of the shaft. When either hole is near the pole of the voltage coil, forces set up by the alternating field tend to hold the disk in this position.

Step 3

Connect meter and take "as found" readings. The connections for testing are as follows: The current circuit of the standard is connected in series with the loading device and current coil of the meter under test, and the voltage circuit of the standard is connected in parallel with the voltage coil of the meter under test. When setting up a standard for making a test, the place selected should be reasonably free from vibration and magnetic influence. The meter should be plumb, without tilt, and the standard should be level during the test. The standard must always be reset to zero before starting a test. A reading of the standard is taken at the end of the test, which gives the number of revolutions of the standard pointer. If no correction is to be applied to the standard readings, the percent registration of the watthour meter under test is obtained as follows:

let r = revolutions of meter under test

R = reading registered by the standard

k_H = watthour constant of meter under test

K_H = watthour constant of standard

then, Percent Registration $= \dfrac{k_h \times r}{K_h \times R} \times 100$

The method may be facilitated by introducing an additional symbol, values for which may be given to the tester in tabular form.

Let R_O = the reading the standard should register when the meter under test is correct.

The revolutions of two watthour meters on a given load vary inversely as their disk constants.

$$\frac{R_O}{r} = \frac{k_H}{K_h}$$

Substituting R_O in the equation for percent registration:

$$\text{Percent Registration} = \frac{R_O}{r} \times 100$$

In testing, the number of watthour meter disk revolutions should be sufficient to permit reading whole divisions of the standard register to the degree of accuracy required.

When the watthour meter under test and the standard have the following constants:

meter	k_H	=	0.6
standard	K_H	=	0.12

number of revolutions of the meter under test, $r = 2$

$$\text{then, } R_O = \frac{0.6 \times 2}{1.12} = 10$$

That is, for two revolutions of the meter under test, the standard registration should equal ten revolutions. Assume the standard actually registered 10.16 revolutions.

$$\text{then, Percent Registration} = \frac{10}{10.16} \times 100 = 98.4$$

When a correction is to be applied to the readings of the standard, the percent registration is determined as follows:

let A = percent registration of standard

$$\text{then Percent Registration} = \frac{k_h \times r \times A}{K_h \times R} = \frac{R_o}{R} \times \text{A}$$

Step 4

Examine original condition of meter. The principal features to look at are:

a. Is the disk centered in both permanent magnet gap and electromagnet gap?

b. Is the magnet gap clean?

c. Examine the mesh of first register gear with shaft worm or pinion. This mesh should be between one-third and one-half the depth of the teeth. A deeper mesh may cause binding. A slight amount of play is necessary. Where the pinion or worm is short, or where the worm is cut concave to match the curvature of the worm wheel, the vertical position of the moving element must be such that the center of the pinion or worm is level with the register wheel which it engages.

Step 5

Check register ratio as marked on register to determine if this ratio is correct for the type and capacity of the meter. For instructions on a complete register check see a subsequent discussion under the subheading "Register Testing and Checking."

Check watthour constant (k_H) for the meter type and rating with the correct constant from manufacturers' tables .

Check kilowatt-hour constant (register multiplier). Kilowatt-hour constant

$$= \frac{k_h \times R_r \times R_s}{10{,}000}$$

where: k_H = watthour constant of meter under test
R_R = register ratio
R_S = shaft reduction ratio

Step 6

Adjustments. Since the heavy-load adjustment affects all loads equally, this adjustment should be made first. If, after adjustment at full load, lightload performance is more than about one percent slow, a cause other than maladjustment of the lightload should be looked for. Such cause may be unusual friction or dirt. If the meter is clean and the register mesh is correct, the meter bearings should be suspected. See bearing maintenance for suggestions.

Power-factor adjustment of single-stator meters is usually limited to shop testing. See discussion following for description of adjustments.

Step 7

Record final readings.

Step 8

Seal meter. Return meter to service or stock.

INDUCTION TYPE METER ADJUSTMENTS SINGLE-STATOR

Full-Load Adjustment

This adjustment is made at nameplate rating of voltage and current or at nameplate voltage rating and test amperes (T.A.). The adjustment is made in most meters by varying the effect of the damping flux passing through the disk. This is done in older meters by changing the position of the damping magnets. Similar results are secured in newer meters by varying the amount of flux passing through the disk by means of a shunt, sometimes called a keeper. The change produced in the percentage registration is practically the same on all loads; that is, if the registration is 98 percent at both heavy load and light load, shifting the heavy-load adjustment so as to increase the speed 2 percent will make the meter correct at both loads.

Moving the magnets toward the disk shaft causes the disk to cut the damping flux more slowly and the meter runs faster. Moving the shunt closer to the damping magnet poles causes more flux to pass through the shunt and less through the air gap in which the disk turns, thereby increasing the disk speed.

Lightload Adjustment

This adjustment is normally made at name-plate rating of voltage and 10 percent of rated amperes or test amperes. It is accomplished by varying the amount of lightload compensating torque.

This adjustment is changed by shifting a coil so that its position with respect to the voltage-coil pole is changed. No torque is produced on the disk as long as the lightload coil is symmetrical with the voltage-coil pole. When the lightload coil is shifted, a torque is produced in the disk which will tend to turn the disk in the direction of the shift. The coil is essentially a short-circuited turn of large cross section placed in the air gap above or below the disk so as to embrace part of the voltage-coil flux. Maladjustment of this coil may result in creep. See Chapter 7, "The Watthour Meter." The effect of this torque on meter percentage registration is inversely proportional to the test load; that is, one-tenth as much effect is produced at heavy load as at 10 percent load.

When a meter, after adjustment at full load, is found inaccurate at light load, the cause may be some condition in the meter which should be removed rather than compensated for. In such cases the tester should locate the trouble, making no adjustments unless the meter is still inaccurate after going over all the parts and restoring them to proper condition.

Lag Adjustment

This adjustment is ordinarily made only in the shop. The flux established by the voltage coil of a meter should lag the flux of the current coil by exactly 90 degrees with unity-power-factor conditions for proper metering accuracy with any load power factor. This flux relationship does not exist because of the inherent resistance of the voltage coil. Compensation to obtain the correct relationship is by means of a lag coil or plate on the voltage pole acting with another lag coil on the current poles. The explanation of how this compensation works is described in Chapter 7.

If the compensation is obtained by means of a coil, the adjustment is made by soldering the exposed pigtail ends of the coil so as to lengthen or shorten the over-all length of the coil conductor to change the resistance of the coil. If a lag plate is used, the adjustment consists of shifting the position of the plate under the voltage pole radially with respect to the disk by means of an adjusting screw. On some types of modern meters the lag adjustment is made by punching a lag plate during manufacturers' testing and cannot readily be changed in the field.

The test to determine the lag or phase adjustment is generally made at 50 percent lagging power factor with rated amperes and voltage applied. Fifty percent power factor is used because it can be obtained readily from a polyphase circuit without auxiliary equipment.

MULTI-STATOR INDUCTION TYPE METER TESTS AND ADJUSTMENTS

Multi-stator meter tests follow a procedure similar to that used for single-stator meters. With the multi-stator meter voltage coils connected in parallel and all current coils in series, the procedure is identical to single-stator tests, but additional tests may be made with each individual stator energized. Each separate stator in a multi-stator meter must provide the same disk-driving torque with equal wattage applied to the individual stators for the meter to provide accurate registration when in service. Therefore, the individual stator torques must be balanced and this is the reason for the individual stator tests and the additional torque-balance adjustment on multi-stator meters. Individual stator tests are also useful in determining correct internal wiring of the voltage and current coils of each stator.

The torque-balance adjustment commonly consists of a magnetic shunt in the stator iron, the position of which may be changed to vary the effective stator air-gap flux and hence, the disk torque produced by the stator. This adjustment, as described, varies the driving torque produced by the individual stators, thereby allowing the torque of one stator to be made equal to that produced by a second

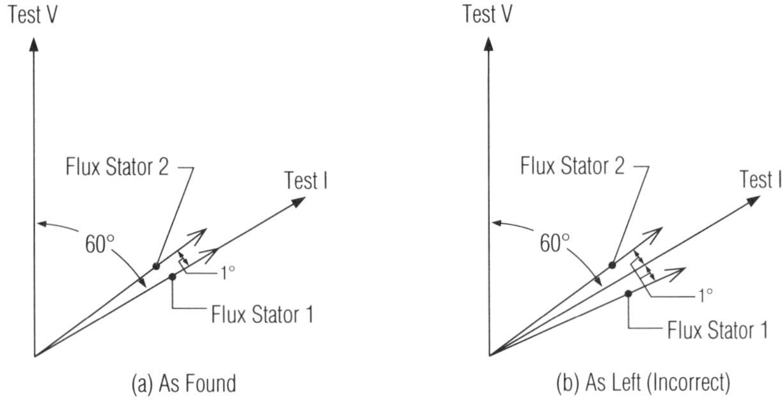

Figure 14-7. Incorrect Adjustment with Lagging Power Factor.

stator. The usual practice is to make the adjustment at unity power factor with rated meter voltage and test amperes. The adjustment may be provided on all stators of a multi-stator meter or, as on some modern meters, it may be omitted on one stator, in which case the torque of the other stator(s) would be adjusted to match the first.

All stators in a multi-stator meter must have a lagging power-factor adjustment to provide the proper flux relationships in each unit. The importance in meter testing of proper balancing of lag adjustments among meter stators is not generally recognized, for not only must the torque-balance adjustment be correct at unity power factor, but the lag adjustments of each individual stator must also agree to provide best meter accuracy under all service conditions.

An example of improper lag balance will illustrate the possible meter errors. Assume a two-stator meter is found running 1.5 percent fast on a series-parallel, lagging-power-factor test caused by a faulty lag adjustment on stator 2. This error is equivalent to a 1 degree phase-angle error in the lag adjustment, as shown in Fig. 14-7a. If instead of correcting the faulty lag adjustment on stator 2, the adjustment on stator 1 is changed, the end result at 100 percent meter accuracy with series-parallel connection will be a 1 degree lead error in the lag adjustment on stator 2 and a 1 degree lag error in the lag adjustment on stator 1, as shown in Fig. 14-7b. With the meter connected to a three-wire delta service, consider its operation on a balanced, unity-power-factor load. Figure 14-8 shows the circuit diagram and metering phasors. Note that in stator 1 the current lags the voltage by 30 degrees and in stator 2 the current leads by 30 degrees. With a maladjusted meter which has power-factor errors of 1 degree lead and lag, one possible circuit connection would, in effect, have both stators registering at a 29 degree angle instead of the proper 30 degrees, or the other circuit connection would cause an effective angle of 31 degrees. In either case, the meter performance error in this application would be approximately 1 percent even though the meter was adjusted to 100 percent performance in the usual series-parallel, single-phase test procedure.

Lightload adjusters may be provided on all stators or only one stator in multi-stator meters. Since the lightload adjustment provides additional disk torque

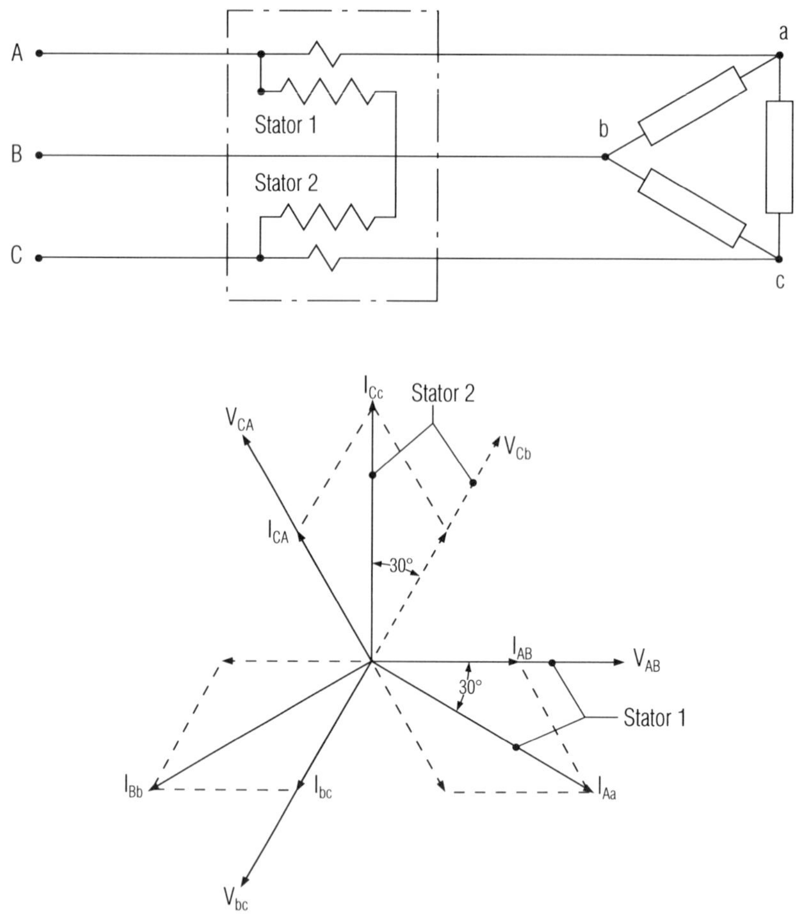

Figure 14-8. Three-Phase, Three-Wire Delta Circuit and Metering Phasors.

dependent on voltage, it is immaterial whether the torque comes from one stator only or all stators as long as all voltage coils are energized. In some meters, there is interdependence between the lightload and lagging-power-factor adjustments so that a change in light load in one stator may affect the lagging power-factor performance of the same stator. If more than one lightload adjuster is provided, it is good test practice when lightload adjustment is required in a series-parallel test, particularly in field tests, to make equal changes with each lightload adjuster.

The full-load adjustment on multi-stator meters, operating on the braking magnets, has an identical effect on all stators. Hence, it cannot be used for torque balance. One or more full-load adjusters may be provided on multi-stator meters.

METER TEST CIRCUITS

Over the years, many arrangements of test circuits have been devised and many forms of test tables are available. Modern test tables show major improvements over many of the older so-called test benches. Figures 14-4 and 14-6 show fundamental schematic diagrams for meter test circuits. Although the circuit arrangements, equipment, and methods of counting revolutions may differ in test tables, fundamentally all the circuits are based on that shown in Fig. 14-6.

Meter Timing and Speed Measurement Methods

Before reviewing more complicated meter test tables, consideration should be given to the automatic timing methods used in various test tables. Two general methods are used, the photoelectric disk revolution counters and the stroboscopic comparison of meter disk speeds.

Photoelectric Counters

Modern test tables do not use a manual voltage switch for applying voltage to the standard meter. Voltage is controlled by a photoelectric counter which contains photoelectric devices and associated equipment for automatically starting and stopping the standard watthour meter. For this method of test, a light beam is directed through the anti-creep holes or reflected from the flag of the disk of the watthour meter under test and illuminates a photodiode or transistor. This illumination causes pulses to be transmitted to the control equipment where they are amplified and used to operate digital displays, which in turn operate relays controlling the voltage circuit of the standard watthour meter. This is done in accordance with a predetermined number of revolutions of the watthour meter disk.

It is important that the start/stop relays have exact or symmetrical reaction times. For this reason, some designs use two relays for the start and stop function, arranged in such a way that they both pull-in to perform their function. Other designs use a single symmetrical reed relay controlled by zero crossing switching circuits to close the relay at zero voltage and open it at zero current. This method of zero crossing switching eliminates the need for arc suppression components used in standard designs which, with time, can cause testing errors.

This method of shop testing eliminates the necessity of manually counting the revolutions of the meter disk and, since the starting and stopping of the standard is automatic, human errors are eliminated.

ROTARY STEPPING SWITCHES

Many older photoelectric counters use rotary stepping switches for the counting, sequencing, and other functions in automatic test boards. For a detailed discussion of rotary stepping switches refer to the Seventh Edition of this *Handbook*, Chapter 15.

WATTHOUR METER TEST TABLES

Among modern tables there are two basically different types, each with many variations of components. One of these, the gang test board, provides connection facilities for a group of meters, all referred to a single standard. The standard, generally by photoelectric means, controls the length of the test run. In the other type, which may be a one- or two-position board, the meter under test controls the length of the test run and the standards are read to find meter performance.

Gang Test Boards

Since the single-stator, self-contained watthour meter normally has a common connection to voltage and current coils, there is always the hazard in gang test boards that unsuspected paths for current-coil or voltage-coil currents may exist. Circuits in which the current coils are shunted by the low-resistance voltage bus should be immediately apparent by meter performance. Other circuits in which the voltage-coil current passes through adjacent meter current coils may be difficult to discover since the voltage-coil current is small. Nevertheless, such a misconnection will cause serious errors at light load.

Figure 14-9 shows two incorrect circuits to be avoided and two methods of isolating voltage and current coil circuits from each other. In Fig. 14-9a, the voltage bus provides a shunt for all meter current coils except the one in the final position. Figure 14-9b shows a circuit which avoids the voltage bus shunt but does so by drawing voltage-coil current through the current coil with resultant meter errors at light loads. Figure 14-9c shows one method of separating voltage- and current-coil circuits by the use of 1:1 voltage transformers at each meter position. An equivalent result may be obtained by matched current transformers at each meter position, illustrated in Fig. 14-9d.

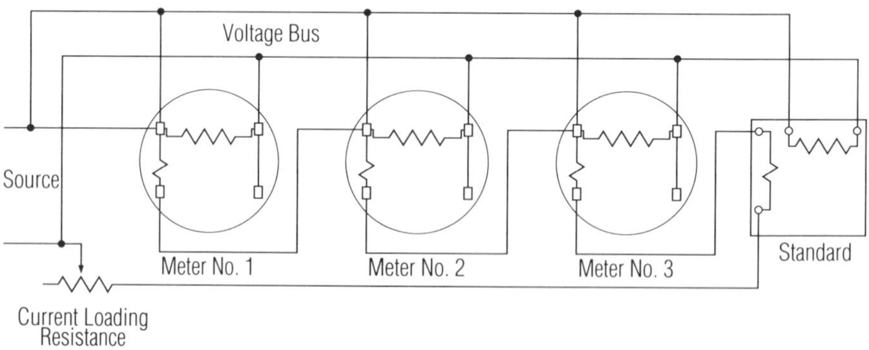

Figure 14-9a. Incorrect Gang Test Board Circuit.

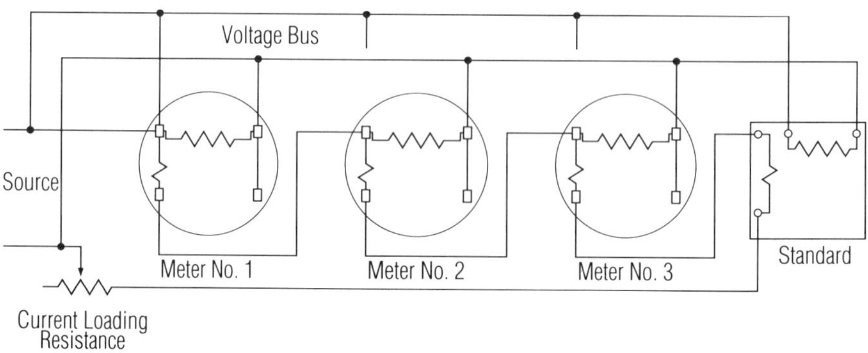

Figure 14-9b. Incorrect Gang Test Board Circuit.

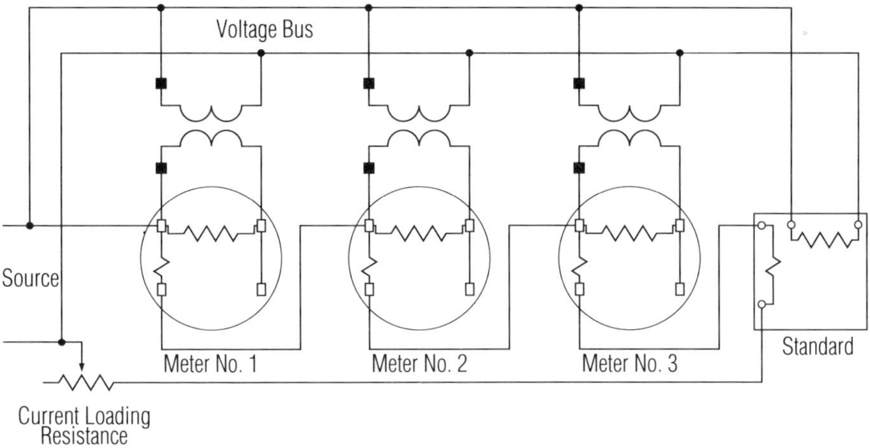

Figure 14-9c. Gang Test Board Circuit with Voltage Transformers.

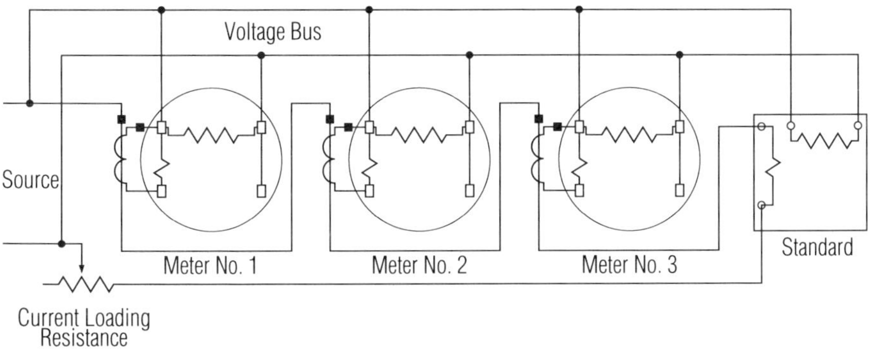

Figure 14-9d. Gang Test Board Circuit with Current Transformers.

The voltage which actuates the standard and each meter being tested must be exactly the same at all meter terminals during tests. Figure 14-9 shows the voltage connections for all meters and the standard tapped from a voltage bus. This bus must be of such size that there is no significant voltage drop between any of the voltage coils of the meters or standard in the testing circuit. Where isolating voltage transformers are used, they must have identical characteristics.

Both meters and standard must always be operating under exactly the same known conditions. Good accuracies cannot be achieved if voltages differ or if there are sneak circuits in the test connection.

The percentage registration of a meter is determined directly at the end of a test by the displacement of the mark on the disk from the reference point. This displacement can be read directly on modern meters with the printed disk divisions or by use of scales similar to that illustrated in Fig. 14-10.

The percentage registration of the meters may also be determined directly from the standard by the use of push-buttons for inching all meters and standard

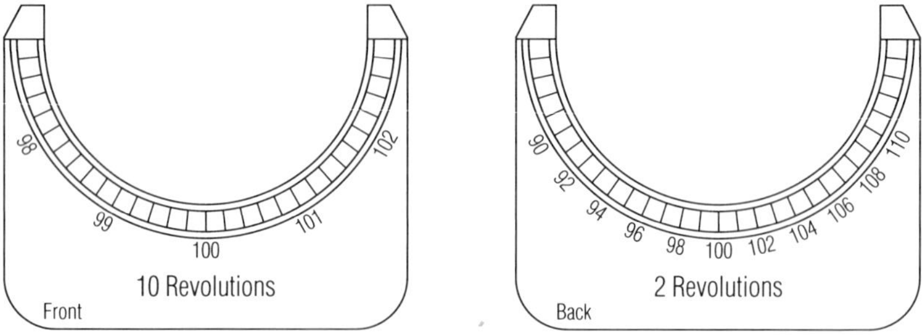

Figure 14-10. Typical Scale Used for Determining the Accuracy of Meters under Test by Start and Stop Method.

either forward or backward to the reference position. This inching forward or backward is done separately for each meter that is out of calibration on the test board.

Figure 14-11 is a schematic wiring diagram of another method of gang testing single-stator meters without opening the voltage gate which utilizes multiple secondary current transformers. This particular board is a gang type which uses a long time run on the meters under test and contains equipment for automatically stopping the meters after a preselected number of watthours have been recorded on the reference meter.

Here the meter accuracies are determined from meter register readings. This type of test, generally limited to full load, does not provide the most precise accuracy readings, but does provide an indication of overall meter performance, including a check on the register ratio, at the load selected.

A trial run, such as can be made on this type of test board, is sometimes used to determine whether new meters run within commercial accuracy limits. At test amperes, the test takes about 30 hours, but the test board is unattended except at the start and end of the run.

Single- or Double-Position Test Boards

Figure 14-12 shows connections designed for testing alternating-current watthour meters, both single-phase and polyphase. This scheme utilizes the Knopp "Uniload" system in which multi-range current and voltage transformers are connected in the current and voltage circuits, respectively, and whose secondaries supply the standard meter, thus permitting the standard to operate at the same current and voltage without regard to the rating of the meters under test.

A modification of this test scheme is the Knopp "Unilink" test board which also uses the "Uniload" principle but which, in addition, permits testing single-stator, three-wire meters without opening the test links. A description of this test method, together with a diagram of connections, may be found in the Eighth Edition of this *Handbook*, Chapter 19.

Figure 14-13 is a schematic wiring diagram of another method of testing single-phase meters without opening the voltage link. This scheme utilizes the TESCO feedback transformer method. The loading current is passed through the

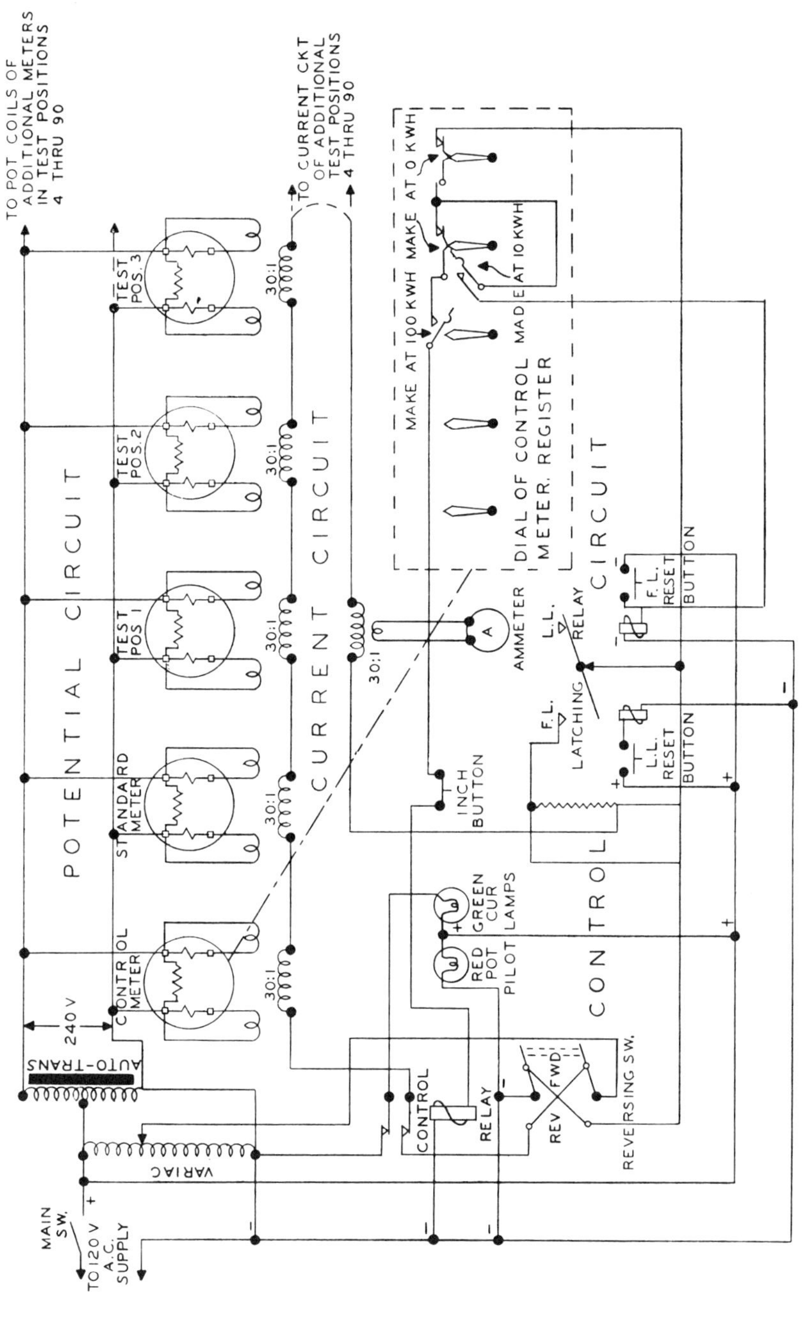

Figure 14-11. Gang Test Board Circuit for Testing with Closed Meter Voltage Gates.

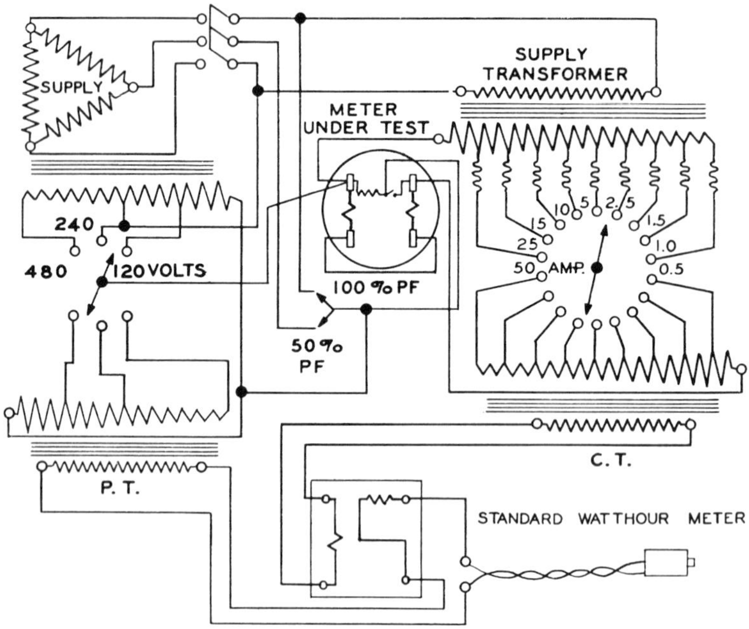

Figure 14-12. Circuit Diagram of Watthour Meter Test Equipment with Knopp "Uniload" Circuit.

feedback transformer twice in identical windings but in opposite directions. Thus, the load current has no effect on the excitation of this transformer, but allows the meter voltage-coil excitation to be obtained and the meter test performed without opening the voltage link.

Figure 14-14 is a schematic diagram of three watthour meters simultaneously tested under live load, an extension of the double-position test procedure. This scheme not only eliminates the need of opening the voltage gate on the meters under test, but also uses only one load, thus reducing the heat dissipation problem. However, use of one load for testing three meters in some ways complicates the connections. The current coils are all connected in series. In actual practice, but not shown in Fig. 14-14, three standards are used for each of the three meter test positions so that all three meter accuracies are registered—heavy load, 50 percent power factor load, and light load—on respective standards for each meter under test.

If all meters had the same current rating and watthour constant, the standard current coils would be in series with the current coils of the meters under test. Since all watthour meters do not have the same current rating and constant, a tapped current transformer is connected in each standard current-coil circuit. These transformers have tapped primary and secondary windings so that the standard current can be varied to make standard disk speed equal meter disk speed, and one revolution of the standard pointer equal one revolution of the meter disk. With this equipment, the standards operate with full load on the 5-ampere coils for all test loads to obtain the best standard accuracy and to reduce the calculation required to determine meter accuracy. During lightload,

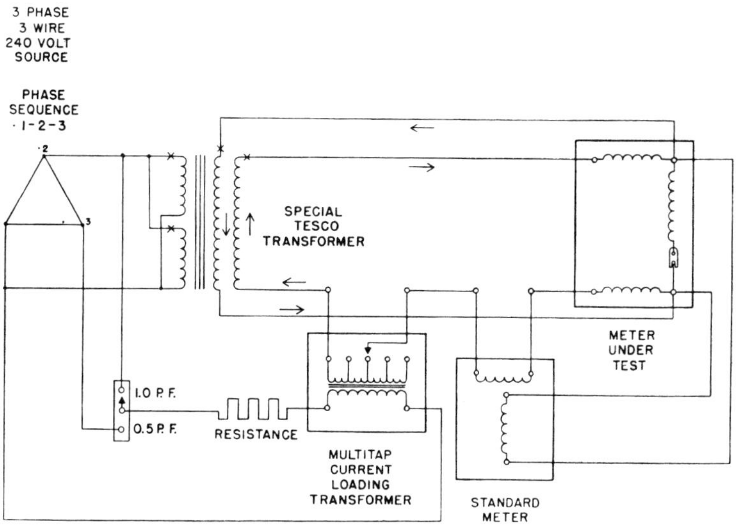

Figure 14-13. Schematic Diagram Showing One Method of Connection for Test with Closed Voltage Link.

test the test current is stepped up to the standard so that its pointer makes ten revolutions to one of the meter.

The schematic diagram shows that each standard voltage coil is in parallel with its respective meter. This is important because looking from right to left on the diagram, meter No. 2 voltage is less than that of meter No. 1 by the amount of voltage drop due to the current going through one current coil of meter No. 1 and the one tapped current transformer for standard No. 1.

The photoelectric counting of disk revolutions may be combined with automatic control equipment to provide automated test functions. The features include photoelectric pickup and counters, automatic load and test sequence selector, and automatic resets for the watthour standards. The heavy load, power factor load, and light load may be run in sequence or any of the three tests may be selected.

Photoelectric watthour meter test equipment is assembled in various ways to test watthour meters. One common assembly consists of a complete test board which includes three portable standard watthour meters, two photoelectric pickups (each of which includes a lamp, a photocell, and supporting stand), and one revolution counter. While the operator is making adjustments on one meter, the photoelectric equipment counts the revolutions of the second meter and indicates the accuracy of heavy load, inductive load, and light load on the three standards, respectively. The photoelectric equipment may also be used with a one-position watthour meter test board if desired.

High-Speed Test Boards

The use of automatic watthour meter testing equipment and the accuracy derived from the equipment has led to further developments in automation for economic watthour meter testing and meter handling procedures. The develop-

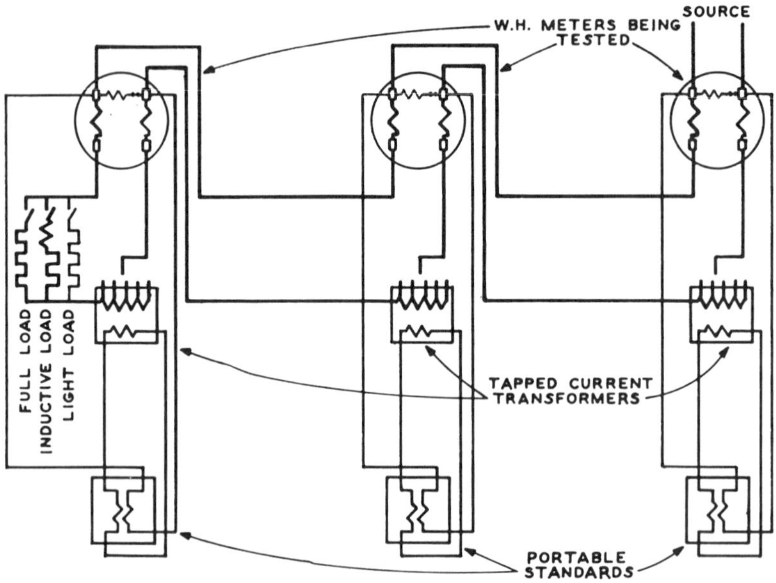

Figure 14-14. Schematic Diagram of Three Watthour Meters Tested under Live Load (Test Amperes). Three Meters Tested in Series. Three Rotating 240 Volt Standards for Each Meter (One for Each Load).

ments include operating circuits for automatically placing meters in test position, starting and sequencing test loads, counting meter revolutions, matching watthour meter revolutions with test meter revolutions for direct reading or print-out readings in percent registration, checking for creep, and three-wire phantom loading transformers or connection to live resistance loads which allow testing without opening the meter voltage links.

Figure 14-15, a schematic diagram, illustrates a one-revolution electronic watthour meter test circuit. The equipment is designed primarily for testing single-stator meters and features the requirement of only one revolution of the disk for each load applied to the meter for meter testing. The equipment includes the necessary loading, load sequencing for desired test loads, and electronic counting and it may be provided with print-out of test results.

The principle is similar to the method making use of the anti-creep holes in the disk or flag, except in this case the standard produces a frequency of about 10 kHz for an input of 600 watts. When the disk of the meter under test rotates so that the light beam shines through the anti-creep hole or reflects from the flag, a pulse triggers the electronic counter.

As shown in Fig. 14-15, the standard frequency pulses become the input to the electronic counter. The electronic counter is connected so that when it is triggered it counts the pulses from the standard. This count is then displayed or, when computers are used, calculated to percent registration. The accuracy figures may also be printed on paper or, in some cases, punched paper tape, magnetic tape, CRTs, or labels.

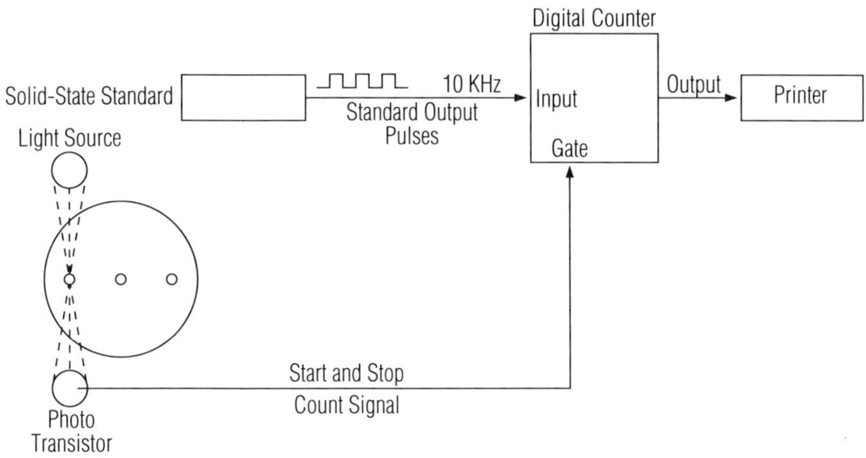

Figure 14-15. Diagram Illustrating a One-Revolution Counter.

In some cases, the front panel has switches which control the current power factor and voltage ranges; in other cases, the test parameters are automatically selected by a computer. Variable-voltage transformers accurately control the voltage, power factor, and current applied to the meter under test. Tests can be omitted or included either by switches or computer selection. The KH selector sets use a condition so that one revolution of the meter being tested is the equivalent of one theoretical revolution of the standard, thereby permitting all tests to be completed in one revolution.

The output of the electronic counter may be displayed on digital units or CRTs; printed on paper; recorded on tape; or transferred directly to a computer for the history file records.

SHOP TEST PROCEDURE

General procedures only are given for testing single-stator and multi-stator meters. Also included are variations in procedures which are dependent upon meter age. No attempt has been made to include details of meter repair, painting, handling, storing, and record-keeping procedures. Such details vary greatly and are subjects for local consideration and decision.

Single-Stator Meters

The following procedure for single-stator meters is typical for a complete shop test. All of the steps listed may not be used by all companies, and the order given may be varied.

Step 1

Check nameplate for wire, phase, volts, and amperes.
Connect standard on proper voltage and current coils.
Open meter voltage link, if necessary.

Connect meter.

Check for creep by applying voltage only.

Step 2

Start "As Found" test.

Record meter number, nameplate data, and reading.

Record "As Found" test results. "As Found" test load points are suggested as follows:

- Heavy load at 100 percent of the current rating or test amperes of the meter; 100 percent power factor
- Light load at 10 percent of the current rating or test amperes of the meter; 100 percent power factor.
- Lagging power factor (if required) at 100 percent of the current rating or test amperes of the meter; 50 percent power factor with lagging current.

Step 3

Remove and clean cover.

Clean meter with compressed air.

Check magnet gaps for iron filings or other dirt.

Check position of disk in air gap.

Step 4

Check insulation by high-voltage test. (This test may be made before "As Found" test according to local company policy.)

Step 5

Remove and examine register.

Check register ratio. (See discussion under a later subheading, "Register Checking.")

Clean register with L. & R. or equivalent cleaning machine, or with brush and cleaning fluid.

Check register constant.

Step 6

Replace bearings, if necessary. Bearing replacement depends upon:

- Light load "As Found" performance slow, or inconsistent.
- Meter on extended test schedule.
- Open-type bearing in very dirty meter "As Found."

If bearing is changed, both jewels and ball or pivot should be replaced.

If meter is noisy, check top bearing.

Top guide pin should be replaced if pin is rusty or bent.

A trouble to be looked for is the bearing in top of disk shaft worn out of round.

Step 7

Remount register on meter.

Check register mesh with disk shaft.

Step 8

Make all necessary adjustments to bring meter accuracy within established company limits.

Check for creep after making adjustments.
Record "As Left" results.

Step 9

Close voltage link if necessary.
Replace cover and seal.

Multi-Stator Meters

All of the foregoing testing procedure applies to multi-stator meters with the addition of several considerations made necessary by the multi-stator meter construction and its application. Multi-stator meters usually have two or three separate stators. It is very important that each individual stator exhibit accurate performance by itself, as well as having good accuracy with all stators combined. This is evident with a three-wire, three-phase circuit metered with a two-stator meter. Here, with a balanced-load power factor of 86.6 percent lagging current, one meter stator is operating at unity power factor and the other at 50 percent power factor, lagging current. This leads to the additional test adjustment of polyphase meters, balancing the performance of the stators to provide the necessary separate stator accuracy.

The balance test is made by connecting all voltage coils in parallel and applying 100 percent of rated current or test amperes of the meter to each current circuit, first at unity power factor and then at 50 percent power factor, lagging current. Calibration of each stator is checked for both currents. In meters which have current circuits which are common to more than one stator, such as the *Z* circuit in two-stator, four-wire wye circuit meters, the common current circuit is not energized during the balance test.

Accuracy limitations for this test are established in accordance with local requirements. If the accuracies are not within the required limits, adjustments are made as follows:

With unity power factor, the torque balance adjusters on the individual stators are used. This additional adjuster allows adjustment of individual stator performance without changing the performance of any other stator. Thus, the individual stator performances may be adjusted to agree within the specified limits at unity power factor. In meters where the torque-balance adjustment is omitted from one stator, the performance of the other stators is adjusted to match that of the first.

With 50 percent power factor, the usual lag adjustment on each stator is used. If the lag adjustment is a fixed factory-made adjustment, the lag balance cannot be easily changed by the meter tester. In meters where the lightload and lag adjustments are not entirely independent in their effects, it must be remembered by the meter tester that a change in lightload adjustment after establishing lag balance may have a detrimental effect on such balance.

After the balance adjustments are made, the "As Left" calibration is made by connecting all voltage coils in parallel and all current coils in series and making tests at the usual light, heavy, and lagging-power-factor loads. If adjustment is required on heavy load, the full-load braking magnet adjuster is used, resulting in an equal effect on the performance of all stators. If lagging-power-factor adjustment is required, equal changes are made with each lag adjuster to maintain as closely as possible proper lag balance. Similarly, required lightload adjustment would be accomplished by equal changes on each stator. However, on meters which have a lightload adjuster on only one stator, this procedure is not possible.

Calibrating Constants

In multi-coil meters the value of one revolution of the meter disk, that is, the K_H, varies with the test connections. The test K_H is sometimes called the calibrating constant. Where the same current passes through more than one full current coil, the calibrating constant can be found by dividing the normal K_H of the meter by the number of current coils in the meter connected in series. The tabulation below may serve to check such calculations.

Calibrating Constants

Meter Type	*Connection*	*Calibrating Constant*
Single-phase, two-wire	All tests	K_H
Single-phase, three-wire	Current windings in series	K_H
Single-phase, three-wire	Individual current windings*	$2K_H$
Three-phase, three-wire	Individual current coil	K_H
Three-phase, three-wire	Current coils in series	$1/2 K_H$
Three-phase, four-wire wye, two-stator	Individual current coil	K_H
	Z coil alone (a double coil)	$1/2 K_H$
	Two individual coils in series	$1/2 K_H$
	Three current coils in series	$1/4 K_H$
Three-phase, four-wire delta, two-stator	Two-wire coil alone	K_H
	Three-wire coil, windings in series	K_H
	Three-wire coil, individual windings	$2K_H$
	Three current coils in series	$1/2 K_H$
Three-phase, four-wire wye, three-stator	Individual current coils	K_H
	Two coils in series	$1/2 K_H$
	Three coils in series	$1/3 K_H$
Three-phase, four-wire delta, three-stator	Individual current coils	K_H
	Two 120-volt stator coils in series	$1/2 K_H$

*In a single-stator, three-wire meter the individual current windings are half-coils.

Procedure Variations

Variations in the preceding test procedures may be followed based on the age and condition of the meters under test.

New meters may not be tested at all before being placed in service, may be sample tested, or may all be tested. However, tests of new meters seldom go beyond the step of taking the "As Found" test if they are found to be within the established accuracy limits.

Meters from service which are to be tested and returned to service are usually subjected to the entire test procedure. Under certain conditions, it may be deemed advisable to take no further tests if the "As Found" test shows that the meter is operating satisfactorily.

Meters from service which are to be retired may have an "As Found" test if such data are required.

TESTING ON CUSTOMERS' PREMISES

Meter testing on customers' premises is often known as "field testing" and will be so called in this chapter.

Even where normal routine testing practice is removing meters for test in the meter shop, field testing is still of importance in connection with complaint or witness testing and with the maintenance of instrument-transformer connected meters. Since there was little uniformity in the method of installation in the early days, and since very large numbers of such installations still exist, the meter tester must have a sound basic knowledge of test connections and methods on which to depend for meeting the unusual problems often encountered.

Customer Relations

On entering a customer's premises to perform a meter test, testers should make their presence and purpose known to someone on the premises. As employees of the utility company, the meter testers are perfectly within their rights in requesting access to the premises and the company meter. They must, however, recognize that the company which they represent is in the position of providing service to the customer and they must, therefore, avoid any action or statement which would be discourteous to, or inconsiderate of, the customer. The tester should be prepared and willing to establish his or her identity as a representative of the company. Where the meter location is in a place accessible to the public and the entrance thereto is open, it should not be necessary to discuss the matter with anyone. It is important, however, that the tester does not interrupt the customer's service without notification. Every precaution should be taken to avoid damaging the property of the customer, and consent should be obtained before making use of any furniture or equipment to assist in making the test. In case of accident resulting in damage to the customer's property, a prompt report should be made to the customer and to the company.

Safety Precautions

It is important that the tester exercise all possible care to avoid accidents. In the interest of safety and the tester's well-being, the prohibitions and suggestions which follow should be continually in mind.

Beware of dogs. If you are bitten, go to a physician at once and then report the injury to your supervisor.

Exercise care when entering customer's premises. Be on the lookout for nails, tripping hazards, low beams, or other overhead projections.

Carefully examine ladders, boxes, and supports expected to carry your weight before making use of them.

If the apparatus to be worked on is in a dangerous condition, or is so located as to be hazardous, a complete report should be made to your supervisor and the location should be passed without doing work.

Do not attempt to make connections until proper light is arranged. A flashlight should be used until your portable lamp is connected.

The use of matches or open flames on customer's premises is prohibited.

Attention should be concentrated on the points where the tester is working; do not attempt to do two things at once.

Only one jumper should be connected at a time. Before connecting a jumper or a test lead, be sure you know where the other end is. If necessary, tape it over or tie it in a safe location. Having connected one end of the jumper, be sure that the final connection of the free end does not create a short circuit. Always check with your voltage tester.

All connections must be made securely to avoid possibility of their dropping or being pulled away from original location.

All wires, jumpers, test leads, instruments, and other equipment should be so placed that they may not be run into or tripped over by passers-by.

Use your voltage indicator to determine whether or not the meter, the meter box, or conduit are alive to ground as a result of insulation failure.

Every tester should be familiar with his or her company's safety rules.

Procedure Preliminary to Test

Before making any connections or in any way disturbing the service meter, the following routine must be followed:

1. Check watthour meter number for agreement with the number given on the test slip.
2. Record the reading of the watthour meter.
3. Enter on the test slip the date, initials of the tester, and the number of the standard watthour meter being used. Make neat and legible records. The work is of little value if the office force cannot read the records.
4. Examine all meter and equipment seals and note condition on test slip.
5. Clear the top of meter of all dust and dirt.
6. Examine the wiring and general condition of the installation for improper or unauthorized connections and possible hazards. When a connection is discovered which apparently was made by an unauthorized person and which might influence meter registration, a report should be made out with a sketch of the connections as found. Do not alter such connections. Do not test meter. Report conditions to your supervisor at the earliest opportunity. If hazardous conditions are found, report them to your supervisor and defer test of this meter until the hazard has been removed.
7. Check voltage and record unusual readings.
8. Note particularly whether or not there is a grounding conductor.
9. Make sure that the grounded or neutral conductor of a two-wire service is properly connected to the voltage and not to the current terminals of the watthour meter.

The first purpose of all field tests is to determine the actual accuracy condition of the meter "As Found," that is, the exact condition the meter is in at the time of test. To meet this requirement, the meter must not be disturbed in a manner which would alter the normal operating condition existing before start of test. The cover of the watthour meter is, therefore, not removed until after completion of the "As Found" test.

As the first step in testing, the rotating standard and load box are set up in a convenient position below and slightly to one side of the meter and with attention to having the standard in a level position and on a stable support.

Test Connections

Before proceeding with making up connections for test, bear in mind that improper and carelessly made connections may produce very serious test errors and, in addition, may be hazardous. Particular care must be observed to ensure that neither the standard nor the meter being tested carries load current other than that supplied or controlled by the load device.

The voltage circuit of the standard must be connected to a point ahead of the meter. It is necessary that the standard be connected in this manner so that neither the standard nor the meter shall record the losses in the voltage circuit of the other and thereby introduce errors at light load.

It is essential that the voltage coils of both the watthour meter and the standard have identically the same voltage impressed upon them. Make sure the voltage switch on the standard is set at the proper voltage. The multiplier for higher than 240-volt tests should be connected in series with the standard voltage circuit. The voltage switch is always on the 240-volt position when a voltage multiplier is used.

Where clip-type connectors are used, they should always be attached to surfaces or studs which give them a good contact area and a solid grip. Clip-type connectors should not be called upon to support the weight of leads.

The initial step in making up connections for test is to open such parts of the customer's service equipment and meter terminal chamber as is necessary to expose line and load connections.

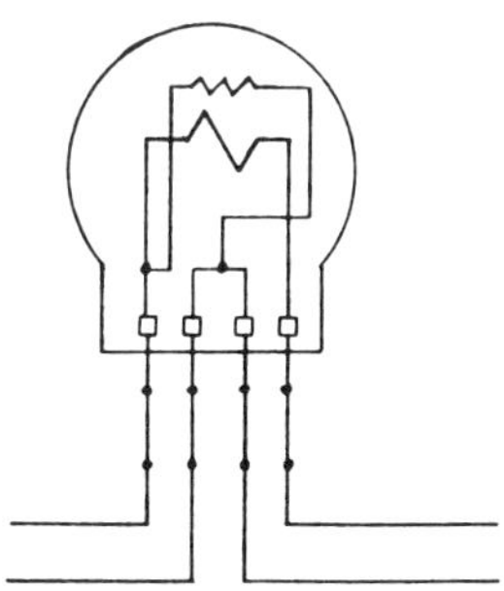

Figure 14-16a. Installed Connection of an A-Base, Two-Wire, Single-Stator Meter.

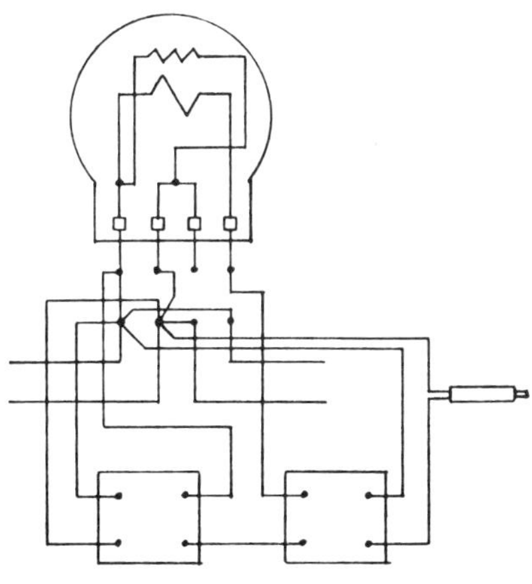

Figure 14-16b. Test Connections for Testing an A-Base, Two-Wire, Single-Stator Meter on Customer's Premises.

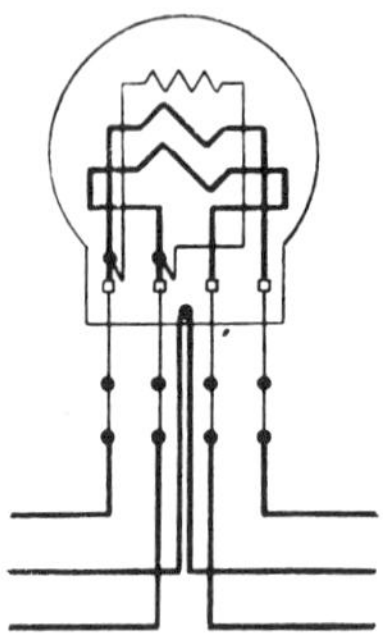

Figure 14-17a. Installed Connections of an A-Base, Three-Wire, Single-Stator Meter.

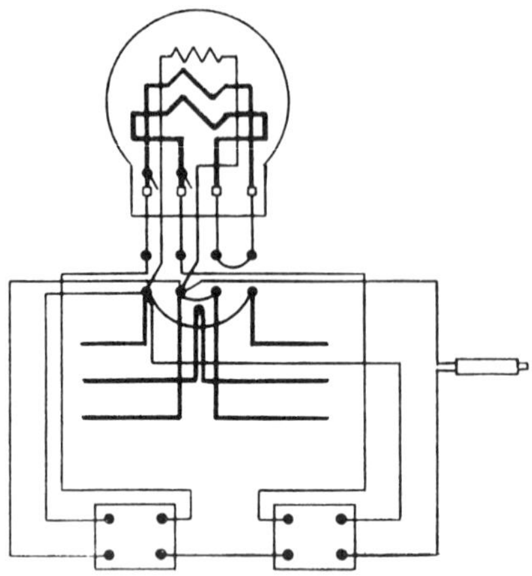

Figure 14-17b. Test Connections for Testing an A-Base, Three-Wire, Single-Stator Meter on Customer's Premises.

Bottom-connected meters, generally installed over a service switch or test switch, may have facilities for by-passing meter and connecting test equipment, In the case of socket meters, the test jack is plugged into the socket and the meter mounted on the test jack. Test connections to the test jack terminals may then be completed without further disturbance of the customer's service.

Basic connections for single-stator meter testing are shown in Figs. 14-16 and 14-17. Test jacks have built-in, by-passing connections so that the customer receives energy as soon as the test jack is inserted in the meter socket. Diagrams of test-jack connections are given in Figs. 14-18 and 14-19. (Certain types of socket are equipped with means to bypass the meter, thus permitting meter test without any momentary interruption of the customer's load.)

Test blocks may be used which allow the meter to be isolated for test without interruption, particularly with high-capacity, bottom-connected meters. Before making bypass connections to a test block it must be noted whether the test block wiring is line-line-load-load or line-load-line-load sequence. In the former case, jumpers must be used from position 1 to 4 and 2 to 3; in the latter case, test links may connect 1 to 2 and 3 to 4 to by-pass the meter. Figure 14-20 shows test connections for a two-stator, three-wire, self-contained meter with test block.

How extensive a meter test procedure is followed depends on many factors, including the interval between scheduled tests, meter retirement policy, type of area and location, history of meter performance, etc. Thus, in some companies meters found within satisfactory accuracy limits are left "As Found" without removing meter covers. Other companies require complete mechanical inspection of all parts of the meter. The steps described in the following are typical of a thorough meter test and inspection.

Field Test Procedure — Single-Stator

1. With test connections in place and the watthour meter voltage coil energized, but with no current in the current circuit, observe whether or not the meter disk creeps. A meter is not considered to creep unless the disk makes a full revolution in ten minutes or less. Intermittent creep due to excessive vibration may also occur. If the meter creeps, note the apparent cause and the time required for one revolution of meter disk.
   (See later discussion of "Meter Maintenance" for causes of creep.)
2. Before disturbing the meter in any way, take "As Found" tests at heavy load (between 60 and 150 percent of nameplate rating or test amperes) and at light load, which should be approximately 10 percent of test amperes. Both tests are made at 100 percent power factor. If the first set of runs shows evidence of excessive errors or improper operation, both tests should include as many runs as may be necessary to obtain a reasonable average of "As Found" conditions. The percent registration or percent error is calculated, where required standard corrections are applied, and corrected percent registrations noted.

Upon completion of "As Found" tests, the top of the meter is cleaned and the meter cover is removed. The register ratio and the marked register constant are checked and noted on the test slip. Next the entire meter assembly is inspected for presence of dust, dirt, paint chips, etc. All such foreign materials must be carefully removed. Particular attention should be given to the gap between the drag magnet and the disk. All particles of dirt in this area must be carefully and thoroughly removed with a non-magnetic cleaner. Check to assure that the mesh between the shaft worm or pinion and the first register gear is of correct depth and that it does not vary as the disk rotates.

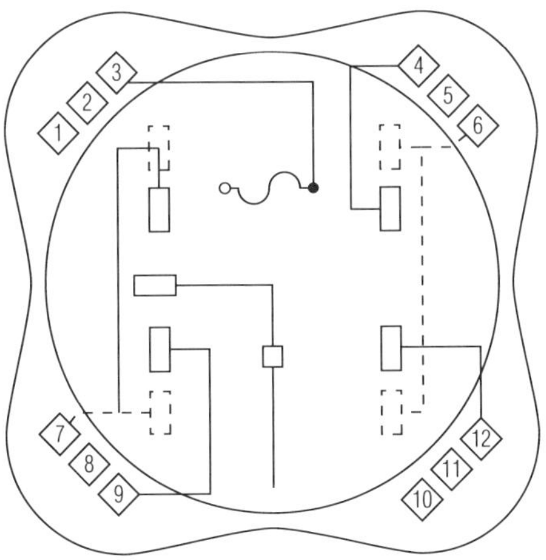

Figure 14-18. Westinghouse Socket Meter Test Jack — Internal Wiring (Rear Blades and Wiring Shown by Broken Lines).

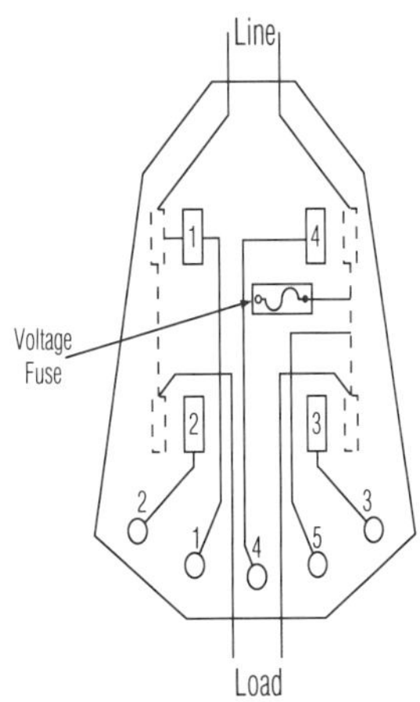

Figure 14-19. States Socket Meter Test Jack — Internal Wiring.

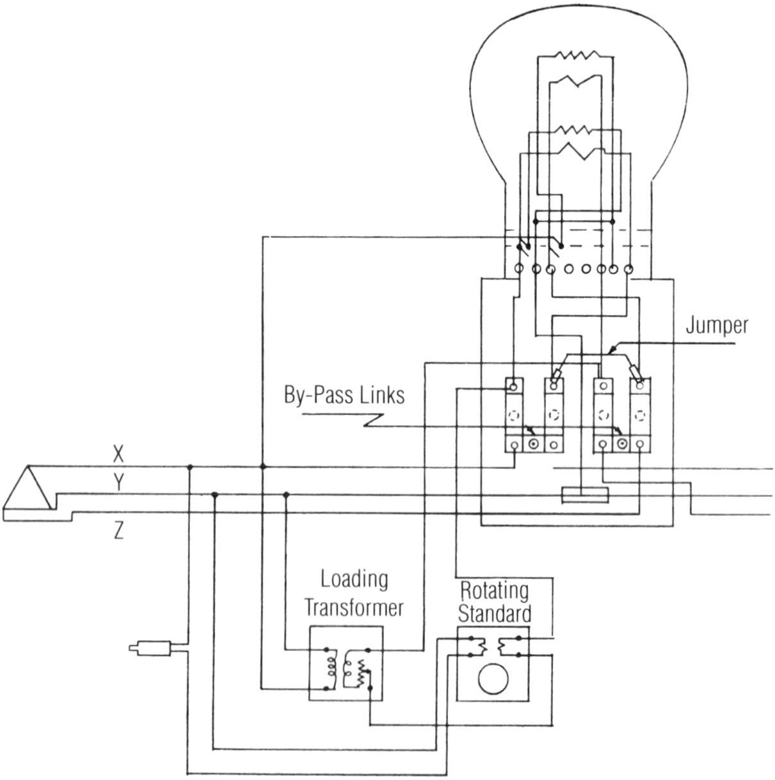

Figure 14-20. Test Connections for Self-Contained, Two-Stator, Three-Wire Meter Using a Loading Transformer. Current Coils Connected in Series and Voltage Coils in Parallel. To Test for Balance of Stators, Test Each Stator Separately.

Feel each dial pointer with the finger to assure that none of them are loose on their shafts and that the shafts are not loose in their pinions.

Check for disk clearances. The vertical play of the shaft should be approximately 3/64ths of an inch. Reference should be made to manufacturers' instructions for this check, particularly for magnetic-bearing meters.

If light-load tests are variable or show the meter to be more than 1 percent slow, no adjustment should be made until a complete examination of the mechanical condition of the meter is made to determine possible causes for excessive friction. Sources of friction might be improper disk clearances, improper mesh between shaft worm and worm gear, faulty upper or lower bearing, or a faulty register.

If light-load results are still in excess of 1 percent slow or show variability, the lower bearing unit should be removed and replaced. If it is found necessary to change the lower bearing, the top bearing should also be removed and, if possible, cleaned with pith and an orange stick. If it is not possible to secure a bright surface on the top bearing pin, it should be replaced.

Adjust meter if necessary. Refer to earlier discussion under the subheading, "Meter Adjustments."

After all inspections, adjustments, and tests have been completed, "As Left" results are noted on the meter test slip. As a final test, the meter must be checked for creep.

In the case of three-wire, single-phase meters, all tests are made with the current coils in series.

It is good practice to make "As Left" tests with the meter in place.

Having completed the "As Left" tests, the temporary connections must be removed in a careful manner. In removing connections, the leads which served to supply the phantom load and voltage circuits must be removed from the line service terminals first. Reconnect the meter line and load conductors in their normal positions and remove jumpers, being careful to reinsulate any section of wiring which has been bared for purpose of test. Finally, replace all seals, verify the fact that service has been returned to normal and that the watthour meter is functioning.

If service has been interrupted, the customer should be so advised, either verbally or by printed form, with a suggestion that his or her electric clocks be reset to correct time.

Before leaving the scene of the test, check your test slip to be sure that all data have been properly entered.

Field Test Procedure — Multi-Stator Meters

In general, the single-stator test procedure applies also to multi-stator meters. Before test connections are made, a check with a voltage indicator is advisable for assurance that the meter case is not alive to ground. With load wires disconnected, a check for creep should be made.

To prepare for test, connect all voltage coils in parallel and all current coils in series. Note that this connection changes the test constant. (See discussion of "Calibrating Constants" under the subheading, "Shop Test Procedure.")

With this series-parallel connection, a multi-stator meter is tested at 1.0 power factor as though it were a single-stator meter. An exception is the four-wire, three-stator delta meter which is commonly tested by individual stators.

In addition to the unity-power-factor tests, multi-stator meters may be tested with 100 percent test amperes at 50 percent lagging power factor.

For lagging-power-factor field tests, the phantom load used may have built-in switching and controls to obtain the desired power factor. If not, the three-phase voltages at the meter installation may be used for this purpose.

When using the service supply to obtain lagging power factor, the phase sequence must be determined in order to make the proper connections. Phase sequence may be determined by means of a phase angle meter or with any of the phase sequence indicators described in Chapter 9.

After phase sequence has been determined, connections for a lag test at 50 percent power factor may be made as follows, for the phasors shown in Fig. 14-21:

When the sequence is 1-2-3:

Use voltage 1-2 with current 1-3, or voltage 2-3 with current 2-1.

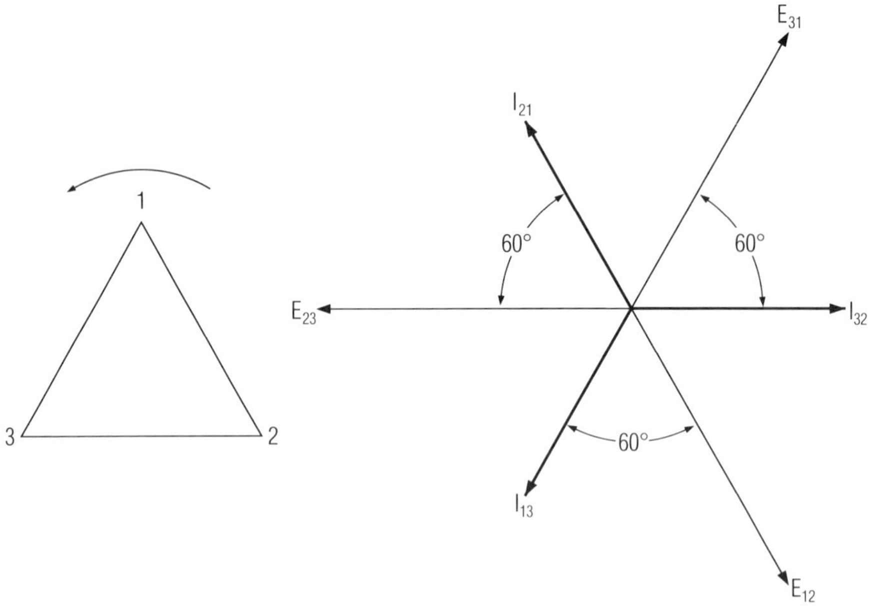

Figure 14-21. Phasor Relations of Voltages and Currents in Lag Test of Multi-Stator Watthour Meter.

When the sequence is 3-2-1:

Use voltage 2-1 with current 2-3, or voltage 1-3 with current 1-2, or voltage 3-2 with current 3-1.

Instrument Transformer Meters

Testers undertaking the test of instruments-transformer-connected meters should be familiar with much of the material discussed in Chapter 11, "Instrument Transformers," where connections and correction factors are covered. The actual test of instrument transformer meters is similar to that of self-contained meters, since for test purposes meters are usually isolated from their associated transformers. Since the meters generally control more revenue than self-contained meters, additional tests may be required and test tolerances may be narrowed.

Meter installations having current and voltage transformers require exceptional care and caution to safeguard personnel from injury through contact with high-voltage primaries or the high voltage developed across an accidentally opened current transformer secondary. If all safety rules are followed, these hazards will be avoided.

Normally, for instrument-transformer meters, a test switch is installed between the transformers and the meter. It is the function of this test switch to short the current transformer secondaries before opening the connections to the meter, and to open the voltage secondary circuit. Where a test switch is of an unfamiliar design, it must be determined that such short-circuiting is effective before opening the switch.

Where test switches are not installed, other means of short-circuiting the current transformer secondaries must be employed. In any case, the short-circuiting connections must be made secure before opening the circuit to the meter. Clip-connection jumpers are not recommended. The use of temporary wire jumpers presents the possibility of leaving the jumpers in place after the test is completed, thus shunting the meter.

Test Procedure

With the meter disconnected from its instrument transformers, proceed to test as a self-contained meter, with multi-stator meters connected with current coils in series and voltage coils in parallel. Additional individual stator tests may also be required for stator balancing. Three-stator delta meters present special problems which are subsequently described.

Particularly where large loads are served from a delta power bank, power factor tests may be required. When power factor adjustments are necessary, they should be made while testing individual stators rather than with series-parallel connection. Lagging power factor values should match the unity power factor performance values in each individual stator so that proper balance is obtained under all conditions of loading. Since, on delta circuits, errors in power factor balance affect meter performance on unity power factor loads, the tolerance for power factor balance should be narrow.

With unusual loads, it is sometimes desirable to make a "running load test," that is, a test using the customer's three-phase load instead of the phantom load. Such tests require two standards for three-wire, three-phase meters, and three standards for four-wire meters, with a standard current coil in series with each meter current coil. The algebraic sum of the standard registrations is used to determine meter performance. Figure 14-22 shows test connections for a running load test on a two-stator, three-wire meter.

When metering large power customers, register constants are particularly important. Primary register constant may be calculated by the standard formula:

$$\text{Primary Register Constant} = \frac{\text{Secondary } k_H \times R_R \times R_S \times \text{CT ratio} \times \text{VT ratio}}{10{,}000}$$

Register ratio and shaft reduction should be examined to make sure that they are correct for the application.

Network Meters

Where meters are installed with current transformers but not with voltage transformers, which is often the case when customers are served from a 120/208 volt network, the test switch voltage blades may present a hazard which must be recognized. In this situation, the voltage supply to the meter is taken from a high-capacity power transformer instead of from a low-capacity instrument transformer. Hence, any accidental short circuit of the voltage conductors can result in a severe arc. Unfortunately, in many test switches the voltage switch blades are alive when open and protrude beyond the test which barriers. Where this is the case, an insulated enclosure to cover the live switch blades will prevent accidents due to falling tools or other metallic objects.

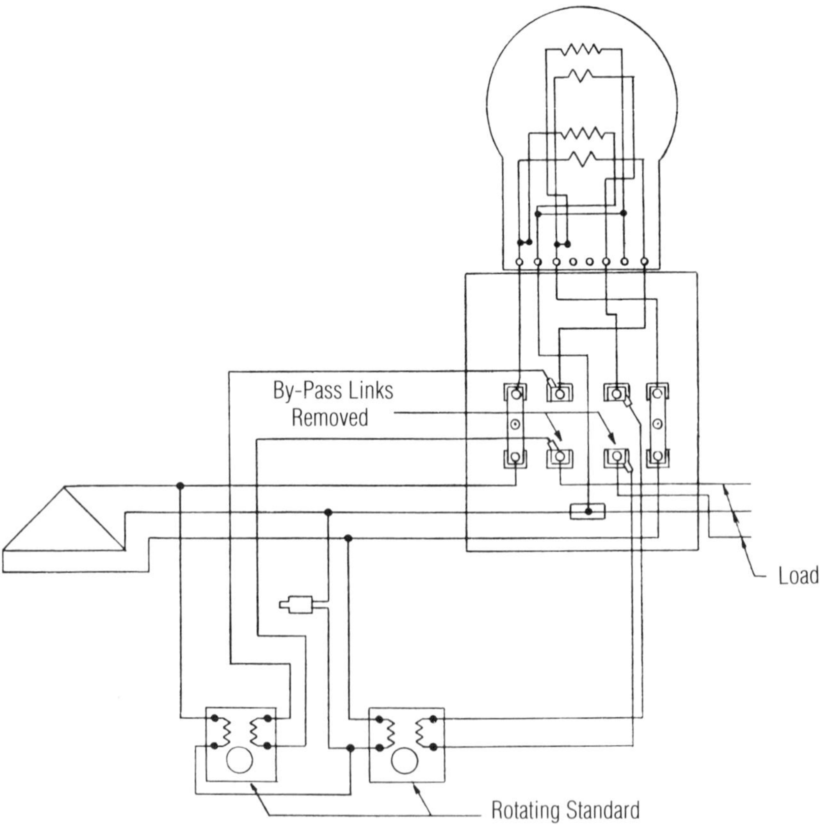

Figure 14-22. Test Connections for Two-Stator, Three-Wire Meter Using Customer's Load and Two Rotating Standards.

Three-Phase, Four-Wire Delta, Two-Stator Meters

This meter comprises one three-wire current circuit at 240 volts and one two-wire current circuit at 208 volts. Except where extreme accuracy is required, this meter may be tested with all current coils in series and voltage coils in parallel. The error introduced by operating the 208 volt coil at 240 volts is generally less than 0.2 percent.

Three-Phase, Four-Wire Delta, Three-Stator Meters

This type of meter is unusual as compared with other meters, especially from a standpoint of testing. Since it consists of one stator with a voltage rating of twice either of the other two stators and a current rating of one-half of either of the other two stators, the usual method of series tests and balance test cannot be used. The common practice is to calibrate each stator independently, although the two like stators can be tested in series and also balanced against each other. A series test can be made by providing the correct ratio current transformer and voltage transformer so that the higher voltage stator can be connected essentially in series-parallel with the two lower voltage stators. This of course requires an accurate step-up voltage transformer and an accurate step-down current transformer. For more detailed discussions, see manufacturers' publications.

Three-Phase, Four-Wire Wye, Three-Stator Meters

The necessity for making separate stator tests on three-stator wye meters in the field and test loads to be used in such tests were investigated by an EEI Meter and Service Committee Task Force and the conclusions were covered in a report dated April 16, 1951.

The following recommendations were made in this report;

1. In calibrating separate stators, test loads should be based on percent rated current, rather than percent rated speed. This is based purely on economics in that fewer adjustments will be necessary under this procedure.
2. In the field, the series test is sufficient except in cases of special investigation. Further details may be obtained by reference to the complete report.

MOBILE SHOP FIELD TESTING

In this system of testing, a large van-type truck or trailer is used to house a compact but very completely equipped meter shop. This mobile shop is moved to a convenient location near the scene of operations and a power supply tapped to it from adjacent company lines. Meter installers then proceed to remove meters from the services in the neighborhood, bypassing each service meter loop. The meters are immediately delivered to the mobile shop where the test crew makes "As Found" tests and high-voltage tests, thoroughly clean the meters, and make adjustments and "As Left" tests. The meter is then returned to service in its initial location, with the same reading that existed at the time of removal. This method of testing has many obvious advantages over the regular house-to-house testing system. Tests are made under nearly ideal conditions; high-voltage tests are possible and a much more thorough job of cleaning the meter can be performed. The very considerable advantage of this system over shop testing methods is that no change of office or meter reading records are involved.

METER TEST BY INDICATING WATTMETER

Load is applied to the meter and watthours are measured by means of indicating instruments and timing devices, such as stop watches or chronographs. The time is usually that required for some convenient and predetermined number of revolutions of the meter under test. The procedure is as follows: the time required for an integral number of revolutions of the meter is measured by a stop watch and the power, in true watts, during the same period is measured by means of indicating instruments. The ratio between the indicated or meter watthours and the true watthours, as determined by the indicating instruments, multiplied by 100 is the percentage registration of the meter under test.

Example:

Let P = true watts (average watts by indicating instruments)

k_H = watthour constant

r = number of revolutions of disk
s = time in seconds for r revolutions

Then, meter watthours = $k_H \times r$
meter wattseconds = $k_H \times r \times 3{,}600$
true wattseconds = $P \times s$

Percentage registration of the meter may then be determined from the following equation:

$$\text{Percent Registration} = \frac{k_H \times r \times 3{,}600 \times 100}{P \times S}$$

A wattmeter is required to measure alternating-current power. This method is generally limited to special meter tests.

When extensive tests to determine performance of meters under all conditions of voltage, current, frequency, and wave form are required, the latest edition of the *Code for Electricity Meters* should be consulted.

METER MAINTENANCE

Causes of Creeping

The causes of creeping may be classified as follows:

a. Incorrect light-load compensation.
b. Vibration.
c. Stray fields, either internal or external.
d. Too high voltage, which has the same effect as overcompensation of light-load adjustment.
e. The voltage circuit being connected on the load side of the meter.
f. Short circuits in current coils.
g. Mechanical disarrangement of the electromagnetic circuit of the meter.

A high-resistance short or ground in the customer's circuit can cause a turning of the rotating element which may be mistaken for creeping; therefore, residence wiring should be isolated from the meter when checking for creep.

If a short circuit is present in the current coils, it will probably be difficult to stop creeping. The creeping, due to current coil shorts, is caused by voltages being induced in the current coils by the voltage magnetic flux and resultant current low in the shorted current coil turns.

Causes of Friction

Friction may be caused by foreign matter, defective bearings, defective registers, or improper alignment of parts interfering with the operation of the rotating element. Friction may also be caused by a meter being out of level.

Particles of iron on other magnetic material cause friction by clinging to the magnet pole-pieces and trailing on the disk. To remove magnetic particles from the magnets, a thin brass or bronze magnet cleaner can be used, but a magnet brush having long bristles is preferable.

Bearings

Magnetic bearings should be inspected to determine if foreign material is present. Since the position of the disk in the air gap depends directly on the magnetic bearing, this characteristic should be checked. Detailed procedure for checking disk position is given in the various manufacturers' publications. The bearing magnets must support the disk and a simple check is to push the disk down gently with a finger to see that it is floating and not resting on the bottom guide pin. Inspection of the guide pin and ring bearings at both ends of the disk shaft may also be desirable. This could be done on a sample basis until experience has been obtained.

New and used meter jewel bearings are usually inspected in the shop. Old jewels are first thoroughly cleaned. After cleaning, the jewels, pivots, and balls are examined with a microscope and those considered unfit for further service are discarded. It is essential when inspecting jewels to provide adequate light in the jewel cups, free from shadows and reflections, to avoid false observations. The inspection of used meter bearing parts is an economic problem. In many cases, such inspection has been eliminated. This practice can be attributed to the use of improved bearing assemblies or the elimination of bearing systems.

Registers

The tester should inspect the register to detect any defects which may prevent its correct registration.

The worm or pinion on the shaft should be examined to see that it meshes properly with the register wheel which it drives. A slight amount of play is necessary to prevent excessive friction.

Where the pinion or worm is short, or where the worm is cut concave to match the curvature of the worm wheel, the height of the moving element should be such that the center of the pinion or worm is level with that of the register wheel which it engages.

For cleaning the pinion or worm, a small stiff brush or a sharpened piece of soft wood may be used.

In some meters, the worm wheel is supported in a separate bracket attached to the meter frame, in which case it is customary to actuate the register by means of a dog on the register engaging with a star wheel or second dog on the worm-wheel shaft. It is important that these parts be in proper alignment.

All the gears on the register must be in mesh and all dial pointers secure. Misplaced pointers should be reset. The tester should record the position of the misplaced pointers. This condition may be an indication of an incorrect train ratio, i.e., one or more of the gears or pinions have the wrong number of teeth.

Register Testing and Checking

An important part of a watthour meter test consists of determining whether the watthour constant, register ratio, gear ratio, and register constant are correct and also bear the correct relation to each other.

The correct watthour constant K_H for any particular size and type of meter can be found from the tables in the manufacturers' section. In order to check that a meter actually has the correct watthour constant, a test of the meter by any one of the methods previously described may be made.

The register ratio R_R may be determined by counting the number of revolutions of the first gear staff of the register which is required for one revolution of the first dial pointer. It is generally checked, however, by comparison with a register of known ratio. In the first method, it is not necessary to take a complete revolution of the first dial pointer, one-tenth of a revolution generally being sufficient. In the second method, the register to be checked may be engaged with a shaft which is driving a register of known ratio, or the registration of the meter under test may be compared with the registration of a meter with known constants.

In order to calculate the gear ratio R_G, it is first necessary to determine the ratio of reduction between the shaft of the rotating element of the meter and the first gear staff of the register—this reduction being referred to as the shaft reduction R_S. If there is a worm drive on the rotating element of the meter, the shaft reduction is:

$$\text{Shaft reduction} = \frac{\text{Number of teeth in first register gear}}{\text{Pitch of worm on rotating element}}$$

If the driving means is a spur gear instead of a worm, the shaft reduction is:

$$\text{Shaft reduction} = \frac{\text{Number of teeth in first register gear}}{\text{Number of teeth in spur gear on rotor}}$$

The gear ratio R_G is equal to the product or the shaft reduction R_S and the register ratio R_R. One revolution of the rotating element of the meter is equal to K_H watthours. The number of revolutions of the rotating element for one revolution of the first dial pointer is equal to the gear ratio R_G and therefore one revolution of the dial pointer will represent:

$$K_H \times R_G \text{ watthours,} \quad \text{or} \quad \frac{K_H \times R_G}{1{,}000} \text{ kilowatt-hours}$$

The numerical value of the one revolution of the first dial pointer of a standard register is 10; therefore, the register constant K_R for kilowatt-hours is:

$$K_R = \frac{K_H \times R_G}{10 \times 1{,}000}$$

$$\text{and } R_G = \frac{K_R \times 10{,}000}{K_H}$$

With these relations established, the value of any one of the factors under consideration can now be expressed in terms of the others.

Let K_H = watthour constant*

R_R = register ratio

R_S = shaft reduction

R_G = gear ratio

K_R = register constant

*Primary watthour constant for meters with instrument transformers.

Then $R_G = R_R \times R_S$

$$K_H = \frac{K_R \times 10{,}000}{R_R \times R_S} = \frac{K_R \times 10{,}000}{R_G}$$

$$R_R = \frac{K_R \times 10{,}000}{K_H \times R_S}$$

$$R_S = \frac{K_R \times 10{,}000}{K_H \times R_R}$$

$$K_R = \frac{K_H \times R_R \times R_S}{10{,}000} = \frac{K_H \times R_G}{10{,}000}$$

For example, assume that the following data are given for a meter and it is desired to verify the correctness of the register constant:

$$K_R = 1$$

$$R_R = 66^2/_3$$

$$K_H = 1.5$$

$$R_S = 100$$

$$K_R = \frac{K_H \times R_R \times R_S}{10{,}000} = \frac{1.5 \times 66^2/_3 \times 100}{10{,}000} = 1$$

The register constant of 1, given in the data, proves to be correct.

In selecting a register of proper ratio for installation on a meter, the formula

$$R_R = \frac{K_R \times 10{,}000}{K_H \times R_S}$$

may be used, but it will be noted that the value for R_R thus obtained depends upon the value selected for the register constant K_R.

There are available mechanical register checking devices for comparison of registers against a master ratio indicator. Any type and make may be checked by providing the required mounting facilities and correct gearing. The device may be operated by a small air turbine which permits flexible speed and reversible operation. The master dial is set at zero and the register or registers to be checked mounted in position. The lowest ratio is checked first by operating the device until the first dial pointer of the register makes one complete revolution. The register ratio is then read directly from the master dial. The register of the next lowest ratio is then checked by continuing the run until its first dial pointer has made one complete revolution and the master dial read, etc.

An alternative method of register checking is by the use of a time-run test in which meters are connected in series with standard meters having known constants and are operated for sufficient periods of time to verify their constants.

This method has some advantage over the use of register checking devices in that the complete performance of the meter is verified as well as establishing the correctness of constants. The disadvantages of this method are the increased time, equipment, and space required. A representative connection diagram of a "time-run" test board is illustrated in Fig. 14-11.

Defective Current and Voltage Coils

A short-circuited turn in the current coil will reduce the effective turns and, consequently, will lower the torque of the watthour meter and hence its speed at or near full load. Induction meters will generally creep when some of the turns are short-circuited and often will be fast on light load and slow on full load. Meters with this defect should be provided with new current coils. On three-wire meters, the test can easily be made by checking the meter registration on each current coil separately.

A short circuit in the voltage coil will change the torque of the watthour meter, hence its speed. A meter with this condition existing will be found out of lag and the voltage coil must be replaced.

WATTHOUR METER TEST DIAGRAMS

This group of diagrams shows single-stator and multi-stator meter test connections. For more detailed meter diagrams, see Chapter 12.

When a special test jack or block is used, be certain that the instructions furnished with the jack or block are followed.

Should both meter and standard run in reverse direction, reverse the leads to the primary of the phantom load. If either the meter or standard runs backward, reverse the voltage leads to the unit that is running backward.

Figure 14-25 shows the portable standard watthour meter used with a voltage multiplier unit. When service meters of higher voltage rating than the rotating standard voltage rating are being tested, it is necessary to use a voltage multiplier. Note that the phantom load primary must have a voltage rating equal to that of the line.

The test diagrams show the necessary test connections required for watthour meter testing and calibration procedures previously outlined. The testing connections shown are for unity power factor testing procedures. For 50 percent power factor testing procedures, a suitable means should be provided for changing the test load to a lagging 50 percent power factor load.

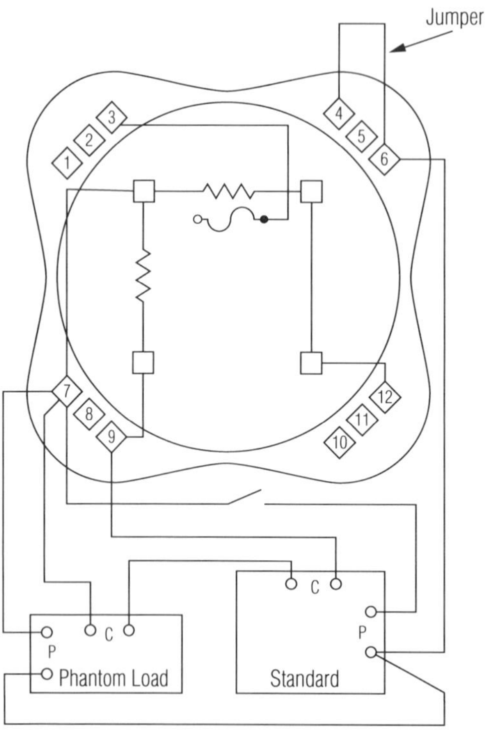

Figure 14-23. Westinghouse Test Jack Connection for Two-Wire, 120 Volt Socket Meter (Refer to Fig. 14-18 for Internal Wiring of Jack).

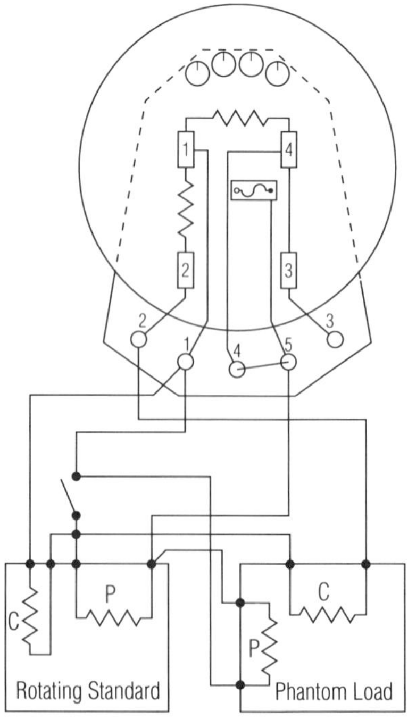

Figure 14-24. States Test Jack Connections for Two-Wire, 120 Volt Socket Meter (Refer to Fig. 14-19 for Internal Wiring of Jack).

Single-Phase, Two-Wire, Single-Stator Meters

See Figs. 14-23, 14-24, 14-25. Also refer to Fig. 14-16.

Single-Phase, Three-Wire, Single-Stator Meters

See Figs. 14-26, 14-27. Also refer to Fig. 14-17.

Three-Phase, Three-Wire, Two-Stator Meters

See Fig. 14-28. Also refer to Fig. 14-20 and 14-22.

Three-Phase, Four-Wire Wye, Two-Stator Meters

See Figs. 14-29 and 14-30.

When the meters are connected for series testing, as indicated in Fig. 14-30, it should be noted that the meter will run four-thirds times as fast when connected in this manner on single-phase as when connected for service on three-

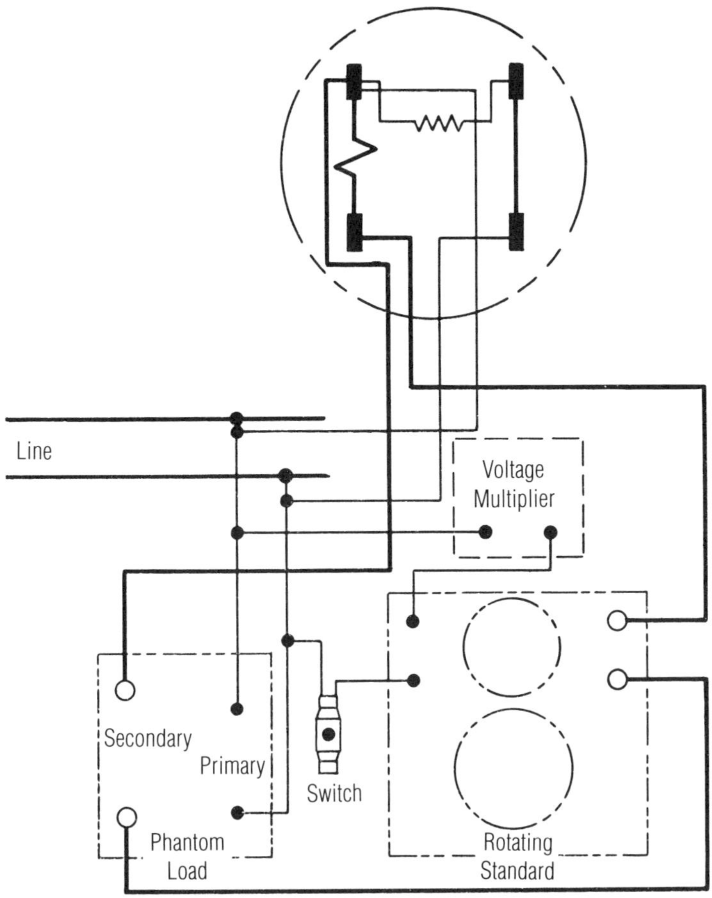

Figure 14-25. Single-Phase, Two-Wire, Single-Stator Meter Showing Voltage Multiplier.

phase with the same magnitudes of current and voltage. This is due to the fact that when operating on a polyphase system with load at unity power factor the current circuit which is common to both meter stators carries current which leads the voltage of one stator by 60 degrees and lags behind the voltage of the other by 60 degrees and thus contributes only half as much torque on a polyphase system as it does when tested on single-phase with a unit power factor load. Since this current circuit contributes one-third of the total driving torque on a balanced polyphase load, the doubling of this by connecting single-phase will result in a speed four-thirds times as great as on polyphase.

Three-Phase, Four-Wire Wye, Three-Stator Meters

See Figs. 14-31 and 14-32.

Three-Phase, Four-Wire Delta, Two-Stator Meters

See Figs. 14-33 and 14-34.

Three-Stator Totalizing Meters

See Fig. 14-35.

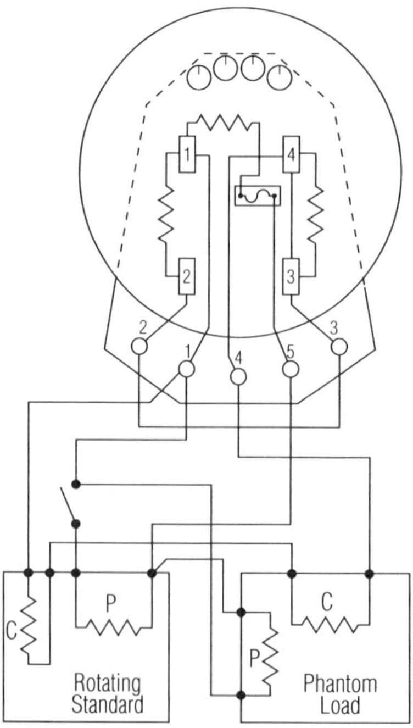

Figure 14-26. States Test Jack Connections for Single-Stator, Three-Wire Meter (Refer to Fig. 14-19 for Internal Wiring of Jack).

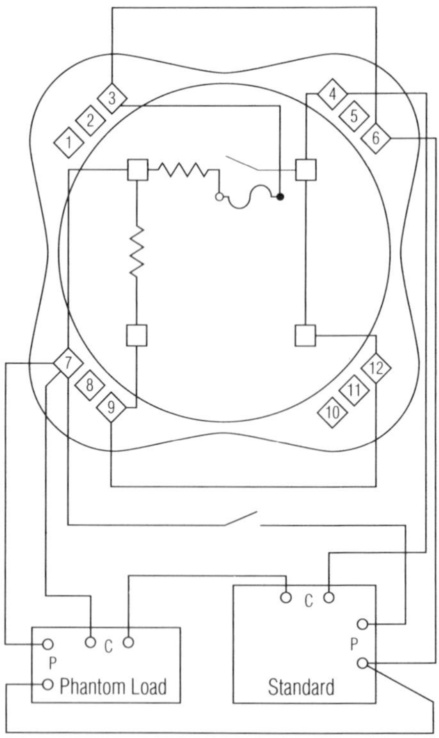

Figure 14-27. Westinghouse Test Jack Connections for Single-Stator, Three-Wire Meter (Refer to Fig. 14-18 for Internal Wiring of Jack).

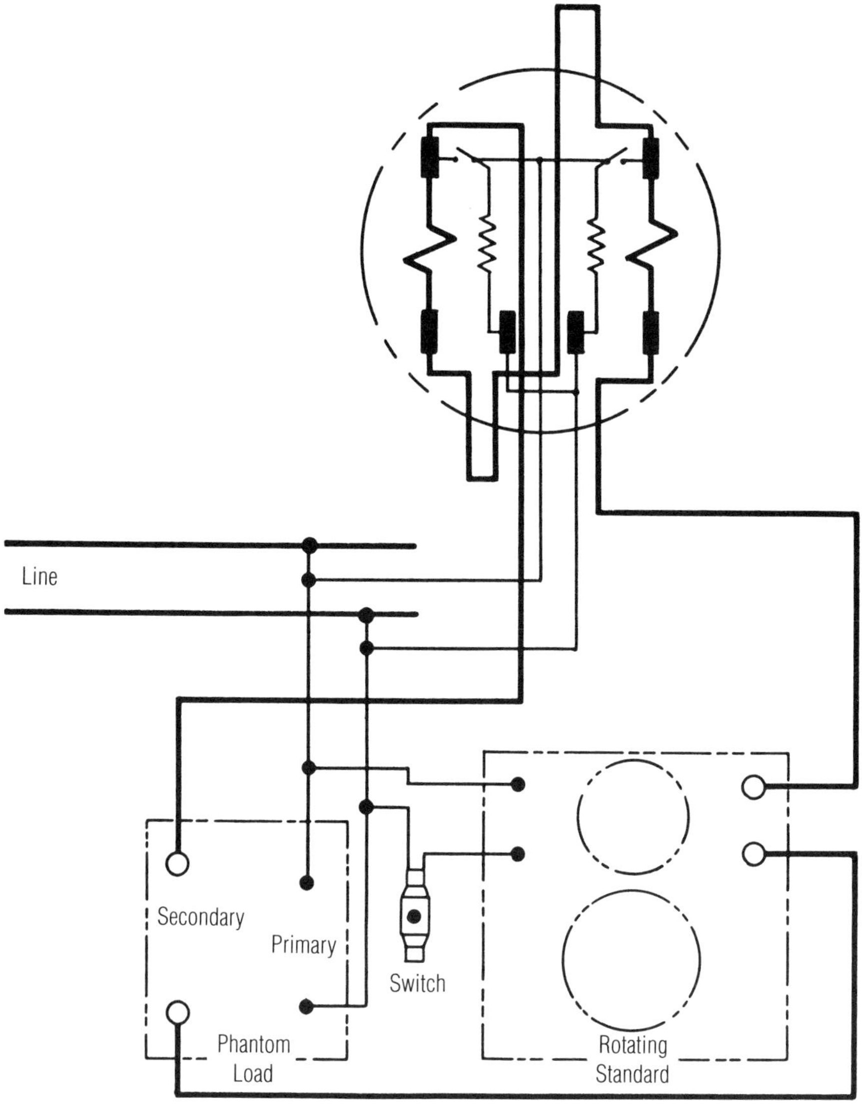

Figure 14-28. Three-Phase, Three-Wire, Two-Stator, Self-Contained Meter with Current Coils in Series

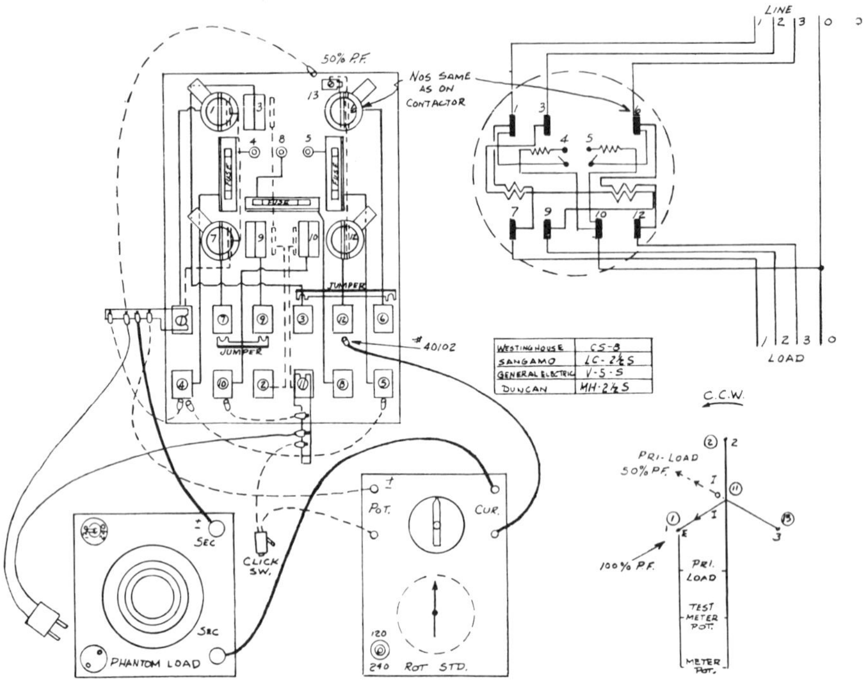

Figure 14-29. States Test Jack Connections for Three Phase, Four-Wire Wye, Two-Stator Meter with Current Coils in Series.

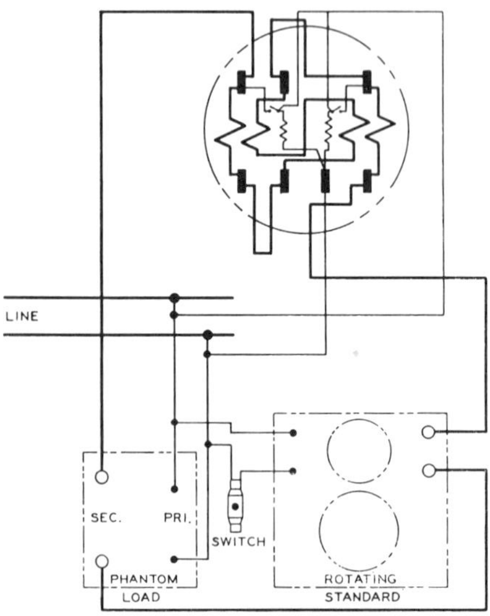

Figure 14-30. Three Phase, Four-Wire Wye, Two-Stator, Self-Contained Meter with Current Coils in Series.

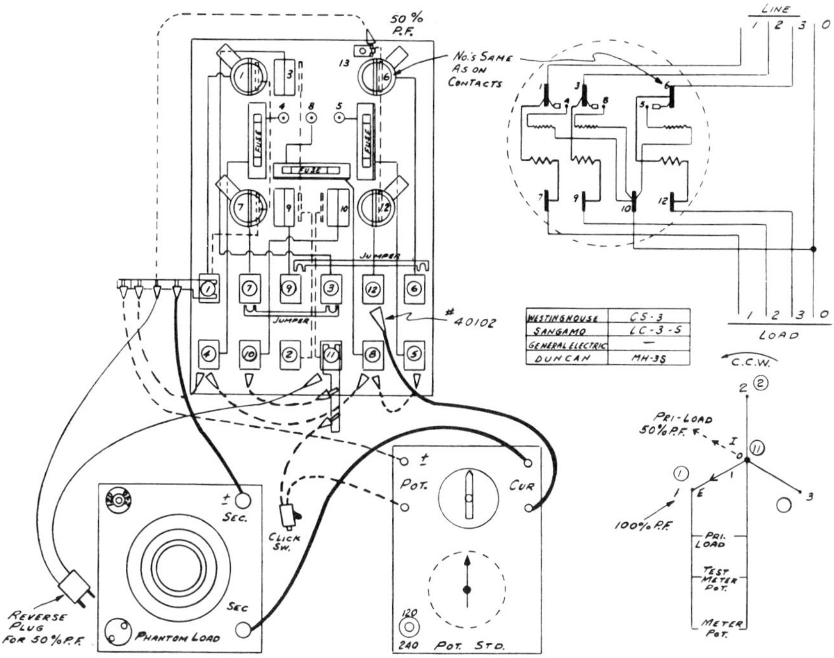

Figure 14-31. States Test Jack Connections for Three Phase, Four-Wire Wye, Three-Stator Meter with Current Coils in Series.

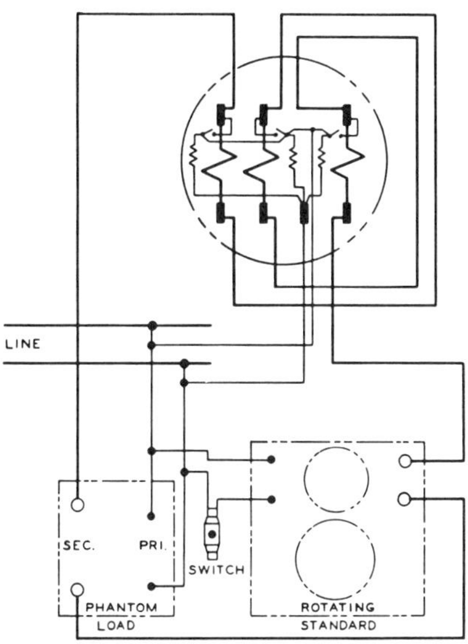

Figure 14-32. Three Phase, Four-Wire Wye, Three-Stator, Self-Contained Meter with Current Coils in Series.

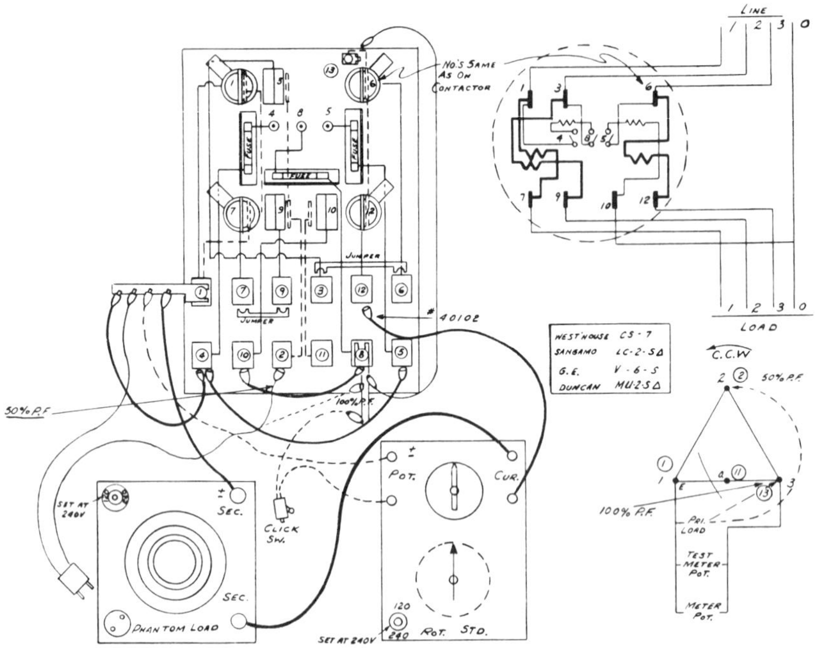

Figure 14-33. States Test Jack Connections for Three Phase, Four-Wire Delta, Two-Stator Meter with Current Coils in Series.

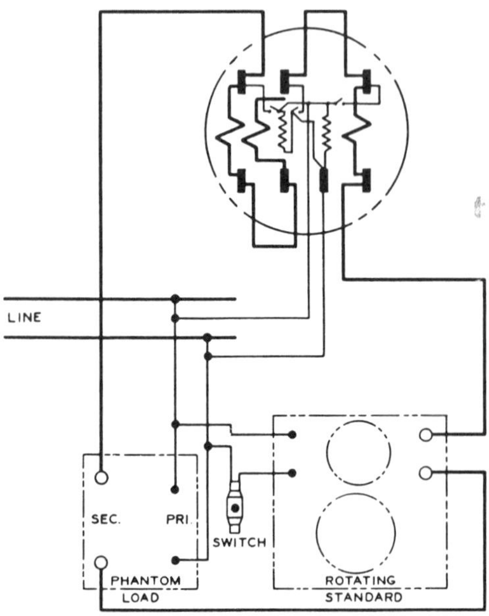

Figure 14-34. Three Phase, Four-Wire Delta, Two-Stator, Self-Contained Meter with Current Coils in Series.

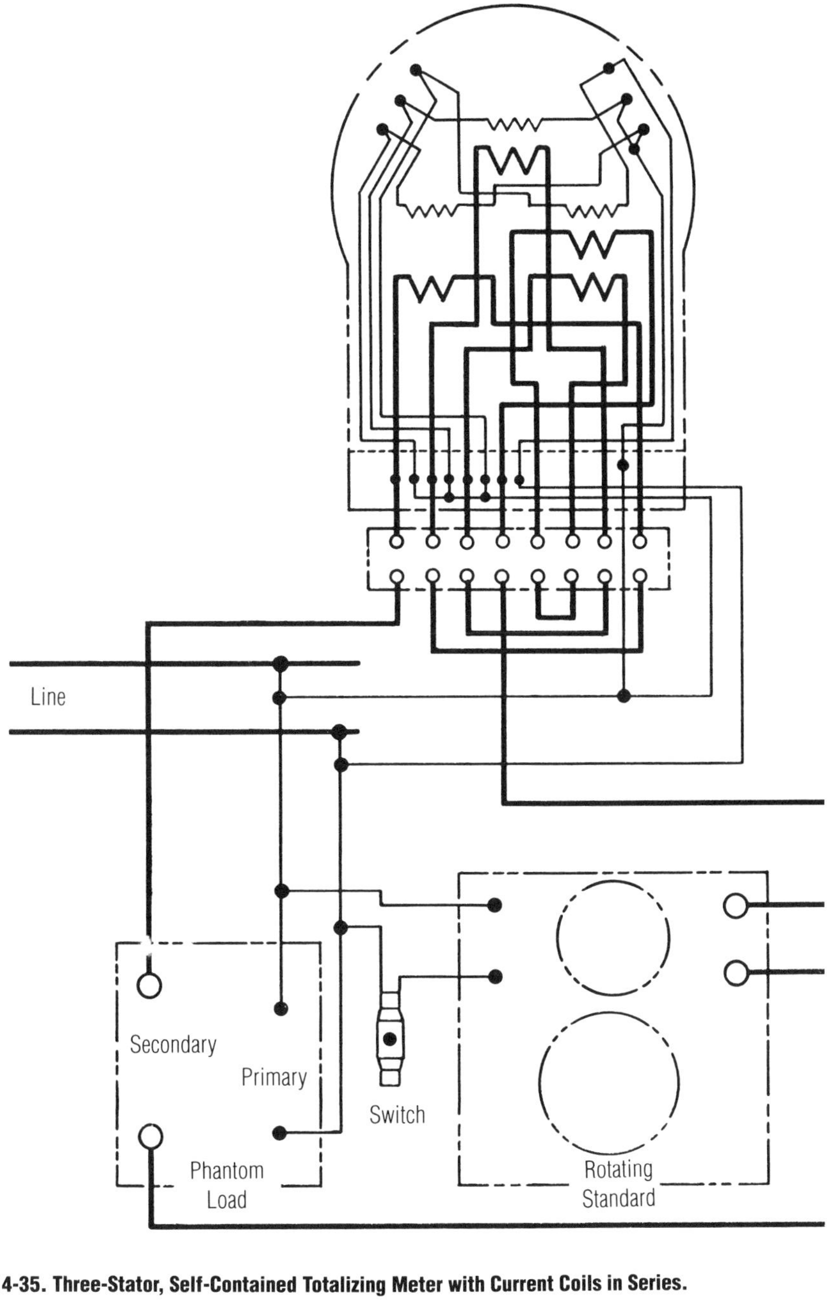

Figure 14-35. Three-Stator, Self-Contained Totalizing Meter with Current Coils in Series.

CHAPTER 15

DEMAND METER TESTING AND MAINTENANCE

The various types of demand meters are described in Chapter 8 and in sections of the *Appendices*. General test procedures and maintenance suggestions will be covered in this chapter, but no attempt will be made to cover all details of adjustments and maintenance. For such details, reference should be made to the manufacturers' instruction books.

MECHANICAL DEMAND REGISTER

A mechanical demand register is designed primarily to replace the conventional register of a watthour meter for the dual purpose of recording kilowatt-hours and the maximum integrated kilowatt demand. In addition to the kilowatt gear train and dials, it includes a gear train driven from the first shaft to provide the demand indication and a synchronous-motor-driven gear train to provide the time interval. The maximum demand may be indicated either by a sweep pointer operating over a semi-circular scale, or by dials similar to the kilowatt-hour dials.

Watthour meters equipped with demand registers are referred to as watthour-demand meters and, with the exception of the register and deeper covers, are identical to watthour meters of the corresponding type.

Principle of Operation

The demand gear train drives a pusher arm which advances the demand indicator in proportion to the speed of the meter disk, which is proportional to the demand.

At the end of a predetermined interval, usually 15, 30, or 60 minutes, the pusher arm is momentarily disengaged from its gearing and returned to zero, either by the motor, a spring counter torque, or by a gravity-driven mechanism. The time interval during which the pusher arm is advanced is controlled by a synchronous motor.

Therefore, a test of a mechanical demand register must satisfy these three questions:

a. Is the advance of the indicator correct?
b. Does the reset operate correctly?
c. Is the time interval correct?

To make clear the distinction between the space interval covered by the advance of the indicator and the demand interval controlled by the timing mechanism, the latter is, in this chapter, called "time interval."

Advancing Mechanism

Since the pusher arm which advances the demand pointer is geared to the watthour meter shaft, the accuracy of the demand indication is dependent upon the accuracy of the watthour meter. Therefore, the watthour meter must be calibrated correctly if the demand indication is to be right. The register ratio of the demand register must be correct for the application.

The register gearing must not impose a heavy or a variable load on the meter. In other words, excessive register friction due to dirt or improper gear mesh must not exist. Where demand meters are located in areas with unusual dust, dirt, or fumes, it is standard practice to clean the demand register thoroughly before reinstallation. If the period between tests is quite long, it is good practice to clean all demand registers as they come into the shop for routine test.

Cleaning methods vary from the use of standard cleaning machines which have a cleaning solution, a rinsing solution, and a drier to the use of ultrasonic cleaners. The cleaning solutions should be nontoxic and rinsing solutions should be acid free and not leave a film on the cleaned parts. Carbon tetrachloride should never be used for cleaning purposes because of the cumulative toxic effects that it may have.

After a register has been cleaned, a thorough examination should be made to detect faulty gears, worn bearing holes, and insufficient or excessive end shake of the various shafts. Particular attention should be given worm gear assemblies because malformation of the gear edges can cause a jerky advance of the succeeding gears.

Clutch

To permit reset of the pusher arm at the end of the demand interval, as well as to permit reset of the maximum demand indicator to zero at time of test, there is a clutch in the demand gear train between the meter disk and the pusher arm.

The clutch usually consists of two flat disks with a felt washer in between, with some means of exerting the proper amount of friction. This may be an adjustable spring. Some registers employ a cam-operated arm to disengage the clutch during the reset operation while others merely slip the clutch under full tension. While the adjustment of clutch tension is not critical, it should be checked to determine that it falls within the limits recommended, particularly if the register has been cleaned or disassembled for overhaul. Tension testing devices are available for this purpose.

The felt friction pads must be dry to insure proper operation and should never be subjected to cleaning fluids. Therefore, if the register is to be put into a cleaning solution, the pads should be removed and then replaced after the register is dry. In many cases, the pads can be removed, without dismantling the clutch assembly, by slitting the pad radially with a very thin knife. New pads may be installed similarly, making certain that the pads, after insertion, lie flat with the disks.

Resetting Mechanism

At the end of the demand interval, the pusher arm must be returned to zero. This requires that a counter-torque be applied to the pusher-arm side of the clutch.

In various types of registers, the counter-torque is supplied by one of three methods, either a spring, direct drive from the timing motor, or by gravity. In the spring-return type, the timing motor, in addition to driving the interval gearing, winds up the return spring during the interval. At the end of the interval, a tripping mechanism releases the spring energy to return the pusher arm to zero. In certain types of spring-return mechanisms, the clutch is simultaneously disengaged, while in others the clutch is allowed to slip but remain engaged. In the latter type, particular attention should be given to correct clutch tension.

On direct-motor-drive reset registers, the timing motor disengages the clutch at the end of the interval. Then, through cam and sector gear mechanisms, the pusher arm is returned to zero by the motor.

The gravity type of return mechanism requires that either a clutch or the demand gearing be disengaged at the end of the demand interval.

There are two main points to be observed concerning the operation of the reset mechanism: first, does the pusher arm return exactly to zero, and second, is the return to zero smooth and within specified time limits for the device?

An error in the zero setting will be reflected at all points on the scale. Most registers have an adjustable zero stop for the maximum pointer which should be checked to see that the pointer is exactly on zero. The pusher arm is provided with a micrometer screw adjustment for coordinating the zero of the pusher arm with the maximum pointer zero position.

A sluggish return to zero may be the result of excessive friction or a decrease in the amount of return power. For various types of registers, the return time, sometimes referred to as the "out-time," varies from a fraction of a second to four seconds for 15 minute interval registers and generally increases proportionally with the length of the time interval. On most registers, no adjustment is provided to change the reset time; however, some do have an adjustment in the cam mechanism.

Sluggishness is more effective and can be detected more readily in the spring-return and gravity-return type of registers. In the spring-return type, a change in torque is possible; however, a change sufficient to effect malfunction is uncommon. In the gravity-return type, since the weight is constant, there can be no change in return force. Therefore, sluggishness can usually be attributed to excessive friction due to dirt or gummy oil, maladjustment of the clutch, or a defective governor mechanism in types where used. Examination of the register should indicate the maintenance required — either cleaning, adjustment of the clutch, or replacement of defective parts.

Timing Motor

Failure of the timing motor is perhaps the most common fault in demand register performance. Under normal conditions, the synchronous motors are as reliable as the system frequency at which they operate. However, excessive friction may impose loads in excess of the motor capability, thus causing the motor to stop or operate intermittently. This means that the demand interval is extended, thus creating a high demand or an off-scale reading. If the condition is due to excess friction in the interval or reset gearing, correction must include removal of the friction by cleaning, adjustments, or replacement of parts as indicated by inspection.

The two most common causes of complete motor failure are the motor running out of oil, and burned-out motor coils. Burned-out coils require replacement of the motor field coils or complete replacement of the motor.

If the motor fails due to loss of lubrication, it must either be reoiled or replaced. If the manufacturer has an exchange program, it is usually more economical to replace the motor.

Certain types of General Electric Co. telechron rotor assemblies were provided with removable oil caps for flushing with a cleaning solution and reoiling. The modern rotors are grease-filled for longer life and do not provide for regreasing. The motor operating unit for the General Electric Type M-60 register is a hermetically sealed unit.

The motors on Westinghouse Type RW and Duncan Type FW and HW registers are identical and oil can be added through the breather hole. However, for complete cleaning and reoiling, it is recommended that the gear case be dismantled by removing the top of the case. After complete cleaning and resealing, the gear case is then filled with the proper amount of oil. The Westinghouse Mark series register motor is provided with a separate oiling hole for convenient maintenance.

The Type A and A-7 motors used in certain Sangamo demand registers can be cleaned and reoiled by completely dismantling the motor. Bearings should be carefully inspected for excessive wear. The Type H motor is much easier to maintain by simply removing the rotor and cleaning and oiling the bearings using a hypodermic needle.

It is recommended that any demand motor that is reoiled or replaced be dated. Although there is no definite period of time that motors will operate correctly without attention, most of them will go five years or longer. Excessively noisy motors or motors that will not operate on a 75 percent rated voltage should be cleaned and reoiled or replaced.

Maximum Demand Pointer

In order to leave an indication of the maximum demand for a billing period, a friction-type pointer is retained in a position representing the maximum advance of the pusher arm that occurred during a demand interval in the period. The friction may be obtained from a friction pad of felt or cork or by use of a silicone grease cup. Where a friction pad is used, there is usually an adjustment by which the compression of the pad may be controlled, which is not required with the silicone grease cup. The friction should be sufficient to prevent moderate vibration from changing the reading, but well below the amount required to slip the driving clutch. Tension testing devices are available for checking and setting the correct pointer friction for the register.

Cumulative Demand Register

Cumulative demand registers in general have all of the operating principles described for indicating demand registers, with an additional feature of retaining the maximum demand reading by adding the kilowatts for the current demand reading period to the accumulated demands of previous reading periods. This is accomplished by adding a gear train to advance dials, when this gear train is manually engaged at the time of reading. Usually a small sweep hand is provided to indicate the current maximum demand. However, the recorded maximum demand for the period is the result of subtracting the previous reading from the reading after reset as indicated on the kilowatt dials.

Test Procedure

Most demand register tests are made to determine the mechanical accuracy of the register only or to determine that, for a marked register ratio and time interval, the demand pointer will give a correct indication for a definite number of revolutions of the first driven gear. The watthour meter on which the register is to be used must be calibrated accurately to give a true indication of demand when in service. Furthermore, the watthour meter should be tested with its demand register energized to insure the best accuracy.

Since there is a definite correlation between the time interval gearing and the demand gear train for any particular register, self-checking devices are available for most types. The register self-checking consists of gearing mounted in a framework, so arranged that when a register is mounted in the framework the gear trains of the register are all locked together through the checker gearing. With the register energized, it will perform its normal function of resetting the register in addition to driving the demand gear train a definite number of revolutions per interval.

Most self-checking devices can be manually operated to run the register more quickly through a demand interval. When operating manually, the checker should be run slowly during the reset cycle to be sure the pusher arm is allowed to return to zero. The true reading that should be indicated may be calculated by counting the number of revolutions of the first gear of the register for the demand interval:

$$\text{kW} = \frac{(\text{Revolution of first gear per time interval}) \times 10}{(\text{Register Ratio}) \times (\text{Marked time interval, in hours})}$$

Usually the self-checking devices are provided with two gear ratios to check two points on the scale. This type of tester verifies the gear train ratios only, and it is necessary to check the timing motor to insure the correct time interval.

Some timing motors can be checked for correct speed by use of a stroboscope light, others by timing one of the slow-speed gears with an accurate stop watch. The time interval may be checked by timing, with an accurate timing device, the interval between two successive resets. It is recommended that motors be checked at a voltage considerably lower than the minimum expected in service.

An overall test of a demand register can best be accomplished by a register-checking device that simulates a watthour meter on a constant load. This consists of a synchronous motor driving a shaft similar to a meter shaft for which the demand register was designed, and studs for mounting the register. With the register motor and checker motor energized from the same source, the register is allowed to operate through one or more complete intervals. Ignoring the demand multiplier, the demand that should be indicated may be calculated as follows:

$$\text{kW} = \frac{(\text{Revolution of first gear per time interval}) \times 10}{(\text{Register Ratio}) \times (\text{Marked time interval hours})}$$

Revolutions of the first gear can be calculated for the time interval, from the checker shaft speed, and the ratio of the first gear reduction. This type of checker is particularly adaptable to shop testing where several such devices may be mounted on a test board. They are usually equipped with two or more speed changes for checking different points on the demand scale. A correct indication verifies that the time interval and the gear ratios are correct.

A manual test may be made on the gear train and zero setting by resetting the interval reset to zero, then advancing the first gear a definite number of revolutions. The kilowatt indication is calculated by the same formula as for mechanical test devices. It is necessary to determine that the time interval is correct as just outlined.

Field test on watthour demand meters may be made by connecting a field standard in series with the watthour meter, as for a regular watthour meter test. With a load applied to produce approximately the desired demand indication, the field standard should be started at the instant of a demand interval reset and stopped at the succeeding reset. The true kilowatts that should be indicated by the demand register can be calculated as follows:

$$\text{True kW} = \frac{(\text{Watthours recorded by standard}) \times N \times 60}{1{,}000 \times (\text{Time interval, in minutes})}$$

N = Number of full current coils in series in the watthour meter

ELECTRONIC DEMAND REGISTER

An electronic demand register is designed to perform the same function as the mechanical demand register, namely, replace the conventional register of a watthour meter for the dual purpose of recording kilowatt-hours and the maximum integrated kilowatt demand. Use of electronics, however, may provide some added benefits not available with the mechanical register. The electronic demand register displays both the total kilowatt-hour consumption as well as the maximum integrated kilowatt demand via a digital display. The programming of this type of register varies and may include an EPROM with a program written by the manufacturer and DOS-based software written by the manufacturer and supplied on diskette.

Principle of Operation

The electronic demand register uses the principle of optical pickup directly from the rotating disk or a small shutter device attached to the shaft of the disk apparatus. The optical pickup feeds pulses to the register, which allow the register to advance the demand display value in proportion to the speed of the meter disk, which is proportional to the demand.

The microprocessor within the electronic register analyzes the pulses fed to it through the optical pickup. At the end of a predetermined interval, usually 15, 30, or 60 minutes, the register calculates the average or integrated maximum kilowatt demand, compares the result with the previous interval, and, if larger, replaces that value. If the resulting demand value is smaller than that of the previous interval, the previous interval value is retained and the new value is discarded. The register then begins analyzing the next predetermined interval, and repeats the aforementioned process.

Therefore, a test of an electronic demand register must also satisfy three questions:

a. Does the register display the correct values upon completion of each interval?
b. Does the reset operate correctly?
c. Is the time interval correct?

Advancing Mechanism

Since the optical pickup which advances the electronic register relies directly on the motion of the watthour meter disk and shaft, the accuracy of the demand indication can only be as accurate as the watthour meter. Therefore, the watthour meter must be calibrated correctly if the demand indication is to be right. Visual inspection of the shutter mechanism on the disk shaft for those meters using that method of pickup, as well as inspection of the holes in the disk for the other method of pickup, should be performed to check the integrity of the mechanical portion of the meter. Refer to the manufacturers' instruction books for more information.

Resetting Mechanism

At the end of the demand interval, the microprocessor contained within the electronic demand register is programmed to return to zero and begin the measurements for the next interval. Many of the electronic demand registers provide visual indication that the interval has come to an end and the next is beginning. This indication consists of an End-of-Interval message displayed on the digital display. The loss of AC power at any time during a demand interval will cause the meter to store the current information into an EPROM where it is held until the meter is again energized. As one example: a meter 25 minutes into a 30 minute interval experiences a power outage. All information is written to the EPROM memory. The meter, when reenergized, will begin its demand interval measurement at the same point within the interval, maintaining the value of demand that had been measured up to the point of the power outage.

Careful attention should be given to the manual demand resets performed for billing. Each of the electronic registers has a demand reset indication that appears on the digital display when a correct billing reset has occurred. The manufacturers' information booklet for each of their registers will give the specific reaction to look for, to determine if a correct reset has occurred.

Test Procedure

The watthour meter on which the register is to be used must be calibrated accurately to give a true indication of demand when in service. Although the electronic demand register does not have any gearing or moving parts, it is still very important that the register be mounted on the meter before the meter itself is tested. The energy required to power the electronic circuitry may be enough to affect the calibration of the meter.

The electronic demand meter to be tested must be programmed to display the maximum demand value. Many of the electronic demand registers have a special test mode which allows a complete test to be performed on the meter without affecting the billing values of the customer. The length of the interval within the test mode may vary from manufacturer to manufacturer and is often programmable as set up by the programmer.

A test voltage only is applied to the meter under test. If the register contains a test mode, it is good practice to enter that mode before any external currents are applied. Visual inspection of the display sequence is necessary to confirm that all segments of the display are functioning properly. If one segment of any of the digits is not functioning, erroneous readings could occur.

The actual test on the meter may be performed by connecting a field standard in series with the watthour meter similar to a regular watthour meter test. Realizing the length of the test mode interval, a desired demand value is established to be "run up" on the meter. With the load applied to produce approximately the desired demand indication, the field standard should be started at the instant of a demand interval reset and stopped at the succeeding reset. The true kilowatts that should be indicated by the demand register can be calculated as follows:

$$\text{True kW} = \frac{(\text{Watthours recorded by standard}) \times N \times 60}{1{,}000 \times (\text{Time interval, in minutes})}$$

N = Number of full current coils in series in the watthour meter

ELECTRONIC TIME-OF-USE REGISTER

An electronic time-of-use register (also referred to as a time-of-day register) is designed to produce time-differentiated billing. The register designates predetermined time periods as being peak and others as non-peak. Some utilities have established a shoulder-peak which may not be as critical as the peak periods, but more so than the non-peak periods.

Time-of-use is typically set up to reward the customer for consumption during the non-peak periods and to discourage consumption during the "shoulder" or peak periods. The reward, or discouraging factor, is accomplished through the use of different billing rates for each time period.

Principle of Operation

The electronic time-of-use register uses the principle of optical pickup to establish pulse information. The pulse information is then analyzed by an on-board microprocessor. The pickup generates pulses directly from the disk rotation and is, therefore, only as accurate as the base meter.

Electronic time-of-use registers contain a microprocessor-based module with a program which is specific to a particular billing rate schedule, as established by the utility. Electronic time-of-use registers are capable of measuring both time-differentiated kilowatt-hours, as well as time-differentiated integrated kilowatt demand. The demand may take the form of the following: block interval demand, cumulative demand, continuous cumulative demand, and rolling demand.

The means of programming this type of register may include: an EPROM programmed by the manufacturer, or DOS-based manufacturer software. Each method of programming provides the register with information pertaining to: the peak, shoulder-peak, and non-peak times; season change dates; daylight savings dates; non-peak holidays; etc. Refer to the manufacturers' instruction books or appendices for more information.

Time is the key word in "time-of-use." Each electronic time-of-use register maintains an ongoing date- and time-keeping function. During AC power outages, the registers rely on battery backup. Currently two types of batteries are used for this function: the non-rechargeable lithium cell and the lead-acid rechargeable cell. The register monitors the battery condition and produces an error code on its digital display if the cell is not adequate. Testing of these cells on a routine basis is advised to maintain proper operation of the register during power outages. Manufacturers' information provides insight as to when the register may require a new cell.

Test Procedure

Periodic testing of the register switch times and special dates is recommended to ensure the correct setup of the programming information. Many of the electronic registers have visual indication of the present rate (peak, shoulder, non-peak). The manufacturers' information on their product may offer additional insight on features that should be tested. Updates to any of the programs to be used with the electronic registers should be verified completely before use in a billing situation.

Careful attention should be given to the manual demand resets performed for billing. Each of the electronic registers has a demand reset indication that appears on the digital display when a correct billing reset has occurred. The manufacturers' information booklet for each of their registers will give the specific reaction to look for, to determine if a correct reset has occurred.

Many of the electronic time-of-use registers may be tested utilizing a "Test Mode" function maintained within the register. This feature disallows any load applied during a test to be applied to the customers bill.

A Test Voltage is only applied to the meter under test. If the register being tested is one of the types that contain a test mode, that mode should be enabled. Visual inspection of the digital display is very important to ensure all segments of the display are functioning properly.

The actual test of the demand portion of the register is performed by attaching a field standard in series with the watthour meter, as for a regular watthour meter test. Some time-of-use registers offer smaller demand intervals for test conditions. The test interval length is used to establish a demand level that is attainable during that time span. With the load applied to produce the calculated demand indication, the field standard should be started at the instant of a demand interval reset and stopped at the succeeding reset. The true kilowatts that should be indicated by the register can be calculated as follows:

$$\text{True kW} = \frac{(\text{Watthours recorded by standard}) \times N \times 60}{1{,}000 \times (\text{Time interval, in minutes})}$$

$N =$ Number of full current coils in series in the watthour meter

RECORDING WATTHOUR DEMAND METER

These meters combine a watthour meter and a recording demand meter. A pen or other recording device is geared directly to the watthour meter shaft causing it to travel over a properly scaled strip chart. At the end of the time interval, the pen is returned to zero. The length of line produced on the chart is proportional to the kilowatt-hours during the specified time interval and is, therefore, proportional to the average kilowatt load for the interval, or the average kilowatt demand. For further discussion of this type of device see the Seventh Edition of this *Handbook*.

PULSE-OPERATED DEMAND METERS

This type of demand meter is essentially a counting device operated by pulses obtained from a pulse initiator found on either a mechanical type meter or one of the electronic type registers. Since each pulse represents a definite number of watthours, the accumulation of pulses for a definite time interval represents the average watts or demand for the time interval. The demand meter may be of either the maximum indicating type or recording type. Depending upon the design, the record may be on a round chart, strip chart, printed or punched into a paper tape, or in the form of pulses on a magnetic tape. More recent technology has found devices where the record may be in the form of pulses collected on an electronic pulse accumulator or solid-state pulse recorder.

Most indicating and chart-type recorders have an internal synchronous motor for establishing the time interval and advancing the chart or tape. The printing type may use an internal timing motor or an external contact-making clock for determining the time interval. Since some advance the indicating or recording mechanism on each pulse received and others on alternate pulses, the multiplying demand constant must be calculated on the basis of watthours per advancing pulse.

$$K = \frac{(\text{Watthours per advancing pulse}) \times (\text{Time intervals per hour})}{1{,}000}$$

TESTS

Shop tests on pulse-operated demand meters are commonly performed by connecting the meters to a controlled pulse source operated by a synchronous motor. The pulse source is so designed that the number of pulses per interval may be varied from a very low value to about two-thirds full-scale value of the meters to be tested. With the interval timing mechanism connected, the meter should be operated through several intervals at varying pulse rates. On indicating types, it is necessary to take readings for each change of pulse rate, while on the recorders, each interval is recorded. If, at the end of the test, readings do not agree with the predetermined pulse rates, the trouble must be located.

Some sources of trouble are as follows:

a. Incorrect time interval. The interval should be checked with an accurate stop watch from reset to reset.

b. Not returning to zero on reset. On recording types, this should be indicated on the chart. However, on indicating and printing types the reset operation must be observed. Most recording types have a register for accumulating the incoming pulses. Over the test period, the accumulation of pulses should check with the sum of the pulses indicated for each interval except that occasionally a pulse may be received while the instrument is resetting. If the meter is returning to zero, low readings usually indicate mechanical fault in the advancing mechanism.

In any test, the legibility of the record should be observed closely and corrected if necessary. This may require a new stylus point on a G-9, cleaning or replacing a pen if used or, replacement or realignment of the printing platen on a printing type. In some cases, realignment of the printing wheels is required.

Any search for trouble on a solenoid-operated, printing-type meter must include inspection for worn or defective parts and proper adjustments.

With electronic pulse accumulators or solid-state pulse recorders, care should be taken to assure the proper configuration program is in place. An improper program will result in incorrect pulse data.

Field testing usually involves more than just a test of the demand instrument, since the main purpose is to determine that the demand of a particular installation is being recorded correctly. The following questions must be answered for a field test.

a. Is the watthour meter calibrated correctly?

b. How many meter revolutions produce one pulse?

c. How many watthours produce one pulse?

d. Is demand meter receiving all pulses sent out?

e. Is time interval correct?

f. Is the proper amount of pulses recorded on the chart or tape?

g. Are pulses properly recorded on the cumulative counter, where applicable?

h. Is demand multiplier correct?

i. Is the demand record legible?

The demand meter test is usually performed at the time of periodic test on the watthour meter.

Since the pulse initiator is mounted in the watthour meter and its correct performance is one of the prime requisites for correct operation of a demand installation, it is generally the first component to be inspected on a field test. This is particularly true if the pulse initiator is a cam-operated contact device. Contact devices, if pitted, should never be filed, but should be dressed with crocus cloth and then cleaned with paper. Slight pitting of the contacts is not serious, but they should be clean and close flatly against each other.

Blade tension should be sufficient to prevent chattering due to vibration, but not so great as to cause excessive friction on the watthour meter. The spacing of cams should be such that the time intervals between successive contacts on a constant meter load are as nearly equal as possible.

After determination that the contact device is satisfactorily adjusted, the watthour meter should be tested and adjusted if necessary.

With the test load connected to the watthour meter, the demand meter should be checked to determine that it is receiving and registering correctly the pulses sent out. One method of doing this is to open the load to the watthour meter, reset the demand meter to zero, and then restart the meter and run it for a definite number of revolutions. The demand meter should then read the correct number of pulses as determined from revolutions per advancing pulse.

The time interval can be checked by timing, with an accurate stop watch, the interval between successive resets.

SOLID-STATE PULSE RECORDERS

Solid-state recorders receive pulses from the pulse initiator in a watthour meter, or the pulse recorder can be "under glass" within the watthour meter.

Pulse data is stored in electronic memory and, therefore, cannot be visually inspected. Instead, the stored pulse data is retrieved by a portable computer with an optical probe, or by a remote computer with dial-up capabilities. Once in a computer, the data may be formatted, printed, and viewed on a monitor. The software which retrieves the data can be supplied by the manufacturer of the solid-state recorder, or by a vendor of general-purpose software supporting several recorders. For a complete description of solid-state pulse recorders, see Chapter 8.

One critical function of an electronic recorder is its clock, which must be both accurate and stable. To test the clock in the recorder, methods described under "Test" earlier in this chapter can be used. If the recorder has an End-of-Interval output pulse, that pulse can be used to test the clock. The length of time between output pulses should exactly match the interval length programmed into the recorder. Another test method, which also tests the interval data recorder, requires pulses from a precision source which are inserted at a known rate. The number of pulses-per-interval can be calculated, and if the End-of-Interval pulse is produced by more or fewer pulses than expected, the timing may be inaccurate.

Other parameters that should be verified during testing are pulse-multiplier factors, calendar settings to change between daylight saving and standard times, and settings for seasonal peak periods.

MAINTENANCE

Solid-state pulse recorders require very little maintenance. Most of these recorders are designed so that the circuit board assemblies in the recorder can be easily replaced. Some recorders use rechargeable batteries which should be checked periodically. Some recorders perform a self-check of the battery and send a low-battery alarm to the device interrogating them. All batteries should be changed at intervals less than their expected shelf life.

Planned replacements of solid-state pulse recorders as part of a maintenance program should be made based on the estimated product-life of solid-state electronics, which is usually 10 to 15 years.

CHAPTER

16

THE METER LABORATORY

SCOPE AND FUNCTIONS

THE DESIGNATION OF "LABORATORY" is very often applied to the meter shop, but for the purpose of this chapter it will be assumed to apply to a separate division of the meter department. The scope of a meter laboratory may include only the certification of portable standard watthour meters and portable instruments, or may be extensive and include such functions as acceptance tests of materials and apparatus, special investigations, and research work. The fundamental responsibility of a laboratory is to obtain accurate measurements.

The basic functions of a meter laboratory are as follows:

1. Calibration and certification of the accuracy of working standards.
2. Calibration and certification of the accuracy of secondary standards by comparison with primary standards. If the primary standards are maintained within the meter laboratory, the comparison is an internal function. In utilities which do not maintain primary standards, the calibration of secondary standards may be performed in other approved laboratories.
3. Certify watthour and varhour secondary standards directly with the National Institute of Standards and Technology—particularly rugged, highly portable solid-state standards.
4. Intercomparison of primary standards. The standard resistors, standard cells, and associated equipment employed to calibrate secondary standards are the laboratory's highest internal authority on the value of electric measuring units and should be intercompared regularly to maintain the highest possible accuracy.
5. Responsibility for regular certification of primary standards. Primary standards should be sent periodically for certification to NIST or to an approved laboratory that has its own primary standards regularly certified by NIST.
6. Acceptance tests and determination of the characteristics of new types or designs of watthour meters, portable instruments, instrument transformers, and other electric measuring devices to determine their suitability for use by the utility company.

7. Special investigations relating to metering or measurement problems. Such investigations may require extensive tests both in the laboratory and in the field under actual operating conditions.
8. In addition to these basic functions, the meter laboratory may include in its scope of operations the repair of electric instruments and related devices. Also, assistance may be given to other sections of the utility company with electric measurements and tests.

FUNDAMENTAL STANDARDS

The basic electric units—the ohm, the ampere, and the volt—have been established by international agreement in conferences attended by scientists from many of the leading nations of the world. Countries such as the United States, England, France, Germany, and Japan maintain bureaus of standardization in which duplicates of these units, or the means for producing the units, are kept under conditions which insure their permanence. As a further check upon the constancy of the units, intercomparison is made periodically.

In the United States, the source of ultimate authority in electric measurements, and our representative in the international system of standardization, is the National Institute of Standards and Technology (NIST) at Gaithersburg, Maryland. Here, basic units are preserved, together with secondary units and high-grade instruments by means of which commercial measuring devices may be compared with these basic units.

Over the past 10 years, NIST has added the certification of secondary varhour standards to its list of services. This addition was the outgrowth of the development and availability of electronic solid-state and multifunction standards, and the inclusion of very accurate and stable var functions in these electronic standards.

Starting with NIST, a comprehensive system of standardization has been developed. Facilities are available whereby each utility may be assured of the accuracy of its measurements, and may have its standards compared with the basic units at Gaithersburg.

The two fundamental standards of electric measurement are the absolute volt and the absolute ohm. Standard cells and standard resistors, referred to as primary standards, are used as a means of maintaining these fundamental standards in a meter laboratory. From these standards, the values of other units of electric measurements can be derived.

With the evolution of electronic standards and their associated high accuracy and exceptional stability, many utility companies which in the past maintained primary standards (standard cells, standard resistors, DC to AC transfer equipment, etc.) have ceased this practice, and instead have their secondary electronic standards periodically certified either by NIST or by a qualified independent laboratory with direct traceability to NIST standards.

STANDARD CELLS

The practical standard of voltage is the cadmium standard cell. This cell is made in two types, the saturated or "Normal" cell, and the unsaturated cell.

The saturated cadmium cell has a relatively large temperature coefficient. Therefore, it is necessary to place this type of cell in a temperature-controlled oil or air bath to eliminate voltage changes caused by varying temperature. With constant temperature, the voltage of a saturated cell is very stable, providing an accuracy of 0.002 percent. Saturated cells are intended as a first-order primary standard and are used only in standardizing other working cells.

The unsaturated cadmium standard cell is the usual working cell. It provides a voltage reference with an accuracy of 0.01 percent. The temperature coefficient of the unsaturated cell is small enough over the normal room temperature range that all temperature influences may be considered as negligible.

To preserve the high accuracy of unsaturated standard cells, certain precautions must be observed. The cells should never be exposed to temperatures below 4° C nor above 40° C. Abrupt changes in temperature should be avoided because they may produce temporary variations in cell voltage of several hundredths of one percent. The stability of unsaturated standard cells can be improved by installing them in a thermally insulated, copper-lined box which smooths out temperature changes and keeps all parts of the cell at the same temperature. Standard cells are not intended to supply any considerable amount of current. A continuous current of 50 μA may be supplied without damage to the cell; 100 μA may be supplied only momentarily; and over 100 μA will cause damage. In any case, standard cells should not be used in circuits where the cell current is continuous. Over a long time period, the voltage of unsaturated cells is less constant than saturated cells, usually decreasing from 40 to 120 μV per year. Therefore, they should be recertified at intervals of one to two years against a group of saturated cells, either at NIST or at another approved standards laboratory.

The standard cells should be intercompared regularly within the meter laboratory on a schedule which may vary from weekly to quarterly, depending upon their use. Two cells should be set aside for use only as reference cells in these intercomparisons. An accurate method of comparison is to oppose the voltages of two cells and measure the difference voltage with a potentiometer. Another, and less accurate, method is to measure the cell voltage directly with a potentiometer which has been balanced to a reference cell. Such intercomparisons are necessary to assure that the voltages of the working standard cells have not suffered temporary or permanent change due to improper use or handling.

POTENTIOMETERS

A potentiometer is an instrument for measuring unknown voltages by comparing them with known voltages. The following is a brief discussion of the principle of potentiometric measurement.

The line *OB* in Fig. 16-1 represents an electric conductor of uniform resistance per unit length, stretched along a scale graduated in uniform divisions. A cell or battery, *W*, causes a steady current of electricity to flow in this conductor. This is called the working current. With current flowing from *O* to *B*, the voltage is higher at *O* than at *B*, and the voltage difference between *O* and *B* is directly proportional to the resistance of the conductor. The resistance being uniform, the fall of voltage per unit length is uniform, and the voltage difference between any two points, as *M* and *M'*, has the same relation to the total voltage difference between *O* and *B* as the scale length between the two points has to the total scale length. Since the voltage falls from *O* to *B*, the point *M*, which is nearer *O*, is at higher voltage than the point *M'*, which is nearer *B*; that is, the polarity of *M* is positive with respect to that of *M'*.

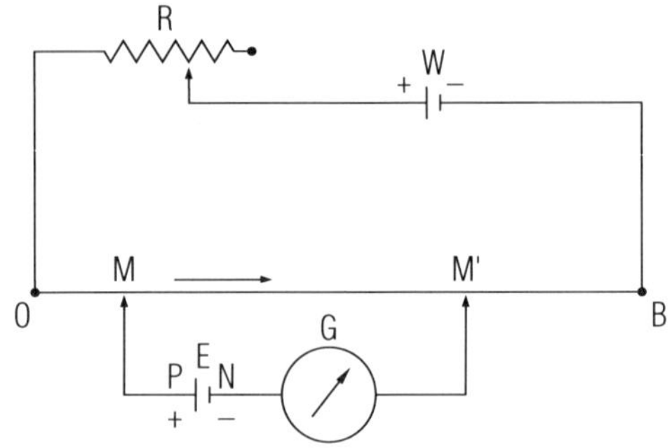

Figure 16-1. Diagram of Elementary Potentiometer Circuit.

A source of steady voltage is represented at *E*, with its positive terminal *P* connected to *M*, and its negative terminal *N* connected to *M'* through a galvanometer. A current will flow in the circuit *MEGM'* if the voltage difference between *M* and *M'* is greater or less than that between *P* and *N*. Since like polarities in *E* and *W* are opposed, the direction of the current will depend on which voltage difference is the greater. The flow of current and its direction will be revealed by the galvanometer, which is a current indicating instrument. If *B* and *M'* are movable contacts on *OB*, they can be adjusted to such positions that the voltage difference between them will be exactly equal to the voltage *E*. With the voltage at *M* the same as that at *P*, and the voltage at *M'* the same as that at *N*, no current will flow in the circuit *MEGM'*, and the deflection of the galvanometer will be zero.

It thus appears that two voltage differences can be opposed in such a manner that one exactly counterbalances the other, and that a galvanometer as an indicator shows when such a condition of balance exists. To employ this principle in measuring an unknown voltage *E*, it is necessary to establish a known voltage difference between *O* and *B*. This can be accomplished by connecting a standard cell in place of *E* with polarities as indicated. If the cell's voltage is, for example, 1.0183 volts, the contacts *M* and *M'* are set to span 1018.3 divisions of the scale, and the current in the resistance *OB* is adjusted by means of the rheostat *R* until the galvanometer indicates the balanced condition. The voltage difference between *M* and *M'* is then 1.0183 volts and the fall of voltage per scale division is 0.001 volts. It is obvious that the voltage of *W* must be higher than that of the standard cell.

With the working current in the potentiometer measuring circuit (*OB*) thus standardized, the standard cell at *E* is replaced by the source of voltage to be measured. When the positions of *M* and *M'* have been adjusted to establish a balance as indicated by the galvanometer, the number of scale divisions between *M* and *M'* represents the number of millivolts in the voltage difference measured.

The fact should be emphasized that when the galvanometer shows a balance, no current flows in its circuit. This shows that no current is taken from the source of voltage which is being measured, which means that the source undergoes no change due to the measurement. Likewise, when the potentiometer measuring circuit is standardized, no appreciable current is drawn from the standard cell; hence, if it is used carefully, the standard cell will last indefinitely because it is required to deliver only very feeble currents, and these only during short and infrequent periods.

The principle described is that of the null method of measurement. The galvanometer in such service is sometimes called the null point indicator. The accuracy of measurement by this method does not depend on the constants of the galvanometer, but on the accuracy of the standard cell voltage and the uniformity of the resistance *OB*. Since the uniformity of the resistance can be determined with exceedingly high precision, the accuracy of measurement depends principally on the constancy of the standard cell voltage, and the method is therefore a primary method.

The precision with which opposing voltages can be balanced depends upon the degree of subdivision of the potentiometer measuring circuit and the sensitivity of the galvanometer used for the null detector. In one type of potentiometer, provision for balancing to a high degree of precision is made by including a portion of the measuring circuit *OB* on a slidewire on which the contact *M'* is continuously adjustable, and the remainder of the circuit in fixed resistances of equal magnitude in a dial switch for adjusting the contact *M* in uniform steps. The scheme of such a circuit is illustrated in Fig. 16-2. This shows also a method for balancing voltage differences in the measuring circuit against the voltage of a standard cell, without disturbing the setting of *M* and *M'*. Fixed connections for the standard cell span a definite portion of the circuit *OB* across which the fall of voltage is equal to the voltage of the standard cell. By means of a double-pole, double-throw switch either the standard cell or the unknown voltage can be put in the circuit through the galvanometer with their respective connections to the measuring circuit. Accuracy of measurement with this type of potentiometer is 0.01 percent.

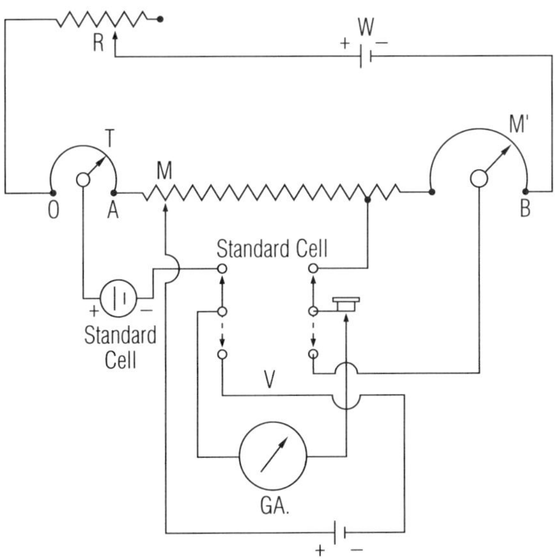

Figure 16-2. Diagram of Improved Potentiometer Circuit.

Potentiometers are usually classified according to the value of the "potentiometer resistance" (*OB* in Fig. 16-2). A potentiometer with about 100 ohms or less total resistance is classified as low resistance and one with a resistance of several hundred or several thousand ohms is classified as high resistance. The low-resistance potentiometer provides greater sensitivity in voltage measurements and is the type usually used in meter laboratory work.

Potentiometers should be certified when new by NIST or the manufacturer. Thereafter, they may be checked and maintained in the meter laboratory in accordance with the instructions furnished by the manufacturer. Potentiometer accuracy is dependent upon the relative values of its various resistors and not on their absolute values. Therefore, a potentiometer is checked by measuring, with a Wheatstone bridge of suitable accuracy, the resistances of the various slidewires and step resistors, and comparing their values to ascertain that they are within the manufacturer's prescribed limits. Potentiometers are usually provided with facilities for conveniently measuring the resistance of each unit.

STANDARD RESISTORS

The potentiometer, standardized against the standard cell, provides an accurate measurement of voltage. The use of current-carrying resistors, or precision shunts, adapts the potentiometer to precise measurement of current. The unknown current is passed through a standard shunt and the voltage drop is measured with the potentiometer. The value of current can then be determined by dividing the voltage drop by the resistance of the standard shunt.

One type of high-precision standard resistor is maintained in a temperature-controlled oil bath for maximum stability and may be certified to accuracies as high as 0.0001 percent. Current ratings are usually low. Standard resistors of a slightly less accurate type which are occasionally used in utility laboratories can be certified to 0.002 percent.

The type of shunt commonly used in the meter laboratory is air cooled and can be used to measure currents up to several thousand amperes. The nominal limit of error is 0.04 percent, but such shunts can be certified by NIST to 0.01 percent. Ratings of shunts commonly used are as follows:

1.0	ohm,	1.0	amperes
0.1	ohm,	15	amperes
0.01	ohm,	100	amperes
0.001	ohm,	300 to 500	amperes

Lower resistance shunts with higher current ratings are also available. The range of current measurements encountered will determine the size of shunts required in a meter laboratory. Where they are used often for a large number of measurements, it is desirable to have two identical sets of shunts.

Care should be taken that standard shunts are not overheated. Both the current terminals and the voltage terminals should be kept clean. Poor contact at the current terminals of large-capacity shunts will produce excessive heating which may permanently change the resistance.

The changes to be expected in resistors of this type when not overloaded are very small. Accuracy may be checked periodically by direct measurement of resistance of the shunt as a four-terminal resistor. This can be done with a precision Kelvin bridge. Intercomparisons can also be made by connecting two shunts in series with each other to a storage battery and rheostat. A Kelvin bridge of the continuously variable ratio type is then connected to the voltage terminals of the two resistors and a balance is established by using a sensitive galvanometer.

VOLT BOXES

When using a potentiometer for calibrating wattmeters or voltmeters, it is usually necessary to measure voltages higher than the maximum range of the potentiometer, which is normally 1.5 volts. For such measurements, it is customary to use a voltage multiplier, known as a volt box. The volt box shown in Fig. 16-3 is arranged for maximum voltages of 3, 7.5, 15, 30, 75, 150, 300, and 750 volts. The respective multipliers are 2, 5, 10, 20, 50, 100, 200, and 500—factors by which the potentiometer reading is multiplied to find the actual voltage. It can be seen that the ratio of resistances is such that full voltage on each range results in the application of 1.5 volts to the potentiometer. Volt box resistances are usually so high (100 to 30 ohms per volt) that only a small current is drawn from the line. At balance, the potentiometer circuit drains no current from the volt box and, consequently, the voltage drop per ohm in the volt box is unaffected by the potentiometer. The nominal accuracy of volt boxes is usually 0.04 percent, but they can be certified by NIST to 0.01 percent.

In using a volt box with the potentiometer, the common lead marked *GR* on the line side and *GR'* on the potentiometer side should be connected to the grounded side of the line. This prevents errors due to leakage currents and, when high voltages are being measured, it protects the operator. When working on a line, neither side of which is grounded, it is best to ground the side of the line which is connected to the *GR* post.

The ratio of resistances of the various ranges in a volt box may be determined by passing a constant current through all of the coils and measuring the drop of voltage across each range by means of the potentiometer. Care must be taken not to exceed the capacity of the potentiometer. Direct measurements of resistance can also be made conveniently with a Wheatstone bridge of sufficiently high accuracy.

WATT STANDARDS

The basic standard for alternating current power has been for many years the electrodynamometer wattmeter. Other devices, such as the Shotter-Hawks and

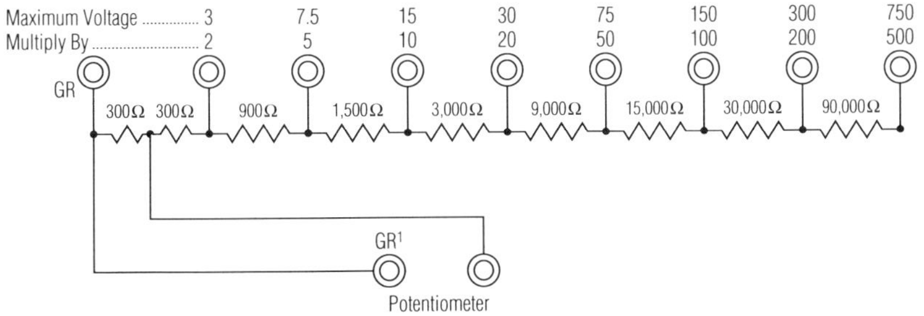

Figure 16-3. Volt Box for Extending Potentiometer Voltage Range, Showing Typical Voltage Ratings and Internal Resistance Values.

Goerz comparators, have come into use in recent years, but the wattmeter is still the most widely used watt standard.

The electrodynamometer wattmeter derives its torque from the reaction between a movable and a fixed coil. The movable coil is connected to the voltage through suitable resistors and the fixed coil is connected in the current circuit. This type of instrument has almost identical accuracy on alternating current over a wide range of power factor in the power frequency range and on direct current, so it is an excellent "transfer instrument" from direct to alternating current. In practice, any difference between DC and AC operation is usually small enough to be neglected. However, for the best possible accuracy, the AC/DC ratio at the frequency, power factor, and test points to be used may be certified by NIST. This difference will normally remain constant and need be determined only once. This difference does not normally exceed a few hundredths of a percent for laboratory wattmeters of good design. In poorly designed or adjusted wattmeters, the DC to AC transfer errors can easily amount to several tenths of a percent. It is desirable, therefore, to have new laboratory-type wattmeters routed through NIST for an AC/DC certification at the time of purchase. Few, if any, utility laboratories have adequate facilities to make an AC/DC ratio test of this type.

The wattmeter is calibrated on direct current using volt boxes, shunts, potentiometers, and standard cells for the highest accuracy. Calibration is obtained from the product of the direct current through the fixed coils and the voltage applied to the moving-coil system. The DC voltage and current may be obtained from entirely independent sources, but as far as the instrument is concerned, it indicates the product of these two quantities and the scale is calibrated in watts. Reversed polarity readings are always taken on direct current and only the average values used to eliminate the effects of residual magnetic fields. The instrument may then be used on alternating current as the standard of AC power, with the application of AC/DC corrections as determined by NIST for highest accuracy.

Instruments of this type may be obtained with certified accuracy of 0.1 percent of the full-scale value without applying any corrections. They may be relied upon to a greater extent with careful handling, an established performance history, and the application of calibration corrections. Wattmeters should be mounted on a solid support to minimize vibration and should be seldom moved and carefully leveled. Connections can be made away from the wattmeter so that it is not necessary to move it at all. Where separate sources are used for current and voltage polarity, terminals of the wattmeter should be connected together, either directly or through a resistor. This "static" tie prevents errors due to electrostatic forces on the moving system. A record should be maintained of all calibrations on the wattmeter to determine any drift in its accuracy. Periodic checks of the wattmeter against the potentiometer may be scheduled at intervals of from two weeks to two months. A large meter laboratory should have at least one spare wattmeter in addition to its working standard.

The wattmeter must be calibrated for all scale readings at which it will be used in watthour standardization. Calibration procedure will usually require two or three people. One person measures the voltage with a potentiometer and volt box and holds it constant with suitable controls. The second adjusts the current to give the desired wattmeter scale reading, and the third measures the current with another potentiometer and shunt. Naturally, the current measurement cannot be made until the voltage and scale reading have been adjusted to their correct values. With a DC voltage source which has automatic regulation, it is possible to use only two people to perform the calibration; one person on a single potentiometer adjusts the voltage to the correct value and then transfers the potentiometer to the current circuit for current measurement.

ALTERNATING VOLTAGE AND CURRENT STANDARDS

The most common standards for alternating voltage and current are electrodynamometer-type indicating voltmeters and ammeters with construction similar to that described for wattmeters. Such instruments also have similar characteristics in that they are of almost identical accuracy on alternating and direct current and may be obtained with accuracies to 0.1 percent of the full-scale reading without applying any corrections. They are used as the references in calibration of AC instruments. For the highest accuracy, the AC/DC ratios should be certified by NIST.

The calibration of electrodynamometer voltmeters is accomplished on direct current using a volt box and potentiometer. Ammeter calibrations require the proper size shunts with the potentiometer. The AC/DC ratios, if significant, should be applied as well as the corrections determined by the mean of reversed DC tests, if the highest accuracy is to be obtained.

TIME STANDARDS

The time standards in current use for watthour standardization are of two types: a high-grade, seconds-pendulum clock and a standard crystal controlled oscillator. Both types of devices may be calibrated by comparison with the time signals from NIST radio stations.

The seconds-pendulum clock is usually equipped with some form of device to obtain time pulses from the pendulum. This may be accomplished by an electric contact attached to the pendulum, photoelectric cell equipment, or other means. Auxiliary equipment is used with the clock pulses to establish or measure the desired time intervals.

Electronic devices for measuring or generating time intervals have been widely used in recent years. The crystal-controlled oscillators used for the time references in these devices provide a high order of accuracy, to less than one part in a million in many cases. Auxiliary components used with the oscillators, however, may reduce the final accuracy. The accuracy of well-maintained electronic or clock time standards used for watthour standardization is usually sufficiently high that any resulting errors are negligible and corrections need not be applied.

PRIMARY WATTHOUR STANDARDS

Although used by fewer and fewer utilities, the watt and time standards previously mentioned can be used as the primary standard in the meter laboratory. It is possible to use these standards to calibrate working portable watthour meter standards by holding the watts constant and running the portable standard for a preselected time interval using auxiliary equipment actuated by the time standard. This procedure is long and becomes tedious if it is necessary to calibrate many working standards. In such a case, it is the usual practice to use secondary watthour standards which are calibrated against the primary standards and which, in turn, are used to calibrate the working standards by a relatively simple procedure.

SECONDARY WATTHOUR STANDARDS

Secondary watthour standards in a large utility meter laboratory usually consist of three special watthour meter standards. Although only one standard may be used, the possibility of unsuspected errors appearing in the value of the watthour must be recognized. Such reference watthour standards are selected to have good electrical characteristics and are maintained so that the best possible performance stability is obtained. To eliminate the possibility of register friction, registers are removed. The standards are installed in a temperature-controlled box which diminishes the effects of varying ambient temperature. They are also usually run continuously to further reduce temperature variations, and also to eliminate any starting and stopping errors during calibrations.

When the reference standards contain a single current coil rated at 5 amperes and 120 volts, precision current and voltage transformers are also used. The precision transformers have multitap primaries with voltage and current ratings corresponding to the values required in all calibrations, thus allowing the reference standards to operate at 5 amperes and 120 volts during all calibrations. Precision transformers can be obtained which have negligible ratio and phase angle errors, so that corrections need not be used. Such transformers should, of course, be certified by NIST. This procedure simplifies the calibration of the reference watthour standards against the primary standards since calibration is normally required at only two points: 5 amperes with 120 volts at unity, and 50 percent lagging current power factors. If corrections are required at other power factors, additional calibration points can be used. Greater reference standard stability is also obtained by this method, because the calibration points operate with full rated current.

The first step in the calibration of the reference watthour standards is to hold the watts constant, either automatically or manually. From this point on, there are several variations in procedure, depending upon the type of time standard and its auxiliary equipment and the design of the watthour standard test equipment. One of two general procedures is usually followed. The time is measured for the watthour standard disk to make a preselected number of whole revolutions, or the disk revolutions, whole and partial, are measured for a preselected time interval. Each reference standard has a photoelectric pickup mounted inside its case which gives an output pulse for each disk revolution. This pickup is used with a photoelectric meter tester to count disk revolutions of the standard. With the photoelectric tester connected to an electronic time interval meter, the time for a selected number of disk revolutions can be measured directly. Using a pendulum clock and chronograph, time pulses from the clock can be recorded on the chronograph chart along with standard disk revolution pulses from the photoelectric tester output. By careful measurement on the chronograph paper chart—a tedious process—the time can again be determined for a selected number of disk revolutions. By including in each reference watthour standard a second photoelectric pickup and special disk, it is possible to obtain a large number of pulses per disk revolution which can be counted by a high-speed electronic counter. Since this combination is equivalent to a mechanical register, it is possible to measure whole and partial disk revolutions for a fixed time interval—the time standard being used to start and stop count on the high-speed counter. Other variations in method are also possible.

Calibration periods for reference standards may vary from weekly to monthly. The corrections used for these standards are usually the average of individual calibrations taken over an extended period of time. Provision should be made for quick, daily intercomparisons among the reference standards to detect any changes in their relative speeds.

The over-all accuracy of the watthour, covering all measurements up to and including the reference watthour standard calibration, can be determined to within ±0.05 percent under ideal conditions with the methods described. To achieve such accuracy, all applicable corrections should be applied and meticulous attention given to measurement techniques.

Solid-state watt and watthour standards are now commercially available. In general, these standards have greater precision and accuracy than induction watthour standards. Modern solid-state standards are programmable multifunction devices which use programmed controls and electronic comparators to eliminate most of the detailed manual activity historically associated with the standardizing process. This is accomplished through the use of electronic gating and control devices, or by using an electronic comparator shown, for example, in Fig. 16-4.

Typically, as with rotating standards, a bank of three electronic standards are maintained. With proper certification traceable to NIST basic standards, these electronic standards serve as both primary and secondary standards for the utility company.

Many companies are now using such solid-state standards as reference standards in the laboratory. Such standards can be shipped to NIST for calibration and may then be used as a laboratory's primary watt and watthour standard, thus

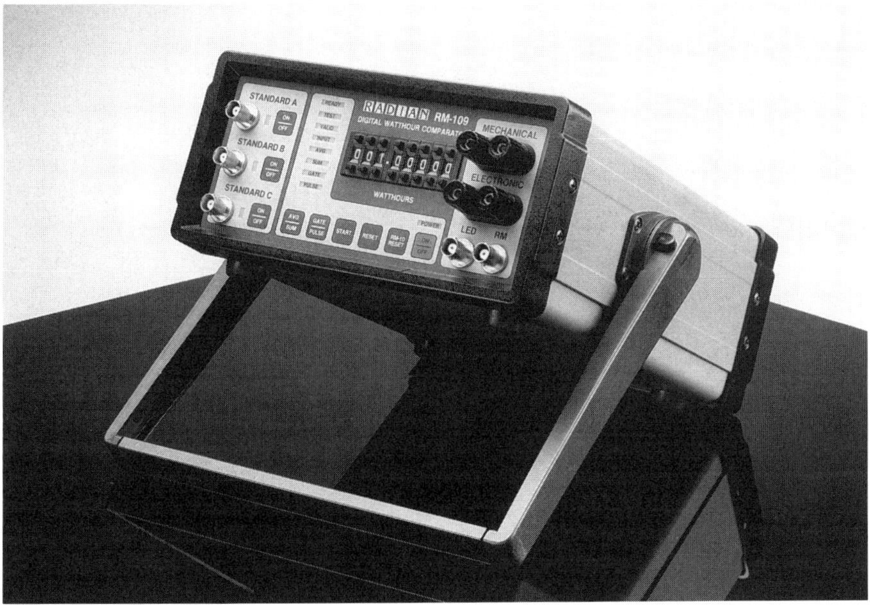

Figure 16-4. Electronic Watthour Comparator.

Figure 16-5. Portable Watthour Standard.

bypassing the DC to AC transfer problem. If two or more such solid-state standards are available and are periodically recalibrated by NIST, this is a very satisfactory method of maintaining a laboratory standard for the watthour.

WORKING WATTHOUR STANDARDS

Working watthour standards are portable standards used in meter test tables or racks or in field test sets as reference standards for testing and calibrating revenue meters.

Electronic portable working standards, such as shown in Fig. 16-5, have nearly eclipsed mechanical rotating standards for new or replacement units. The accuracy of electronic standards is typically 0.05% power factor or better under normal operating conditions, which is more than twice the accuracy of rotating standards. Electronic portable standards also are temperature-compensated over a wide range of ambient temperatures, and function within published specifications regardless of physical position.

Most modern electronic portable standards are auto-ranging on current, voltage, or both. This greatly simplifies their use as standards, and eliminates the possibility for damage to the standard by the accidental application of excessive voltage or current.

At least one electronic portable standard now available has a current tunnel feature which permits passage of a current-carrying conductor of up to 150 amperes continuous through an aperture in the standard. This allows testing, at most nominal load levels, of Class 200 meters without an external current transformer.

All presently available electronic working portable watthour standards are physically compatible with test tables and with field test sets which were designed for older rotating standards.

Both single and polyphase solid-state portable watthour standards are now available for field use. These can provide greater precision, higher accuracy, and increased resolution without increasing the time of a test.

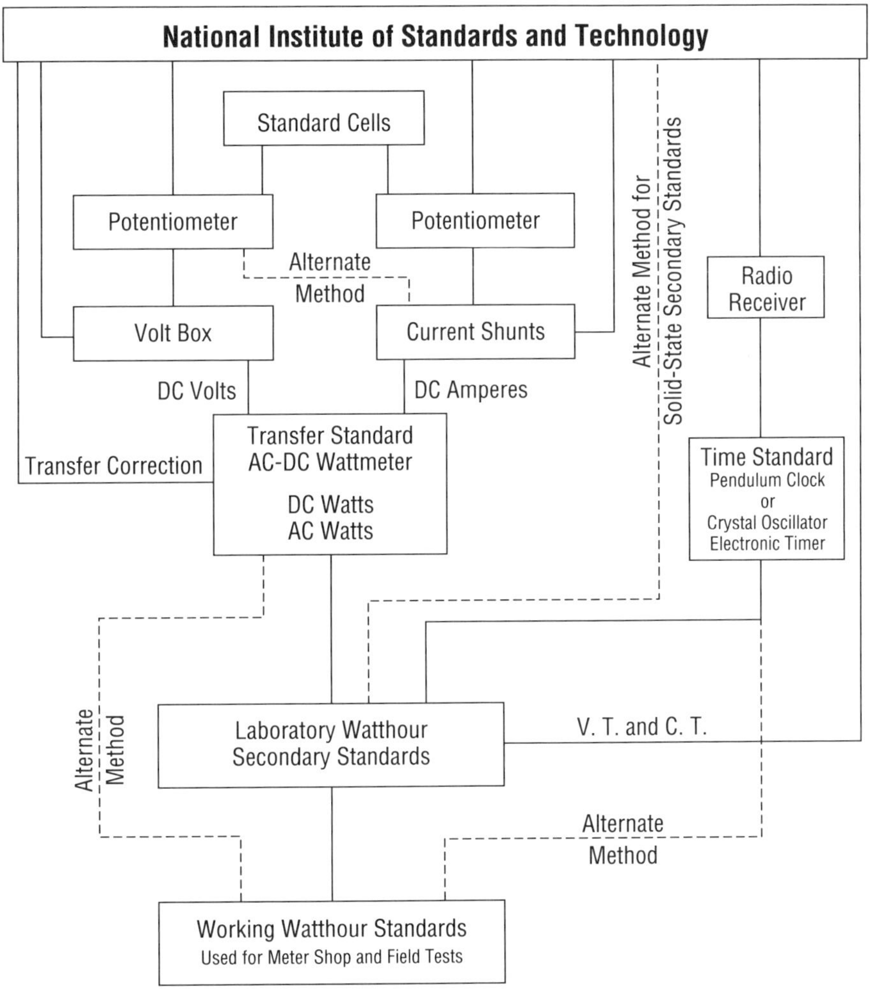

Figure 16-6. Simplified Block Diagram of a Typical Laboratory Chain of Standardization for the Determination of the Watthour.

THE CHAIN OF STANDARDIZATION

Figure 16-6 is a typical block diagram of the chain of standardization from NIST to the working standard watthour meters. It illustrates the interrelations of the equipment and methods discussed. Numerous variations of the basic procedures are possible. In some cases, certification of equipment may be obtained from sources other than NIST, such as manufacturers or other laboratories having primary standards certified by NIST. Large utility laboratories having certified precision resistance standards, ratio boxes, and bridges may have the capability of calibrating their own potentiometers, volt boxes, and shunts. If a certified and temperature-controlled bank of saturated standard cells is available, it is possible to calibrate the unsaturated standard cells. In all cases, the calibration of the equipment used to establish the watthour should be traceable to NIST.

The accuracy in the chain of standardization is limited by the combined accuracy of calibration involved in all previous steps. Theoretically, the total error could equal the sum of all previous errors. In actual practice, there will generally be some cancellation of errors. High accuracy is not the result of coincidence.

ACCURACY RATIOS

To reduce the effect of accumulated errors in a chain of standardization, each higher standard in the chain should also have higher accuracy. Ideally, each standard should be 10 times more accurate than the one it is to test. A ten-to-one accuracy ratio in each step cannot be maintained, however, since in a six-step chain the over-all accuracy ratio would be 10^6, or a million to one, and the accuracy requirements of the higher steps would greatly exceed the capabilities of the best available standards.

In practice, accuracy ratios in each step in the range of 2:1 to 5:1 are the best that can be normally realized in the present state of the art. With these lower accuracy ratios, great care is required to eliminate, or correct for, errors in each step.

RANDOM AND SYSTEMATIC ERRORS

Small residual errors are always present in any measurement. These errors may be divided into two general types: random or systematic. Random errors occur without any apparent pattern in a series of repeat measurements with the same equipment. Random errors are easily detected and may be analyzed statistically. The standard deviation, sigma, of a series of measurements containing random errors is a good indication of the precision of the measurement. (The calculation and use of the standard deviation, sigma, is well covered in books on statistical analysis.)

Systematic errors are much more difficult to detect or analyze. These are errors that either may be fixed or may vary in a recognizable pattern. If variable, they may be detected as a systematic change in the results of a series of repeated measurements. The pattern may be a trend in one direction with time or a periodic function of time, or it may be a function of temperature or some other variable. Systematic errors which remain constant are the most troublesome of all, as they cannot be readily detected in a series of repeated measurements.

Fixed systematic errors are detected and eliminated by making cross checks using a different method or different equipment, or both. If the results agree, the

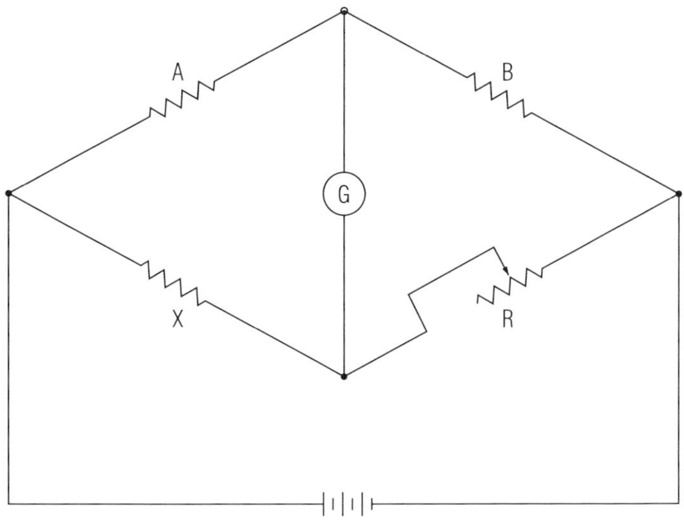

Figure 16-7. Wheatstone Bridge Circuit.

probability of undetected systematic error is small, since it is unlikely that each test involved identical systematic errors. If the results do not agree, systematic error is present in one or both measurements. Fixed systematic errors may be due to an unknown change in calibration of any piece of equipment, to unknown voltage drops in test circuits, to unknown magnetic coupling between circuits, or innumerable other causes.

CROSS CHECKS BETWEEN LABORATORIES

To eliminate the possibility of systematic errors, and thereby establish confidence levels for accuracy at each level of standardization, suitable cross checks must be made. Much of this cross checking may be done within a single laboratory if some alternate equipment is available; but complete duplication of all equipment is seldom economically feasible. To overcome equipment limitations, cross checks may be made with other laboratories. Two utility laboratories may easily make cross checks of their watthour standards by transporting carefully calibrated working portable watthour meter standards from one laboratory to the other. The standards should be rechecked upon return to the home laboratory to be sure they have not changed due to transportation. Such cross checks, or round robin checks, involving several laboratories, provide information of great value to all laboratories participating. Good agreement in such interlaboratory cross checks provides assurance that accuracy is being maintained, while a lack of agreement will often uncover unsuspected systematic errors.

RESISTANCE MEASURING INSTRUMENTS

In the laboratory, resistances are usually measured by bridge methods. Because not only laboratory workers but metering personnel, as well, may have occasion to use such bridges, the two most common types of bridge circuits, the Wheatstone and Kelvin, will be discussed.

Although both the potentiometer and the bridge employ the null balance principle and have much else in common, they differ radically in the manner of balancing. In the potentiometer, a balance is effected between the voltage to be measured and a measurable portion of the voltage gradient along a conductor; but in the bridge, a balance is obtained between equivoltage points in two circuits generally connected to one common voltage. In the bridge, balance indicates that the relative resistance values of the subdivisions of the circuits fulfill certain conditions.

The operating principle of the Wheatstone bridge may be explained as follows: Fig. 16-7 shows the elementary Wheatstone bridge circuit in which *A*, *B*, and *R* are accurately known resistances and *X* is the resistance to be measured. When using the bridge, the ratio of *A* to *B* and the value of *R* are adjusted until the galvanometer, *G*, shows no current flowing. With this condition:

$$X = \frac{A}{B} R$$

The Wheatstone bridge is, in general, suited to measuring resistances from about 1 to 100,000 ohms. If the unknown resistance is intermediate in value between the limits stated, errors to be expected with a first-class bridge are of the order of 0.01 to 0.05 percent.

For the measurement of resistances under 1 ohm, there exist a number of laboratory methods capable of greater precision than the Wheatstone bridge, but the best circuit for general work is that of the Kelvin (Thomson) bridge. The Kelvin bridge differs importantly from the Wheatstone bridge in the means employed to nullify the effects of the unknown and inconstant resistances of connecting leads: a system of double ratio arms, whereby the resistances of connecting leads are placed in series with high-resistance ratio arms, and not with the resistance under measurement or with the standard of reference. Because of the

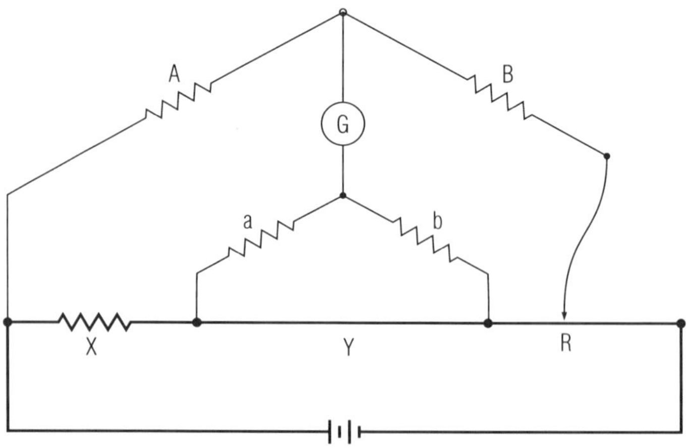

Figure 16-8. Kelvin Bridge Circuit.

D'Arsonval instruments will not agree on a DC source containing significant ripple. Rectifiers must be extremely well regulated if they are to be used for potentiometer work.

Regardless of the accuracy of the equipment used, accurate electric measurements cannot be made with unstable or distorted sources of power.

LABORATORY OPERATION

Successful operation of a standards laboratory requires the possession and maintenance of certified standards of suitable types, highly trained laboratory personnel who understand the art and science of accurate measurement, and the availability of sufficient time and money for the regular program of periodic standardization and cross checking.

CHAPTER 17
METER READING

THE ACCURATE READING of meters is an operation of major importance in any electric utility company, not only from a revenue standpoint but also in the promotion of good customer relations. The meter reader makes many personal contacts with customers in gaining access to meters and is often the only employee seen by the customer. With this in mind, most electrical utility companies choose, for their meter reading personnel, employees who are conscientious and have the natural attributes of friendliness, courtesy, and a neat appearance.

A meter reader's initial training course usually includes the fundamentals of good public relations, familiarization with the types of metering equipment likely to be encountered, and thorough explanations of the terms used in customer billing. It is important that meter readers be familiar enough with their company to be able to channel a customer's requests and questions to the proper departments.

Meter reading is carried on under a carefully planned program and meter routes are arranged in proper reading order. Readings are taken on a prearranged schedule to make billing periods as nearly equal as practicable and to provide an even flow of work for other operations, such as billing, auditing, and collecting. To prevent undue annoyance to customers and to expedite the work, meter reading records usually bear notations showing the exact locations of meters, means of access, keys needed, and notes of any unusual conditions, such as physical hazards and bad dogs.

If it is necessary to enter a customer's premises, it is good practice for readers to announce themselves courteously and produce proper credentials and identification when requested. Meter readers can be of considerable help to their company in reporting irregularities, such as changes in customers, meters without a reading record, vacant buildings, stopped meters, unsealed meters, unmetered service, and any other condition that might adversely affect customer billing or the quality of service rendered.

The principal duty of a meter reader is to obtain the meter readings and to make certain that these readings are entered on the correct reading records by verifying addresses and meter numbers. A good meter reader will enter the readings on mark sense cards or meter reading sheets in a clear and precise manner.

HOW TO READ A WATTHOUR METER

There are three styles of kilowatt-hour meter register faces in general use. One has individual dial circles as shown in Fig. 17-1, another has interlocking dial circles as shown in Fig. 17-2, and a third style of register uses cyclometer-type dials.

Registers with dial circles have either four or five dials, five dials being provided to avoid a dial multiplier of 10 and the possibility of a register "turn-over" during the normal billing period. Adjacent pointers rotate in opposite directions and are so geared for travel that the pointer on the right will make one complete revolution while the one next to it on the left makes one-tenth of a revolution. When a pointer is between two figures, the smaller figure is the one to use as a part of the reading.

Figure 17-1. Conventional 5-Pointer Kilowatt-Hour Dial.

Figure 17-2. Conventional 4-Pointer Kilowatt-Hour Dial with Overlapping Circles.

Figure 17-3. Kilowatt-hour Register Showing Reading of 0562.

Figure 17-4. Kilowatt-hour Register Showing Reading of 2198.

A watthour meter is read from right to left by reading all dials and recording the reading on a meter reading form in this same sequence. The reason for reading the dials from right to left is that the right-hand dial governs the one on its left in each instance. With all pointers at zero, and a dial multiplier of 1, one clockwise revolution of the unit's dial pointer will indicate a reading of 10 kilowatt-hours on the register. A complete counterclockwise revolution of the 10's dial pointer will indicate a reading of 100 kilowatt-hours on the register and so on. In reading the dials the procedure is like reading 1's, 10's, 100's, and 1,000's.

Each pointer must complete a revolution to advance the pointer located at its left by one division. Therefore, in deciding upon the reading of a pointer, the pointer before it (to the right) must be consulted. Unless this pointer has reached or passed the 0, that is, completed a revolution, the pointer in question has not completed the division on which it may appear to rest. For this reason, accuracy and rapidity are gained by reading the meter from right to left.

A simple analogy can be made with a watch. When the hour hand is near 8 and the minute hand is at 11, it is not yet 8 o'clock, but it is 7:55 and, obviously, it will not be 8 o'clock until the minute hand has advanced to 12.

Figures 17-3 and 17-4 show examples of typical watthour meter readings.

To obtain the use in kilowatt-hours over a period, it is necessary to subtract the previous reading from the present reading. When the dial multiplier is 1, the difference will be the number of kilowatt-hours used between the two readings. When the dial multiplier is other than 1, the difference between the readings following subtraction must be multiplied by the given dial multiplier to obtain the kilowatt-hours used. Dial multipliers of 1 generally are not shown, but those of other than 1 are shown on the dial faces.

Double- or two-rate registers employ two sets of dials and two complete register mechanisms which are automatically switched into gear with the moving element shaft at predetermined times. These two-rate registers are generally used in conjunction with off-peak water heating rates.

DEMAND METERS

Demand readings are usually in terms of either kilowatts, kilovolt-amperes, or kilovars, the quantity measured being indicated on the dial face or the name plate of the instrument. The demand device may be incorporated within a watthour meter, may be associated with a separate watthour meter, or may be independent of any watthour meter (a separate demand meter). Demand readings may be either indicating or recording. In reading a demand meter of the sweephand type, it is important that the reading be taken, as closely as is practicable, from directly in front of the meter. This is to avoid errors in reading due to parallax.

Figure 17-5. Watthour Demand Register, Indicating Type.

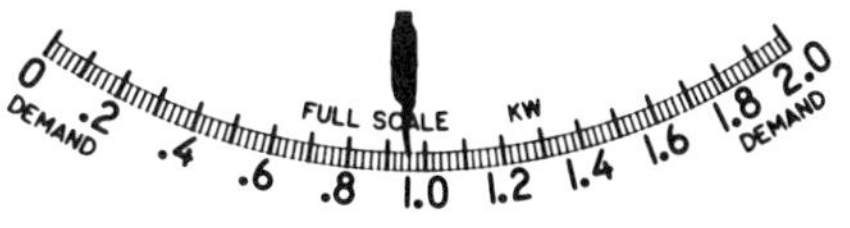

Subdivision Value: 0.5
Indicated Value: 13.5

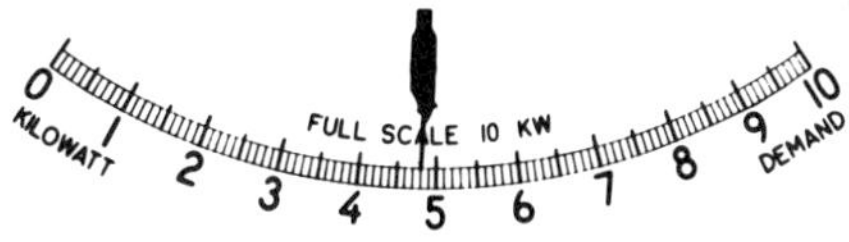

Subdivision Value: 0.1
Indicated Value: 4.8

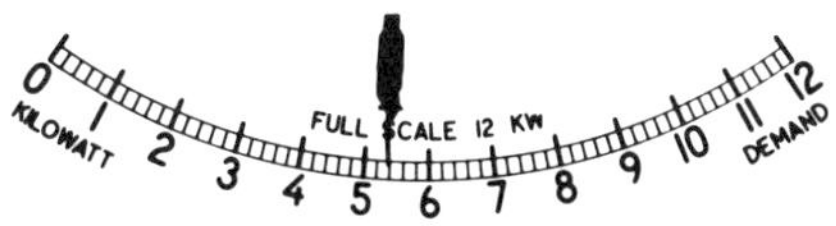

Subdivision Value: 0.2
Indicated Value: 5.4

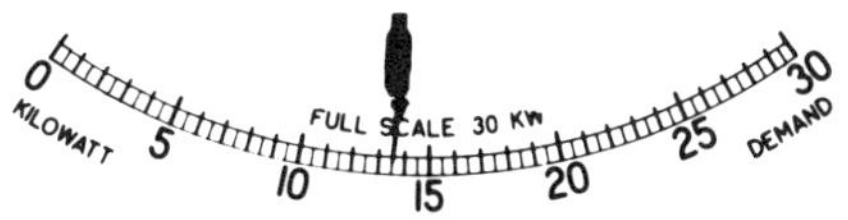

Subdivision Value: 0.02
Indicated Value: 0.96

Figure 17-6. Watthour Demand Scales Showing Various Subdivision Values.

Indicating Demand Meters

Indicating demand meter registers fall into four basic types.

Sweephand type of Pointer (Mechanical-Type Register)

How to Read—The position of the sweephand is read and the pointer reset to zero. A register of this type is shown in Fig. 17-5. The maximum demand is determined by the position of the pointer on the scale in a manner similar to reading a voltmeter or a speedometer. The value of the smallest subdivision on the scale may be found by dividing the first numeral by the number of divisions between this numeral and zero. When the number of subdivisions between the marked or numbered divisions is ten, the reading of the pointer is the value of the marked division plus the value of the subdivisions in tenths of the major division. The demand value shown in Fig. 17-5 is 1.45 kilowatts. Fig. 17-6 shows four demand scales with different subdivision values.
How to Reset—This type of register has a clutch in the advancing mechanism and therefore both the indicating pointer and the pusher finger may be returned to zero by a counterclockwise movement of the reset knob. The wire reset should not override the pointer, as the resulting spring action may cause a false demand indication.

Sweephand Type of Pointer Showing Maximum Demand, Plus a Sweephand Pusher Pointer, Showing Current Demand (Thermal-Type Register)

How to Read—The maximum demand is read in the same manner as the mechanical-type register.

How to Reset—The pusher pointer is not driven through a clutch as in the mechanical register, but is directly attached to the moving element and must not be forced back to zero.

Therefore, it is customary to record the position of the maximum pointer and then reset the maximum pointer back to, or slightly depressed beyond, the indication of the pusher pointer.

Dial Type of Pointers (or Cyclometer) Showing Cumulative Readings

How to Read—This type of register has, in addition to four or five kilowatt-hour dials, four similar dials at the bottom for registering maximum demand. These demand dials are read in the same manner as the kilowatt-hour dials. To obtain a reading of the maximum demand on this device, it is necessary to operate a mechanism attached to the meter cover. The kilowatt demand for the billing period is then automatically added to these dials and a record is then made of the new readings. The dial pointers are stationary between reading periods. A solid black index line equal in height to the demand dials is used to denote the decimal point.

How to Reset—When resetting cumulative demand registers which are motor operated, firmly push the reset plunger against the reset knob until the knob starts to rotate. Do not hold the plunger against the knob for the completion of the accumulation since this may jam the register. For a meter equipped with a reset lock device, insert the key, turn to the left one eighth of a turn and push the key forward causing the plunger on the lock to push against the reset knob. When the knob starts to rotate, release the pressure on the key and rotate it back one-eighth of a turn to its normal position and remove the key.

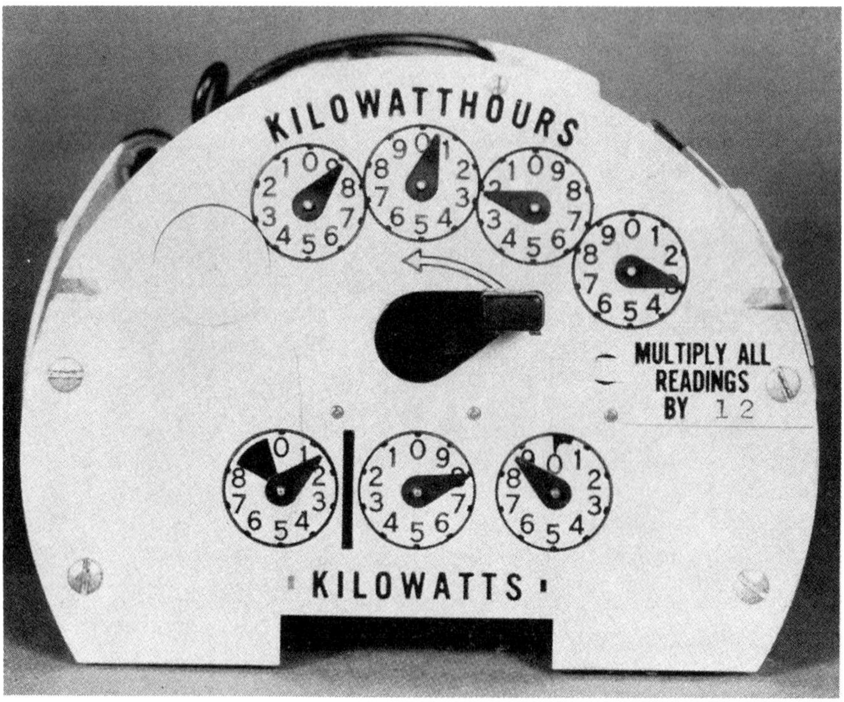

Figure 17-7. Watthour Demand Register, Non-Cumulative Dial Type.

Dial Type of Pointers, Non-Cumulative

How to Read—This register has three demand dials similar in appearance to the kilowatt-hour dials, Fig. 17-7. These dials are located at the bottom of the register and are read in the same manner as the kilowatt-hour dials. The pointers are reset to zero at each reading. A solid black index line equal in height to the demand dials is used to denote the decimal point.

How to Reset—The dial pointers are reset to zero by turning the reset lever about one-fourth of a turn in a counterclockwise movement; then the lever is returned to its normal position.

Sealing

Following the operation of a reset device or operating mechanism it is necessary that the device be sealed or locked and that it remain so until the next reading. Some companies require that old seals be turned in, both for their salvage value and in the interest of good housekeeping and safety on the customer's premises.

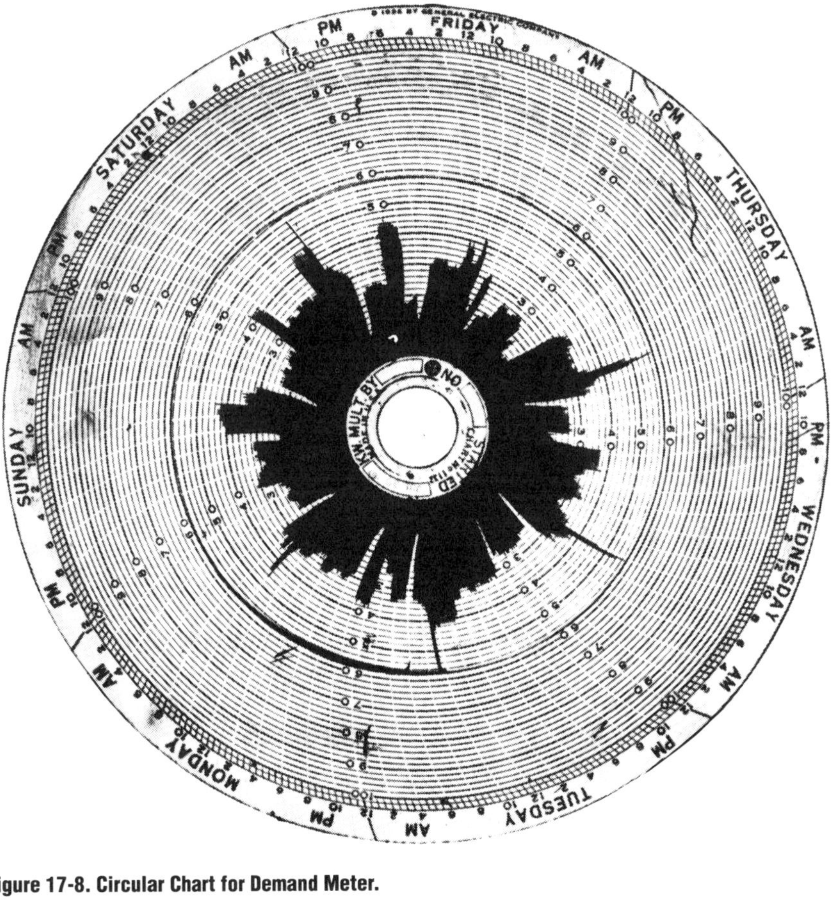

Figure 17-8. Circular Chart for Demand Meter.

External Demand Meter, Indicating Type

This type is separate from the watthour meter, but in some instances is electrically operated by pulses from one or more watthour meters. This meter is read in the same manner as the indicating demand register, shown in Fig. 17-5. The external demand meter may also be of the cumulative type.

Recording Demand Meters

Recording demand meters may be of either the circular-chart or stripchart type. In either case, it is generally the practice to remove the chart at the time of reading, to write in the time of removal on the chart, to identify it with the meter with which it was used, to record the reading of any pulse totaling device, and to return the chart to the office for determination of demand. Important considerations in setting a new chart are that it be set accurately on time and correctly zeroed.

Some demand meters are direct reading, having scales or charts so designed that the instrument has a constant of 1. In most cases, however, it will be necessary to multiply the indication by the constant shown on the meter or billing record to obtain the demand in the desired unit. The value of each subdivision on the chart may be found by dividing the first numbered position by the number of divisions between this numeral and zero. When kilowatt-hour and kilowatt readings are obtained from the same meter, the multiplying constant for each may or may not be the same, so each should be verified before being used to determine the quantities to be billed.

External Demand Meter, Circular-Chart Type

The kilowatt demand for each demand interval is recorded upon a circular chart, Fig. 17-8. The time of day and the day of the week are indicated on this chart which is driven by a time mechanism. The chart may be either a 7-,16-, or 32-day type. Some of these charts have inverted scales with zero at the outer edge of the chart.

How to Read—The maximum kilowatt demand is obtained by finding the highest value recorded for the demand interval, which may be 5, 15, 30, or 60 minutes, depending upon the rate established. This determination is generally made for a monthly billing period. When the highest kilowatt value is found, the time of day and the day of the week or month in which it occurred can be determined and stated for billing purposes.

How to Remove Chart—To remove a chart, raise the stylus (pointer) until it swings back past the stylus-mounting-plate center and is automatically held out of the way. Then unscrew the knurled thumb nut which holds the used chart in place.

How to Change Chart—Before putting a new chart in place, read the dials of the totalizing register and make a record of the multiplying constant (kilowatt-hour constant) printed on the face of the register. At the same time, record the register reading of the watthour meter in conjunction with which the demand meter is used.

Then put the new chart in place, setting it so the proper time of day coincides with the pointer found on the case at the top of the chart. See that the edge of the chart is properly in position beneath all the guide clips. After the chart is adjusted, securely tighten the knurled thumb nut which holds the chart in place and lower the stylus to make contact with the chart.

How to Set on Time—Before putting a new chart in position on a mechanical-type demand meter, check the time-setting disk, a part of which appears through the irregularly shaped opening in front of the case, visible when the chart is removed. If the time is incorrect, turn the time-setting disk in the upward direction by the finger until the proper minute of the interval appears opposite the index.

A thermal-type recording demand meter has no interval adjustments.

External Demand Meter, Strip-Chart Type

The operation of this type meter is similar to that of the combined watthour and demand meter of the stripchart type.

Combined Watthour and Demand Meter, Strip-Chart Type

The demand device is mounted within the same enclosure with the watthour meter and mechanically operated by the watthour meter. A strip-type chart is used and generally covers a period of one month. The determination of demand is the same as for the circular chart type.

DEMAND AND TIME-OF-USE METERS WITH ELECTRONIC DISPLAYS

Electronic meters use digital displays to indicate power readings and other information. The display can show a series of readings, automatically stepping through each parameter at a programmed rate, for example, 5 or 10 seconds per step. This allows the meter reader to record each reading, either by writing it on a meter card or by keyboarding it into a hand-held terminal. In addition to measured values, the display can show time, date, and perform an All-Segments test to verify the proper operation of all display elements. Figure 17-9 shows an All-Segments test. Some electronic registers display prompts such as 01 or 02, which identify the parameters being displayed. Figure 17-10 shows an electronic display which incorporates prompts.

Figure 17-9. Electronic Display, All-Segments Test.

Figure 17-10. Electronic Display, with Prompts.

MAGNETIC TAPE PULSE RECORDERS

The following steps describe the tape-change procedure for the Westinghouse-type WR pulse recorder series, and serves to describe the typical operation for most pulse recorders employing the Westinghouse-style cartridge.

Removal from Recorder

Note the time to see if an interval is near completion. Do not change the cartridge at the end of an interval. If two minutes or less remain, wait until the end of that interval and approximately two minutes of the next interval have passed so that the translator and computer operator know from the reported start and stop times the exact time of the first and last time pulse and that none are omitted.

The "dead band" (black portion of interval) of ± two minutes indicates the period to avoid changing the cartridge.

The steps given below should be followed in the sequence given to remove the cartridge from the recorder:

1. Unseal the recorder and open cover.
2. Record appropriate data from initiator register along with day, date, recorder time (avoid +two minute "dead band"), and actual time, if different on the data card.
3. Press the back edge of cartridge toward the capstan and lift cartridge away from locking posts. Once clear of the locking posts, the reel lock spring will push the cartridge out somewhat. Remove from recorder.
4. Place cartridge with data card in place into a plastic bag before putting into carrying case.
5. Forward to central location for processing.

Inserting the Cartridge

Upon initial receipt of new recorders, the reflective area (silver area approximately four inches long) will be positioned so that it may be seen when inspecting the cartridge. The reflective area, after translation, will be on the take-up reel and therefore not visible when returned to a field recorder. The steps for proper installation of the cartridge should be followed in the sequence given below:

Note: On recorders equipped with an outage indicator, the interval pointer must be turned by hand to two minutes prior to the end of the interval and allowed to complete the interval under recorder power after application of AC power in order to energize the outage relay. Failure to do so will result in an outage pulse being recorded prior to the first interval.

1. If recorder time and actual time differ, set the recorder time before inserting cartridge. The recorder time may be set simply by turning the interval knob. If the clock is off several hours, the hour pointer may be moved by hand since it is a slip fit on the hour shaft. The interval pointer should be on zero when the hour hand is moved. With the interval pointer on zero, the hour pointer should be set on the hour and the remaining portion set using the interval knob in order to maintain synchronism of the two hands.
2. Inspect the tape guide, recording head, capstan, and bias magnet for cleanliness. If dirty, use a soft cloth and gently wipe until clean.

   Caution: Start recorder at an odd time (12:03) so the translation person knows the exact starting time of the first time pulse.

3. Insert the cartridge into the recorder (aluminum side out) between the dial plate and the middle plate. This requires pushing the pressure roller firmly against the capstan such that the cartridge seats itself on the two locating and locking posts. When properly positioned, marks at the top and bottom of the cartridge will align with markings on the middle plate.
4. Slowly turn the capstan clockwise, observing that the take-up drive spring is functioning properly. Turn the capstan until there is no slack tape between the capstan and the take-up reel.

Handling Tape Reel

If the reel and tape is removed from the cartridge for any reason, the flanges should not be pressed such that abusive contact is made with the tape. Pressing the flanges against the tape could cause edge damage that would result in lost data.

Cleaning the Recording Head and Guides

Magnetic tapes have a ferromagnetic oxide coating bonded to one side of the plastic base material. Some oxide or dust may accumulate on the recording head, guide, and bias magnet. These components should be checked for cleanliness at each tape change. For cleaning, a lint-free non-abrasive cloth (or chamois) slightly dampened with common ethyl alcohol is satisfactory.

Accidental Erasure

Magnetic tapes should not be exposed to magnetic fields of high intensities. Fields of this type may be generated by electric motors, transformers, and other electrical equipment. Bulk spacing is effective in reducing the possibility of accidental erasure. For example, one inch spacing is adequate shielding from a field produced by 400 amperes in a conductor.

Tape Splices

Magnetic tape used with these recorders must be splice-free. If it is necessary to salvage certain information from a broken tape, use splicing tape as a temporary measure until the data can be translated. Then discard the tape.

Tape Translation

After tapes are removed from recorders, they are delivered to a central location for translation. There, a tape translator converts pulses recorded on tape to units such as KWH and KQH. If the translation is successful, the metered information is sent to the utility's central billing computer system; if the translation is unsuccessful, a manual analysis of the data may be required and can lead to sending a meterman to the recorder location to effect repairs. After the metering information has been transferred from tape to the billing system, tapes are erased and re-used.

REGISTER CONSTANTS

The register constant is also known as the dial constant, dial multiplier, or reading multiplier. It is the factor by which the register reading is multiplied to obtain the total registration.

Each disk revolution represents a definite value of the units being measured. The register gearing converts these disk revolutions to the units used for customer billing. For small capacity domestic meters, the register commonly reacts directly in the billing units.

Changing the gearing between the dial pointers and the disk shaft introduces a constant. If the value of this gearing is doubled, the speed of the register pointers is halved and the reading must be multiplied by two to obtain the true value.

On modern high-capacity meters, particularly with bi-monthly billing, a four-dial register may not have the capacity to indicate the customer's use during the billing period. That is, the register may "turn over." This difficulty is usually avoided by increasing the capacity of the register.

One way to obtain this increased capacity is to add an extra dial at the left of the four dials commonly used. This means that the capacity of the register is increased tenfold and, instead of a maximum reading of 9,999 kilowatt-hours, there is a maximum reading of 99,999 kilowatt-hours. Another solution is to change the register ratio or gearing and introduce a multiplying factor or constant of 10. In this way, the capacity of the register is also changed from 9,999 kilowatt-hours to 99,999 kilowatt-hours.

Many factors enter into the decision as to which method is used. Among these factors are the procedures and billing machines used by the billing group which might require major modification to accept five digit readings. On the other hand, constants other than one must be carefully controlled and checked if errors are to be avoided.

Register constants are also introduced when meters are supplied from instrument transformers. If the meters are supplied for use with instrument transformers of specified ratios, the register constant is an even multiple of 10, that is, 10, 100, 1,000, etc. If transformer-rated meters are not supplied for specified transformers, then the register will read directly for the meter alone (considered as self-contained), and the register constant will be the current-transformer ratio or the product of the ratios of the current and potential transformers when both are used.

Meter manufacturers have recently made available demand registers with a scale and gear shifting design which provides full-scale deflection at 50 percent and 100 percent of meter capacity. These registers provide flexibility for future load growth with better reading accuracy at smaller loads.

AUTOMATIC RETRIEVAL OF DATA FROM SOLID-STATE PULSE RECORDERS

It is possible to retrieve information electronically from certain solid-state pulse recorders. If the recorder has a modem and is connected to a telephone line, the recorder can connect with the central office periodically, and transmit the data it has stored. If the recorder does not have a modem or a telephone line, a portable computer with modem can be taken to the site to automatically collect the information.

Some solid-state recorders have an optical port for transmitting readings through an optical probe, to a portable computer or hand-held device.

When this data is returned to the central office, it can be processed by software provided by the recorder manufacturer, or by general software which has been developed for processing data from a variety of recorders.

ELECTRONIC METER READING

The use of electronic devices within meters opens up new options for accurately retrieving larger volumes of data from meters. Various methods for automating the meter reading process are evolving, with several different technologies being developed.

New electronic meters have the ability to record a wide variety of consumption information. As described in earlier chapters, electronic meters can measure consumption by time-of-use, record peak consumption values, and collect load survey data. Emerging meter reading technologies make it easier for the utility to accurately collect data on all these items.

Hand-Held Data Entry Terminals

Hand-held terminals are carried to the field, and meter readings are entered via a keyboard on the face of the terminal. The meter reader can verify each reading as entered, by checking the digital display on the terminal. Some terminals are programmed to guide the meter reader through the daily route, and to advise him of special conditions related to the reading of each meter. At the end of the route, data collected is automatically downloaded to a computer, where it can be used immediately by the billing department.

Hand-held terminals, specially designed for reading utility meters, are available from several manufacturers. General purpose computers, such as laptop and palmtop computers, are also used by electric utilities.

Hand-held terminals can increase the productivity of the meter reader, and reduce errors, because meter readings are entered in the field and loaded into the central computer without further human involvement.

Hand-Held Terminals, Automatic Data Entry

Some meters and some hand-held terminals have the ability to transfer readings from meters into terminals automatically. Instead of manually entering data via a keyboard, the meter reader uses an optical probe to read in the data. Meter data is sent through an optical port on the meter directly to the hand-held terminal. This technique virtually eliminates human reading errors.

REMOTE METER READING

With remote meter reading, no on-site meter reader is required at all. These systems automatically connect meters in the field with the utility's computer system using telephone lines, radio transmissions, or electric power lines.

Remote Meter Reading, Using Telephone Lines

Telephone Dial-Out System

With a telephone dial-out system, the utility's computer initiates outbound calls to the meters. With this approach, the utility has the flexibility to poll meters at any time. It is a convenient solution for capturing initial readings, final readings, re-readings after a high-bill complaint, and testing for special conditions such as peak or nighttime consumption.

Dial-out meter reading systems require assistance from the local telephone company. To call customers without ringing their telephones, the telephone company injects special tones into the phone call, signals originally intended to test phone lines without disturbing customers. These test signals are created at the phone company central office and are not available to the public. With these signals, a meter reading can be taken even when the phone is disconnected.

To place outbound phone calls, the utility host computer maintains a directory of phone numbers of all customers. This can be a challenge with the increasing popularity of unlisted phone numbers and the frequent turnover of phone listings.

Telephone Dial-In System

With a telephone dial-in system, each meter calls the utility at specified times and reports its information. The meter, in effect, calls home once a month to report the latest readings. During that phone call, the host computer can send back to the meter the time for the next call, verify the meter's clock and calendar, test for other conditions, and update the meter module software if necessary. Dial-in systems typically use existing customer telephone lines. Each meter acts like a telephone extension on the line, or is the only instrument when installed on a dedicated line. Dial-in systems are installed and operated without special assistance from the local telephone company.

Dial-in systems usually place calls at night when the line is not likely to be in use. If the line is busy, or becomes busy during the call, the dialer in the meter hangs up and tries again later. The phone calls can be to a toll-free number, so there is no per-call cost to the customer. When no phone exists, the utility can pay to install a phone line specifically for meter reading. This investment can be cost-effective for reading meters installed in remote locations, or for meters which are read frequently.

Electronic meters can be programmed to dial-in if a special event occurs such as an out-of-range reading (suggesting a damaged circuit), suspected meter tampering, or a low battery in the meter.

Figure 17-11 (a). Meter Equipped for Remote Reading by Radio.

Figure 17-11 (b). Hand-held Terminal for Remote Reading by Radio.

Dial-in systems generally rely on customer phone lines and are vulnerable to breaks in that service. Some of these systems do not work when the customer's phone is disconnected.

Remote Meter Reading, Using Radio

With a radio-based system, each meter is adapted with a compact, low-power transmitter-receiver which, upon request, broadcasts a radio signal containing meter reading data.

These meters are polled by a mobile transmitter-receiver which collects meter readings from many meters, and carries the readings back to the central office. At the end of the day, the readings are loaded directly into the utility's billing computer. The mobile transceiver can be in a hand-held terminal, or can be a higher-powered unit installed in a truck which drives down streets, polling meters, and recording replies. Hand-held terminals can read several hundred meters in a day. Truck-mounted units can read thousands of meters in a day.

Figure 17-11a shows a watthour meter adapted for remote reading by radio. A small semicircular circuit board with a transceiver (difficult to see in the photo) has been installed immediately under the glass cover. The portable terminal shown in Fig. 17-11b can read meters at a distance of several hundred feet.

Radio-based meter reading is fast and generally reliable. Some meter installations are inside metal buildings rendering them inaccessible for two-way transmissions. These site-related problems can often be solved by re-locating the meter antenna.

Remote Meter Reading, Using Power Lines

Electric power lines can be used for the transmission of meter data. Information collected by the meter is encoded, then sent through the power lines to special receivers.

Since power lines are electrically "noisy," their use for reliable transmission of data is usually restricted to short distances, or to installations where the power lines have been specially conditioned.

CREDITS

INDEX

A

B

J

K

L

M

N

O

P

Q

R

S